SYMPOSIA OF THE ZOOLOGICAL SOCIETY OF LONDON
NUMBER 34

The Biology of Hystricomorph Rodents

(The Proceedings of a Symposium held at the Zoological Society of London on 7 and 8 June, 1973)

Edited by

I. W. ROWLANDS

*Wellcome Institute of Comparative Physiology,
Zoological Society of London,
Regent's Park, London*

and

BARBARA J. WEIR

*Wellcome Institute of Comparative Physiology,
Zoological Society of London,
Regent's Park, London*

Published for

THE ZOOLOGICAL SOCIETY OF LONDON

by

ACADEMIC PRESS
1974

ACADEMIC PRESS INC. (LONDON) LTD.

24/28 Oval Road

London NW1

U.S. Edition published by

ACADEMIC PRESS INC.

111 Fifth Avenue,

New York, New York 10003

Library of Congress Catalog Card Number: 74-5683

ISBN: 0-12-6133334-4

PRINTED IN GREAT BRITAIN BY
BILLINGS & SONS LTD., GUILDFORD

CONTRIBUTORS

AMOROSO, E. C., *A.R.C. Institute of Animal Physiology, Babraham, Cambridge, England* (p. 447)

ASIBEY, E. O. A., *Department of Game and Wildlife, P.O. Box M239, Accra, Ghana* (pp. 161, 251)

BUGGE, J., *Department of Anatomy, The Royal Dental College, DK 8000, Aarhus C, Denmark* (p. 61)

EISENBERG, J. F., *National Zoological Park, Smithsonian Institution, Washington, D.C., U.S.A.* (p. 211)

GEORGE, WILMA, *Lady Margaret Hall, Oxford, England* (pp. 79, 143)

HEAP, R. B., *A.R.C. Institute of Animal Physiology, Babraham, Cambridge, England* (p. 385)

ILLINGWORTH, D. V., *A.R.C. Institute of Animal Physiology, Babraham, Cambridge, England* (p. 385)
Present: Division of Steroid Endocrinology, Department of Chemical Pathology, University of Leeds, England

KLEIMAN, D. G., *National Zoological Park, Smithsonian Institution, Washington, D.C., U.S.A.* (p. 171)

LAVOCAT, R., *Laboratoire de Paléontologie des Vertebrés, Université des Sciences et Techniques du Languedoc, Place Eugene-Bataillon, Montpellier, France* (p. 7)

LAZARUS, N. J., *Diabetes Research Unit, Wellcome Foundation, Temple Hill, Dartford, England* (p. 417)

NEVILLE, R. J. W., *Diabetes Research Unit, Wellcome Foundation, Temple Hill, Dartford, England* (p. 417)

PEARSON, O. P., *Museum of Vertebrate Zoology, University of California, Berkeley, California, U.S.A.* (p. 109)

PERRY, J. S., *A.R.C. Institute of Animal Physiology, Babraham, Cambridge, England* (pp. 249, 333)

ROBERTS, CHRISTINE M., *Wellcome Institute of Comparative Physiology, Zoological Society of London, Regent's Park, London NW1 4RY, England* (p. 333)

ROWLANDS, I. W., *Wellcome Institute of Comparative Physiology, Zoological Society of London, Regent's Park, London NW1 4RY, England* (pp. 131, 303, 361)
Present: Department of Anatomy, University of Cambridge, Downing St, Cambridge

SIMPSON, G. G., *5151 East Holmes Street, Tucson, Arizona 85711, U.S.A.* (p. 1)

TAM, W. H., *Wellcome Institute of Comparative Physiology, Zoological Society of London, Regent's Park, London NW1 4RY, England* (p. 363)
Present: *Department of Zoology, University of Western Ontario, London 72, Ontario, Canada*

WEIR, BARBARA J., *Wellcome Institute of Comparative Physiology, Zoological Society of London, Regent's Park, London NW1 4RY, England* (pp. 79, 113, 265, 303, 417, 437)
Present: *The Journal of Reproduction and Fertility, 7 Downing Place, Cambridge, England*

WOOD, A. E., *Amherst College, Amherst, Massachusetts, U.S.A.* (p. 21)
Present: *20 Hereford Avenue, Cape May Court House, New Jersey 08210, U.S.A.*

ORGANIZERS AND CHAIRMEN

ORGANIZERS

I. W. ROWLANDS and BARBARA J. WEIR, on behalf of the Zoological
Society of London

CHAIRMEN OF SESSIONS

O. P. PEARSON, *Museum of Vertebrate Zoology, University of California,
Berkeley, California, U.S.A.*

J. S. PERRY, A.R.C. *Institute of Animal Physiology, Babraham,
Cambridge, England*

I. W. ROWLANDS, *Wellcome Institute of Comparative Physiology,
Zoological Society of London, Regent's Park, London NW1 4RY,
England*

G. G. SIMPSON, *5151 East Holmes Street, Tucson, Arizona 85711, U.S.A.*

PREFACE

The first seeds of the events leading to this symposium were sown when one of us (I.W.R.), already working on the laboratory guinea-pig, read papers by Pearson and Mossman in 1949 on reproduction in the mountain viscacha and the Canadian porcupine respectively. These papers stirred the imagination and aroused the desire, which further acquaintance has in no way diminished, to learn more about these and other hystricomorph rodents. Some years later, at the Wellcome Institute, this work was extended by I.W.R. and Dr R. B. Heap, a former colleague at the A.R.C. Institute of Animal Physiology, Babraham, to a study of reproduction in the coypu. At the same time, as part of a bigger programme to study unusual aspects of mammalian reproduction at the Wellcome Institute, the World Health Organization agreed to support some work on hystricomorphs, which Dr Devra Kleiman undertook on the behaviour and reproduction of acouchis; we are glad to note that her interest has been maintained and extended.

The hystricomorph invasion of the Wellcome Institute really started, however, with the arrival of chinchilla, agoutis, acouchis, coypu and B. J. W. from Cambridge! Our original team was completed by Dr W. H. Tam who studied the endocrinology of pregnancy in some of these newly acquired animals. Several years later and after two expeditions to South America, African porcupines, tuco-tucos, cuis, degu, chozchoris, casiragua, Jamaican hutias, wild guinea-pigs, Patagonian and Peruvian mountain viscacha, and plains viscacha appeared at the Institute, not, fortunately, all at once. The maximum number at any one time has been about 450 of nine species. We should like to think that none has been an extravagance. An extraordinary amount of information has been derived from our studies of these animals and complements that of the other devotees of hystricomorphs who were able to attend this Symposium.

We are grateful to the Zoological Society of London for asking us to organize the scientific programme, and to edit the papers and the discussion they evoked. We are equally grateful to all Chairmen, participants and contributors, and we should thank, especially, Professor E. C. Amoroso for his conscientious attendance as our rapporteur. Miss Unity McDonnell made all the house arrangements and members of the Institute staff recorded the discussions on tape.

The programme fell readily into three basic sections, the last one—physiology—being inevitably the largest since most of the work centred around the colonies of hystricomorphs at the Wellcome Institute, which is primarily a laboratory for the study of mammalian reproduction. The substance of all but one of the papers in these proceedings was presented at the meeting. The exception, the paper dealing with the origin of the domestic guinea-pig, was withheld from the programme through lack of space and time, but as the guinea-pig and its ancestry were constant reference points throughout our deliberations, we decided to include it, even though one of our Chairmen had indicated that its origins were uncertain.

There have been a few editorial problems since the Symposium was based on a multi-disciplinary approach, but authors have been very willing to assist us in the interpretation and clarification of obscure points. We hope that our efforts will lead each of us to a greater awareness of the current status of work in disciplines not our own.

While speaking to participants during the course of the meeting, it became obvious that we are not alone in feeling a fascination for hystricomorph rodents. The reason for this seems to be the extraordinary ability of members of the group to be different from each other and from any one else, and their motto could well be "you name it, we do it, but not as others do".

I. W. Rowlands
Barbara J. Weir

September 1974

CONTENTS

Chairman's Introduction: Taxonomy
GEORGE GAYLORD SIMPSON

What is an Hystricomorph?
R. LAVOCAT

The Evolution of the Old World and New World
Hystricomorphs
ALBERT E. WOOD

The Cephalic Arteries of Hystricomorph Rodents

J. BUGGE

Hystricomorph Chromosomes

WILMA GEORGE and BARBARA J. WEIR

CONTENTS   xiii

Chairman's Introduction: Ecology and Behaviour
OLIVER P. PEARSON 109

The Tuco-Tuco and Plains Viscacha
BARBARA J. WEIR

Mountain Viscacha
I. W. ROWLANDS

Notes on the Ecology of Gundis (F. Ctenodactylidae)

WILMA GEORGE

The Grasscutter, *Thyronomys swinderianus* Temminck, in Ghana

E. O. A. ASIBEY

Patterns of Behaviour in Hystricomorph Rodents

DEVRA G. KLEIMAN

The Function and Motivational Basis of Hystricomorph Vocalizations

J. F. EISENBERG

Chairman's Introduction: Reproductive Physiology

J. S. PERRY

Reproduction in the Grasscutter (*Thryonomys swinderianus* Temminck) in Ghana

E. O. A. ASIBEY

Reproductive Characteristics of Hystricomorph Rodents

BARBARA J. WEIR

CONTENTS   xvii

Functional Anatomy of the Hystricomorph Ovary

BARBARA J. WEIR and I. W. ROWLANDS

Hystricomorph Embryology

CHRISTINE M. ROBERTS AND J. S. PERRY

CONTENTS

Chairman's Introduction: Endocrinology

I. W. ROWLANDS

The Synthesis of Progesterone in some Hystricomorph Rodents

W. H. TAM

The Maintenance of Gestation in the Guinea-pig and other Hystricomorph Rodents: Changes in the Dynamics of Progesterone Metabolism and the Occurrence of Progesterone-binding Globulin (PGB)

R. B. HEAP and D. V. ILLINGWORTH

CONTENTS xix

Hystricomorph Insulins

R. W. J. NEVILLE, BARBARA J. WEIR and
NORMAN R. LAZARUS

Notes on the Origin of the Domestic Guinea-pig

BARBARA J. WEIR

Concluding Remarks

E. C. AMOROSO 447

Symp. zool. Soc. Lond. (1974) No. 34, 1–5

CHAIRMAN'S INTRODUCTION: TAXONOMY

GEORGE GAYLORD SIMPSON

5151 East Holmes Street, Tucson, Arizona 85711, U.S.A.

The rodents constitute the most diverse and most numerous order of mammals. Taxonomic difficulties tend to increase with diversity, so that this may also be expected to be an unusually difficult order. This expectation is fulfilled, and there are taxonomic problems in the Rodentia at every level from subspecies upward. In these introductory remarks I shall confine myself to problems at high taxonomic levels, although there are equally interesting questions at lower levels as will doubtless appear in the course of this Symposium.

It is true that in what now seem to us days of happy innocence, major difficulties were not noticed or were brushed aside. We may ourselves immediately brush one aside by noting that the lagomorphs, long included in the Rodentia, are now universally excluded. In 1855, Brandt divided the rodents, or the other rodents, into three suborders: Sciuromorpha, Myomorpha, and Hystricomorpha. Up until our own days that arrangement has continued overtly or covertly to dominate much thinking about rodents.

In 1899 Tullberg retained Brandt's three groups, but he separated the "Bathyergomorphi," which were, however, united with the "Hystricomorphi" in a taxon "Hystricognathi". The "Myomorphi" and "Sciuromorphi" were similarly united in the "Sciurognathi." Although some radically different arrangements had been proposed in the meantime (notably by Miller & Gidley, 1918, and Winge, 1923–1924), Ellerman's monumental work of 1940–1941 has exactly the same higher taxa as Tullberg, with only insignificant semantic changes.

Although I noted that the arrangement was becoming increasingly unsatisfactory, I retained Brandt's division of rodents into three suborders in my classification published in 1945. The dissatisfaction was expressed, in part, by leaving a number of groups *incertae sedis*. That system has been quite widely followed, although my classification of more than 30 years ago (written in and before 1942) obviously now needs radical revision, which may be forthcoming from other hands before long. Meanwhile, students of recent mammals in particular frequently continue to divide rodents into Sciuromorpha, Myomorpha, and Hystricomorpha. For example, Anderson (1967) did so even though he put the names in quotation marks. That is also the usage in the most recent textbook of mammalogy (Vaughn, 1972). The author says he is going to use these names in a morphological sense, without taxonomic status, but in fact he goes on to use them as if they were taxonomic.

As this Symposium is devoted to hystricomorphs (see p. 4) let us especially consider what, if anything, this term has come to mean. I shall try to do so without excessive trespassing on the territory of the following speakers. Although the term hystricomorph has a literal meaning no more explicit than 'porcupine-like' (with reference to the Old World porcupines), it was from the beginning applied to resemblance in one respect only: the jaw musculature, or, still more explicitly, the arrangement of origins of the muscles of the masseter complex. The great majority of living rodents do indeed show variations on three general masseteric patterns. (There is in fact a fourth, present in only one living genus, *Aplodontia*, but universal in early and primitive rodents.) As is well known, the main feature of the hystricomorph pattern is that a massive part of the masseteric complex, the masseter medialis, originates principally on the rostrum and hence passes to its insertion on the mandible (or dentary) through a greatly enlarged infraorbital foramen medially to the zygoma.

The use of Hystricomorpha as the name of a taxon including all the rodents with an hystricomorph masseteric complex soon came to have the evolutionary, phylogenetic implication that such an arrangement arose only once and that all the animals that have it inherited it from a common ancestor. The term hystricognath similarly indicates a mandibular structure highly but not exclusively correlated with an hystricomorph masseteric complex. The less common use of Hystricognathi as a taxonomic term came to have evolutionary implications similar to but different from those of Hystricomorpha.

There remains the problem whether the rodents that are anatomically hystricomorph really do constitute a natural (by which I mean an evolutionary) taxon. It has been mostly the specialists on fossil rodents who have attacked this problem. That is understandable because fossils of early rodents surely can be expected to cast considerable, perhaps even conclusive, light. Nevertheless, it must be remembered that the fossil record of the rodents is still very incomplete. Among mammalian orders, only the Chiroptera have a lower ratio of families and genera known as fossils to those now living. That may be in part but not altogether a true reflection of the evolution of the order. That is, the Rodentia may in fact have diversified more steadily and more extensively into the Recent. It is also noteworthy that the rate of discovery of fossil rodents is now higher than ever before and still accelerating, owing to an increase in the number of interested palaeontologists and use of modern mass methods of collecting small mammalian fossils.

The late Swiss palaeontologist Samuel Schaub, a leading student of fossil rodents along with his predecessor H. G. Stehlin, discarded the concept of hystricomorphy altogether and proposed a new system based primarily on a theory of molar evolution (Stehlin & Schaub, 1951; Schaub, 1953). Schaub proposed a suborder Pentalophodonta which included all the hystricomorphs of other authors but gave them some strange bedfellows. His Palaeotrogomorpha, one of the two infraorders of Pentalophodonta, included not only the Old World hystricomorphs and hystricognaths, the bathyergids among them, but also the beavers and a mixture of other groups that are anatomically sciuromorph or myomorph. He believed that the older groups of South American rodents arose from palaeotrogomorph ancestors with teeth like the European Oligocene genus *Theridomys*, but that once in South America these rodents evolved in complete isolation into a large natural group that he called the infraorder Nototrogomorpha. In it he placed all the South American anatomical hystricomorphs and no other rodents. He refused to speculate on how their theridomyid ancestor reached South America.

The essentials of Schaub's classification were followed in both of the great French treatises, the *Traité de zoologie* by Grassé and Dekeyser (1955) and, not surprisingly, the *Traité de paleontologie* by Schaub himself (1958). Elsewhere, however, few students of rodents have adopted either Schaub's theory of molar evolution or the taxonomic inferences that he based on it. In fact his system seems merely to evade what is generally considered the crucial question. That question is quite simply whether the anatomically hystricomorph rodents are a natural evolutionary taxon. The question plainly is not answered by theoretical derivation of all these rodents from an early Tertiary European genus or family, itself doubtfully hystricomorph, supposedly ancestral also to a variety of other rodents which are definitely not hystricomorph.

No one doubts that all the South American rodents in question had a common ancestry somewhere and sometime. Although there is still some question as to just what lesser taxa are to be included, no one doubts that there is also a natural Old World higher taxon to which the name Hystricomorpha properly belongs, whether or not they are to be given sole possession of that name in formal classification.

The question also has an important biogeographic aspect. It now presents itself as a choice between these alternatives:

1. the hystricomorph-like South American rodents were of remote

(Eocene) North American ancestry and their resemblances to Old
World Hystricomorpha evolved separately. In that case they are not
taxonomically Hystricomorpha but are best designated as Caviomorpha,
a name decidedly preferable to Nototrogomorpha;

2. the hystricomorph-like South American rodents were of remote
(likewise Eocene) African ancestry among the Old World Hystricomorpha. In that case they may be included taxonomically in the
Hystricomorpha, although their separation as a group at a lower
hierarchical level would still merit consideration.

There I stop because our first two speakers represent these two views
and are better able than anyone else to present and to evaluate the
evidence, especially from the palaeontological point of view. They will
be followed by two taxonomic studies of Recent rodents, one on cephalic
arteries, following the classic lines of anatomical enquiry, and the
other on chromosomes, using a more recently developed sort of taxonomic evidence.

Appendix

[These comments were made by Dr Simpson at the end of the Symposium
but are complementary to his introductory remarks, and so they have been
included here. Eds.]

I should like to make a general remark, which is to some extent an appendix
to my introductory comments because when I had the privilege of speaking
first yesterday, I did not know just what the trend would be. This is a matter of
nomenclature and, therefore, may seem to be purely semantic but I think it has
rather more fundamental significance. The term 'hystricomorph' is a perfectly
good anatomical term so it would seem quite proper to use it for a symposium
of this sort. I observe however that, with the exception of one palaeontologist,
everyone here is assuming that it is a taxonomic term, in other words that it is
the name of an evolutionarily or phylogenetically defined group. The implication
is that the Hystricomorpha are a group of unified origin including some African
forms and some South American forms. I do not want to argue either for or against
that view here because it does not really matter for the point that I am about to
make. Leaving aside for this purpose again the Erethizontidae, we all agree, I
believe, that all the other forms which here have been called uniformly hystricomorphs and occur in South and Central America do form a natural group, even
if one refers them to a broader group called the Hystricomorpha. Whether we
think of this as a separate suborder, or whether we think of it as an infraorder of
hystricomorphs, at least it is a natural group for which the name Caviomorpha
is rather generally used, universally amongst palaeontologists, and I think it will
be more and more used amongst zoologists. I also note that most of the hard facts
that have been presented here refer not to the Hystricomorpha in a taxonomic
sense as a whole unit but only to the Caviomorpha. I think that communication
would be clearer, and perhaps even our thinking on the matter would be clearer,
if when we meant Caviomorpha we said Caviomorpha, and when we meant Hystricomorpha in that broader sense, we said Hystricomorpha, and said it only then.

REFERENCES

Anderson, S. (1967). Introduction to the rodents. In *Recent mammals of the world. A synopsis of families*: 206–209. Anderson, S. & Jones, J. K., Jr (eds). New York: Ronald Press.

Brandt, J. K. (1855). Beiträge zur näheren Kenntnis der Säugetiere Russland's. *Zap. imp. Akad. Nauk* (6) **9**: 1–365.

Ellerman, J. R. (1940–1941). *The families and genera of living rodents*. London: British Museum.

Grassé, P.-P. & Dekeyser, P. L. (1955). Ordre des rongeurs. In *Traité de zoologie* **17**: 1321–1573. Grassé, P.-P. (ed.). Paris: Masson.

Miller, G. S., Jr & Gidley, J. W. (1918). Synopsis of the supergeneric groups of rodents. *J. Wash. Acad. Sci.* **8**: 431–448.

Schaub, S. (1953). Remarks on the distribution and classification of the "Hystricomorpha". *Verh. naturf. Ges. Basel* **64**: 389–400.

Schaub, S. (1958). Simplicidentata (= Rodentia). In *Traité de paléontologie* **6** (2): 658–818. Piveteau, J. (ed.). Paris: Masson.

Simpson, G. G. (1945). The principles of classification and a classification of mammals. *Bull. Am. Mus. nat. Hist.* **85**: I–XVI, 1–350.

Stehlin, H. G. & Schaub, S. (1951). Die Trigonodontie der simplicidentaten Nager. *Schweiz. palaeont. Abh.* **67**: 1–385.

Tullberg, T. (1899). *Ueber das System der Nagethiere: eine phylogenetische Studie*. Upsala: Akademischen Buchdruckerei.

Vaughn, T. A. (1972). *Mammalogy*. Philadelphia etc.: W. B. Saunders.

Winge, H. (1923–1924). *Pattedyr-Slaegter*. Copenhagen: H. Hagerups.

Symp. zool. Soc. Lond. (1974) No. 34, 7–20

WHAT IS AN HYSTRICOMORPH?

R. LAVOCAT

*Laboratoire de Paléontologie des Vertebrés,
Université des Sciences et Techniques du Languedoc,
Place Eugène-Bataillon, Montpellier, France*

SYNOPSIS

Only rodents which are hystricognathous as well as hystricomorphous should be retained
in the suborder Hystricomorpha. It is proposed that this systematic unit is referred to
subsequently by the name first used for the group by Tullberg (1899), the Hystricognathi.

Many osteological features, not in themselves depending one upon the other, are
discussed to determine the common plan. The relevance of other facts such as tooth
structures, brain fissures, fetal membranes, circulation and endoparasites, is considered.
The systematic unity of the suborder needs a common origin from a single stem. That
origin is found in fauna from the African Eocene, from which members migrated across
the South Atlantic to South America by rafts. This would have been possible if the two
continents were as close as theories of continental drift imply.

INTRODUCTION

The use of the word 'hystricomorph' is rather equivocal since it can
have a structural and a systematic meaning, but the two are not necessarily the same.

The structural term 'hystricomorph' refers to a particular condition
of the infraorbital region in which the infraorbital foramen is large and
accommodates the passage of an important portion of the masseter
muscle. Such a condition is typically found in *Hystrix*, and is believed
to be present in all members of the suborder Hystricomorpha (in the
wide sense of Simpson, 1945) although secondarily reduced in the
Bathyergidae. However, the enlarged infraorbital foramen is not the
only diagnostic character of the Hystricomorpha, and there are other
groups of rodents which can be said to be hystricomorphous but are not
'hystricomorphs'. It is this fact which has, in the past, led to confusion
over the term 'hystricomorph'.

My friend, A. E. Wood, and I both agree that the following taxa are
not 'hystricomorphs' although they are hystricomorphous: the Anomaluroidea, Ctenodactyloidea, Dipodoidea, Pedetoidea and Theridomorpha. The basis on which these groups are excluded from the Hystricomorpha is that they are all sciurognathous whilst the true 'hystricomorphs' are also hystricognathous, although there are many other

significant differences. All rodents of the suborder Hystricomorpha have an hystricognathous mandible and this feature is very rare outside the suborder, although hystricomorphy appears in several unrelated groups of rodents. The sciurognathous (Fig. 1D) and hystricognathous (Fig. 1A, B, C) conditions are, in fact, more typical at higher systematic levels than the hystricomorphous characteristics. This distinction is less noticeable for the Pedetoidea than for the others but there are many other differences from the 'hystricomorphs' which become obvious when the characteristics of a true 'hystricomorph' are examined.

The suborder Hystricomorpha is an assemblage of rodents which not only possess the hystricomorphous condition but also share several other characters which I believe to be important in deciding affinities.

THE SUBORDER HYSTRICOGNATHI

Since the hystricognathous condition is not only the first and most easily recognized characteristic of members of the so-called suborder Hystricomorpha, but also the most diagnostic, I have recently decided to resurrect the name Hystricognathi for this taxon. The suborder Hystricognathi was established by Tullberg (1899) and I believe it to contain exactly the same families now as then, as well as having fewer equivocal connotations.

I began a detailed osteological study of rodent skulls two years ago and I consider the following features of the head are important:

(a) the mandibular structure which may be sciurognathous or hystricognathous (see Fig. 1);

(b) the infraorbital opening which may be myomorphous, sciuromorphous or hystricomorphous;

(c) the shape of the zygomatic arch which is partly correlated with the structure of the infraorbital region;

(d) the structure of the pterygoid region which may be flat, excavated, blind or open in front;

(e) the relationships between the palatine and maxillary bones in the floor of the orbito-temporal cavity;

(f) the alisphenoid bones which may be higher or lower than the adjacent bones;

(g) the position of the lacrimal;

(h) the interconnections between the parietal, squamosal, occipital and mastoid;

(i) the structure of the middle ear, particularly of the promontorium, fenestrae and ossicles;

(j) characters of the dentition.

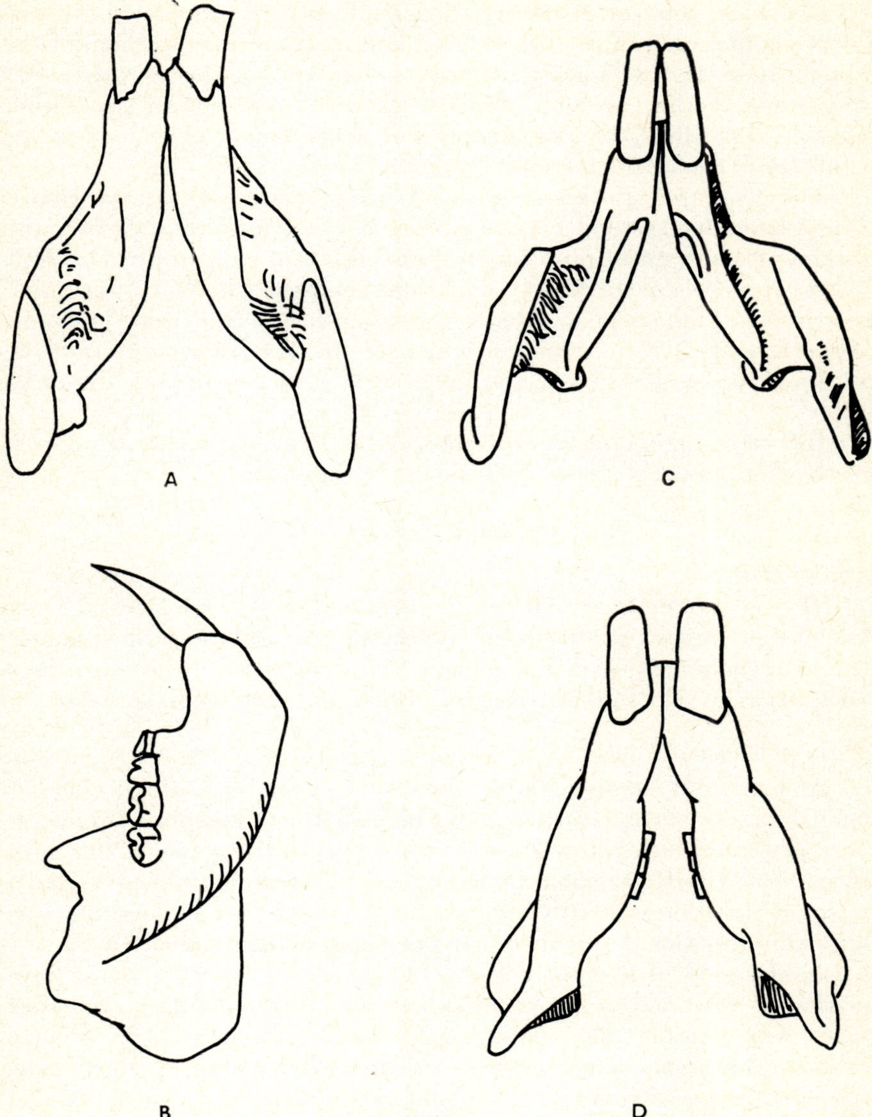

FIG. 1. Diagrammatic representation of the hystricognathous (A, B and C) and sciuro-gnathous (D) condition of the lower jaws of rodents. A and B. *Bathyergoides*; A, ventral view; B, medio-lateral view of the left half of the jaw: C, *Bathyergus*, ventral view (after Ellerman, 1940): D, *Rhizomys*, ventral view (after Tullberg, 1899). In the sciurognathous condition the angle is in the same line as the incisival anterior part of the jaw. In the hystricognathous condition, the angle is in a latero-external position, being extreme in *Bathyergus* and more typical of the general condition in *Bathyergoides*.

Other anatomical characteristics, such as the proportions of the palatine foramina, of the paroccipital and mastoid processes and of the bony palate and its rectangular or triangular shape, are clearly less important as they are subject to variation in lower taxa. Yet other characteristics like the arrangment and development of sinuses in the skull are extreme adaptations.

The geographical distribution of the Hystricognathi is such that a complete understanding of the suborder can be gained not only from the solving of anatomical problems but also those of geography. The New World Hystricognathi (the Caviomorpha) and the Old World Hystricognathi (the Phiomorpha) are now separated by the width of the South Atlantic and, if the systematic unity of the Hystricognathi is to be accepted, one must solve the problem of how the separation arose and when.

Before trying to answer this question, I shall examine in detail each of the anatomical structures I believe are significant.

Cranial osteology

Hystricognathy

This character is shared by all members of the suborder. To my knowledge, hystricognathy is found only in one other described genus, the Mongolian Oligocene *Tsaganomys*. This genus was believed to be a bathyergid by Matthew & Granger (1923), but this seems unlikely.

Hystricomorphy

At first sight, it seems impossible to consider the Bathyergidae as being a family of the Hystricognathi because the infraorbital foramen is very small and only rarely gives way to a part of the masseter in recent forms. But the Hystricognathi do have an evolutionary history, and a study of the Miocene Bathyergoidea clearly shows that the foramen was originally functional and there has been a secondary reduction.

It seems evident that the ancestors of the hystricognath rodents were neither hystricomorph nor sciuromorph, but protrogomorph, and that there was a progressive enlargement of the foramen. The African Phiomorpha clearly demonstrate, to my mind, that by the Middle Eocene the move towards enlargement had already started in the ancestral stock and subsequently progressed at different speeds in the various lines. *Kenyamys mariae* (Lavocat, 1973), from the Kenya Miocene has a more advanced dentition but, nevertheless, a more primitive infraorbital structure than has *Phiomys*, the main Oligocene genus of Fayum. In the Miocene *Bathyergoides* the infraorbital foramen, similar to that of *Kenyamys*, clearly shows that the masseter was attached only to the rim of the foramen and did not pass through to the muzzle.

In the South American Deseadan Oligocene, *Platypittamys* is found having a rather small infraorbital foramen alongside such genera as *Cephalomys* (Wood & Patterson, 1959) and *Sallamys*, *Incamys* and *Branisamys* (Hoffstetter & Lavocat, 1970) which have an extensively enlarged foramen. The condition in *Platypittamys* can be considered either as the persistence of a primitive state or as a stage in the secondary reduction of the foramen like that occurring in the Bathyergidae. It is clear from the evidence of the African fossil fauna that there is not necessarily a correlation between the evolutionary development of the foramen and that of the teeth.

Zygomatic arch

The shape of this arch is not the same in all rodents which are hystricomorphous. Of the non-hystricognaths, the Anomaluridae and Dipodidae have a thin ascending ramus and the Pedetidae has a robust arch, similar to that in *Paraphiomys*. *Paraphiomys*, like most of the hystricognaths, has a wide ascending ramus of the jugal and a peculiar shape to the inferior border of the arch. The Bathyergidae do not fit this pattern, perhaps because of their adaptations to a fossorial habit, and some of the caviomorphs are also discrepant but at present the *Paraphiomys* model is the most typical one for comparison.

Pterygoid region

Tullberg (1899) recognized the importance of this region in rodents because it is the place of insertion for the internal and external pterygoid muscles. Several patterns can be described amongst rodents. The basic formation found in all genera of Hystricognathi is a deep fossa with an anterior opening. In the Bathyergidae, at least, Tullberg (1899) has shown that this opening allows the pterygoid muscle to elongate and become inserted in the orbito-temporal fossa, and eventually in the cranial cavity. To my knowledge, a similar opening is found outside the Hystricognathi only in the Geomyidae and *Spalax*. There is good evidence to suggest that the opening in *Spalax* is a secondary adaptation of the normally blind fossa of the Muroidea. I have recently found a similar arrangement in the Oligocene genus *Tsaganomys* from Asia.

Maxillary–palatine relationships

In rodents, the relationship between the maxillary and palatine bones is peculiar because of the unique position of the teeth in relation to the orbito-temporal fossa. Two main arrangements are found. In one, the floor of the orbito-temporal fossa is composed equally of the maxillary in front and the palatine behind. In the other, the palatine bone is practically excluded from the fossa and can be seen only as a small apophysis behind the tooth row. It is this second pattern which has been found in all groups of the Hystricognathi, and occurs only in *Castor* and the Ctenodactyloidea outside the suborder (Lavocat, 1971). The presence of this arrangement can, therefore, be accepted as a characteristic of the Hystricognathi.

Alisphenoid

Landry (1957) has discussed the relationships of the alisphenoid with adjacent bones. The alisphenoid either has a vertical branch which reaches to the top of the skull, or is limited to the basicranium. The latter is common to all members of the Hystricognathi, although it is also found in other groups.

Lacrimal

In *Paraphiomys*, *Diamantomys*, *Thryonomys* and brachyodont caviomorphs, the lacrimal bone is posterior to the infraorbital foramen and the lacrimal canal is vertical. This is assumed to be the basic pattern. Modifications which are probably a consequence of hypsodonty are seen in the Caviidae and Chinchillidae, but further study is needed to clarify this point.

Occipital–mastoid relationships

The particular interrelationships of the squamosal, occipital and mastoid bones were thought by Tullberg (1899) to be typical of the Hystricognathi. Landry (1957) argues that similar connections can be found in other groups but it seems to me that this particular feature, together with all the others common to the Hystricognathi, gives it a diagnostic value.

Auditory region

The structure of the auditory region, particularly of the middle ear, is, in my opinion, important and significant. There are many reasons for assuming that this region is particularly stable and can be representative, and studies of the middle ear of the Theridomorpha from the Eocene and Oligocene have provided good evidence to support this contention (Lavocat, 1967). When studying the Miocene rodents from Kenya, I was lucky enough to find a very well preserved middle ear of *Diamantomys*, which proves to be very similar to that of *Thryonomys* and the South American genus, *Lagostomus*. A brief survey of the structure of the middle ear in the Hystricognathi indicated that there was

considerable homogeneity thoughout the suborder, and my colleague,
J. P. Parent, is now investigating the problem in detail.

Dental characters

Up to this point, I think it is true to say that there is good agreement
between the views of A. E. Wood and myself, but we clearly differ in
our interpretation of the dentition. In my opinion, the dental structures
observed in the Thryonomyoidea, Bathyergoidea, Hystricoidea and
Caviomorpha are the result of the evolution of a fundamental pattern
initiated by some ancestral Eocene African rodent.

Cheek tooth crests

The basic crest pattern of an upper molar (see Fig. 2) consisted of an

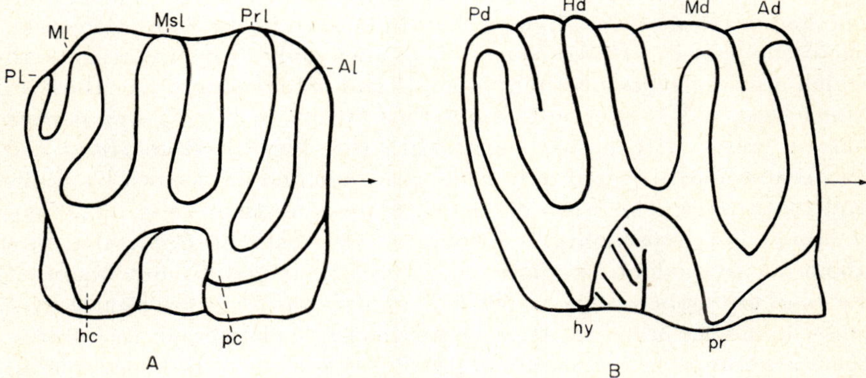

Fig. 2. Diagrammatic representation (greatly enlarged) of an upper (A) and lower
(B) molar of *Simonimys*, a phiomorph from the East African Miocene. The arrows
indicate the anterior side of the tooth.

A. Upper molar showing typical pentalophodont structure with an anteroloph
(*Al*), protoloph (*Prl*), mesoloph (*Msl*), metaloph (*Ml*) and posteroloph (*Pl*) and a protocone (*pc*) and hypocone (*hc*). The internal end of the metaloph is not connected to the
hypocone but to the posteroloph, midway along this loph. In many forms the mesoloph does not reach the external border of the tooth but only to the antero-internal
border of the metaloph.

B. Lower molar with four well developed lophids, the anterolophid (*Ad*), mesolophid (*Md*), hypolophid (*Hd*) and posterolophid (*Pd*), and a protoconid (*pr*) and hypo-
conid (*hy*). The mesolophid may be considered to be the posterior arm of the protoconid.

anteroloph, a protoloph, a rather variable mesoloph, a metaloph, whose
lingual end reached midway between the labial and lingual ends of the
posteroloph rather than the hypocone, and a posteroloph. The size of
the mesoloph has proved to be very variable; in some African genera it

B

is very short, in others it reaches the external border of the tooth, and yet in others there is no loph at all. A comparative study of the fossil and recent Thryonomyoidea indicates that there is also a tendency for the metaloph to become reduced, vestigial and finally to disappear. Reduction of the mesoloph is not necessarily correlated with that of the metaloph. For example, in *Paraphiomys* the small and narrow valley between the metaloph and posteroloph disappears in M3 and the metaloph is not then distinguishable but the mesoloph reaches the labial border of the same tooth, while stopping mid-way along the metaloph in M1 and M2. Using *Paraphiomys* as a basis, it is then easy to understand the teeth of juvenile, recent *Thryonomys* which have only three main crests. On the external half of the anterior slope of the posteroloph of this genus there is usually a small fourth crest which is undoubtedly the last remnant of the metaloph.

Structures which are directly related to the ancestral pentalophodont pattern of the Phiomorpha can be seen in the South American Oligocene genera. *Protosteiromys* is known only from isolated teeth which are somewhat brachyhypsodont and astonishingly like those of *Simonimys*, a Miocene genus from East Africa, in having five perfect narrow crests in its upper cheek teeth (see Fig. 2). *Branisamys* and *Incamys* were discovered only a few years ago and are known by skulls and tooth rows. The teeth are more hypsodont than those of *Protosteiromys* but hypsodonty is moderate. The teeth of *Branisamys* are about as hypsodont as those of *Paraphiomys pigotti*, but those of *Incamys* are more so. The upper molars of *Branisamys* are clearly five-crested, and the M2 of the type shows this particularly well. In *Incamys* there is an imperfect structure on the anterior slope of the posteroloph, and I interpret this as a metaloph, bearing in mind the vestigial crests which can be found in the recent African *Thryonomys*.

The interpretation of the teeth of *Sallamys pascuali* is not instantly apparent. By careful investigation, one can see that the third crest in M1 is connected to the middle of the posteroloph, as is the metaloph in the Phiomorpha, and there is a very, very small crest apparent at the place where the mesoloph should be present. The structure of M3 is reminiscent of that of *Paraphiomys* and M2 is intermediate.

Considering again the African rodents, *Bathyergoides* represents a good intermediate between the typical ancestral structure and the extremely simplified teeth of the recent Bathyergidae. The molars of *Bathyergoides* are already fairly simple; many of the connections between the crests have disappeared and the crests themselves are more or less composed of tubercles. But there are some samples in which the connections have been preserved and these provide a good guide for relating the dentition of the Bathyergoidea to the ancestral pattern. The teeth of the Hystricidae are easier to interpret. They are tuberculate and, as shown by Ellerman (1940), are similar to those of *Dasyprocta* from South America. The underlying crests appear after wear and then the resemblance to the Phiomorpha pattern can be seen. Descriptions of the teeth of *Atherurus* (Lavocat, 1973) show that their pentalophodont pattern can be derived directly from that of several phiomorphs in which the mesoloph extends to the labial side of the crown. The short fourth crest, which stops halfway across the tooth, is connected to the posterior cingular crest and is an exact reproduction of the disposition to be seen in *Paraphiomys*, *Epiphiomys* and others.

In the lower cheek teeth of the Phiomorpha (Fig. 2B), it is usually possible to distinguish an anterolophid, an hypolophid and a postero-lophid. The presence of a mesolophid or of a posterior arm of the protoconid is highly variable and one can find molars with three or four lobes in the same species. However, I think that the teeth of all the Hystricognathi can be explained using the same basic pattern in spite of their variability.

Number and replacement of premolars

An upper P3 or dP3 and an upper P4 or dP4 are found in several Oligocene and Miocene genera of Africa. *Metaphiomys*, *Diamantomys*, *Andrewsimys*, *Epiphiomys* and *Simonimys* have a P3 or dP3 and *Phiomys* has a dP4 and P4. All other genera seem to have at least a permanent dP4. Of the South American Deseadan genera, *Branisamys* and *Incamys* had dP4 and P4, and the former also had dP3 or P3. This suggests that the ancestral hystricognath animals probably had these characteristics also. Wood (1968) has shown that there was a tendency in the Phiomorpha to retain the milk tooth dP4 as a permanent tooth, although such retention was not present at the beginning of the Oligo-cene in *Phiomys*. In the Hystricidae there is still a normal replacement of dP4 by P4 which has been used as an argument against the Hystri-cidae being descendants of some African Miocene phiomorphs. But there are some indications that an ancestral phiomorph reached the area which is now Pakistan at about the Middle Eocene. The Hystricidae would then have evolved in this region throughout the Tertiary and their first appearance in the Siwaliks would then be explained. At that Eocene epoch, such a phiomorph ancestor must have acquired already the essential characteristics of the Hystricognathi (those of the Therido-morpha were already obvious in Europe at that time) but the normal replacement of the teeth which then existed may have been retained without change in only this branch of the Phiomorpha.

The Hystricidae are in fact as similar to the Phiomorphs of Africa as are the Caviomorpha, which probably separated from the main stem at about the same time.

Other characters

Little is really known about the postcranial osteology in rodents. Nevertheless, the tibiofibular shape and its inter-relationships are similar in *Paraphiomys* and *Thryonomys* in Africa and *Platypittamys* and *Erethizon* in America.

There is a tendency for the frontal sinus to become expanded. Even in the Hystricidae, the expression of this tendency is highly variable; *Hystrix* has a very enlarged sinus, and *Atherurus* has a smaller, though evident, one, which is rather similar to that of *Thryonomys*. A greatly enlarged sinus is also found in some South American genera such as *Coendou*.

In general, the brains of rodents are lisencephalic, but even the early workers had noticed that the brains of *Thryonomys* and several caviomorph genera showed signs of incipient fissuration (Beddard, 1892). I have studied several natural braincasts from the Miocene phiomorphs, *Paraphiomys* and *Diamantomys*. When these were compared with an artificial cast of a recent *Thryonomys* they were found to be almost exactly the same. So we can assume that fissuration is a very ancient character inherited in the Caviomorpha and Phiomorpha from their common ancestor(s).

According to Tullberg (1899) the presence of a sacculus urethralis is typical of the Hystricognathi, a similar, but vestigial (?), structure being found only in the Ctenodactyloidea.

Mossman & Luckett (1968) have considered the fetal membranes of some caviomorphs and the African genus *Bathyergus*: "It seems unlikely that the peculiarly hystricomorph nature of fetal membrane morphogenesis in *Bathyergus* could be the result of convergent evolution, since this system is so obviously free from adaptations to external environment. Therefore we conclude that the Bathyergidae are closely related to New World hystricomorphs."

The tendency to a spiny pelage is also found in the Hystricognathi. This is a relatively minor characteristic and, of course, the extreme adaptation seen in *Hystrix* is not reached by most hystricognaths, or even by all hystricids, but it is remarkable that the Erethizontidae are also very spiny. The fur of the African *Thryonomys* is relatively spiny to touch, and according to Ellerman (1940), a spiny pelage is found in several genera of the Echimyidae of South America.

There are also reported (Woods, 1972) to be great myological similarities between *Thryonomys* and some caviomorphs. Even if it is assumed that these may not be greatly significant factors they at least do demonstrate that myology cannot be used to argue against a fundamental unity of the Hystricognathi.

Bugge's (1971, 1974) work on the cephalic arteries also indicates that there is no justification for separating widely the Old World and New World Hystricognathi.

UNITY OF THE HYSTRICOGNATHI

A definition of an animal taxon is not only anatomical; it should also be historical and evolutionary. A true systematic unit must be homogeneous in its origin and composition and by its evolution. So before answering our question of what is an hystricomorph, I shall consider what I believe may have been the evolutionary history of these rodents.

At the beginning of the Eocene there was a population of very primitive mammals living in Africa. They were derived partly from animals which had arrived from Eurasia, and partly, perhaps, from caenozoic mammals already in Africa. Representatives of the Paramyidae or Sciuravidae would have been among this population and by about the Middle Eocene some of them would have begun their evolution towards the Anomaluroidea and Pedetoidea whilst the others would have started to develop the incipient characteristics later to be found in the Hystricognathi. I believe this to be true by analogy with the European Theridomorpha, in which we know that the main osteological features of the suborder were established as early as the Middle Eocene. Such an ancient origin in Africa is also essential to explain the subsequent history of the group. Furthermore, in the Middle Eocene of Pakistan, there exists a tooth which I think, rightly or wrongly, is of a phiomorph pattern; this early record could represent the migration of the Hystricoidea group. If the genus *Tsaganomys* of the Oligocene in Mongolia is some day proved to be an hystricognath rather than cylindrodontid (although I do not really suppose this is very probable) then an Africa–Pakistan–Mongolia migration would be, to my mind, the only possible explanation of its presence. This part of the story is hypothetical and there are important and large gaps.

The events between Africa and South America are easier to determine. The probability of finding so many coincident characteristics in two unrelated infraorders is so unlikely, to my mind, that any explanation other than close genetic affinities resulting from a common origin seems entirely impossible (Lavocat, 1969). So we have to postulate a migration of a well established but primitive member of the ancestral group from Africa to South America. The migration must have been by raft between the Middle Eocene and Lower Oligocene. The Oligocene is the limit because the caviomorphs were already well established and reasonably diversified by the Deseadan. There are quite big differences of opinion amongst geophysicists about the width of the South Atlantic at that time but there is a consensus that the eastern part of South America could have been on the same meridian as the western part of Africa. Specialist studies of oceanic currents (Lacombe, personal communication) indicate that the current which starts from the south of Africa and passes northwards along the African coast, then across the Atlantic to South America, was probably active by the Oligocene. It is evident that raft travel cannot have been easy, and the filter to animal migration was very effective, but experience has shown that rodents are very quick to take a chance in such situations.

The invocation of migration is not unsupported. Two students of Professor Chabaud of the French Museum of Natural History have been studying the nematodian parasites of rodents. Durette-Desset (1971) has shown that the genus *Paraheligmonella* (F. Heligmosomidae) infests members of the Echimyidae. A related, but more primitive genus, *Heligmonella*, is found infesting *Thryonomys*, while the Pudicinae, which are related to *Paraheligmonella* but are more advanced, infest the echimyids, capromyids, erethizontids and dasyproctids. Different species of *Evaginuris* (F. Oxyurinae) have been found infesting *Hystrix*, *Erethizon*, *Coendou* and *Dinomys* (Quentin, 1973). Both these authors feel that this distribution of the nematode parasites can only be explained by a migration of animals from Africa to South America and this evidence gives further support to a theory which is itself the only one to fit a number of facts.

Wood & Patterson (1970) have criticised the migration hypothesis, but I have not changed my mind; indeed, I think I have found more arguments to support it.

DEFINITION OF THE HYSTRICOGNATHI

The Hystricognathi are rodents whose earliest ancestors were living in the Middle or Early Eocene period in Africa. During the Eocene, some of these ancestral rodents migrated to Asia and others to South America. The mandible is always hystricognathous (Fig. 1); the hystricomorph infraorbital foramen is primitively fairly wide but is secondarily reduced in the Bathyergidae; the pterygoid fossa is deep and opens mostly in the anterior region of the orbito-temporal cavity but secondarily in the cranial cavity; the alisphenoid is low; there is generally a distinct interparietal; the palatine is seen only at the back of the orbito-temporal floor; there is no transverse ductus; the occipital, mastoid and squamosal bones are peculiarly interconnected in the lateral occipital part of the skull; the lacrimal primitively has a vertical canal; the middle ear structures are similar to those of *Diamantomys* and *Lagostomus*; early fissurations are present in the brain; the teeth have evolved from a pentaloph condition (Fig. 2) with an anteroloph, protoloph, variable mesoloph, metaloph connected backwards to the posteroloph, and a posteroloph in the upper teeth, and an anterolophid, variable mesolophid and/or posterior arm of the protoconid, metalophid and posterolophid in the lower teeth; frequently there is a dP3 or P3; there is always P4 or dP4; there is a tendency for the development of frontal sinuses and spiny fur; there is a common plan of the cephalic arterial system; a sacculus urethralis is present; the fetal membranes are of the same basic pattern; and there are affinities of the genera of endoparasites.

ACKNOWLEDGMENTS

I am indebted to the Zoological Society of London for the kind invitation, conveyed by the Organizers, to participate in this symposium.

REFERENCES

Beddard, F. E. (1892). On the convolutions of the cerebral hemispheres in certain rodents. *Proc. zool. Soc. Lond.* **1892**: 596–613.

Bugge, J. (1971). The cephalic arterial system in New and Old World hystricomorphs and in bathyergids with special reference to the systematic classification of rodents. *Acta anat.* **80**: 516–536.

Bugge, J. (1974). The cephalic arteries of hystricomorph rodents. *Symp. zool. Soc. Lond.* No. 34: 61–78.

Durette-Desset, M. C. (1971). Essai de classification des Nématodes Héligmosomes. Corrélations avec la paléobiogéographie des hôtes. *Mém. Mus. natn. Hist. nat. Paris* N.S. **69**: 1–126.

Ellerman, J. R. (1940). *The families and genera of living rodents* **I**: 1–689. London: British Mus. (Nat. Hist.)

Hoffstetter, R. & Lavocat, R. (1970). Découverte dans le Déséadien de Bolivie de genres pentalophodontes appuyant les affinités africaines des rongeurs caviomorphes. *C. r. hebd. Séanc. Acad. Sci., Paris* **271**: 434–437.

Landry, S. O. (1957). The interrelationships of the New and Old World hystricomorph rodents. *Univ. Calif. Publs Zool.* **56**: 1–118.

Lavocat, R. (1967). Observations sur la région auditive des Rongeurs Théridomorphes. In *Problèmes actuels de paléontologie, Colloq. int. Cent. nat. Rech. sci.* **163**: 491–501.

Lavocat, R. (1969). La systématique des Rongeurs Hystricomorphes et la dérive des continents. *C. r. hebd. Séanc. Acad. Sci., Paris* **269**: 1496–1497.

Lavocat, R. (1971). Affinités systématiques des Caviomorphes et des Phiomorphes et origine africaine des Caviomorphes. *Anais Acad. bras. Cienc.* **43**, suppl.: 515–522.

Lavocat, R. (1973). Les rongeurs du Miocène d'Afrique Orientale. **I**. Miocène inférieur. *Trav. Mém. Inst. E.P.H.E. Montpellier* **1**: 1–284.

Matthew, W. D. & Granger, W. (1923). New Bathyergidae from the Oligocene of Mongolia. *Am. Mus. Novit.* No. 101: 1–5.

Mossman, H. W. & Luckett, W. P. (1968). Phylogenetic relationships of the African mole rat *Bathyergus janetta*, as indicated by the fetal membranes. *Am. Zool.* **8**: 806.

Quentin, J. C. (1973). Affinités entre les Oxyures parasites de Rongeurs Hystricidés, Erethizontidés et Dinomyidés. Intérêt paléobiogéographique. *C. r. hebd. Séanc. Acad. Sci., Paris* **276**: 2015–2017.

Simpson, G. G. (1945). The principles of classification and a classification of mammals. *Bull. Am. Mus. nat. Hist.* **85**: 1–350.

Tullberg, T. (1899). Ueber das System der Nagetiere; eine phylogenetische Studie. *Nova Acta R. Soc. Scient. Upsala* **18**: 1–154.

Wood, A. E. (1968). The African Oligocene Rodentia. *Bull. Peabody Mus. nat. Hist.* **28**: 23–105.

Wood, A. E. & Patterson, B. (1959). The rodents of the Deseadan Oligocene of Patagonia and the beginnings of South American rodent evolution. *Bull. Mus. comp. Zool. Harv.* **120**: 279–428.

Wood, A. E. & Patterson, B. (1970). Relationships among hystricognathous and hystricomorphous rodents. *Mammalia* **34**: 628–639.

Woods, C. A. (1972). Comparative myology of jaw, hyoid, and pectoral appendicular regions of New and Old World hystricomorph rodents. *Bull. Am. Mus. nat. Hist.* **147**: 115–198.

Symp. zool. Soc. Lond. (1974) No. 34, 21–60

THE EVOLUTION OF THE OLD WORLD
AND NEW WORLD HYSTRICOMORPHS

ALBERT E. WOOD

*Amherst College, Amherst, Mass., U.S.A.**

SYNOPSIS

Among hystricomorphous rodents are the South American Caviomorpha, the African
Phiomorpha and the Old World Hystricidae, all hystricognathous. The first two groups
appear in the fossil record in the early Oligocene, the third in the late Miocene. The
primitive caviomorph had a small infraorbital foramen with little or no penetration by
the masseter muscle, four-crested upper and lower molars with clearly distinguishable
cusps and normal replacement of deciduous teeth, but was hystricognathous and had
multiserial incisor enamel. The primitive phiomorph differed in having five-crested
molars and deciduous teeth that were usually not replaced. The hystricids, with normal
replacement of deciduous teeth, could not have been descended from any known phiomorph. Five sciurognathous and hystricomorphous families or superfamilies are known,
which can have no relationships with the hystricognathous groups or each other, and
must have evolved independently. The only known Eocene rodents that could possibly
have been ancestral to the hystricognathous groups are from the United States and
Mexico. The distribution of the features, sometimes considered characteristic of a suborder "Hystricomorpha", is reviewed, and the extreme amount of parallelism in the
evolution of these features, no matter what taxonomic arrangement one adopts, is
emphasized. Since parallelism is a strong indication of relationship, this supports a
common ancestry for the hystricognaths, but they must have been independently derived
from non-hystricomorphous and at most sub-hystricognathous northern hemisphere
ancestors that had been evolving independently in the Old and New Worlds since the
late Palaeocene or early Eocene.

INTRODUCTION

In the previous paper, Lavocat (this volume, pp. 7–20) has set up a
definition of what we both believe should be included in a discussion of
hystricomorph rodents. We are in agreement that certain of these forms
—the South American Caviomorpha, the African Phiomorpha (including the Bathyergidae) and the Old World Hystricidae—are probably
related to each other, but we disagree as to the date of their latest
common ancestor and the manner in which they attained their present
distribution. We also disagree as to the stage of evolution that had been
achieved by this common ancestor.

The term hystricomorph refers to the migration of the origin of the
masseter medialis through the greatly enlarged infraorbital foramen

* Present address: *20 Hereford Ave., Cape May Court House, New Jersey 08210 U.S.A.*

onto the snout in front of the zygomatic arch. Hystricomorphy, sciuromorphy and myomorphy were derived from the ancestral protrogomorphous condition (Fig. 1). Because this feature (hystricomorphy)
occurs in many rodents not related to the three groups mentioned
above, I do not like to use hystricomorph as a taxonomic term. As
Lavocat has indicated, the Theridomyoidea from the Eocene and Oligocene of Europe, the Ctenodactyloidea from the Eocene to the Pliocene
of Asia and the Miocene to the Recent of Africa, and the Anomaluridae
and Pedetidae from the Miocene to the Recent of Africa, are hystricomorphous. So are the Dipodoidea, even though they are rarely considered
to be related to any of the other forms. This second group of rodents
possess what is termed a sciurognathous mandible, in which the angular
process arises from the ventral surface of the incisive alveolus, and lies
in the plane of that alveolus (although the process itself may be curved
laterally or mesially).

The palaeontologic evidence has convinced both Lavocat and me that
none of these sciurognathous groups has anything to do with the Caviomorpha, Phiomorpha and Hystricidae, which are all hystricomorphous
and hystricognathous, i.e. the angular process of the mandible arises
from the lateral surface of the incisive alveolus, and the two are not co-
planar.

The origin of these hystricognath groups remains a major problem for
which there is no agreed solution. The Caviomorpha appear in the early
Oligocene of South America; the Phiomorpha in the early Oligocene of
Egypt; the Hystricidae in the late Miocene of southern Asia and the late
Miocene or early Pliocene of Europe and North Africa (Wood & Patterson, 1970: 632). In each case, there is no trace of possible ancestors within
the area where they subsequently evolved. A half century or more ago,
the problem was solved by deriving the African and South American
forms from the European Eocene or Oligocene Theridomyidae, which
were assumed to have migrated over a variety of land bridges (Wood,
1950). Later work has shown that there were no land bridges across the
South Atlantic over which Eocene rodents might have wandered. The
recent resurgence of interest in the possibility of continental drift, now
under the name of plate tectonics, has suggested to some authors
(Lavocat, 1969, 1971; Hoffstetter & Lavocat, 1970; Hoffstetter, 1972)
that the origin of the South American Caviomorpha could be explained
by raft transportation across the South Atlantic from Africa in the late
Eocene when the continents were assumed to have been much closer
than now. The fact that the South Atlantic was demonstrably at least
3000 km wide in the late Cretaceous (Maxwell *et al.*, 1970) and, therefore, presumably even wider in the late Eocene, has been considered of

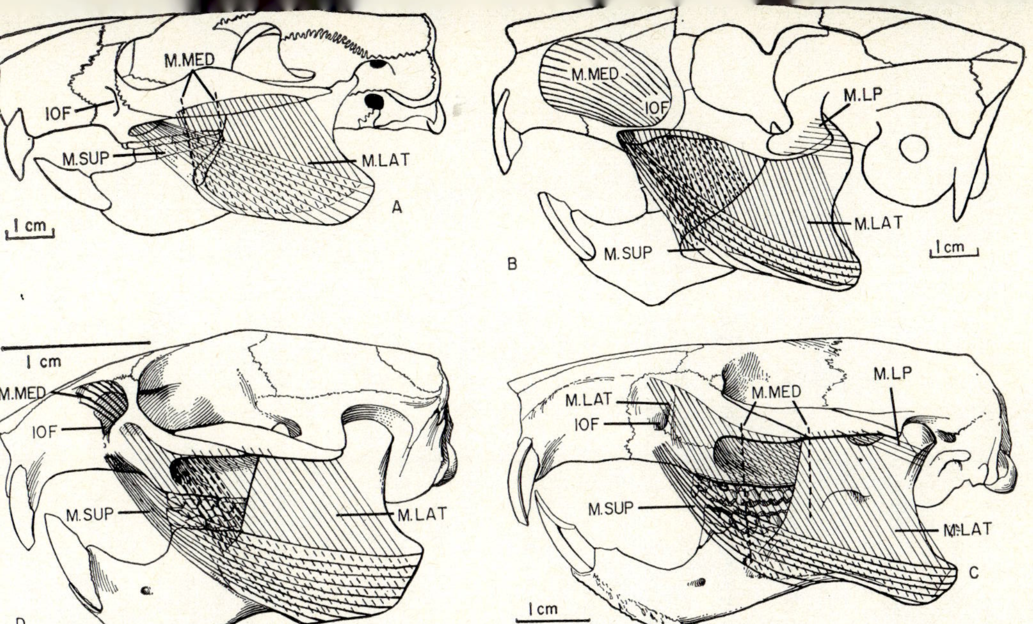

Fig. 1. Restoration of the four types of masseter muscles in rodents. Abbreviations: IOF = infraorbital foramen; M LAT = masseter lateralis, dashed portions lying beneath masseter superficialis; MLP = masseter lateralis profundus, pars posterior, deep division; M MED = masseter medialis, its concealed course indicated by dashed lines; M SUP = masseter superficialis.

A. Protrogomorphous: the primitive type with the origin of the masseter limited to the zygoma (*Paramys delicatus*; middle Eocene of Wyoming).

B. Hystricomorphous: the origin of the masseter medialis has migrated forward, through a greatly enlarged infraorbital foramen, and occupies much of the side of the snout; the masseter lateralis profundus, pars posterior, deep, has differentiated and inserts on the ascending ramus of the jaw behind the condyle (*Neoreomys australis*; middle Miocene of Patagonia).

C. Sciuromorphous: the origin of the masseter lateralis has spread upward in front of the zygoma, lateral to the infraorbital foramen, almost to the top of the snout (*Eutypomys thomsoni*; middle Oligocene of South Dakota).

D. Myomorphous: a combination of the hystricomorphous and sciuromorphous types (*Eumys elegans*; middle Oligocene of South Dakota).

minor importance by Hoffstetter and Lavocat, although the possibility of a transatlantic crossing still seems to me to be as improbable as it ever did (Wood, 1950).

There are, unquestionably, great similarities between the Caviomorpha and Phiomorpha; so many, in fact, that apparently no one in his right mind could doubt that the two groups had a special relationship. This, therefore, poses the dilemma that I shall discuss in more detail.

THE HYSTRICOGNATHS

The beginnings of the Caviomorpha

The Caviomorpha, or South American hystricognaths, are very clearly a natural group, having had either a common ancestor that invaded South American from elsewhere, or a small group of closely related ancestors, who reached South America from the same homeland at essentially the same time. Relying on Occam's Razor, I shall accept the first alternative, but there is evidence suggesting the second possibility.

There is no trace of rodents in the quite diverse faunas of any of the known South American Eocene deposits, unfortunately largely limited to Patagonia. Rodents are abundant in fossiliferous deposits of areas where they were in existence; therefore, I believe it to be certain that rodents were not present in Patagonia during Mustersan (late Eocene) time (for geologic terms, see Table I). Their subsequent history convinces me that they had not yet established themselves in South America at this time, but that they arrived during the several-million-year hiatus between Mustersan and the early Oligocene Deseadan. In the Deseadan deposits of Patagonia, the rodents are a diverse group, represented by nine genera, referred to six families (Wood & Patterson, 1959; Patterson & Pascual, 1968). They were, however, much more closely related to each other than this splitting would indicate, and an early Oligocene taxonomist would probably have placed them all in a single family. They are referred, however, to the living families of which they are the early stages. The overall picture presented by the diversity of these animals is somewhat like that of the living murids of Australia, and, I believe, indicates a similar recent (less than four or five million years old) invasion of the continent. Recently, three additional rodent genera of Deseadan age have been described from Bolivia (Hoffstetter & Lavocat, 1970). One of these represents the early stages of another modern family, not so far reported from Patagonia, and a second belongs to a distinct subfamily of a family present in Argentina. Hoffstetter *et al.* (1971: 2217) have reported a second Bolivian Deseadan occurrence, with representatives of two other families, previously known from Patagonia. In Bolivia, as in Patagonia, the rodents are, in individuals, the most abundant order represented in the collections.

This early Oligocene diversity, with no late Eocene South American ancestors, is what one would expect if the rodents reached South America during the latest Eocene or earliest Oligocene. There were no real competitors in the rodent niche, and they diversified very rapidly (an evolutionary explosion) and quickly became a dominant element in the fauna.

Among the Deseadan caviomorphs, some genera are obviously specialized, in that the cheek teeth have lost all traces of the ancestral cusps and have become very hypsodont. A few of the known Deseadan

TABLE I

Geological terms used in the text

Epochs of the Caenozoic	South America	North America	Europe	Asia	Africa
Pleistocene				SIWALIKS	
Pliocene					
Miocene				Chinji	Kenya and SW Africa
Oligocene	Deseadan		QUERCY Aquitanian		Fayum
Eocene	Mustersan	Washaskie	Lutetian		
Palaeocene					

caviomorphs had low-crowned cheek teeth, still showing distinct, rounded cusps in the crests of both uppers and lowers. All that is known of the dental evolution of rodents—indeed, of all mammals—clearly indicates that these genera are the more primitive, and that the ones that do *not* show identifiable cusps are the more advanced.

Rounded cusps (including, in some cases, the conules of the upper cheek teeth, which in the development of lophate rodent teeth, are always the first cusps to disappear) are present in *Platypittamys*

(Fig. 2), in *Deseadomys arambourgi* (Wood & Patterson, 1959, Figs 4 & 5) and *Xylechimys obliquus* (Patterson & Pascual, 1968, Fig. 2). These three genera are referred to the Octodontidae and Echimyidae, the most primitive families of the Caviomorpha, and are at least as brachydont as any other genera of Deseadan caviomorphs.

Another very important feature of all these obviously primitive caviomorphs is that there is never, in any of them, any suggestion that there were more than four crests in the molars. Hoffstetter & Lavocat (1970: 174) and Lavocat (1971: 516) categorically deny the primitive nature of *Platypittamys*, but give no evidence to support their point of view. The separate cusps of the teeth are only slightly less obvious in the Bolivian echimyid *Sallamys* in which Hoffstetter & Lavocat (1970) identified as a mesoloph, and hence a fifth crest, a structure in M^1 that seems to me to be the lingual end of an interrupted metaloph, as can be seen in their picture of M^{2-3} of the same specimen (see Discussion, Fig. 1a).

In all of these genera, the premolars were distinctly less complex than the molars and this is also a primitive character. In *Deseadomys*, *Xylechimys* and *Sallamys* there clearly was a change from the primitive condition and an active trend towards the loss of the second crest on the lower molars (see Wood & Patterson, 1959: Fig. 5, *D. arambourgi* and Fig. 7, *D. loomisi*; Patterson & Pascual, 1968: Fig. 2, *X. obliquus*; Patterson & Wood, in preparation: *S. pascuali*).

Among other Deseadan rodents, the lower molars are generally four-crested (Wood & Patterson, 1959: Fig. 19, *Cephalomys*; Fig. 26, *Asteromys*; Figs 31 & 32, *Protosteiromys*; Hoffstetter & Lavocat, 1970: Fig., lowers of *Branisamys* and *Incamys*), whereas the uppers are more complex. In the upper molars of the low-crowned *Protosteiromys medianus* (see Discussion, Fig. 1b and Wood & Patterson, 1959: Fig. 30), there are three anterior crests and a partly subdivided posterior crest. There usually are rounded enlargements at the buccal ends of the second and third crests, the remnants of the paracone and metacone, demonstrating that these crests are the protoloph and metaloph, respectively. The partly subdivided posterior crest, must be, therefore, the posteroloph which is in the process of giving rise to a new crest (the neoloph) that ultimately comes to occupy the space between the metaloph and the posteroloph (Wood & Patterson, 1959: 288, 333–336, 380). A similar indication is given by the upper molars of *Cephalomys*, seemingly formed of five crests after wear, but in which there were only four crests in the virgin state (Wood & Patterson, 1959: Fig. 15A).

Skulls or partial skulls of Deseadan caviomorphs are not abundant; a number are known from Bolivia, but only three from Patagonia, of

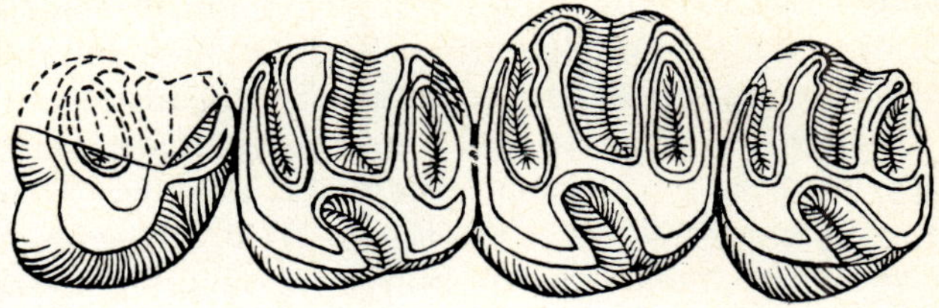

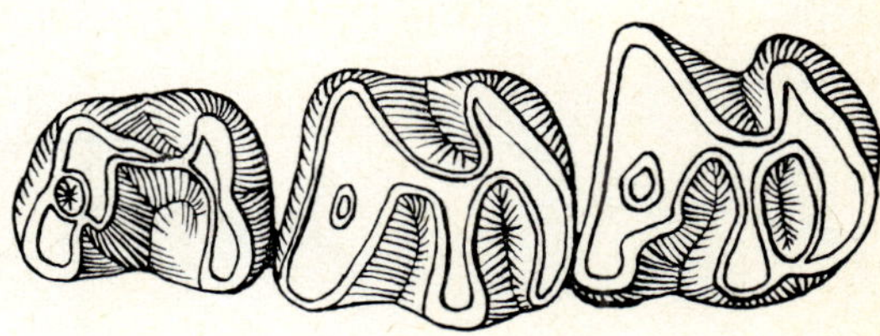

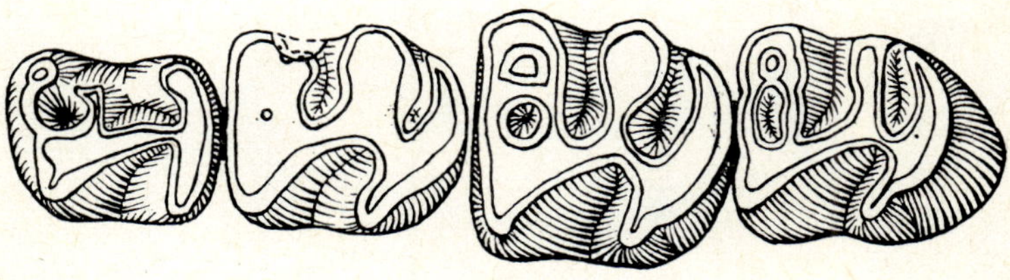

Fig. 2. Upper (A, B) and lower (C, D) cheek teeth of the early Oligocene caviomorph *Platypittamys brachyodon* from Patagonia, showing the primitive caviomorph tooth pattern, × 15·9. Anterior end of B to the right; anterior ends of A, C and D to the left.

A. Left P⁴ — M³ (American Museum of Natural History 29601; P⁴ restored from AMNH 29600).

B. Right P⁴ (AMNH 29600).

C. Left P₄ — M₂ (AMNH 29601).

D. Right P₄ — M₃ (AMNH 29600).

(From Wood, 1949, Fig. 3)

which two belong to *Platypittamys*. For a caviomorph the *Platypittamys*
skulls (Fig. 3) have a very small infraorbital foramen and unlike Landry
(1957: 93) I have never been able to detect any expansion of the masseter
medialis through the foramen and onto the face, although the muscle
may have penetrated the foramen a very short distance. All other known
Deseadan caviomorph skulls (*Cephalomys* from Patagonia and *Branisamys*, *Incamys* and *Sallamys* from Bolivia) show full penetration of the
large infraorbital foramen by the masseter medialis, although there is
some variation in just how far forward the muscle reaches. This would
seem to be evidence for the development of hystricomorphy within the
Caviomorpha after they reached South America (Patterson & Pascual,
1972: 278), which would rule out any possible direct descent of the Caviomorpha from the African Phiomorpha.

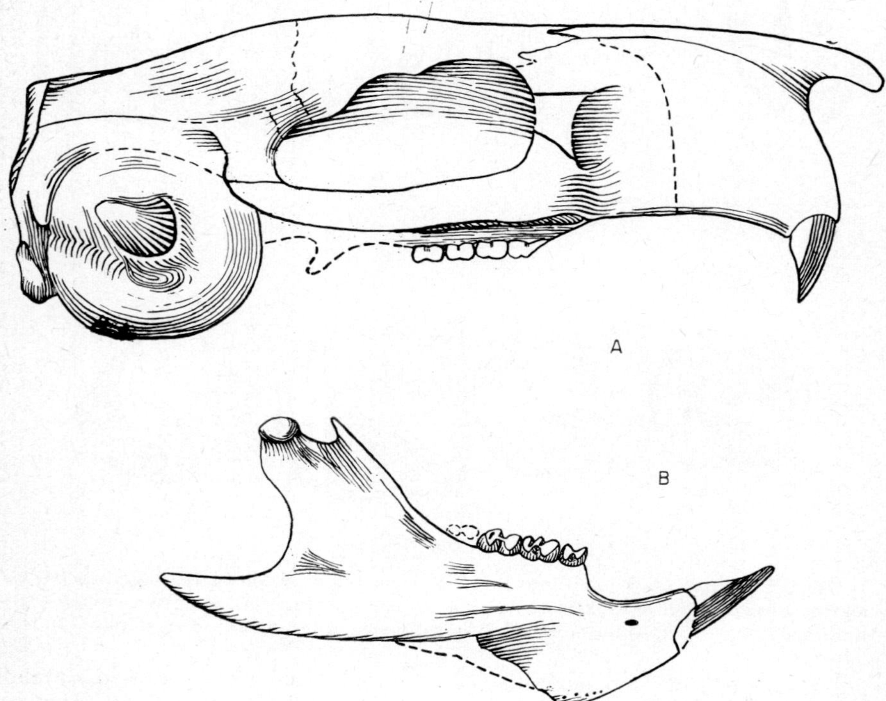

Fig. 3. Skull and lower jaw of the early Oligocene caviomorph, *Platypittamys
brachyodon*, from Patagonia, × 3.
   A. Right side of skull (American Museum of Natural History 29601).
   B. Lateral view of right lower jaw (AMNH 29601) partly restored from AMNH
29600.
(From Wood, 1949, Fig. 2)

There seems to me to be no question as to the fact that the animals with brachydont cheek teeth, rounded cusps, identifiable conules on the upper molars and a small infraorbital foramen are primitive, whereas those with high-crowned cheek teeth, cusps inseparable from the crests, no conules, and large infraorbital foramina are more specialized. Furthermore, the primitive characters are concentrated in animals referable to the two primitive caviomorph families, the Echimyidae and Octodontidae. The four-crested pattern, which is also found in these more primitive caviomorphs, must therefore also be the primitive pattern for the group. Hoffstetter & Lavocat (1970: 174) and Lavocat (1971: 516) disagree completely with this view, and accept the primitiveness of the pentalophodont pattern, for no very valid reason that I can see, unless they have taken the view of Schaub (in Stehlin & Schaub, 1951) that five-crested rodent teeth are always more primitive than four-crested ones. Hoffstetter & Lavocat (1970) stated (in my opinion, in error) that "les deux genres montrant la stucture la plus complète ont des dents brachyodontes, donc plus conservatrice." It is impossible to determine from their text to which two genera they refer, although I would assume one to be *Protosteiromys*. The occurrence of its four-crested teeth in association with the other clearly primitive characters implies that the cheek teeth are just as primitive as they appear to be.

Among the Deseadan caviomorphs, dP^4_4 are always replaced, in the normal mammalian fashion, by P^4_4, as is the case in all post-Deseadan caviomorphs except the Echimyidae. There is a completely normal replacement of deciduous teeth in the three known Deseadan echimyid genera, *Deseadomys*, *Xylechimys* and *Sallamys*. However, in all later echimyids (Wood & Patterson, 1959: 301), the deciduous teeth are retained throughout life and the permanent premolars are suppressed. This retention of the deciduous teeth was, therefore, not a characteristic of the ancestral caviomorph, but was a feature that evolved rapidly and independently in several phyletic lines (Patterson & Pascual, 1968) within the Echimyidae only in South America.

Where the jaws are adequately preserved, there is always a typical hystricognathy and a backward expansion of the ascending ramus, just behind and below the condyle, which marks the insertion of the pars posterior, deep division, of the masseter lateralis profundus (Woods, 1972: 127–128).

From this review a picture emerges of the ancestral caviomorph, the animal that presumably invaded South America. The angle of the lower jaw arose lateral to the incisive alveolus—hystricognathy—as in all known caviomorphs. There was a distinct pars posterior of the masseter lateralis profundus. The infraorbital foramen was somewhat enlarged,

probably having a vertical diameter about a third of that of the snout
(Fig. 3A), but with little or no penetration by the masseter medialis.
That penetration and enlargement of the foramen may proceed *pari
passu* is shown by the sequence in the European Eocene and Oligocene
pseudosciurids *Protadelomys* and *Adelomys* (Fig. 4), as pointed out by

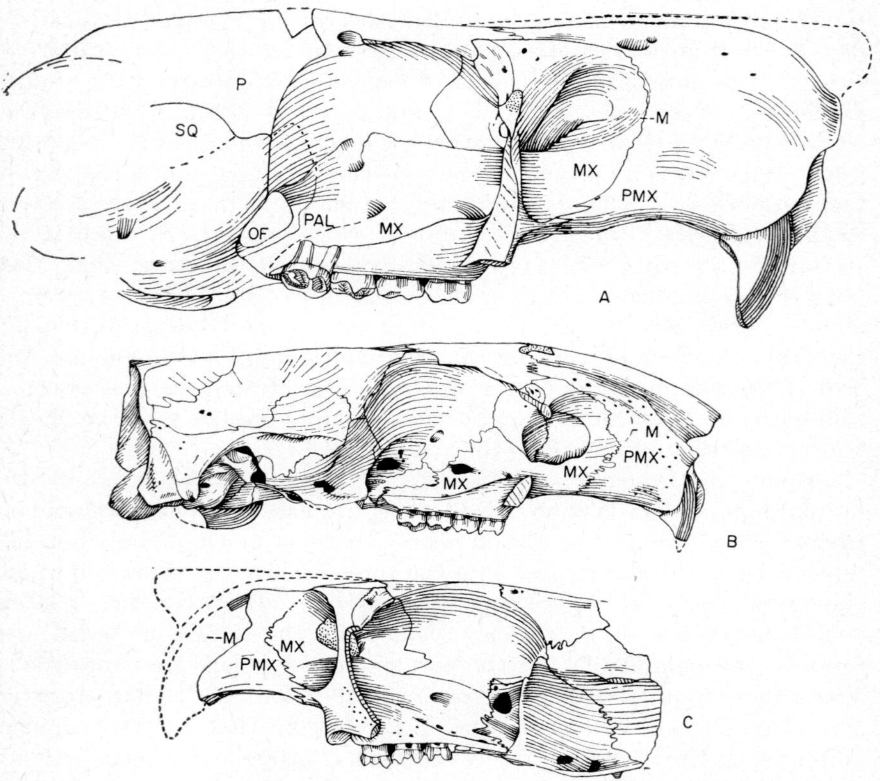

Fig. 4. Lateral views of skulls of European pseudosciurid theridomorphs showing
progressive enlargement of the infraorbital foramen and concomitant forward movement
of the origin of the masseter medialis.

A. Right side of skull of *Protadelomys cartieri* (Naturhistorisches Museum, Basel,
EK 245), Lutetian of Egerkingen, Switzerland, × 4·75. Vertical diameter of foramen
about a third of the height of snout.

B. Right side of skull of *Sciuroides* sp. (Basel, QT 756), late Eocene of Quercy,
France, × 1·93. Vertical diameter of foramen half of the height of snout.

C. Left side of skull of *Adelomys* sp. (Musée de Montauban), late Eocene of Quercy,
France, × 2·4. Vertical diameter of foramen ¾ the height of snout.

Abbreviations: M = anterior end of fossa for origin of masseter medialis (approximate
on C); MX = maxilla; OF = optic foramen; P = parietal; PAL = palatine; PMX = premaxilla; SQ = squamosal.

Hartenberger (1968). Presumably, as in all caviomorphs, the ancestral cheek teeth were P^4_4, M^{1-3}_{1-3}, and probably dP^3 was present (as well as dP^4_4) in the ancestral stage, and there was no tendency for the retention of the deciduous teeth. An alveolus for dP^3 of *Branisamys* is shown by Hoffstetter & Lavocat (1970) and an unerupted and partly resorbed dP^3 seems to be present in one specimen of *Sallamys*. A dP^3 is present also in *Incamys* but no statement can be made about this tooth, except that it is not preserved among the rodents of the Patagonian Deseadan. The cheek teeth were brachydont and the molars were four-crested, with the cusps still clearly differentiated. The incisor enamel was presumably multiserial, as it is in all caviomorphs that have been checked (Wahlert, 1968: 16–17).

There are no known fossils anywhere in the world that precisely fit this description. The closest approaches are to be found among certain North American Eocene rodents, including the Reithroparamyinae, *Prolapsus* and *Protoptychus*, some of which had four-crested cheek teeth. Hystricognathy is incipient in the Palaeocene and Eocene Reithroparamyinae (Wood, 1962a: 117, 122, 136 and Figs 41E, 46B) and is fully developed in *Prolapsus* (see Fig. 5; Wood, 1972, 1973). There are the beginnings of associated hystricomorphy and hystricognathy in *Protoptychus* (Wahlert, 1973) and probably in the reithroparamyine *Rapamys* (Wood, 1962a: 148 and Fig. 52A) from the late Eocene. As far as known, these forms all have pauciserial incisor enamel. These fossils were all found in the United States. The ancestors of the Caviomorpha, if they came from the north, presumably were living in Middle America, where Tertiary fossil vertebrates are almost unknown (Patterson & Pascual, 1972: 252–257 and Table 2). Recent studies of Eocene to early Oligocene rodents from south-western Texas (Wood, 1973, 1974) and from Guanajuato in central Mexico (Black & Stephens,

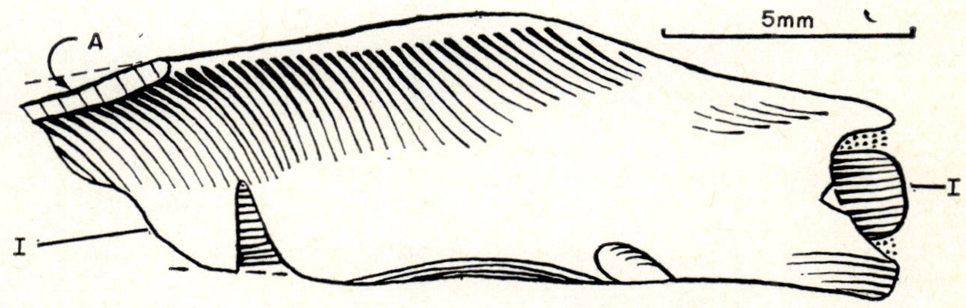

FIG. 5. Ventral view of lower jaw of the hystricognath *Prolapsus sibilatoris* (Texas Memorial Museum 41372–179), mid-Eocene of Texas, showing angular process (A) arising entirely lateral to the plane of the incisor (I–I), × 5·6. (From Wood, 1972, Fig. 1)

TABLE II

The distribution of characters supposedly indicative of hystricomorph relationships

Taxon	Hystrico-morphous	Hystrico-gnathous	Deep division, pars posterior, masseter lateralis profundus	Fusion of malleus and incus	Multiserial incisor enamel	Sacculus urethralis
Caviomorpha	+	+	+	±[1]	+	+
Phiomorpha	+	+	+	+	+	+
Bathyergidae	—[2]	+	+	+	+	+
Hystricidae	+	+	+	+	+	+
Theridomorpha	+	—	—	—	—	
Anomaluridae	+	—	—	—	—	—
Pedetidae	+	—	—	±[3]	+	—
Ctenodactylidae	+	—	—	+	+	±[4]
Reithroparamyinae	?—[5]	±[6]	±[7]		—	
Prolapsus		+			—	
Protoptychus	+	+				
Dipodoidea	+	—		—	—	
Eutypomyidae	—	—	+		—	—

+ = present; — = absent; blank indicates unknown.
[1] Except in *Proechimys, Echimys, Octodon, Spalacopus, Aconaemys.*
[2] Probably secondarily reduced.
[3] Not fused but so closely appressed that little or no motion is possible between them.
[4] A structure is present which has been interpreted as an incipient or a degenerate sacculus urethralis.
[5] None except, possibly, *Rapamys.*
[6] Incipient hystricognathy is a diagnostic character of the subfamily.
[7] There is evidence of this muscle in some members of the subfamily.

1973) suggest that there was a varied Middle American rodent fauna, including at least two hystricognathous lines (*Prolapsus* from Texas and *Guanajuatomys* from Mexico). All of the North American and Middle American possible pre-caviomorphs possessed low-crowned cheek teeth, and most of them had four-crested molars. Although few if any of the currently known possible pre-caviomorphs from Mexico or the United States could have been directly ancestral to the Caviomorpha, the group as a whole does have the necessary features (except for the absence, as far as we are aware, of multiserial incisor enamel), and is the only group of rodents known from the Eocene of any part of the world of which this is true.

The beginnings of the Phiomorpha

The Phiomorpha are the African hystricognathous rodents. Wood (1955) originally established the Phiomyidae as a separate family of early Tertiary African rodents. Lavocat (1962) showed that they were of considerable importance as the beginning of an African radiation, previously hinted at by the work of Andrews (1914), Stromer (1926) and Hopwood (1929). Thaler (1966) raised the group to subordinal rank (not named), for which Lavocat (1969: 1496) later coined the term 'Phiomorphes'. Wood (1968) showed that the Phiomorpha were abundant and dominant rodents in the African Oligocene, and Lavocat (1962) demonstrated that they had become highly diversified by the Miocene. The Miocene invasion of Eurasian rodents greatly reduced the phiomorph diversity, and their present survivors are the relict *Thryonomys* and *Petromus* (Wood, 1962b), each the end stage of a line already recognizable in the Oligocene (Wood, 1968). Finally, I believe that Lavocat is showing (1973) that the burrowing Bathyergidae are a native African group, of phiomyid ancestry, in which hystricomorphy has been secondarily reduced or lost.

The phiomorphs are almost exclusively African. They have been reported elsewhere only from the Miocene of Chios (Tobien, 1968: 54), and the late Miocene of the Siwaliks of Pakistan (Black, 1972: 243–246, Fig. 11), whence Hinton (1933) described the genus *Paraulacodus* unillustrated and with an inadequate description that precluded the certain determination of its affinities.

The Fayum phiomyids are the earliest described African rodents, although Savage (1969: 69; 1971: 220) has cited isolated incisors from the late Eocene or early Oligocene of Libya. He informs me (R. J. G. Savage, personal communication, 6 February 1973) that these are most probably early Oligocene. In the Fayum, there is the same type of diversity among the rodents as in the Deseadan—all the rodents are clearly related, with enough differentiation possibly to have reached the level of there being two recognizable subfamilies. Here, however, all five Oligocene genera were referred by Wood to an ancestral family, the Phiomyidae, rather than being placed in descendant families. This was because, in South America, the Deseadan rodents were described after all the later stages were known, whereas in Africa the description of the Fayum rodents preceded that of the rich East African faunas. The internal diversity, therefore, seems to have been about the same on the two continents; this suggests, in view of the total absence of known possible competitors in both continents, that the phiomorphs had been in Africa about as long, before Fayum time, as the caviomorphs had been in South America before the Deseadan.

The Oligocene phiomorphs show, perhaps even more clearly than do the Oligocene caviomorphs, that there had been but a single ancestral stock that invaded Africa. A study of their cheek teeth shows that two genera (*Phiomys* and *Phiocricetomys*) have very brachydont cheek teeth in which at least the major cusps are still present as rounded enlargements of the crests; two other genera (*Paraphiomys* and *Metaphiomys*) are somewhat higher crowned than *Phiomys* and are fully crested, with the cusps absorbed into the crests, although the four primary cusps are still identifiable after wear; and in one more genus (*Gaudeamus*) in which the teeth are fully crested, the lowers are considerably higher crowned than in the other genera, and the uppers are beginning to develop unilateral hypsodonty. Skulls are unknown among Oligocene phiomyids; maxillary fragments of *Metaphiomys* (Wood, 1968: Fig. 6B) suggest that there was a fully hystricomorphous infraorbital foramen; those of *Gaudeamus* (Wood, 1968: Fig. 14B) indicate a rather smaller infraorbital foramen, possibly no larger than that of the Patagonian *Platypittamys*. All the phiomyids are fully hystricognathous, with the angle well lateral to the incisive alveolus. The incisor enamel is multiserial, as in the living genera (Wahlert, 1968: 17). The portion of the jaw immediately behind and below the condyle is not preserved in any available specimen from the Fayum (Wood, 1968: Figs 3C, 7), or in any of the Miocene material from Southwest Africa, described by Stromer (1926: Pls 41 & 42). I suspect, however, that there probably already was a deep division of the pars posterior of the masseter lateralis profundus as in living phiomorphs.

The analysis of the ancestry of the phiomyids begins with the assumption, as in the caviomorphs, that brachydont are more primitive than hypsodont teeth. In one brachydont genus, *Phiocricetomys*, known from a single lower jaw, there are only three cheek teeth, the cusps are unusually rounded, and there are extensive cingula (Wood, 1968:

Fig. 16), all of which seem to be specializations. The other brachydont genus with clearly recognizable cusps, *Phiomys*, is, I believe, the most primitive phiomorph genus known (Fig. 6). It could have been structurally (or even genetically) ancestral to the other phiomorphs, and should give clues as to the tooth pattern of the ancestral stock. Within the Fayum deposits, there seems to be an evolutionary trend from *Phiomys* leading to *Paraphiomys*; *Metaphiomys* must have been derived from

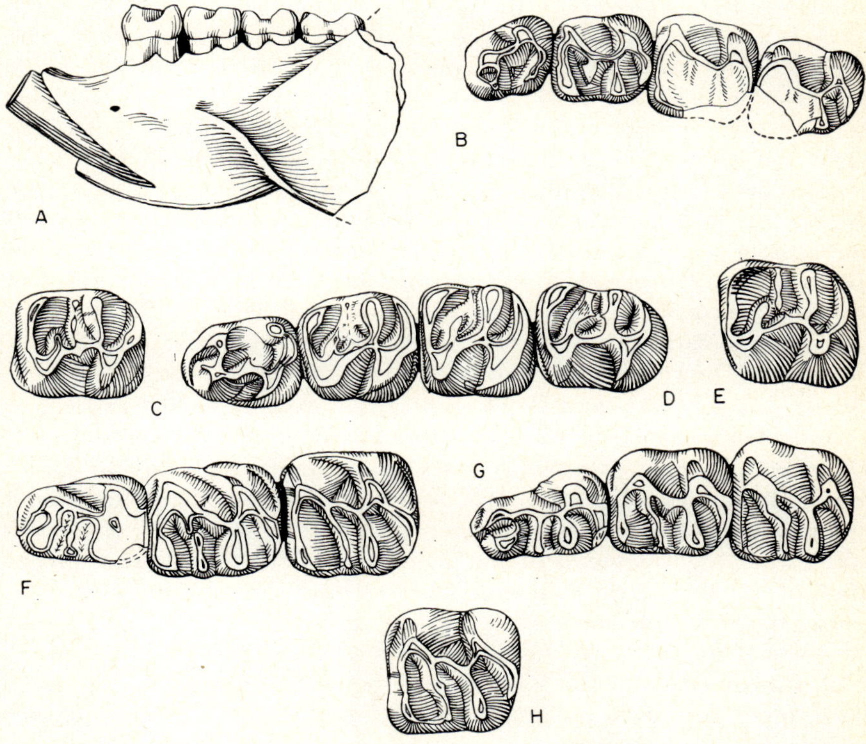

FIG. 6. Jaw and lower teeth of the early Oligocene phiomorph *Phiomys andrewsi* from Egypt, showing the primitive phiomorph tooth pattern. Anterior ends of all figures to the left. Jaw × 4·4, teeth × 8·8.

A. Lateral view of left lower jaw (Yale Peabody Museum 18071).
B. Right $P_4 - M_3$ (American Museum of Natural History 13275).
C. Left M_1 (YPM 18060).
D. Left $P_4 - M_3$ (YPM 18071).
E. Left M_2 (AMNH 13274).
F. Right deciduous $P_4 - M_3$ (YPM 18064).
G. Right deciduous $P_4 - M_2$ (YPM 13271).
H. Right M_2 (YPM 18076).

(From Wood, 1968, Fig. 1)

something very similar to *Phiomys andrewsi*, and the most divergent genera, *Phiocricetomys* and *Gaudeamus*, probably had an ancestor not too different from *P. andrewsi* (Wood, 1968: Fig. 17).

Within *Phiomys*, there are specimens in which the cheek teeth, both uppers and lowers, are five-crested, with a well developed mesoloph or mesolophid. In the lower teeth, the posterior two crests (Fig. 6C to H) are clearly the hypolophid and the posterolophid of rodents in general (Wood & Wilson, 1936: Fig. 2A). The anterior crest of the lower molars runs from the metaconid to the protoconid along the front of the tooth (Fig 6C to H), and seems to be what Wood & Wilson (1936: 390) termed a 'metalophulid I'; the anterior cingulum has nothing to do with this crest (Wood, 1968; Figs 1D, 2B, 3A & B, 8–10, 16D). The posterior arm of the protoconid forms what is, primitively, an independent crest, but it soon joins with part of the mesolophid, after which both the mesoconid and its attachment to the mesolophid are lost (Wood, 1968: Figs. 1 & 2: 38 & 39). This is followed by reduction and loss of the posterior arm of the protoconid, so that *Phiomys lavocati*, *Phiomys paraphiomyoides*, *Paraphiomys* and *Gaudeamus* have only three crests in the lower molars.

Upper molars are not abundant in the Fayum collections. Specimens are known, however, for *Phiomys andrewsi*, *Phiomys paraphiomyoides*, *Metaphiomys schaubi* and *Gaudeamus*. In all except the last (which, as stated above, is one of the most specialized of the Fayum rodents), these teeth have five transverse crests, the anteroloph, protoloph, mesoloph, metaloph and posteroloph, which occasionally have additional complications. In contrast to the more primitive caviomorphs, the metacone, when identifiable, is at the buccal end of the fourth crest, rather than of the third crest as in e.g. *Platypittamys* (Fig. 2) and *Protosteiromys* (Wood & Patterson, 1959: Fig. 30).

Thus, if one can use the usual methods of palaeontological interpretation, the ancestors of the Fayum rodents had five transverse crests in both upper and lower molars, the third crests being the mesoloph and mesolophid, respectively.

Another feature of great importance in connection with the phiomyids is the progressive retention of the deciduous teeth and the suppression of the permanent premolars. Three permanent lower premolars (one in *Gaudeamus*, exposed by the removal of dP4, that might never have erupted even if the animal had lived; and two of *Phiomys andrewsi*) and 58 deciduous lower premolars are known from the Fayum. Two specimens of dP^3 and 11 of dP^4 are known in *Metaphiomys schaubi*, one dP^4 from either *Phiomys* or *Metaphiomys*, and there is an upper premolar of *Gaudeamus*, which I have been unable to determine with certainty as either dP^4 or P^4. Thus, the Fayum phiomyids were well on the way toward suppressing the permanent premolars, although they occasionally still replaced them. The tendency to suppress replacement was continued in their Miocene and later descendants, there being no trace of permanent premolars in any of the rodents from Southwest Africa (Stromer, 1926: Pls 41 & 42) or in *Thryonomys* and *Petromus* (Friant, 1945: 202–204; Wood, 1962b: 314–320, Figs 1 & 2).

The retention of the deciduous teeth and the presence of five crests in both upper and lower cheek teeth separate the phiomyids rather sharply from the caviomorphs. The phiomyids are so advanced in these characters that it would be impossible for them to have given rise to the ancestral caviomorph; a derivation of the phiomyids from the caviomorphs might theoretically be possible, but it would have to involve the *de novo* development of the mesocone, mesoconid, mesoloph and mesolophid, structures for which there is not the slightest evidence in any of the Caviomorpha.

There are numerous European Eocene rodents with five-crested cheek teeth; among these are the Theridomyoidea, which are also hystricomorphous but show no signs of hystricognathy. Of all known Old World Eocene rodents, the theridomyids are the most logical ancestors for the phiomyids. However, among the theridomyids, the incisor enamel is either of the pauciserial or universal type (Korvenkontio, 1934: 116–117), but never of the multiserial type found in the Phiomorpha. Because of the structure of the angle and that of the incisor enamel, I cannot help but agree with Lavocat (1955) that the Theridomyidae could not have been the ancestors of the phiomyids. There are, however, no other Old World rodents known that could have been such ancestors. We know enough of the Eocene rodents of Europe strongly to suspect that the phiomorph ancestors came to Africa via southwestern Asia (an area whose Eocene rodent fauna is utterly unknown). The numerous similarities between the Phiomorpha and the Caviomorpha suggest that the two were related through a common ancestry and that these ancestors belonged to a stock related either to the North American Reithroparamyinae, to the North American *Prolapsus*, or to some of the other possible North American hystricognaths.

The Hystricidae

The Old World porcupines must fit into the story somewhere; there is at present no good evidence as to how. Lavocat believes them to have been derived from the phiomorphs; I very much doubt this. The fact that the hystricids agree with the Phiomorpha and Caviomorpha in a number of characteristics (Table II) is, of course, the basis for lumping all of these animals in a 'Suborder Hystricomorpha', or, more accurately, a 'Suborder Hystricognathi', following Tullberg (1899).

The palaeontological history of the hystricids is poorly known, which implies that their earlier evolution occurred in an area where the Eocene and Oligocene is likewise poorly known. They appear in the late Miocene Chinji Zone of the Siwaliks (Black, 1972: 246–247, Fig. 11), and in southeastern Europe and northern Africa either about the same time, or slightly later. They clearly were invaders in Europe; I see no evidence that the same was not true of Africa. Our absence of knowledge permits the assumption that they may have originated either on the Indian sub-continent or farther east in Asia. The cheek tooth pattern seems to be, basically, the five-crested pattern so common among rodents (Stehlin & Schaub, 1951), and which was present in the earliest of the Phiomyidae. Most authors, however, disagree with Schaub (1958: 691–694) that this pattern of itself is indicative of close relationships, as it recurs independ-ently too many times among the rodents. The development of a multi-plicity of cuspules over the crowns of the teeth is clearly a specialization (see Wood & Patterson, 1959: 412–414) although this has been contested by Landry (1957: 85). One feature, however, clearly separates the Hystricidae from the Phiomorpha—there is never any tendency for the retention of the deciduous teeth in the hystricids, as P^4_4 appears in a perfectly normal manner (Wood & Patterson, 1970: 632–633). Therefore, if the hystricids and the phiomorphs had a common ancestor, it was more primitive than any known member of the latter group. It is entirely possible that this hystricid ancestor was one of the hypothetical southern Asiatic sub-hystricomorphous, sub-hystricognathous forms postulated above as possible phiomorph ancestors.

THE SCIUROGNATHOUS HYSTRICOMORPHS

Finally, I should like to summarize something of the history of sciurognathous hystricomorphs, even though both Lavocat and I are sure that these animals have nothing do with the history of the hystri-cognathous forms discussed above.

The Theridomyoidea

The Theridomyoidea, with two families, the Pseudosciuridae and the Theridomyidae, are a European group known from the mid-Eocene to the Oligocene. The development of hystricomorphy from protrogomorphy (Fig. 1) can be clearly seen within the pseudosciurids (Fig. 4, and Hartenberger, 1968: Fig.; 1969: Fig. 2). In *Protadelomys cartieri* from the middle Eocene (Lutetian) of Egerkingen, the infraorbital foramen is small for an hystricomorph, being only about a third of the height of the snout, and the masseter medialis has moved only a very short distance through it onto the snout, just reaching the premaxillary-maxillary suture (Fig. 4A). In the late Eocene *Sciuroides* sp. from Quercy (Basel, QT 756), the masseter reached slightly forward of the premaxillary-maxillary suture (Lavocat, 1951, Pl. 11, Fig. 2), and the foramen is enlarged to about half the height of the snout (Fig. 4B). And, finally, in a skull of *Adelomys* sp. (Mus. Montauba n),also from Quercy, the foramen is three-quarters the height of the snout, and the masseter medialis arose as far forward as the anterior part of the premaxilla (Fig. 4C). The skull in all theridomyids is fully hystricomorphous. There is never any indication of there being a pars posterior, deep division, of the masseter lateralis profundus, or of hystricognathy, in any pseudosciurid or theridomyid. The upper cheek teeth are pentalophate, with well developed mesolophs; the lowers may be either four- or five-crested (Stehlin & Schaub, 1951: 22–39, 207–216, Figs 21–35, 311–323). In the primitive condition, the incisor enamel was pauciserial, but, among the Theridomyidae, it became progressively uniserial (Korvenkontio, 1934: 116–117).

The best available evidence is that the Pseudosciuridae arose in Europe from early Eocene microparamyine paramyids (Wood, 1962a: 170, 248; Hartenberger, 1968: 1819; Schmidt-Kittler, 1971: Fig. 46) and that the theridomyids were descended from the Pseudosciuridae (Lavocat, 1951: 47, 70). There is no evidence that either family ever spread out of Europe or that the group left any descendants (Lavocat, 1955).

The Ctenodactylidae

The Ctenodactylidae are at present an African hystricomorphous family. Because of this distribution and their hystricomorphy, they have sometimes been considered to be related to the hystricognaths. However, the ctenodactylids are not known in Africa before the Miocene (and then not in the Miocene of Southwest Africa), but occur in the Siwaliks of India in the late Miocene to early Pliocene (Wood, 1937a: 73; Black, 1972: 239–243). Furthermore, ctenodactylids were abundant in the Oligocene of Mongolia (Matthew & Granger, 1925) and the Oligocene and Miocene of Kansu (Bohlin, 1946: 75–147), and have been traced back, in central Asia, into the middle Oligocene of Kazakhstan (Shevyreva, 1971a: 81–85), where they are represented by two genera, one known only from the lower jaw, but the other, *Terrarboreus*, having an enlarged infraorbital foramen, presumably with muscle penetration, although the snout is badly damaged (Shevyreva, 1971a: 81–82, Fig. 7). The ctenodactylids are probably derived from forms such as *Advenimus* from the late Eocene of Mongolia (Dawson, 1964) or *Tamquammys* from the Eocene of Kazakhstan (Shevyreva, 1971b), and, presumably, by way of Asiatic intermediates (N. S. Shevyreva, personal communication, 5 May, 1973), ultimately from early Eocene paramyids (Lavocat, 1962: 289). This independent history of the ctenodactylids is important because it demonstrates the presumed parallel acquisition not only of hystricomorphy, but also of the fusion of the malleus and incus, of multiserial incisor enamel, and of the development of a structure in the urogenital system that has been interpreted as being either a degenerate or an incipient sacculus urethralis (Wood & Patterson, 1959: 414, footnote).

The Anomaluridae

These African hystricomorphous but sciurognathous rodents are first represented in the Miocene of Africa, although R. Lavocat (personal communication) has seen a specimen of the very characteristic ulna of an anomalurid supposedly from the Fayum Oligocene of Egypt. A careful search of the Yale collections, however, has not uncovered this bone, and it must be presumed to have been lost. Nothing is known as to when, where, or from what source the anomalurids arose. They seem to have no particular relationship with any other African group. Their cheek teeth are reminiscent of those of the theridomyids, but Lavocat, who has studied both the anomalurids and the theridomyids much more extensively than I have, assures me that there are so many differences they can have nothing special in common.

The Pedetidae

These hystricomorphous, sciurognathous African rodents are known from the Miocene to the Recent of Africa and from the Miocene of Chios (Tobien, 1968: 54). Until relatively recently, nothing has been known of the detailed structure of the cheek teeth, even in the Miocene (MacInnes, 1957: Figs 6-8; Plate, Figs 2–3), except that there were two lobes in each tooth. More recently, Wood (1965b) has studied the unworn cheek teeth of a newborn modern *Pedetes* and Lavocat & Michaux (1966) reported on unworn lower teeth of the Miocene *Megapedetes*. The cheek teeth of *Pedetes* are made up of two transverse lobes, each of which is formed of

three cusps, closely united with each other and widely separated from
the cusps of the other loph (Wood, 1965b: Fig. 1). The lower teeth of
Megapedetes are much more brachydont, and are similar in pattern to
those of *Pedetes*, but show slightly greater complexity, seeming to be
bilobed but with an additional cusp at the middle of the buccal side
(Lavocat & Michaux, 1966).

There apparently is no premolar replacement in *Pedetes*, and as this
ever-growing tooth erupts before the first molar (Wood, 1965b: 422), it
would seem to be a retained deciduous tooth. In *Parapedetes* from the
Miocene of Southwest Africa, Stromer (1926: 130) reported the normal
replacement of the deciduous premolars by $P^4/_4$. The deciduous teeth
(clearly recognizable by their widely divergent roots) are quite high
crowned, suggesting that this is a preliminary step toward the suppression of replacement. Neither MacInnes (1957) nor Lavocat & Michaux
(1966) comment on deciduous teeth in *Megapedetes*. However, since the
premolar figured by these last authors is more worn than the molar
(Lavocat & Michaux, 1966), it presumably is dP4. Partly for this
reason, I suspect that some of the cusp and crest homologies presented
by Lavocat & Michaux (1966: 1677) are in error. The pedetids, then,
would seem to have paralleled the phiomorphs and the echimyids in the
suppression of premolar replacement, but at a later date than either
of the other groups.

In addition to being hystricomorphous, the pedetids have multiserial incisor enamel (Korvenkontio, 1934: 166) and the malleus and
incus, although not fused, are so closely appressed that there is little
or no possibility of movement between them (Wood & Patterson, 1959:
293). The pre-Miocene history of the pedetids is a complete mystery,
and there are no rodents known from any part of the world that look
to me like possible relatives.

The Dipodoidea

The Dipodoidea (Dipodidae and Zapodidae) are just as hystricomorphous as are any of the other rodents discussed above, and again
are sciurognathous. But, in contrast to the last four groups, all of which
have often been considered members of a 'Suborder Hystricomorpha',
the Dipodoidea are almost universally recognized as having been independent of the hystricognaths in their entire history, whoever else may
or may not be their relatives. They appear in the Middle Oligocene of
Europe and Asia and the early Miocene of North America (Schaub,
1958: 788); possibly related forms occur in the late Eocene of southern
California and the earliest Oligocene of west Texas (Wood, 1974).

POSSIBLE ANCESTRY OF HYSTRICOGNATHS

The North American Eocene subfamily Reithroparamyinae of the Paramyidae are, potentially, of considerable importance in solving this problem. The genera referred to this subfamily are all incipiently hystricognathous (Landry, 1957: 82; Wood, 1962a: 117) although most are protrogomorphous (Fig. 1). *Franimys* is represented by a skull, jaw and associated skeletal material from the late Palaeocene of Wyoming, the earliest known rodent skull or jaw (Wood, 1962a: 140–147; date as corrected by Wood, 1974: 16–17). In another reithroparamyine, *Rapamys*, the very fragmentary material available suggests that the infraorbital foramen was considerably larger than in other paramyids (Wood, 1962a). In at least some of the Reithroparamyinae (Wood, 1962a: Fig. 46F), the pars posterior, deep, of the masseter lateralis profundus seems to have begun to expand backward, below and behind the condyle, as in hystricognathous rodents. This is reported elsewhere, so far as I am aware, only in the sciuromorphous and sciurognathous North American Oligocene genus *Eutypomys* (Fig. 1 and Wood, 1974).

Prolapsus, from the Eocene (perhaps middle Eocene) of southwest Texas, is the earliest rodent from anywhere in the world known to have been completely hystricognathous (Wood, 1972, 1973). Its tooth morphology is not close to that of any other rodent known, but *Prolapsus* might have been derived from the North American Sciuravidae. A new genus, *Guanajuatomys*, from central Mexico and described by Black & Stephens (1973), also seems to have been clearly hystricognathous, but with a quite different type of cheek teeth from that of *Prolapsus*. Black & Stephens referred it merely to the Paramyidae ("Ischyromyidae" of those authors).

These rodents must, of necessity, be given very serious consideration as possible ancestors of the Caviomorpha, as they are the only Eocene rodents known from any part of the world that show either complete or incipient hystricognathy. Hoffstetter (1972) opposes this point of view, without citing any reason for neglecting these, the only known Eocene hystricognaths, and without having seen any of the material. The fact that *Franimys*, the only Palaeocene rodent represented by more than isolated teeth, is incipiently hystricognathous indicates that this modification of the angle of the lower jaw occurred very early in rodent history. Unfortunately, reithroparamyines are apparently unknown outside North America. Two species of the subfamily were reported, on the basis of isolated teeth, from Europe (Michaux, 1964), but one species was later made the type of a new genus placed in a new (and non-hystricognathous) paramyid subfamily, the Ailuravinae, and the other species was stated to be generically indeterminable (Michaux, 1968: 155, 173). Reithroparamyines have not been reported from Asia.

The Protoptychidae are a family from the late Eocene of North America comprising a single described genus and two described species. Two skulls (one edentulous) are on record, but no lower jaws have as yet been described. The skull has an enlarged hystricomorphous infraorbital foramen, and the scar of the masseter medialis can be clearly seen on the snout (Wahlert, 1973). Specimens from slightly earlier in the Eocene (Washakie), probably referable to *Protoptychus* or certainly closely related, have hystricognathous lower jaws. The upper cheek teeth are four-crested and have no trace of a mesoloph, mesostyle or mesocone. The nature of the incisor enamel is unknown. The known material of the genus presumably lived too far north (Wyoming and northern Utah) to have been ancestral to the Caviomorpha; the molariform P^4 and the greater penetration of the infraorbital foramen by the masseter medialis than in *Platypittamys* also presumably indicate that *Protoptychus* was not directly ancestral to the Caviomorpha, although they are probably related (Wahlert, 1973).

The known North American rodents that are fully hystricognathous or hystricomorphous, then, are all too specialized in one way or another to have been ancestral to the Caviomorpha; the Reithroparamyinae could have been ancestral, but they are very much more primitive than any caviomorph, and a large series of hypothetical intermediates would have to be assumed. However, these rodents are the only Eocene rodents known from any part of the world that seem to have been members of a taxon that might have been immediately ancestral to the Caviomorpha.

The ancestry of the Phiomorpha and Hystricidae could readily be sought in Asiatic reithroparamyines, but we have no evidence that such animals existed. However, the presence of incipient hystricognathy in the late Palaeocene *Franimys* indicates that this condition developed early enough in rodent history for animals with sub-hystricognath lower jaws to have been among the early rodent immigrants to Europe and Asia. If such hypothetical ancestors are not accepted, we must conclude that no potential Old World ancestors of the Phiomorpha and Hystricidae have yet been described.

I think that Lavocat and I are in essential agreement as to the type of rodent that gave rise to the Phiomorpha and Hystricidae. We differ most in regard to the source for the invasion of South America and the date of the invasion of Africa. There is, of course, no direct evidence available on this latter point at the present time. The best clue, it seems to me, comes from South America, where the known rodent diversity by Deseadan time was equal to or greater than the known African diversity by Fayum time. In South America, we can be sure that the rodents arrived during the late Eocene and, therefore, it seems a reasonable assumption that they reached Africa not far from the same time.

DISTRIBUTION AMONG RODENTS OF FEATURES SUPPOSED TO CHARACTERIZE THE 'HYSTRICOMORPHA'

General

There are, in my opinion, no taxonomically useful characteristics that are diagnostic for a group that includes only the Caviomorpha, Phiomorpha (including the Bathyergidae) and Hystricidae. Most features are found in these and some other rodents, or usually (but not always) in these forms (Table II). I shall discuss some of these characters briefly.

Hystricomorphy

The original feature used as the diagnosis of a suborder, the Hystricomorpha (Waterhouse, 1839; Brandt, 1855) was the enlarged infraorbital fenestra, through which passes the bulk of the masseter medialis between its origin on the snout and its insertion on the mandible. Hystricomorphy is present in all Caviomorpha except *Platypittamys*, in the Hystricidae, and in all Phiomorpha except most of the Bathyergidae. In the Bathyergidae the fenestra is greatly reduced, and the muscle penetrates the foramen only slightly in *Cryptomys* and not at all in other genera. If Lavocat (1973) is correct that the bathyergids were derived from phiomorphs, this must be a secondary reduction of the foramen and of the muscle. Hystricomorphy also characterizes the Theridomyoidea, Anomaluridae, Ctenodactylidae, Pedetidae, Dipodoidea, Protoptychidae, African glirids, and possibly the paramyid *Rapamys* (Table II; Wood, 1962a: 148), in most or all of which the feature seems to have developed independently. That is, hystricomorphy is a grade (Wood, 1965a) that has originated at least eight times among the rodents, possibly nine times (if *Rapamys* was an hystricomorph) and, I believe, at least once more (in the Caviomorpha, where *Platypittamys* shows an ancestral non-hystricomorphous condition). Hystricomorphy therefore cannot be used to demonstrate relationships among forms that possess it.

Hystricognathy

The most important character common to the rodents that we are discussing is the development of hystricognathy (Table II). This is found in all four of the groups in question (Caviomorpha, Phiomorpha,

Bathyergidae and Hystricidae), but in none of the other hystricomorphous groups that have sometimes been considered as possibly related (Theridomyoidea, Anomaluridae, Pedetidae and Ctenodactylidae). However, hystricognathy is clearly present in the middle Eocene genus *Prolapsus* from southwest Texas (Wood, 1972, 1973), in the late Eocene *Protoptychus* (Wahlert, 1973), and in the Eocene *Guanajuatomys* from central Mexico (Black & Stephens, 1973); it is incipiently present in other North American Eocene rodents. If hystricognathy is a diagnostic character of a Suborder Hystricognathi, these North American forms must be included, as the only demonstrable Eocene members of the suborder. If they are ruled out, hystricognathy must, as a result, be a character that has evolved several times independently in the North American Eocene and it cannot be considered a diagnostic feature of any importance until it is proven how many more times it evolved independently.

Masseter lateralis profundus, pars posterior, deep division

A very general feature of the rodents we are considering is the expansion of the mandible, immediately behind and below the condylar process, to form an area for the insertion of the masseter lateralis profundus, pars posterior, deep division (Woods, 1972: 127). This expansion occurs in the Caviomorpha, Phiomorpha, Bathyergidae and Hystricidae (Table II), and is incipiently developed in at least some of the Reithroparamyinae (Wood, 1962a: 136–137 and Fig. 46F) and in some members of the peculiar North American Oligocene and Miocene family Eutypomyidae, a family that is both sciuromorphous and sciurognathous, which may be related to the beavers, but which certainly had nothing in common with any of the rodents we are now discussing (Wood, 1974). That is, this development of the deep division of the masseter lateralis profundus, pars posterior, must have occurred independently at least twice, even if it is assumed that all the hystricognathous forms inherited it from a common reithroparamyine ancestor.

Fusion of malleus and incus

In most hystricognathous rodents, the malleus and incus are fused. Since no one has proposed any selectively advantageous functional basis for this condition, it is generally assumed to be a clear indication of close relationship among these forms (Landry, 1957: 16). However, the ossicles are not fused in at least some adult specimens of *Proechimys, Echimys, Octodon, Spalacopus* and *Aconaemys* (Wood & Patterson, 1959: 292–293). These are all members of the Octodontidae and Echimyidae, the

C

two most primitive of the caviomorph families. This, it seems to me, indicates that the fusion of the ossicles has been acquired, independently, several times *within* the Caviomorpha. Furthermore, the two bones are fused in the Ctenodactylidae (Cockerell, Miller & Printz, 1914) and so closely appressed (although with no fusion) in *Pedetes* that there seems to be no possible motion between the bones (Table II). Even if it were postulated that the lack of fusion of the ossicles in some echimyids and octodonts resulted from a number of independent cases of secondary separation of the ossicles (a most unlikely prospect), fusion must never-theless have occurred independently at least twice (ctenodactylids and the others), and be about to occur a third time (*Pedetes*). It seems much more probable that the considerable number of primitive caviomorphs with separate ossicles indicates that fusion has occurred at least half a dozen times among the Caviomorpha, independently of any other groups. This character is, unfortunately, of little use from the palaeontological point of view, since the only published record of ear ossicles in a fossil hystricomorphous rodent is that in *Theridomys* (Lavocat, 1967), in which the bones are not fused.

Multiserial incisor enamel

Another feature whose adaptive significance has been rarely discussed is the modification of the incisor enamel, that occurred in all lines of rodents about the end of the Eocene. In primitive rodents, the incisor enamel was of a type termed 'pauciserial' (Korvenkontio, 1934: Wahlert, 1968). From this, there have been two types of modification, one to give what is termed 'uniserial' and the other 'multiserial'. Wilson (1972: 220) considered that these changes represented an "im-provement of incisor strength to a critical point which allowed rapid development of a more powerful chewing and gnawing mechanism", and believed that the change in the incisor enamel preceded and per-mitted the forward migration of the origin of the masseter muscle. Docu-mentation of the exact stage at which the enamel and muscle changes occurred is poor; however, in *Protadelomys*, *Sciuroides* and *Adelomys* (Fig. 4) the muscle shifts preceded modification of the incisor enamel; in the ischyromyid *Titanotheriomys* (Wood, in press) and, in the Cavio-morpha, the reverse may have been the case.

As shown in Table II, all the Caviomorpha, Phiomorpha, Bathyer-gidae and Hystricidae have multiserial enamel, as do also the Pedetidae and Ctenodactylidae, although there is considerable variation in the details of the enamel structure among the Oligocene and Miocene ctenodactylids (Bohlin, 1946: 143–146, Pls 4–5). No instances of multi-serial enamel have yet been reported from the Eocene. That is, this condition must have developed independently at least three times, and four times if I am correct that the Caviomorpha were of North American origin. Uniserial enamel, which is as complicated as is multiserial, was apparently acquired independently by the Theridomyidae, Ischyromyidae, Sciuridae, Aplodontidae, Gliridae, Castoridae, Anomaluridae and myomorphs, at the very least.

Sacculus urethralis

A most peculiar structure, of unknown evolutionary significance, is the sacculus urethralis, an outgrowth of the male urogenital system. This is found in all living phiomorphs, bathyergids, hystricids, and in almost all living caviomorphs. It is, however, absent in *Lagostomus*, which almost convinced Dathe that this genus had nothing to do with the other South American hystricomorphous rodents (Dathe, 1937: 54). In addition, there is a structure in the ctenodactylids that is either an incipient or degenerate sacculus urethralis (Table II). That is, this structure, unfortunately not found in fossils, must have developed at least twice, unless its absence in *Lagostomus* is primitive, in which case it must have evolved several times independently within the Caviomorpha.

Other common features

Discussion of similarities in tooth structure and in the retention of deciduous teeth is omitted because, as I have indicated above, these characteristics did not appear among the earliest members of the various groups.

One of the facts that has convinced many workers, including Lavocat and Woods, of the unity of the Hystricomorpha is the great similarity that occurs between some of the New and Old World forms in many features of their anatomy. This led Ellerman (1940) to refer *Thryonomys* and *Petromus* to the American echimyids and octodontids, which he considered subfamilies of the Octodontidae. A little later, Ellerman, Morrison Scott & Hayman (1953) considered *Petromus* and *Thryonomys* to be subfamilies (along with the Echimyidae) of the Octodontidae. This, of course, would require one or two invasions of Africa from South America after the separation of the octodontids and echimyids. Fortunately, this difficult palaeogeographic feat is no longer required, the work by Wood (1968) and Lavocat (1973) having shown that *Petromus* and *Thryonomys* were clearly derived from the African Oligocene Phiomyidae, and that their similarities to the South American forms (in so far as they differ from their Oligocene or Eocene ancestors) must have been due to parallelism. Woods' (1972) detailed myological studies indicated great similarity, almost to the point of identity, in the cephalic and fore-limb myology between *Thryonomys* and *Petromus* on the one hand, and various South American genera, although the two groups cannot, on any assumption, have had a common ancestor more recent than the late Eocene, and none of the genera in question seems close, morphologically, to the primitive condition for the rodents of its own continent. Furthermore, Woods (1972: 189) pointed out that *Erethizon* is very different, myologically, from the other Caviomorpha and the Phiomorpha, which he felt indicated that the erethizontids were widely separated from other caviomorphs.

A number of other features have been used to indicate relationships, especially between the Caviomorpha and Phiomorpha. These include the fusion of the tibia and fibula (Landry, 1957: 17–20). This frequently occurs; however, fusion is absent in *Platypittamys* (Wood, 1949: 35), the only Oligocene form whose skeleton is known from either South America or Africa, and seems certainly to have evolved independently on the two continents.

The fetal membranes would seem to provide an entirely independent line of evidence as to relationships (Mossman, 1937: 200–303). Mossman & Luckett (1968) indicated that there is similarity, almost to the point of identity, in the fetal membranes of the caviomorphs and of *Bathyergus*, which is not merely a general, overall similarity, but extends to all significant details of structure. Luckett (1971: 167) indicated that *Hystrix* has the same type of fetal membranes as does *Bathyergus*. Fischer & Mossman (1969: 102) concluded that *Pedetes* is a somewhat specialized sciuromorph. They further stated that the "remarkable resemblance in all known details of the fetal membranes between *Pedetes* and *Ctenodactylus* argues for their close phylogenetic relationship ..." (Fischer & Mossman, 1969: 103). Luckett (1971) observed that *Anomalurus* is fundamentally like *Pedetes* and *Ctenodactylus* in its fetal membranes. Since the separate ancestry of *Ctenodactylus* can be traced back to the middle Eocene and since, therefore, it seems highly unlikely that a common ancestor of *Ctenodactylus*, *Anomalurus* and *Pedetes* existed later than early Eocene, this similarity of the fetal membranes implies that these structures either exhibit the same parallelism that occurs elsewhere among the rodents or that they are exceedingly conservative; this would suggest that the caviomorph-phiomorph split could also date from the early Eocene. Moreover, similarities in the fetal membranes of the lemurs and perissodactyls led Mossman (1937: 204 and Fig. 7) to conclude that these two groups were closely related. I feel that the caviomorph-phiomorph relationship is unquestionably closer than that of the lemurs and perissodactyls (after all, they belong to the same order), but the conclusions about lemur–perissodactyl and *Ctenodactylus-Anomalurus-Pedetes* relationships force me to question the value of fetal membranes in determining anything except extremely broad caviomorph-phiomorph relationships.

CONCLUSIONS

In conclusion, if the similarities of the Caviomorpha and Phiomorpha are due neither to retention of characters from a common ancestor nor to a transatlantic migration by one or the other, they must be due to parallelism. Some of the similarities (such as those between *Thryonomys* and *Petromus* and the caviomorphs, reported by Woods (1972) or the independent retention of the deciduous teeth in both groups) *must* be parallelisms. I have given reasons why I believe that many of the other resemblances are also parallelisms. Some might question whether such great parallelism is not too extensive to be believed. However, what most students of rodents believe to be dental parallelisms occur over and over within the order. Many years ago (Wood, 1937b), I pointed out the extreme parallelism that occurs in various lines of the North American heteromyid and geomyid rodents—parallelisms so great that identification of jaws with teeth as to family or subfamily is often impossible if the geologic age is not known. Subsequent work has demonstrated that the parallelism was even greater than I had believed, and that it extends to all parts of the anatomy where there is adequate fossil representation (Rensberger, 1971). The different lines within these two families did exactly the same things, over and over again, but at slightly different times. I have also commented on the selective pressures acting on the rodents to bring about a forward migration of a part of the masseter, resulting in numerous parallel developments of the sciuro-morph and hystricomorph conditions. Hystricomorphy, as stated above, must have developed independently at least eight times (and I believe at least once more), and sciuromorphy at least a half dozen times.

Landry (1970: 366) apparently misunderstood some of my previous remarks. I was *not* implying that rodents had a restricted evolutionary potential (which is obviously not the case), but rather that the "patterns of rodent cheek teeth may be fitted into a very limited number of types", and that, in the cheek teeth, apparently "only a relatively small number of primary modifications of the primitive ground plan were possible" (Wood, 1937b: 171). These statements were correct. I also stated (Wood, 1935: 249) that "there is far and away a greater amount [of parallelism] in this order than in any other order of mammals, and perhaps as much as in all other placentals". I would now consider this a gross understatement. The rodents were restricted in the possible directions of evolution by the initial development of evergrowing incisors and the associated structures, and, when there was selective pressure for a forward migration of the origin of the masseter, there apparently were only two available methods for bringing this about— the sciuromorph and the hystricomorph methods, myomorphy being a combination of the other two (Fig. 1; Wood, 1959: 171). There were a myriad detailed variations on a limited number of basic patents; this is why rodent phylogeny is so difficult to unravel.

One of the clearest expositions of the principles to follow in studying rodent evolution was presented by Lavocat (1962: 288):

(1) each of the three principal zygomasseteric types or of the two main types of angle could have developed independently several times in different parts of the world;

(2) groups that have been geographically isolated for a very long time, even when they show close anatomical resemblances, should be presumed to have independent origins in the absence of proof to the contrary;

(3) groups of both primitive and advanced structure, especially in regard to dentition, can be considered related if other characters are homogeneous, and

(4) groups that resemble each other in only two sets of characters— orbital region, mandibular structure, or dental structure—should be considered independent.

When the selective pressures acting on a group of abundant and successful animals are heavy, and the number of alternative solutions are limited, the same results may arise many times independently, as in the case of the migration of the masseter. On the other hand, the fact that similar results develop independently may be considered indicative of a common genome, producing identical phenotypic results by similar genetic methods (Wood, 1937b: 175). This, then, would suggest a common ancestral background for the Caviomorpha, Phiomorpha, and presumably, the Hystricidae, but all the available evidence seems to me to point to this common ancestral background having been late Palaeocene or early Eocene at the latest, and all the available evidence suggests that these common ancestors were neither fully hystricognathous nor even incipiently hystricomorphous. Since most of these common ancestors are still unknown, it seems to me very premature to try to establish a taxon to include these ancestral forms and all their descendants. These conclusions are in very close agreement with those of Woods (1972: 191) who concluded that the Caviomorpha and

Phiomorpha had a common paramyid ancestor, but that the two groups had most probably evolved independently from ancestors living in the tropical and semitropical portions of North America and Asia.

<h2 style="text-align:center">REFERENCES</h2>

Andrews, C. W. (1914). On the lower Miocene vertebrates from British East Africa, collected by Dr. Felix Oswald. *Q. Jl geol. Soc. Lond.* **70**: 163–186.

Black, C. C. (1972). Review of fossil rodents from the Neocene Siwalik Beds of India and Pakistan. *Palaeontology* **15**: 238–266.

Black, C. C. & Stephens, J. J., III (1973). Rodents from the Paleocene at Guanajuato, Mexico. *Occas. pap. Mus. Texas Tech. Univ.* No. 14: 1–10.

Bohlin, B. (1946). The fossil mammals from the tertiary deposit of Taben-Buluk, Western Kansu. Part II: Simplicidentata, Carnivora, Artiodactyla, Perissodactyla, and Primates. *Palaeont. sin.* (n.s. C,) No. 8b: 1–259.

Brandt, J. F. (1855). Beiträge zur näheren Kenntniss der Saugethiere Rüsslands. *Mem. Acad. Imp. Sci. St. Petersbourg* (6) **9**: 1–365.

Cockerell, T. D. A., Miller, L. I. & Printz, M. (1914). The auditory ossicles of American rodents. *Bull. Am. Mus. nat. Hist.* **33**: 347–378.

Dathe, H. (1937). Ueber den Bau des männlichen Kopulationsorganes beim Meerschweinchen und anderen hystricomorphen Nagetieren. *Morph. Jb.* **80**: 1–65.

Dawson, M. R. (1964). Late Eocene rodents (Mammalia) from Inner Mongolia. *Am. Mus. Novit.* No. 2191: 1–15.

Ellerman, J. R. (1940). *The families and genera of living rodents* I: 1–689. London: Brit. Mus. (Nat. Hist.).

Ellerman, J. R., Morrison-Scott, T. C. S. & Hayman, R. W. (1953). *Southern African mammals 1758 to 1951; a reclassification.* London: Brit. Mus. (Nat. Hist.)

Fischer, T. V. & Mossman, H. W. (1969). The fetal membranes of *Pedetes capensis*, and their taxonomic significance. *Am. J. Anat.* **124**: 89–116.

Friant, M. (1945). La formule dentaire des Rongeurs de la famille des Thryonomyidae. *Revue Zool. Bot. afr.* **38**: 200–205.

Hartenberger, J.-L. (1968). Les Pseudosciuridae (Rodentia) de l'Eocène moyen et le genre *Massillamys* Tobien. *C. r. hebd. Séanc. Acad. Sci., Paris* **267**: 1817–1820.

Hartenberger, J.-L. (1969). Les Pseudosciuridae (Mammalia, Rodentia) de l'Eocène moyen de Bouxwiller, Egerkingen et Lissieu. *Palaeovertebrata* **3**: 27–61.

Hinton, M. A. C. (1933). Diagnoses of new genera and species of rodents from Indian Tertiary deposits. *Ann. Mag. nat. Hist.* (10) **12**: 620–622.

Hoffstetter, R. (1972). Origine et dispersion des Rongeurs hystricognathes. *C. r. hebd. Séanc. Acad. Sci., Paris* **274**: 2867–2870.

Hoffstetter, R. & Lavocat, R. (1970). Découverte dans le Déséadien de Bolivie de genres pentalophodontes appuyant les affinités africaines des Rongeurs Caviomorphes. *C. r. hebd. Séanc. Acad. Sci., Paris* **271**: 172–175.

Hoffstetter, R., Martinez, C., Mattauer, M. & Tomasi, P. (1971). Lacayani, un nouveau gisement bolivien de Mammifères déséadiens (Oligocène inférieur). *C. r. hebd. Séanc. Acad. Sci., Paris* **273**: 2215–2218.

Hopwood, A. T. (1929). New and little-known mammals from the Miocene of
    Africa. *Am. Mus. Novit.* No. 344: 1–9.
Korvenkontio, V. A. (1934). Microskopische Untersuchungen an Nagerincisiven
    unter Hinweis auf die Schmelzstruktur der Backenzähne. Histologisch-
    phyletische Studie. *Suomal. eläin-ja Kasvit. Seur. van. eläin. Julk.* **2**: i–xiv,
    1–274.
Landry, S. O. (1957). The interrelationships of the New and Old World hystrico-
    morph rodents. *Univ. Calif. Publs Zool.* **56**: 1–118.
Landry, S. O. (1970). The Rodentia as omnivores. *Q. Rev. Biol.* **45**: 351–372.
Lavocat, R. (1951). *Révision de la faune des mammifères oligocènes d'Auvergne et
    du Velay.* Paris: Éditions "Science et Avenir":
Lavocat, R. (1955). Sur les relations systématiques entre les Théridomyidés
    et divers rongeurs d'Afrique. *C. r. hebd. Séanc. Acad. Sci., Paris* **240**: 634–635.
Lavocat, R. (1962). Réflexions sur l'origine et la structure du groupe des rongeurs.
    In: Problèmes actuels de Paléontologie. *Colloq. int. Cent. nat. Rech. sci.* **104**:
    287–299.
Lavocat, R. (1967). Observations sur la région auditive des rongeurs thérido-
    morphes. In: Problèmes actuels de Paléontologie. *Colloq. int. Cent. nat. Rech.
    sci.* **163**: 491–501.
Lavocat, R. (1969). La systématique des Rongeurs hystricomorphes et la dérive
    des continents. *C. r. hebd. Séanc. Acad. Sci., Paris* **269**: 1496–1497.
Lavocat, R. (1971). Affinités Systématiques des Caviomorphes et des Phiomorphes
    et origine africaine des Caviomorphes. *Anais Acad. bras. Cienc.* **43**, Suppl.:
    515–522.
Lavocat, R. (1973). Les rongeurs du Miocène d'Afrique Orientale. **I.** Miocène
    inférieur. *Trav. Mém. Inst. E.P.H.E. Monpellier* **1**: 1–284.
Lavocat, R. & Michaux, J. (1966). Interprétation de la structure dentaire des
    rongeurs africains de la famille des pédétidés. *C. r. hebd. Séanc. Acad. Sci.,
    Paris* **262**: 1677–1679.
Luckett, W. P. (1971). The development of the chorio-allantoic placenta of the
    African scaly-tailed squirrels (Family Anomaluridae). *Am. J. Anat.* **130**:
    159–178.
MacInnes, D. G. (1957). A new Miocene rodent from East Africa. *Fossil Mammals
    Afr.* No. 12: 1–36.
Matthew, W. D. & Granger, W. (1925). New creodonts and rodents from the
    Ardyn Obo formation of Mongolia. *Am. Mus. Novit.* No. 193: 1–7.
Maxwell, A. E., von Herzen, R. P., Hsü, K. J., Andrews, J. E., Saito, T., Percival,
    S. F., Jr, Milow, E. D. & Boyce, R. E. (1970). Deep sea drilling in the South
    Atlantic. *Science, N.Y.* **168**: 1047–1059.
Michaux, J. (1964). Diagnoses de quelques Paramyidés de l'Eocène inférieur de
    France. *C. r. somm. Soc. géol. Fr.* **1964** (4): 153.
Michaux, J. (1968). Les Paramyides (Rodentia) de l'Eocène inférieur du Bassin de
    Paris. *Palaeovertebrata* **1**: 135–193.
Mossman, H. W. (1937). Comparative morphogenesis of the fetal membranes and
    accessory uterine structures. *Contr. Embryol.* No. 26: 126–246.
Mossman, H. W. & Luckett, W. P. (1968). Phylogenetic relationship of the
    African mole rat (*Bathyergus janetta*) as indicated by the fetal membranes.
    Am. Zool. **8**: 806.
Patterson, B. & Pascual, R. (1968). New echimyid rodents from the Oligocene
    of Patagonia, and a synopsis of the family. *Breviora* No. 301: 1–14.

Patterson, B. & Pascual, R. (1972). The fossil mammal fauna of South America. In *Evolution, mammals and southern continents*. 247–309. A. Keast, F. C. Erk & B. Glass (eds), Albany, N.Y.: State Univ. of New York Press.

Patterson, B. & Wood, A. E. (in preparation). The caviomorph rodents from the early Oligocene of Bolivia and their place in caviomorph evolution. *Bull. Mus. comp. Zool. Harv.*

Rensberger, J. M. (1971). Entoptychine pocket gophers (Mammalia, Geomyoidea) of the early Miocene John Day Formation, Oregon. *Univ. Calif. Publs geol. Sci.* **90**: 1–209.

Savage, R. J. G. (1969). Early Tertiary mammal locality in southern Libya. *Proc. geol. Soc. Lond.* **1969** (1657): 167–171.

Savage, R. J. G. (1971). Review of the fossil mammals of Libya. *Symposium on the Geology of Libya, Fac. Sci. Univ. Libya*: 217–225.

Schaub, S. (1958). Simplicidentata (= Rodentia). In *Traité de Paléontologie*, **6** (2): 659–818. J. Piveteau (ed.). Paris: Masson et Cie.

Schmidt-Kittler, N. (1971). Odontologische Untersuchungen an Pseudosciuriden (Rodentia, Mammalia) des Alttertiärs. *Abh. bayer. Akad. Wiss.* (Math.-Naturwiss. Kl.) (n.f.) **150**: 1–133.

Shevyreva, N. S. (1971a). Novie sredneoligocenovie grizuni Kazakhstana i Mongolii. [New middle Oligocene rodents of Kazakhstan and Mongolia.] *Trudỹ paleont. Inst.* **130**: 70–86. [In Russian]

Shevyreva, N. S. (1971b). Pervaya naxodka eocenovikh grizunov v SSSR. [The first find of Eocene rodents in the USSR.] *Soobshch. Akad. Nauk. gruz. SSR* **61**: 745–747. [In Russian]

Stehlin, H. G. & Schaub, S. (1951). Die Trigonodontie der simplicidentaten Nager. *Schweiz. palaeont. Abh.* **67**: 1–385.

Stromer, E. (1926). Reste Land- und Süsswasser-bewohnender Wirbeltiere aus den Diamantenfeldern Deutsch-Sudwestafrikas. In: E. Kaiser's *Die diamanten-wüste Südwestafrikas* **2**: 107–153. Berlin: Dietrich Reimer.

Thaler, L. (1966). Les rongeurs fossiles du Bas-Languedoc dans leurs rapports avec l'histoire des faunes et la stratigraphie due Tertiaire d'Europe. *Mém. Mus. natn. Hist. nat., Paris* (N.S.) (C) **17**: 1–295.

Tobien, H. (1968). Paläontologische Ausgrabungen nach jungtertiären Wirbeltieren auf der Insel Chios (Griechenland) und bei Maragheh (NW-Iran). *Jb. Verein. "Freunde Univ. Mainz"* **1968**: 51–58.

Tullberg, T. (1899). Ueber das System der Nagethiere, eine phylogenetische Studie. *Nova Acta R. Soc. Scient. upsal.* (3) **18**: 1–514.

Wahlert, J. H. (1968). Variability of rodent incisor enamel as viewed in thin section, and the microstructure of the enamel in fossil and recent rodent groups. *Breviora* No. 309: 1–18.

Wahlert, J. H. (1973). *Protoptychus*, a hystricomorphous rodent from the late Eocene of North America. *Breviora* No. 419: 1–14.

Waterhouse, G. R. (1839). Observations on the Rodentia, with a view to point out the groups, as indicated by the structure of the crania, in this order of mammals. *Ann. Mag. nat. Hist.* (N.S.) **3**: 90–96, 184–188, 274–279, 593–600.

Wilson, R. W. (1972). Evolution and extinction in Early Tertiary rodents. *Proc. int. geol. Congr.* **24** (Sec. 7): 217–224.

Wood, A. E. (1935). Evolution and relationships of the heteromyid rodents with new forms from the Tertiary of western North America. *Ann. Carneg. Mus.* **24**: 73–262.

Wood, A. E. (1937a). Fossil rodents from the Siwalik beds of India. *Am. J. Sci.* (5) **34**: 64–76.

Wood, A. E. (1937b). Parallel radiation among the geomyoid rodents. *J. Mammal.* **18**: 171–176.

Wood, A. E. (1949). A new Oligocene rodent genus from Patagonia. *Am. Mus. Novit.* No. 1435: 1–54.

Wood, A. E. (1950). Porcupines, paleogeography, and parallelism. *Evolution* **4**: 87–98.

Wood, A. E. (1955). A revised classification of the rodents. *J. Mammal.* **36**: 165–187.

Wood, A. E. (1959). Are there rodent suborders? *Syst. Zool.* **7**: 169–173.

Wood, A. E. (1962a). The early Tertiary rodents of the family Paramyidae. *Trans. Am. phil. Soc.* (N.S.) **52**: 1–261.

Wood, A. E. (1962b). The juvenile tooth patterns of certain African rodents. *J. Mammal.* **43**: 310–322.

Wood, A. E. (1965a). Grades and clades among rodents. *Evolution* **19**: 115–130.

Wood, A. E. (1965b). Unworn teeth and relationships of the African rodent, *Pedetes. J. Mammal.* **46**: 419–423.

Wood, A. E. (1968). The African Oligocene Rodentia. Part II of: Early Cenozoic Mammalian Faunas, Fayum Province, Egypt. *Bull. Peabody Mus. nat. Hist.* **28**: 23–105.

Wood, A. E. (1972). An Eocene hystricognathous rodent from Texas: its significance in interpretations of continental drift. *Science, N.Y.* **175**: 1250–1251.

Wood, A. E. (1973). Eocene rodents, Pruett Formation, southwest Texas; their pertinence to the origin of the South American Caviomorpha. *Texas Meml Mus. Pearce–Sellards Series* **20**: 1–40.

Wood, A. E. (1974). Early Tertiary vertebrate faunas, Vieja Group, Trans-Pecos Texas: Rodentia. *Bull. Texas Meml Mus.* **21**: 1–112.

Wood, A. E. (in press). The Oligocene rodents *Ischyromys* and *Titanotheriomys* and the content of the family Ischyromyidae. *Life Sci. Contr., R. Ont. Mus.*

Wood, A. E. & Patterson, B. (1959). The rodents of the Deseadan Oligocene of Patagonia and the beginnings of South American rodent evolution. *Bull. Mus. comp. Zool. Harv.* **120**: 279–428.

Wood, A. E. & Patterson, B. (1970). Relationships among hystricognathous and hystricomorphous rodents. *Mammalia* **34**: 628–639.

Wood, A. E. & Wilson, R. W. (1936). A suggested nomenclature for the cusps of the cheek teeth of rodents. *J. Paleont.* **10**: 388–391.

Woods, C. A. (1972). Comparative myology of jaw, hyoid, and pectoral appendicular regions of New and Old World hystricomorph rodents. *Bull. Am. Mus. nat. Hist.* **147**: 115–198.

Simpson (*Tucson*): Doctors Wood and Lavocat have been speaking on essentially the same subject and they have asked that their papers be discussed together. They have not only discussed the same subject but also have done so on the basis of much the same evidence. Both have a background of specialization on early Holarctic fossil rodents, mostly but not exclusively Nearctic for Wood and mostly but not exclusively Palaearctic for Lavocat. Even more directly pertinent to the present symposium, each has made detailed studies of both African and South American fossil rodents of the group here being called Hystricomorpha, including what both call Caviomorpha in South America and Lavocat calls Phiomorpha in Africa.

In view of that near duplication in knowledge, it is, I think, quite fascinating that they have reached such different conclusions. The disagreement is not, indeed, quite as complete as might appear. I take it that they agree that the South American and African rodents in question both arose from closely related, if not from exactly the same, Holarctic ancestors. They agree that the earliest South American rodents reached that continent by a sweepstakes route or waif dispersal while it was isolated by oceanic barriers. They differ as to whether that dispersal was from North to South America before the present land connection existed or from Africa to South America after the present South Atlantic Ocean existed but before it was quite as wide as it now is. If the former is true, the assumption is that the New and Old World groups evolved in parallel from a common ancestry; if the latter, there was a single origin and the New World groups were derived from, and hence were more closely related to, the Old.

Hence the taxonomic alternatives to which I alluded in my introductory remarks, and we can now discuss them with the authoritative data provided by our friendly antagonists, Doctors Wood and Lavocat.

As I am the only other palaeontologist directly involved in this programme and as I have long been concerned with the history of the South American fauna as a whole, I cannot forbear from opening the discussion myself with a brief remark on the place of rodents in that history. No rodents occur in the earliest known South American mammalian faunas, those believed to be of Palaeocene and Eocene age. If they were present at all, they were then so few or so local as not to appear in the now rather rich fossil collections of those ages. Then with apparent suddenness they appear in abundance and in some but still quite limited variety in the next age or subepoch, designated as Deseadan and placed with considerable uncertainty in the early Oligocene of the world time scale.

Before rodents reached South America, or before they appear in the fossil record there, their ecological niches were apparently filled in part by a variety of marsupials. It is a curious and probably a significant fact that the most rodent-like of all known South American marsupials, a small animal known as *Groeberia*, occurs in a somewhat peculiar fauna from Mendoza, Argentina, which is just slightly older than the South American faunas in which true rodents appear. *Groeberia* has no known descendants, and after it the placental rodents took over. They expanded and diversified enormously thoughout the rest of the Caenozoic, the age of Mammals. All were descendants of those early immigrants from—where?—until toward the end of the Caenozoic when other groups of rodents, notably cricetids, moved in, quite certainly from North (including Central) America. Thus, the tendency for the marsupials in South America to become pseudo-rodents was nipped in the bud so to speak by the invasion of true rodents and their explosion there.

WEIR (*London*): Professor Wood and Professor Lavocat, may I ask why you have both ignored Antarctica as a possible route between the two continents?

WOOD (*Amherst*): I'll answer that first. I have two reasons for neglecting Antarctica. One is that I don't know anything about Antarctica, but the other one is that, according to all the people who are now drawing plate tectonics maps with great freedom, any connections of South America with Antarctica led not to Africa but to Australia at the time with which we are concerned. I think it is perfectly clear that any connection between South America and Australia through Antarctica was broken before the rodents reached Patagonia because no rodents got into Australia at that time. So I think we can neglect it unless you want them swimming the southern oceans.

LAVOCAT (*Montpellier*): I will say something about the problem of the relations of South America and Africa. I think it is very difficult for geophysicists to know the exact relationships between the two continents in the Middle Eocene times. Many scientists think that in the Middle Eocene the meridian of the eastern part of America was the same meridian as the one of the place that is now Dakar. At the moment I feel we have no real proof of more proximity. The separation of the two continents could have been approximately 600 miles. And we must also remember that there was no land at that time between North and South America. Professor Lacombe, who is a specialist in ocean currents, suggests that if Africa was in the same relative position to the equator then as now, there would be currents from Africa, but between North America and South America, and without a motor it is extremely difficult to cross a current, and I don't think that even very well evolved rodents had a motor on the raft. It is difficult to know how far rafts with animals can travel but it seems that recently some Frenchmen sailing in the ocean found a raft with trees and five rodents 1000 kilometres from any coast.

Another problem is that *Platypittamys* is not the only animal known in the Oligocene of South America. There are other rodents from the Deseadan which have been found recently, and among them there is one (*Sallamys*) which has teeth rather similar to those of *Platypittamys* (see Wood, 1974; Fig. 2), and I will ask my friend Dr Wood how he explains this.

WOOD: This is very simple, I accept Dr Lavocat's picture (Fig. 1a) as being 100% accurate for M1. However, on M2 and M3 the structure is like that shown in Fig. 1a and I think it is quite clear that what he considers to be a mesoloph in this tooth is simply the labial part of the metaloph which has become separated from the buccal part in M1 of *Sallamys*, because *Sallamys* is not quite as primitive in my opinion as *Platypittamys* is. But this is the kind of thing you run into, of course, when you are worrying about details of tooth structure. This is why many non-palaeontologists say 'forget about teeth because anybody can interpret them any way he wants to.' I would like at this time also to mention another Deseadan rodent, *Protosteiromys* (Fig. 1b). In *Protosteiromys*, as in *Platypittamys*, we can still see the rounded paracone and metacone at the ends of the second and third crests but the fourth crest is beginning to subdivide. At least I say it is beginning to subdivide; Dr Lavocat would say that what I call the metacone is really the mesoloph which has gotten to look like a metacone in a most unmesoloph-like way. Obviously, what we need now are a few slightly earlier fossils from South America to settle the argument for us. At the moment we don't have them.

LAVOCAT: Yes. You see the problem is extremely technical and Professor Wood and I could argue all day about metalophs, mesolophs and posterolophs because of our different interpretations. In the Oligocene of Africa there are samples which have very similar structures to those found in *Sallamys*. They are not strictly identical, but you can interpret these animals of Africa in the same way as *Sallamys*. Professor Wood himself interprets the African genera with a metaloph, which is exactly as I described it, that is, with this posterior part reaching the posteroloph. And what I could say also is that I have animals of the Miocene of Africa in which the structure of the teeth is exactly like those of *Protosteiromys* in the Oligocene. We have no proof that the teeth of *Platypittamys* are strictly primitive. The proof could be supposed to be given by the fact that the infraorbital foramen is not really enlarged but, as I mentioned in my lecture, *Kenyamys* has one of the most primitive infraorbital foramina and the teeth are more evolved than those of *Phiomys*.

WEIR: Could I ask you both to draw side by side your versions of the relationships between these two groups, the Caviomorpha and Phiomorpha.

WOOD: I would do it this way:

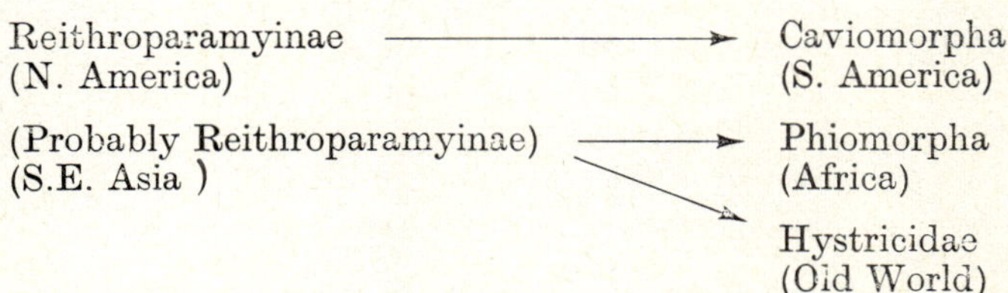

At the present time I feel a little uncertain about trying to unite these two reithroparamyine groups into a single taxon, since the Asiatic group

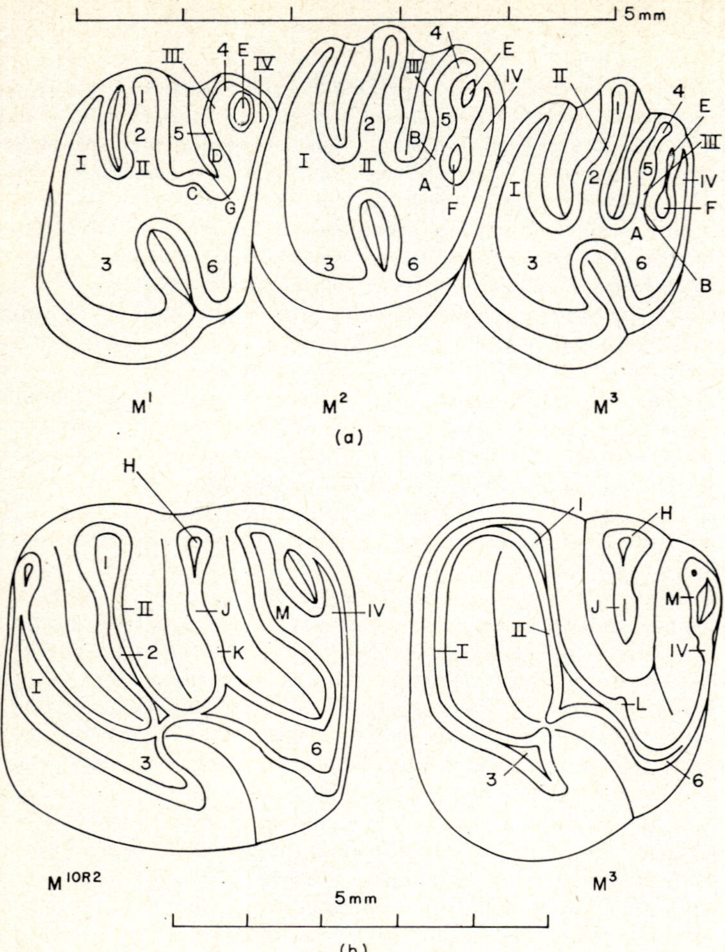

Fig. 1. Left upper molars of Oligocene caviomorphs *Sallamys* and *Protosteiromys* showing agreement and disagreement between Lavocat and Wood in the identification of various parts. Anterior ends to the left. Scales as shown.

(a) *Sallamys pascuali*, redrawn after Hoffstetter, R. & Lavocat, R. (1970). *C. r. hebd. Séanc. Acad. Sci. Paris*, **273**: 172–175.
(b) *Protosteiromys medianus*, redrawn after Wood, A. E. & Patterson, B. (1959). *Bull. Mus. comp. Zool. Harv.* **120**: 279–428: Fig. 30.

I = anteroloph; II = protoloph; III = metaloph; IV = posteroloph; 1 = paracone; 2 = protoconule; 3 = protocone; 4 = metacone; 5 = metaconule; 6 = hypocone.

(a) Wood's interpretation of *Sallamys*:

A–B is part of metaloph connecting the metaconule and hypocone.
C is the same as A.
D is the same as B, but separated from its primitive connection to the hypocone at C.
E is a valley between the external parts of metaloph and posteroloph.
F is a valley between internal parts of metaloph and posteroloph.
G is homologous with F.

is completely hypothetical and I am not sure how to define the characteristics of a taxon that is half hypothetical. If we ultimately establish that
this phylogeny is correct, it is possible that we would then have a suborder
Hystricognathi including all of these groups. We would then probably
have to establish a different definition to fit this suborder Hystricognathi
because the kind of definition that Dr Lavocat was talking about is valid
only for the later descendants. But it seems to me that the tremendous
amount of parallelism going on here is an indication of relationships between
the three modern groups, because I believe it is true that close parallelism
is a strong indication of close genetic relationships. One reason why I was
willing to accept all the parallelism that there must have been in my ideas
is that I had an opportunity many years ago to study the North American
heteromyids and geomyids and I was impressed by the overwhelming
parallelism that there must have been between these two families and
their subfamilies in their evolution in North America. I have more recently
discovered that I had grossly under-estimated the parallelism. It is impossible now, unless you know the geologic horizon from which a fossil

---

Lavocat's interpretation (based partly on the comparison with *Paraphiomys pigotti*
(Lavocat, R. (1973). *Trav. Mém. Inst. E.P.H.E. Montpellier* 1: 1–284: Pl. 26, stereo-
photo 5) and with *Diamantomys*:

A–B is not a metaloph but a mesoloph which reaches the metaloph at its labial end;
the crest 4-5-B-A is therefore a composite one.

C is the same as A but is an incipient mesoloph and not a metaloph.

E is a valley between the metaloph and posteroloph.

F, because of the interpretation of A–B, is a valley between the mesoloph and posteroloph.

G is homologous with F.

Two interpretations can be proposed of the unique valley E–F in M³.

1. As has probably happened in *Paraphiomys*, the valley is the result of an enlargement of F; E and a distinct metaloph have disappeared, and the crest labelled III is,
therefore, only an enlarged mesoloph.

2. It is possible that E opens lingually into F. Thus the crest labelled III remains a
combination of a lingual mesoloph and a labial metaloph, and the connection of the
metaloph with the posteroloph has disappeared, as it does in the M3 of *Diamantomys*.

(b) Wood's interpretation of *Protosteiromys*:

H is the metacone.

J is the metaconule.

K is the same as B.

L is the same as C, i.e. the remains of an interrupted metaloph.

M is considered to be a newly developed crest, a neoloph formed by separation from
the front of the posteroloph.

Lavocat's interpretation of *Protosteiromys* is based partly upon comparison with
Simonimys (a Miocene phiomorph from Kenya) in which the M¹ structure is identical
with that of *Protosteiromys* (Lavocat, 1973: Pl. 42, Fig. 3):

H is the mesocone.

H-J-K is the mesoloph.

K is the same as B (mesoloph).

L is the same as C of *Sallamys*, but both represent an interrupted mesoloph, and not
a metaloph.

M is considered to be the true metaloph.

member of one of these forms comes, to identify the family, subfamily, or tribe to which it belongs. Each of about 10 lines within the two families did exactly the same things, evolutionarily speaking, in tooth structure, in skull structure, in limb structure, at slightly different times. Rodents can be exceedingly parallel.

LAVOCAT: Well, this is what I think the relationships are:

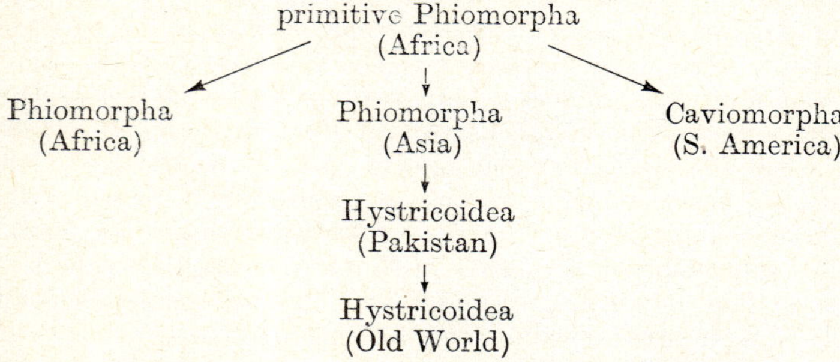

At one time I held exactly the same views as Professor Wood about the relationships between the Caviomorpha and Phiomorpha. I had even written these conclusions in my paper about East African Miocene rodents, (Lavocat, 1973*) and it was only three weeks afterwards that I had to change my mind. There is no one character important enough to distinguish between a South American caviomorph and an African phiomorph rodent, except perhaps the teeth. If you have only one, two or three structures similar, you can postulate separation for a long time, but here you have practically every feature in common. In *Thryonomys* and all the skulls that I have studied of South America I cannot discover any real difference in the osteology. The same is true for all the other characters known. These are sometimes shared by other groups but in fact you cannot find any important character which is not shared by the Caviomorpha and Phiomorpha. I think the separation of the groups with a completely different anatomy began in the Middle Eocene. The Theridomorpha have a really distinct structure in the Middle Eocene. I assume for that reason that it was the same for the root of the Hystricognathi. We know, from the discoveries of Dr Savage, that there were rodents in the late Eocene of Africa [see personal communication to Wood, 1973, this symposium, p. 33]. Phiomorphs must have been present in the Middle Eocene of Africa and this must also have been true for the anomalurids and pedetids, since neither is known anywhere else. When you see which are the fundamental differences between the different sorts of Phiomorpha, their common ancestors must have been much earlier than the Oligocene and there is nothing that will fit this out of Africa except the Hystricidae, which I think is an off-shoot of the Phiomorpha in the Middle Eocene.

* Lavocat, R. (1973). *Trav. Mém. E.P.H.E. Inst. Montpellier*. **1**: 1–284.

Symp. zool. Soc. Lond. No. 34, 61–78

THE CEPHALIC ARTERIES OF HYSTRICOMORPH RODENTS

J. BUGGE

Department of Anatomy, The Royal Dental College, DK 8000, Aarhus C, Denmark

SYNOPSIS

Sixty-four corrosion specimens from 17 species of hystricomorphs are used to determine the cephalic arterial pattern. The species include New and Old World forms and some bathyergids. All the superfamilies and most of the families of the suborder Hystricomorpha (*sensu lato*) are represented and the patterns of the cephalic arteries are considered a useful character for determining the taxonomic relationships of hystricomorph rodents. The resulting suggestions are compared with other formal classifications.

INTRODUCTION

I have been studying the cephalic arterial system of many mammals for several years and have found that it may serve as a character of importance for taxonomic purposes. Some studies have involved insectivores and primates (Bugge, 1972), but most have concerned the order Rodentia (Bugge, 1970, 1971a,b,c).

Until the middle of the present century the hystricomorphs were considered to constitute a natural entity. In recent years, one of the most important taxonomic problems has concerned the relationship between the New and the Old World hystricomorphs. Morphologically, they form a single entity, whereas it has been difficult on the basis of phylogeny and palaeogeography to link them in a single suborder (see Wood, 1950, 1955; Wood & Patterson, 1959).

I have attempted to revise the systematic classification of the New and Old World hystricomorphs on the basis of the cephalic arterial system, with regard to continental drift and rotation in the early Tertiary, and with reference to earlier classifications based on other organ systems.

MATERIALS AND METHODS

The 17 species and 64 specimens used in this investigation are detailed in Table I. Fresh or frozen heads were treated by a plastic injection technique followed by partial corrosion as described by Bugge (1963).

TABLE I

The hystricomorph rodents investigated and their suggested classification in the suborder **Hystricognathi** *Tullberg, 1899*

Superfamily	Family	Subfamily	Species	Common name	No. used
Infra- or suborder **Caviomorpha** Wood & Patterson, 1959 (but excluding Erethizontidae)					
Cavioidea	Caviidae	Caviinae	*Cavia porcellus*	Guinea-pig	8
			Galea musteloides	Cuis	4
		Dolichotinae	*Dolichotis patagona*	Mara	2
	Dasyproctidae		*Dasyprocta aguti*	Agouti	3
Chinchilloidea	Chinchillidae		*Chinchilla laniger*	Chinchilla	6
			Lagostomus maximus	Plains viscacha	6
Octodontoidea	Capromyidae		*Myocastor coypus*	Coypu	5
	Octodontidae		*Octodon degus*	Degu	5
	Ctenomyidae		*Ctenomys talarum*	Tuco-tuco	4
	Echimyidae		*Proechimys guairae*	Casiragua	4
Infra- or suborder **Erethizontomorpha** new infra- or suborder					
Erethizontoidea	Erethizontidae		*Coendou prehensilis*	Prehensile-tailed porcupine	1
Infra- or suborder **Hystricomorpha** Brandt, 1855 (or **Phiomorpha** Lavocat, 1972)					
Hystricoidea	Hystricidae		*Hystrix cristata*	Crested porcupine	1
			Hystrix leucura	White-tailed Indian porcupine	2
Thryonomyoidea	Thryonomyidae		*Thryonomys swinderianus*	Cane rat	2
Bathyergoidea	Bathyergidae		*Bathyergus suillus*	Cape dune mole-rat	2
			Cryptomys natalensis	Natal mole-rat	4
			Heterocephalus glaber	Naked mole-rat	5

RESULTS

Basic pattern

The definitive cephalic arterial system in all mammals is assumed to have been derived from a primary basic pattern (Fig. 1A), comprising the internal carotid artery (ci), the external carotid artery (ce) and the stapedial artery (st) with its supraorbital (rs), infraorbital (ri) and mandibular branches (rm). In this study, special emphasis has been placed on the contribution of the internal and external carotid arteries and the stapedial artery to the cephalic arterial system.

There is evidence from ontogenetic studies that this is the primitive pattern (see Grosser, 1901; Tandler, 1902; Hofmann, 1914; Struthers, 1930; Padget, 1948). However, the internal carotid artery seems originally to have been divided into a medial and a lateral stem (Matthew, 1909; Van Valen, 1966). Of these two presumptive original stems, the lateral has apparently been retained in certain mammals, such as insectivores and primates, and the medial in others, like lagomorphs and those rodents which have not entirely lost an internal carotid artery. The primary area supplied by the internal carotid artery (ci) is part of the brain and the eyeball (oi–c), whilst that supplied by the external carotid artery (ce) is the tongue (li) and face (fa, ts, tf). Of the three branches of the stapedial artery (st), the supraorbital branch (rs) supplies the dura (mm) and the extrabulbar part of the orbit (l, f, e), the infraorbital branch (ri) supplies the upper jaw, and the mandibular branch (rm) supplies the lower jaw.

Alterations in this basic pattern are presumed to have occurred by obliteration of certain parts of the original system in connection with the development of a varying number of up to six anastomoses, designated *a 1–a 6* (Fig. 1B). These occur in various combinations in hystricomorph rodents and all other mammals, and as permanent (*a 1* and *a 3*) or temporary anastomoses (*a 2, a 4, a 5* and *a 6*) during human embryogenesis (see Padget, 1948; De la Torre & Netsky, 1960). Apart from a possible existence of anastomosis *a 1*, hardly any of the anastomoses shown in Fig. 1B can have been developed in the primitive placentals from the early Tertiary.

Of the six anastomoses shown in Fig. 1B, the most important in hystricomorphs are anastomosis *a 3'*, connecting the proximal part of the external carotid artery system (ce) with the mandibular branch (rm) and forming the first part of the maxillary artery (m^1), and anastomosis *a 2* between the infraorbital branch (ri) and the orbital part of the supraorbital branch (rs), forming the external ophthalmic artery (oe). By means of these anastomoses the external carotid artery

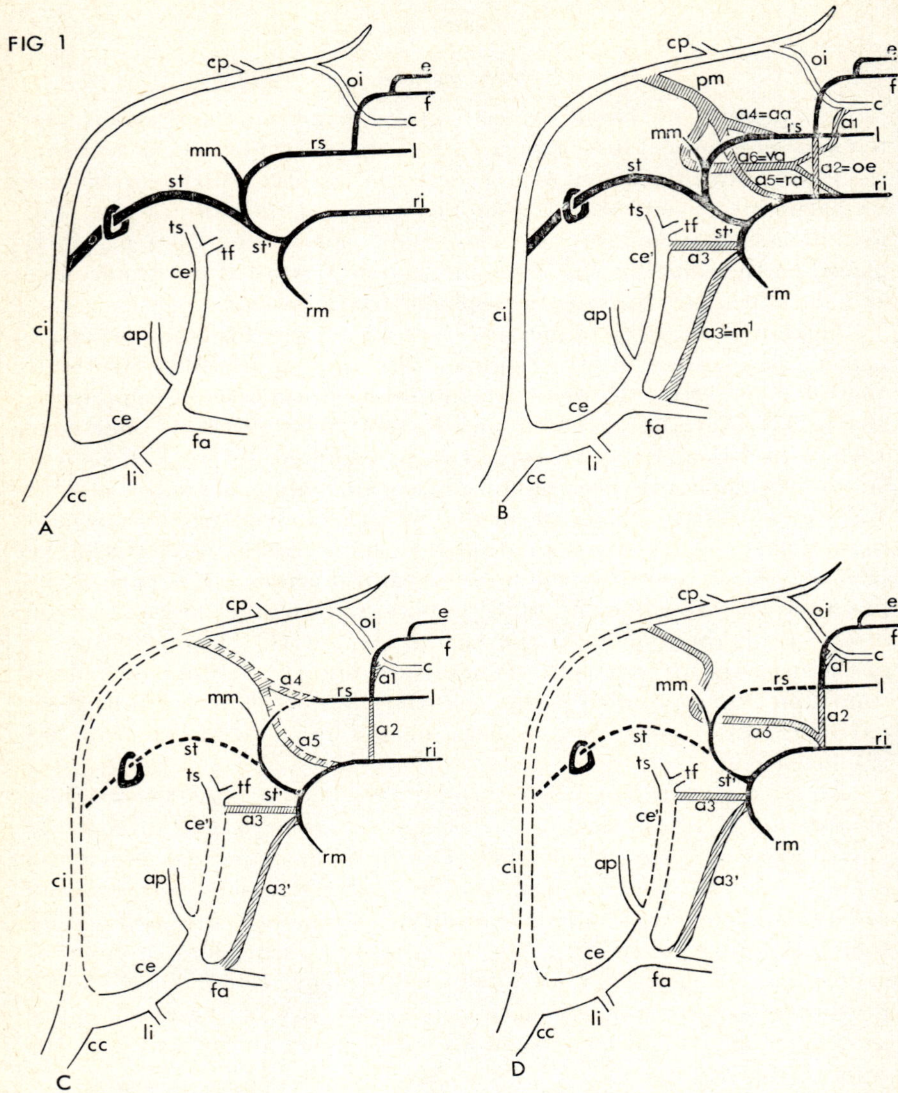

Fig. 1. The cephalic arterial pattern in New and Old World hystricomorphs. A: Basic pattern. B: Basic pattern and six possible anastomoses (*a 1–a 6*). C: Cavioidea; *Cavia porcellus* (minus *a 4*) and *Dolichotis patagona* (minus *a 3* and *a 5*). D: Cavioidea; *Galea musteloides* (minus *a 3*) and *Dasyprocta aguti*.

White: external–internal carotid system; *black*; stapedial artery system; *hatched*: anastomoses; *dashed*: only temporarily developed; *one side dashed*: postembryonically inconstant.

aa = arteria anastomotica (*a 4*); ap = a. auricularis posterior; c = a. ciliaris (proximal (ce) annexes the whole stapedial area of supply, that is, the upper and lower jaw (ri and rm), the dura (mm) and the extrabulbar part of the orbit (l, f, e); the stapedial artery stem (Figs 1C–D and 2A–D, st) and the central part of the supraorbital branch (rs) become obliterated. Anastomosis *a 1*, connecting the extrabulbar arteries (Fig. 1B, f, e) with the bulbar arteries (c) of the orbit, is also important. By means of this anastomosis the external carotid artery also annexes the supply of the eyeball (c), and the internal ophthalmic artery (oi) from the circulus arteriosus in most cases disappears. An anastomosis, *a 3*, sometimes develops between the distal part of the external carotid artery (ce′) and the mandibular branch (rm). In hystricomorphs (Figs 1C–D and 2A–D) this forms a common trunk for the superficial temporal artery (ts) and the transverse facial artery (tf); the external carotid artery (ce) then disappears distal to the departure of the posterior auricular artery (ap). Anastomoses *a 4*, or anastomotic artery (aa), and *a 5*, or anastomotic branch (ra), which connect the distal end of the internal carotid artery (ci) with the supraorbital (rs) and infraorbital (ri) branches respectively, occur irregularly and are unimportant for the systematic classification of hystricomorph rodents. Anastomosis *a 6*, the distal end of which forms the Vidian artery (va), connects the distal end of the internal carotid artery (ci) with the ciliary artery (c) or the infraorbital branch (ri) or both. It occurs in a few hystricomorphs but contributes only slightly to the cephalic pattern of supply. The common trunk (Fig. 1B, pm) for anastomoses *a 4*, *a 5* and *a 6* is thought to represent the vestige of a "persistent primitive maxillary artery" (De la Torre & Netsky, 1960).

Hystricomorph patterns

Cavioidea Kraglievich, 1930

In the cavioids examined (see Table I), the internal carotid artery, the tympanic part of the stapedial artery stem and the mid-part of the supraorbital branch (Fig. 1C–D, ci, st, rs) are lacking. The brain is

---

part = *a 1*); cc = a. carotis communis; ce = a. carotis externa (proximal part); ce′ = a. carotis externa (distal part); ci = a. carotis interna; cp = a. communicans posterior; e = a. ethmoidalis; f = a. frontalis; fa = a. facialis; l = a. lacrimalis; li = a. lingualis; m^1 = first part of the maxillary artery (*a 3′*); mm = a. meningea media; oe = a. ophthalmica externa (*a 2*); oi = a. ophthalmica interna; pm = a persistent primitive maxillary artery; ra = ramus anastomoticus (*a 5*); ri = r. infraorbitalis; rm = r. mandibularis; rs = r. supraorbitalis; st = a. stapedia; st′ = distal part of the stapedial artery stem; tf = a. transversa faciei; ts = a. temporalis superficialis; va = Vidian artery (distal part of anastomosis *a 6*). (Reproduced by permission from Bugge, 1971c).

supplied by the vertebral artery, assisted by the external carotid artery (see below), while the stapedial area of supply has been annexed by the external carotid artery system, and up to five of the following anastomoses are developed: *a 1, a 2, a 3, a 3', a 4, a 5, a 6* (Fig. 1B).

By means of anastomosis *a 3'* between the proximal part of the facial artery (Fig. 1C–D, fa) and the mandibular branch (rm), the external carotid artery (ce) supplies the upper and lower jaws (ri, rm), and the dura (mm) (*via* the distal part of the stapedial artery stem, st', and the proximal part of the supraorbital branch, rs). The whole orbit, including the extrabulbar part (l, f, e), originally belonging to the stapedial (st) area of supply, and the bulbar part (c), originally supplied by the internal carotid artery system (ci), is annexed by the external carotid artery (ce) *via* anastomoses *a 2* and *a 1* respectively. Anastomosis *a 1* is extremely short, and the common stem for the frontal and ethmoidal arteries (Fig. 1C–D, f + e) and the internal ophthalmic artery (oi) almost coalesce where they cross. Although the external carotid artery has taken over the supply of the eyeball (c) by means of anastomosis *a 1*, the internal ophthalmic artery (oi) persists and is very well developed. The external carotid artery system (ce) makes in this way a considerable contribution to the supply of the anterior part of the circulus arteriosus. This circuit is considerably stronger distally than it is proximally to the junction with the internal ophthalmic artery, in which the flow is reversed. With the exception of *Dolichotis*, the internal ophthalmic arteries (oi) of cavioids are connected by a strong interorbital artery and there is an additional anastomosis, *a 3* (Fig. 1C–D), forming a common trunk for the superficial temporal artery (ts) and the transverse facial artery (tf); the external carotid artery (ce) disappears distal to the departure of the posterior auricular artery (ap). Anastomosis *a 4* in *Dolichotis* and anastomosis *a 5* in *Cavia* occur irregularly (Fig. 1C). Anastomosis *a 6* (Fig. 1D) is well developed in *Dasyprocta* and *Galea*. In the former the vertebral artery system contributes by this anastomosis to the supply of the upper jaw, whereas in *Galea* the flow of this anastomosis is reversed, and the external carotid artery contributes to the supply of the mid-portion of the circulus arteriosus.

Thus, in the cavioids, the presence of the interorbital artery and anastomoses *a 3, a 4, a 5* and *a 6* is variable. However, as in all other hystricomorphs, the proximal part of the stapedial artery stem (Fig. 1C–D, st) and the central part of the supraorbital branch (rs) are obliterated, the whole stapedial area of supply being annexed by the external carotid artery (ce) by means of the anastomoses *a 3'* and *a 2*. The supply of the bulbar part of the orbit (c) is taken over by the external carotid artery system *via* anastomosis *a 1*, the flow in the internal ophthalmic artery (oi) being reversed. In addition, the internal carotid artery (ci) is obliterated, and the brain becomes supplied not only by the vertebral artery but also from the external carotid artery system by the well developed internal ophthalmic artery (oi). The contribution of the external carotid artery to the supply of the anterior part of the brain in this way has not been found in any mammals other than cavioids.

Chinchilloidea Kraglievich, 1940

In the two chinchillids investigated (see Table I) the internal carotid artery and the stapedial system are obliterated to the same extent as in cavioids (Fig. 2A, ci, st) and the brain is supplied by the vertebral artery alone, the internal ophthalmic artery (oi) having disappeared in the chinchilla and become attenuated in the plains viscacha. The external carotid artery system (ce) has taken over the entire stapedial area of supply together with the bulbar part of the orbit by means of anastomoses $a\ 3'$, $a\ 2$ and $a\ 1$. There is no interorbital artery.

Octodontoidea Simpson, 1945

In all the octodontoids examined (see Table I) the internal carotid artery (Fig. 2B, ci) is lacking and the stapedial artery system (st) is obliterated to the same extent as in cavioids and chinchilloids. The brain is supplied as in the chinchilloids by the vertebral artery alone while the stapedial area of supply is taken over by the external carotid artery system (ce), the following four anastomoses being developed: $a\ 1$, $a\ 2$, $a\ 3$, $a\ 3'$. Octodontoids are distinguished from chinchilloids by the development of anastomosis $a\ 3$ as well as anastomosis $a\ 3'$. This forms the common trunk for the superficial temporal artery (Fig. 2B, ts) and transverse facial artery (tf).

Erethizontoidea Simpson, 1945

A persistent carotid canal was reported by Hill (1935) in *Erethizon epixanthum* and in the present study of *Coendou prehensilis*, the internal carotid artery was found to be well developed (Fig. 2C, ci). The stapedial artery stem (st) and the mid-part of the supraorbital branch (rs) are obliterated as in all caviomorphs. The brain is supplied by the internal carotid artery and vertebral artery, while the whole stapedial area of supply together with the bulbar part of the orbit is taken over by the external carotid artery system (ce) by means of three anastomoses ($a\ 1$, $a\ 2$, $a\ 3'$) as in the chinchilloids. The internal ophthalmic artery (oi) has become obliterated.

FIG 2

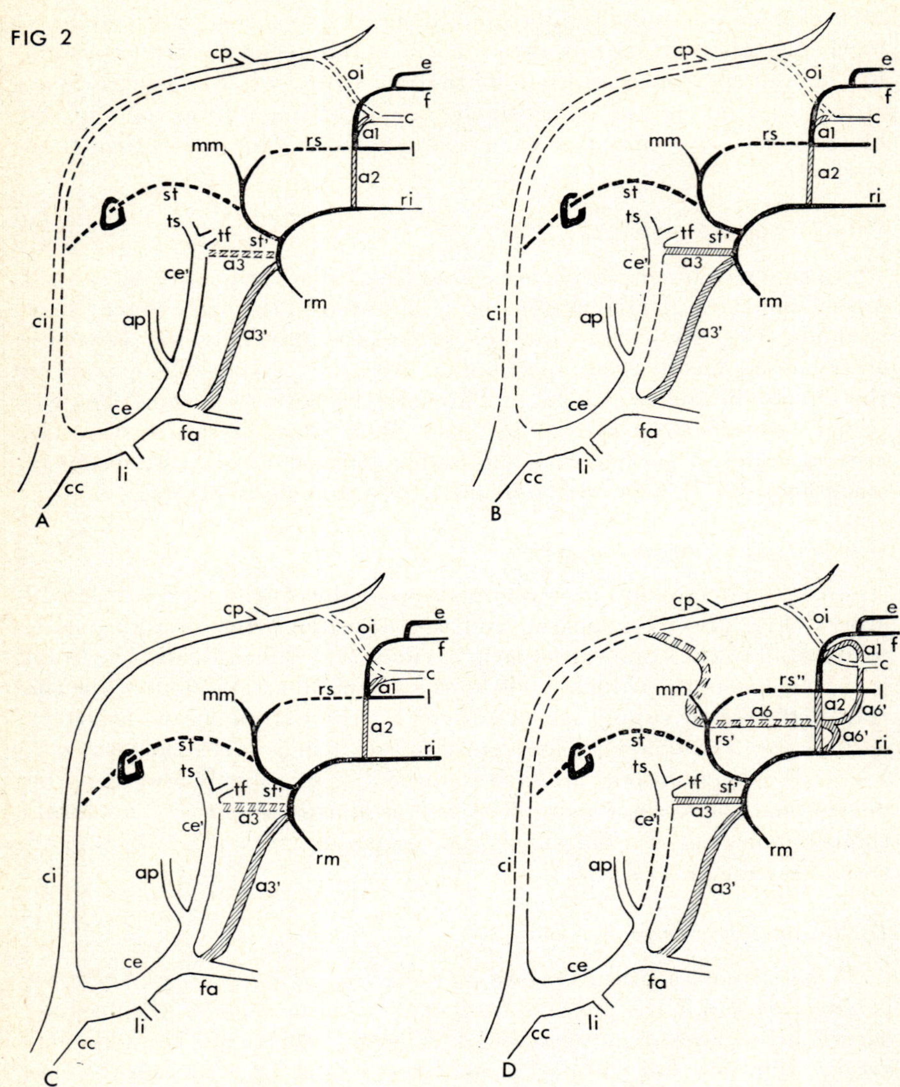

FIG. 2. The cephalic arterial pattern in New and Old World hystricomorphs. A: Chinchilloidea; *Chinchilla laniger* and *Lagostomus maximus* (plus weak oi). B: Octodontoidea; *Myocastor coypus*, *Octodon degus*, *Ctenomys talarum* and *Proechimys guairae*; Thryonomyoidea; *Thryonomys swinderianus* and Bathyergoidea; *Bathyergus suillus*, *Cryptomys natalensis* and *Heterocephalus glaber*. C: Erethizontoidea; *Coendou prehensilis*. D: Hystricoidea; *Hystrix cristata* and *Hystrix leucura*. See legend to Fig. 1 for key to symbols and abbreviations.

Hystricoidea Gill, 1872

In the two hystricoids studied (see Table I), the internal carotid artery (Fig. 2D, ci), the stapedial artery stem (st) and the supraorbital branch (rs) are obliterated to the same extent as in the caviomorphs. The brain is supplied by the vertebral artery alone, as in chinchilloids and octodontoids, while the external carotid artery system has taken over the entire stapedial area of supply. Five anastomoses (a 1, a 2, a 3, a 3' and a 6') are developed. The upper and lower jaws (Fig. 2D, ri and rm) and dura (mm) are supplied by the external carotid artery by anastomosis a 3' from the proximal part of the facial artery (fa). Anastomosis a 3 forms the common trunk for the superficial temporal artery (ts) and the transverse facial artery (tf) only. Most of the orbit is also supplied by the external carotid artery system (ce) by means of anastomoses a 2, a 1 and a 6'. The infraorbital branch (Fig. 2D, ri) annexes the supply of the extrabulbar part of the orbit (l, f, e) by means of anastomosis a 2, lateral to the 1st and 2nd trigeminal branches. It also annexes the supply to the bulbar part (c) by the distal part of anastomosis a 6 (= a 6') medial to the 1st trigeminal branch. During this process, the proximal part of anastomosis a 6 and the internal ophthalmic artery (oi) usually become obliterated. Anastomosis a 1 is weak and only establishes a connection across the muscle cone between the bulbar (c) and the extrabulbar artery system (l, f, e). The medial ophthalmic arteries (a 6'), forming the proximal part of the ciliary arteries (c), are connected by an interorbital artery through the right and left optic foramina.

Thryonomyoidea Wood, 1955 and Bathyergoidea Osborn, 1910

In the species listed in Table I the cephalic arterial pattern is identical to that of octodontoids (Fig. 2B), except that the lacrimal artery (l) in the bathyergoids becomes secondarily annexed by the buccolabial artery.

DISCUSSION

From a palaeogeographical point of view it seems reasonable to divide the classical rodent suborder Hystricomorpha Brandt, 1855, into several relatively independent lines, whereas from a morphological standpoint the New and Old World hystricomorphs together with the bathyergoids seem to form a natural entity, because they have many common features, including the cephalic arterial system, which are difficult to explain as pure parallelism. For example, there is remarkable similarity between the cephalic arterial systems of the four South American octodontoid families investigated and those of the African thryonomyids and bathyergoids (Fig. 2B), which are also placed in the same superfamily by Simpson (1945). There is, however, a considerable difference between the cephalic arterial pattern of cavioids (Fig. 1C–D) and that of other caviomorphs from the New World (Fig. 2A–B), especially with respect to the brain supply. Apart from the anastomosis *a 6'*, the cephalic arterial system of the hystricids (Fig. 2D) is very similar to that of the other Old World hystricomorphs including the bathyergoids. Although a specific contribution is made by the external carotid artery system to the supply of the brain in cavioids (Fig. 1C–D) *via* a persistent internal ophthalmic artery (oi), the only important departure from the general cephalic arterial pattern in New and Old World hystricomorphs and the bathyergoids is the persistent internal carotid artery (Fig. 2C, ci) in New World porcupines (Erethizontoidea). This is undoubtedly the original condition, retained from the common, early Tertiary ancestor of all hystricomorphs.

The placing of the erethizontoids in a new independent infra- or suborder, Erethizontomorpha, rests primarily on the unusual cephalic pattern of supply (Fig. 2C). They undoubtedly left the common caviomorph stem before the obliteration of the internal carotid artery took place, that is, before the separation of the other caviomorph superfamilies in the early Oligocene (see Table II) since reacquisition of a primitive character must be considered extremely unlikely. The erethizontoids most probably evolved independently some time in the last half of the Eocene. Their isolated position in relation to other caviomorphs is also emphasized by their external parasites (Vanzolini & Guimaraes, 1955), serology (Moody & Doniger, 1956) and osteology and myology (see Wood & Patterson, 1959).

Whilst the present investigation confirms the isolated position of the erethizontoids, the cephalic arterial system together with numerous other morphological features seems to presuppose a relationship between the other caviomorphs and the Old World hystricomorphs and bathyergoids across the South Atlantic. As the major objection to placing the New and Old World hystricomorphs in the same suborder has been the seemingly impassable barrier created by the South Atlantic (see Wood, 1950, 1955; Wood & Patterson, 1959), the research of recent years into continental drift and rotation is very important. Such research reveals that although the South Atlantic existed in the Eocene, it was not then at its present formidable width (Thenius, 1971).

The detailed investigations of Lavocat (1969) have revealed that all Old World hystricomorphs (i.e., thryonomyids, petromurids, bathyergids and hystricids) can be traced to the Lower Oligocene phiomyids from Fayum (Table II), and that the similarity between primitive

TABLE II

*A possible evolutionary relationship between the New and Old World hystricomorphs**

OLD WORLD HYSTRICOMORPHS incl. BATHYERGOIDS				NEW WORLD HYSTRICOMORPHS		
HYSTRICOMORPHA sensu stricto or PHIOMORPHA				ERETHIZONTOMORPHA, CAVIOMORPHA excl ERETHIZONTOIDEA		
HYSTRICOIDEA	THRYONOMYOIDEA	BATHYERGOIDEA		ERETHIZONTOIDEA		
ci: 0; st: 0	ci: 0; st: 0	ci: 0; st: 0		ci: +; st: 0		ci: 0; st: 0

Columns (left to right): HYSTRICIDAE [A+EA], PETROMURIDAE [A], THRYONOMYIDAE [A], BATHYERGIDAE [A], ERETHIZONTIDAE [NA], CAVIOIDEA [SA], OCTODONTOIDEA [SA], CHINCHILLOIDEA [SA].

Time scale (left column): QUART.; TERTIARY — PLIOCENE (U, M, L); MIOCENE (U, M, L); OLIGOCENE (U, M, L); EOCENE (U, M, L).

HYSTRIX (E)
ci: 0; st: 0 *(MIOCENE M)*

BATHYERGOIDES (A)
ci: (0); st: (0) *(MIOCENE L)*

PHIOMYS (A)
ci: (0); st: (0) *(OLIGOCENE L)*

PROTOSTEIROMYS (SA)
ci: +; st: (0) *(OLIGOCENE L)*

PLATYPITTAMYS (SA)
ci: ?; st: (0) *(OLIGOCENE L)*

?······INVADING PARAMYID (A) *(EOCENE U)*

?

PHIOMYID (primitive) (A): ?·····················► ? INVADING (SA)? *(EOCENE M)*
ci: +; st: +

INVADING PARAMYID (A)·········· *(EOCENE)*
ci: +; st: +

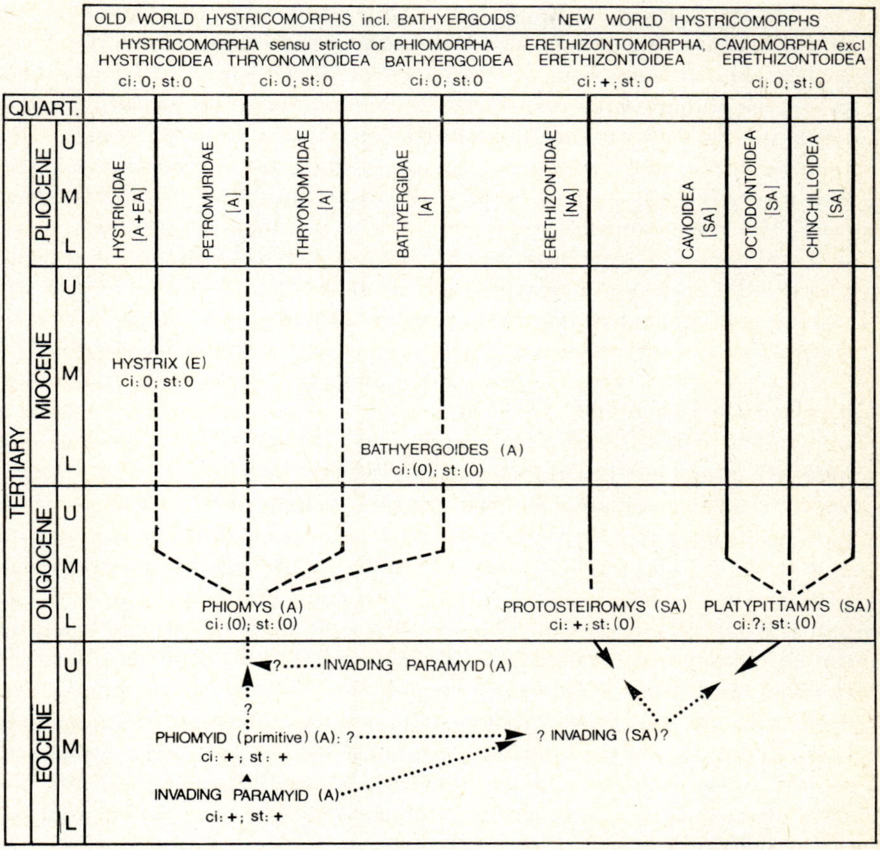

* Based on the cephalic arterial system with special reference to the persistence of the internal carotid artery (ci) and the stapedial artery (st) and partly on Wood & Patterson (1959); Wood (1968); Lavocat (1969, 1972).

A = Africa; E = Europe, EA = Eurasia; NA = North America; SA = South America; U = Upper; M = Middle; L = Lower.

Solid lines represent reasonably certain, broken lines probable, and dotted lines possible relationships.

+ = artery present; 0 = artery absent; (+) = artery presumably present; (0) = artery presumably absent; ? = uncertainty.

(After Bugge, 1971c).

caviomorphs from the Lower Oligocene of South America and the oldest phiomyids from the same period in North Africa is so striking that it is difficult to explain as simple parallelism. Owing to the great uniformity of the Lower Oligocene phiomyids from Fayum, Wood (1968) finds it most likely that their ancestor, one of the Old World paramyids, reached the African continent about the beginning of the Upper Eocene and quickly differentiated in the phiomyid direction (Table II). These earliest known Oligocene phiomyids can, however, hardly have been primitive enough with respect to the cephalic arterial pattern to have been ancestral to the erethizontoids (Table II). Their presumed close relationship with thryonomyoids makes it most likely that the stapedial and internal carotid arteries were already obliterated (Table II). However, the erethizontoids have retained a particularly well developed internal carotid artery, and its persistence may be explained as a result of a double invasion of South America: an early invasion by Old World paramyids which gave rise to the erethizontoids, and a later invasion by African phiomyids from which the other caviomorphs have evolved.

It seems, however, more probable that the striking morphological similarity between Old World and most New World hystricomorphs is due to the African and South American continents both having been invaded some time during the Eocene by Old World paramyids with a very primitive cephalic arterial pattern (Table II). In Africa, these paramyids gave rise *via* the phiomyids to the Old World hystricomorphs and bathyergoids, and in South America to erethizontoids and other caviomorphs. The closely related Eocene ancestors presumably had many common genes which, in conjunction with a uniform selective pressure, have resulted in the marked morphological similarity between the Old and New World hystricomorphs, despite long geographical separation (Wood, 1950). Although no genuine paramyid remains have been found in South America, this may be because large areas still remain to be explored and the paramyids rapidly diversified into the stem groups of the recent caviomorph superfamilies after their invasion of the continent.

R. Lavocat (personal communication) also agrees that the internal carotid artery was absent in the Lower Oligocene phiomyids from Fayum, and that these cannot therefore be directly ancestral to the erethizontoids, but unlike Wood (1968) he considers the phiomyids to have differentiated in Africa as early as the Mid-Eocene (Table II). As these presumptive early African phiomyids must be considered to be close to the original paramyids, it is reasonable to assume that the cephalic arterial system was primitively developed. These phiomyids could have been ancestral to the Old World hystricomorphs and bathyergoids *via* the Oligocene Fayum phiomyids (Table II), and have differentiated into erethizontoids and caviomorphs after their invasion of South America. On the basis of the assumption that all the classical hystricomorphs and bathyergoids stem from the same primitive Mid-Eocene African phiomyids, Lavocat (1972) collects them into one suborder, Hystricognathi Tullberg, 1899, which is divided into two infraorders, Caviomorpha Wood & Patterson, 1959, and Phiomorpha Lavocat, 1972, comprising the New and Old World hystricomorphs respectively. However, the persistent internal carotid artery in the erethizontoids suggests that they left the common caviomorph stem long before the true caviomorphs were diversifying into the recent superfamilies. They probably evolved independently of the other New World hystricomorphs after leaving the presumptive common Mid-Eocene phiomyid ancestor (Table II). The New World hystricomorphs should thus undoubtedly be divided into two infraorders, the Erethizon-tomorpha (new infraorder) and the Caviomorpha of Wood & Patterson (1959) but excluding the erethizontoids.

The collection of all the classical hystricomorphs and the bathyergoids into a single suborder, based on the assumption of the existence of a common primitive African phiomyid ancestor as proposed by Lavocat (1972) but divided into three rather than two infraorders, seems to be the most obvious classification when the results of morphological, palaeontological and palaeogeographical investigations of recent years are taken into account. This classification would also explain the remarkable morphological similarity of the cephalic arterial system between the Old World and most of the New World hystricomorphs, especially between the African thryonomyids and bathyergoids and the four South American octodontoid families, and accounts for the presence of a well developed internal carotid artery in the erethizontoids.

If, however, as supposed by Wood (1968), the African phiomyids evolved at the end of the Eocene and gave rise only to the Old World hystricomorphs, it is most natural to assume that the erethizontoids and the other caviomorphs evolved directly from paramyids which invaded the South American continent some time during the latter half of the Eocene. If the erethizontoids, caviomorphs and Old World hystrico-morphs evolved directly from the same or closely related Old World paramyids, the possibility of a subdivision of the classical hystrico-morphs into three suborders should be admitted. The present investigation confirms the classification of Simpson (1945) at the superfamily level although the thryonomyids (and petromurids) are removed from the octodontoids. This modification, however, is made from a zoo-geographical and not a morphological point of view because the cephalic arterial pattern in thryonomyids and South American octodontoids is identical. The echimyids, for example, which are placed by Landry (1957) among the erethizontoids, lack an internal carotid artery and with respect to the cephalic arterial system, are identical to the rest of the classical octodontoids. The dasyproctids, which are placed by Landry (1957) and Wood & Patterson (1959) among the octodontoids and chinchilloids respectively, are characterized by the same highly specific cephalic pattern of supply as the caviines and dolichotines.

SUMMARY

The cephalic arterial system of 17 species of New and Old World hystricomorphs has been studied from 64 specimens. They represent all the classical superfamilies and most of the families. The cephalic arterial pattern in the New World porcupines (Erethizontoidea) is more primitive than in other hystricomorphs examined because the internal carotid artery is still present. This persistence undoubtedly reflects the original condition of the early Tertiary ancestor of the hystricomorphs. There is a specific contribution of the external carotid artery to the supply of the brain in the cavioids *via* a persistent internal ophthalmic artery. Apart from these two features, the most remarkable result of the investigation is the demonstration that the cephalic arterial pattern of the four South American octodontoid families examined and that of the African thryonomyids and bathyergoids are identical and very similar to the patterns in the New World chinchilloids and Old World hystricoids.

The cephalic arterial patterns are used to suggest a systematic classification which takes into account earlier classifications based on other organ systems, recent investigations of fossil material and current theories of continental drift and rotation in the early Tertiary. If it is correct, as is assumed by R. Lavocat (personal communication), that all the hystricomorphs have evolved from primitive Mid-Eocene African phiomyids, it seems reasonable, if the cephalic arterial pattern is taken into account, to place all hystricomorphs in one suborder, which would be divided into three infraorders, one for the erethizontoids, one for the South American caviomorphs and one for the Old World hystricomorphs and the bathyergoids. However, if as assumed by Wood (1968), the African phiomyids gave rise only to the Old World hystricomorphs, the cephalic arterial pattern suggests that the erethizontoids and the other South American caviomorphs evolved directly from, presumably, Old World paramyids which invaded the South American continent some time during the Eocene. The erethizontoids, the South American caviomorphs and the Old World hystricomorphs and bathyergoids should perhaps then be given the status of independent suborders.

Acknowledgments

I wish to express my gratitude to Dr I. W. Rowlands and Dr B. J. Weir, Wellcome Institute of Comparative Physiology, Zoological Society of London, for important supplementary material for this investigation. My participation in the symposium was supported by a grant from the Danish State Research Foundation.

References

Brandt, J. F. (1855). Untersuchungen über die craniologischen Entwickelungs-stufen und die davon herzuleitenden Verwandtschaften und Classificationen der Nager der Jetztwelt, mit besonderer Beziehung auf die Gattung *Castor*. *Mém. Acad. Sci. St. Petersb.* (Sci. mat. phys. nat.) (6) **1**: 125–336.

Bugge, J. (1963). A standardized plastic injection technique for anatomical purposes. *Acta anat.* **54**: 177–192.

Bugge, J. (1970). The contribution of the stapedial artery to the cephalic arterial supply in muroid rodents. *Acta anat.* **76**: 313–336.

Bugge, J. (1971a). The cephalic arterial system in mole-rats (*Spalacidae*), bamboo rats (*Rhizomyidae*), jumping mice and jerboas (*Dipodoidea*) and dormice (*Gliroidea*) with special reference to the systematic classification of rodents. *Acta anat.* **79**: 165–180.

Bugge, J. (1971b). The cephalic arterial system in sciuromorphs with special reference to the systematic classification of rodents. *Acta anat.* **80**: 336–361.

Bugge, J. (1971c). The cephalic arterial system in New and Old World hystrico-morphs, and in bathyergoids, with special reference to the systematic classification of rodents. *Acta anat.* **80**: 516–536.

Bugge, J. (1972). The cephalic arterial system in the insectivores and the primates with special reference to the Macroscelidoidea and Tupaioidea and the insectivore-primate boundary. *Z. Anat. Entwickl.-Gesch.* **135**: 279–300.

De la Torre, E. & Netsky, M. G. (1960). Study of persistent primitive maxillary artery in human fetus: some homologies of cranial arteries in man and dog. *Am. J. Anat.* **106**: 185–195.

Gill, T. (1872). Arrangement of the families of mammals with analytical tables. *Smithson. misc. Collns* **11** (230): 1–98.

Grosser, O. (1901). Zur Anatomie und Entwickelungsgeschichte des Gefässsystemes der Chiropteren. *Arb. anat. Inst., Wiesbaden* **17**: 203–424.

Hill, J. E. (1935). The cranial foramina in rodents. *J. Mammal.* **16**: 121–129.

Hofmann, L. von. (1914). Die Entwicklung der Kopfarterien bei *Sus scrofa domesticus*. *Morph. Jb.* **48**: 645–671.

Kraglievich, L. (1930). Diagnosis osteológicodentaria de los géneros vivientes de la subfamilia "Caviinae". *An. Mus. nac. Hist. nat. B. Aires* **36**: 59–96.

Kraglievich, L. (1940). Los roedores extinguidos del grupo Neoepiblemidae. *Obras Geol. Paleont., La Plata* **3**: 739–764. (Cit. *Simpson* (1945)).

Landry, S. O. (1957). The interrelationships of the New and Old World hystrico-morph rodents. *Univ. Calif. Publs Zool.* **56**: 1–118.

Lavocat, R. (1969). La systématique des rongeurs hystricomorphes et la dérive des continents. *C. r. hebd. Séanc. Acad. Sci., Paris* (D) **269**: 1496–1497.

Lavocat, R. (1972). Miocene rodents of East Africa and Oligocene rodents of Bolivia. 20*th Symp. vertebr. Palaeont. comp. Anat.*

Matthew, W. D. (1909). The Carnivora and Insectivora of the Bridger Basin, Middle Eocene. *Mem. Am. Mus. nat. Hist.* **9** (6): 291–567.

Moody, P. A. & Doniger, D. E. (1956). Serological light on porcupine relationships. *Evolution, Lancaster, Pa.* **10**: 47–55.

Osborn, H. F. (1910). *The age of mammals in Europe, Asia and North America.* New York: Macmillan Co.

Padget, D. H. (1948). The development of the cranial arteries in the human embryo. *Contr. Embryol.* **32**: 205–261.

Simpson, G. G. (1945). The principles of classification and a classification of mammals. *Bull. Am. Mus. nat. Hist.* **85**: 1–350.

Struthers, P. H. (1930). The aortic arches and their derivatives in the embryo porcupine (*Erethizon dorsatum*). *J. Morph.* **50**: 361–392.

Tandler, J. (1902). Zur Entwicklungsgeschichte der Kopfarterien bei den Mammalia. *Morph. Jb.* **30**: 275–373.

Thenius, E. (1971). Zum gegenwärtigen Verbreitungsbild der Säugetiere und seiner Deutung in erdgeschichtlicher Sicht. *Natur. Mus., Frankf.* **101**: 185–196.

Tullberg, T. (1899). Ueber das System der Nägethiere. Eine phylogenetische Studie. *Nova Acta R. Soc. Scient. upsal.* (3) **18**: 1–514.

Van Valen, L. (1966). Deltatheridia, a new order of mammals. *Bull. Am. Mus. nat. Hist.* **132**: 1–126.

Vanzolini, P. E. & Guimaraes, L. R. (1955). South American land mammals and their lice. *Evolution, Lancaster, Pa.* **9**: 345–347.

Wood, A. E. (1950). Porcupines, paleogeography, and parallelism. *Evolution, Lancaster, Pa.* **4**: 87–98.

Wood, A. E. (1955). A revised classification of the rodents. *J. Mammal.* **36**: 165–187.

Wood, A. E. (1968). Early Cenozoic mammalian faunas Fayum Province, Egypt, Pt. 2: The African Oligocene Rodentia. *Bull. Peabody Mus. nat. Hist.* **28**: 23–105.

Wood, A. E. & Patterson, B. (1959). The rodents of the Deseadan Oligocene of Patagonia and the beginnings of South American rodent evolution. *Bull. Mus. comp. Zool. Harv.* **120**: 279–428.

DISCUSSION

SIMPSON (*Tucson*): I am sorry this beautiful new evidence supports questionable points in my classification that I wrote in 1942, because I have since abandoned it.

BUGGE (*Aarhus*): I beg your pardon, sir, but the cephalic arterial system in the South American octodontid families and the African thryonomyids and bathyergids is absolutely identical.

SIMPSON: That won't be the first time that a scientist has become more and more wrong as time went on rather than more and more right. The second point is that the idea of two different invasions of South

America strikes a palaeontologist as quite strange and it does seem to be contrary to the palaeontological consensus. In fact, I think everyone up until now has assumed that the caviomorphs, that is the South American morphologically hystricomorph rodents, are the result of a single invasion.

BUGGE: I wouldn't say that there were two invasions, but I have discussed this problem with Professor Thenius in Vienna and he wouldn't exclude that there could have been, and nobody could do so. I also think that there has been one single invasion in South America, but then it must have been a rodent with so primitive a cephalic arterial system that it could have been a possible ancestor of both the erethizontids and the true South American caviomorphs. This early ancestor of all the New World hystricomorphs must have had an internal carotid artery. Therefore, it could be these very primitive phiomyids which Professor Lavocat has thought could have existed in Africa in the Middle Eocene. Of course, it is also possible that Africa and the New World were invaded by some paramyids which were closely related and, therefore, had many common genes which in conjunction with the uniform selective pressures could have resulted in similar appearances.

SIMPSON: The only other point I want to make is to point out to some of you who are not familiar with the bio-geographic implications of all this, that we all agree that South America at the time of the invasion, or the invasions, of rodents was an island continent. That wherever these rodents came from, they came across a considerable sea barrier.

WOOD (*Amherst*): The palaeontological evidence somewhat suggests that there might have been two invasions of South America, one by the ancestral erethizontid and one by everyone else. If so, I feel absolutely sure the two ancestors came from the same place and at about the same time, and had differentiated not long before they invaded South America. But it seems to me, just because of what Dr Simpson said, that South America was an island continent and that it was difficult for animals to get there, it is much simpler to assume a single invasion which was of animals sufficiently primitive to have been ancestral to whatever is the most primitive form, and then very rapid evolution in South America. Of the latter I feel absolutely certain, because we have no late Eocene rodents, but we have them all over the place in the early Oligocene; they must have differentiated very rapidly.

SALE (*Nairobi*): I am not very clear as to what are the postulated selective forces altering the evolution of the cephalic arteries.

D

BUGGE: I have speculated very much about this problem. I have investigated about 500 rodents and 100 insectivores and primates and I have seen that their stapedial artery obliterates when the bulla and pars petrosa are nearly connected and when the chewing musculature is very strongly developed, and I think there is some mechanical explanation for obliteration of the stapedial artery. But I am not able at this moment to say anything about why the internal carotid artery is obliterated in some cases and not in others.

SALE: Well, if we know so little about the selective forces, are we justified in assuming that a character once obliterated cannot reappear later on? It seems to me that if the selective forces change, producing a situation, an evolutionary context which was formerly present, it is quite possible with characters of this sort that they can reappear.

BUGGE: I wouldn't think so. It is quite clear that the internal carotid artery and the stapedial artery are primitive characters found in all the early Tertiary mammals. Another thing is that these arteries pass through canals in the cephalic bones and when these arteries are obliterating these canals are also obliterated. Therefore, I would think it most unlikely that such osseous canals should re-open for the internal carotid artery or the stapedial artery. It would be helpful if palaeontologists would in the future comment on these osseous canals as well as the teeth.

SIMPSON: I finally can't resist adding one point which refers rather to an earlier comment during this morning's session, that palaeontologists would be delighted to work with things other than teeth, and do so when they have things other than teeth, but usually all they have is teeth so they have to make do with what they have.

Symp. zool. Soc. Lond. (1974) No. 34, 79–108

HYSTRICOMORPH CHROMOSOMES

WILMA GEORGE

Lady Margaret Hall, Oxford, England

and

BARBARA J. WEIR

Wellcome Institute of Comparative Physiology,
Zoological Society of London, Regent's Park, London, England

INTRODUCTION

Several attempts have been made to use karyotype patterns to elucidate phylogenetic relationships on the assumption that chromosome morphology is less subject to superficial adaptive changes than are other features of an organism. Matthey (1958) surveyed possible evolutionary changes of karyotypes within the Eutheria, based on earlier studies (Matthey, 1952, 1956) of the generic relationships of myomorph rodents. The sciuromorph genera *Tamias* and *Eutamias* have been distinguished by karyotypic differences (Nadler, 1964). A detailed study of the subfamily Herpestinae (Fredga, 1972) followed a survey by Wurster & Benirschke (1968a) and Wurster (1969) on karyotypic evidence for phylogenetic relationships of families within the order Carnivora. Similar studies have been reported on insectivore (Gropp, 1969; Borgaonkar, 1969) and primate (Egozcue, 1969) family relationships.

It is clear from these reports that karyotypes are more useful at the specific and population level than they are for the determination of relationships within higher taxa. Nevertheless, it was felt that a comparison of known hystricomorph rodent karyotypes might be of some help in elucidating possible phylogenetic relationships. The Hystricomorpha is considered here in its widest sense to include the New World (caviomorph) species and those of the Old World, although karyotypic information is only available for one African species. The following suggestions are based on general karyotype patterns and they are intended to be provocative rather than the basis for yet another taxonomy of this group of rodents.

TABLE I

Synopsis of karyotypic features of rodents of the suborder Hystricomorpha

Family and species	$2n$	NF* $2(m + sm + sa) + a + X$	NF* $2(m + sm) + (sa + a) + X$	Autosomes $m + sm$	Autosomes sa	Autosomes a	X-chromosome size	X-chromosome type	X-chromosome rank	Y-chromosome size	Y-chromosome type	Rank of marker chromosomes	References
Hystricidae													
Hystrix cristata	66	114	98	30	16	18	med.	m	1	—	—	—	D. H. Wurster, personal communication
Hystrix cristata	60	118	98	36	20	2	med.	m	1	micro	a	?	Renzoni, 1967
Erethizontidae													
Erethizon dorsatum	42	80	78	34	2	4	big	sm	1	med.	sm	—	Benirschke, 1968
Coendou rothschildi	74	86	82	6	4	62	med.	sm	1	micro	a	—	D. H. Wurster, personal communication
Caviidae													
Cavia aperea	64	128	118	52	10	0	small	sm	3	micro	m	—	George *et al.*, 1972
Cavia porcellus	64	128	118	52	10	0	small	sm	3	micro	m	—	Cohen & Pinsky, 1966; George *et al.*, 1972
Galea musteloides	68	136	130	60	6	0	med.	sm	1	small	m	—	George *et al.*, 1972
Dolichotis patagonum	64	128	114	48	7	0	med.	m	1	micro	m	—	Wurster, *et al.*, 1971
Dasyproctidae													
Dasyprocta aguti	64	126	116	48	12	2	med.	m	1	micro	m	—	Fredga, 1966
Dasyprocta variegata	64	126	116	50	10	2	med.	m	1	micro	sm	—	Hungerford & Snyder, 1964
Myoprocta acouchy	62	122	118	54	4	2	med.	sm	1	micro	a	—	Fredga, 1966
Octodontidae													
Octodon degus	58	116	110	50	6	0	med.	m	1–2	micro	sm	8	Fernández, 1968; George & Weir, 1972b
Spalacopus cyanus	58	116	112	52	4	0	med.	m	2	micro	sm	10–12	Reig *et al.*, 1972
Aconaemys sp.	58	116											O. A. Reig, personal communication
Octodontomys gliroides	38	68	60	20	8	8	med.	m	6	micro	sm	12	George & Weir, 1972b
Ctenomyidae													
Ctenomys talarum	48	78	64	10	32	8	med.	m	7	micro	a	15	Reig & Kiblisky, 1969
Ctenomys torquatus	68	98	94	24	4	38	med.	m	2	small	sa	—	Reig & Kiblisky, 1969
Ctenomys tuconax	62	122	88	24	34	2	med.	m	2	—	—	—	Reig & Kiblisky, 1969
Ctonomys minutus	50	78	64	12	14	22	med.	m	6	micro	a	18	Reig & Kibilsky, 1969
Ctenomys porteousi	48	84	74	24	10	12	big	m	1	micro	a	19	Reig & Kiblisky, 1969
Ctenomys latro	42	50	46	2	4	34	med.	m	11	small	a	18	Reig & Kiblisky, 1969
Ctenomys magellanicus	36	68	58	20	10	4	?small	m	13	?small	a	10	Reig & Kiblisky, 1969
Ctenomys tucumanus	28	56	46	16	10	0	med.	m	7	small	sa	10	Reig & Kiblisky, 1969
Ctenomys opimus	26	52	50	22	2	0	med.	m	6	small	m	10	Reig & Kiblisky, 1969
Ctenomys occultus	22	44	40	16	4	0	?med.	m	?	?	?	?	Reig & Kiblisky, 1969

Echimyidae													
Proechimys guairae	46	74	70	24	2	18	med.	*sa*	3	micro	*m*	9	Reig *et al.*, 1970; George & Weir, 1973
Proechimys guayannensis	40	58	54	14	4	20	small	*a*	?	small	*a*		Patton & Gardner, 1972
Proechimys hendeei	32	60	52	20	8	2	small	*a*	8	small	*a*	10	Patton & Gardner, 1972
Proechimys semispinosus	30	56	50	20	6	2	small	*a*	6	micro	*a*	?	Patton & Gardner, 1972
Proechimys longicaudatus	28	52	46	18	6	2	small	*a*	7	micro	*a*	9	Patton & Gardner, 1972
Proechimys cherriae	26	44	42	16	2	8	?			?		?	Reig *et al.*, 1970
Proechimys brevicauda	24	44	42	16	4	2	small	*a*	7	micro	*a*	5	Patton & Gardner, 1972
Capromyidae													
Capromys pilorides	40	68	62	20	6	12	med.	*sm*	5	small	*a*	13	Hsu & Benirschke, 1971
Geocapromys brownii	88	136	114	26	22	38	med.	*a*	4	micro	*m*	18	George & Weir, 1972a
Myocastor coypus	42	84	84	40	0	0	med.	*m*	4	micro	*a*	11	Fredga, 1966
Chinchillidae													
Chinchilla laniger	64	128	126	60	2	0	med.	*m*	1	micro	*sm*	2	Nes, 1963; Fredga, 1966
Lagidium boxi	64	128	126	60	2	0	med.	*m*	1	small	*sm*	2	George & Weir, in preparation
Lagidium peruanum	64	128	126	60	2	0	med.	*m*	1	micro	*sm*	2	George & Weir, in preparation
Lagostomus maximus	56	112	112	54	0	0	med.	*sm*	2	micro	*sm*	2	Wurster *et al.*, 1971; George & Weir, in preparation
Cuniculidae													
Cuniculus paca	74	90	88	12	2	58	med.	*m*	1	micro	*sm*	1	Fredga, 1966
Cuniculus taczanowski	42	80		36	0	3	med.	*m*	1	?		?2	Gardner, 1971
Hydrochoeridae													
Hydrochoerus hydrochaeris	66	106	106	38	0	26	med.	*m*	1	small	*a*	1	Wurster *et al.*, 1971

* Two values are given according to the method of calculation from the chromosome types; *m*, metacentric, *sm*, submetacentric, *sa*, subacrocentric, *a*, acrocentric.

KARYOTYPE CHARACTERISTICS

The karyotype of 41 of the 180 or so species of hystricomorphs has been examined (see Table I). This proportion is greater than for myomorph (16%) or sciuromorph (11%) rodents (Matthey, 1969). All the hystricomorph families of Simpson's (1945) classification are represented in Table I except the Dinomyidae, Abrocomidae, Thryonomyidae and Petromyidae. The Bathyergidae, Ctenodactylidae and Pedetidae, left *incertae sedis* by Simpson, are also excluded.

The diploid number of the chromosomes ranges from $2n = 22$ in *Ctenomys occultus* to $2n = 88$ in *Geocapromys brownii* (Fig. 1). This range is similar to that for cricetine rodents (Matthey, 1958; Gardner, 1971) but the distribution is more skewed. The mode is at $2n = 56\cdot6$ for hystricomorphs and is considerably higher than that for Myomorpha ($2n = 48$: Matthey, 1952, 1956) or Sciuromorpha ($2n = 38$: Nadler, 1964, 1966a,b, 1969; Nadler & Hughes, 1966; Hoffman & Nadler, 1968), or for eutherian mammals as a whole ($2n = 48$: Matthey, 1958, 1969).

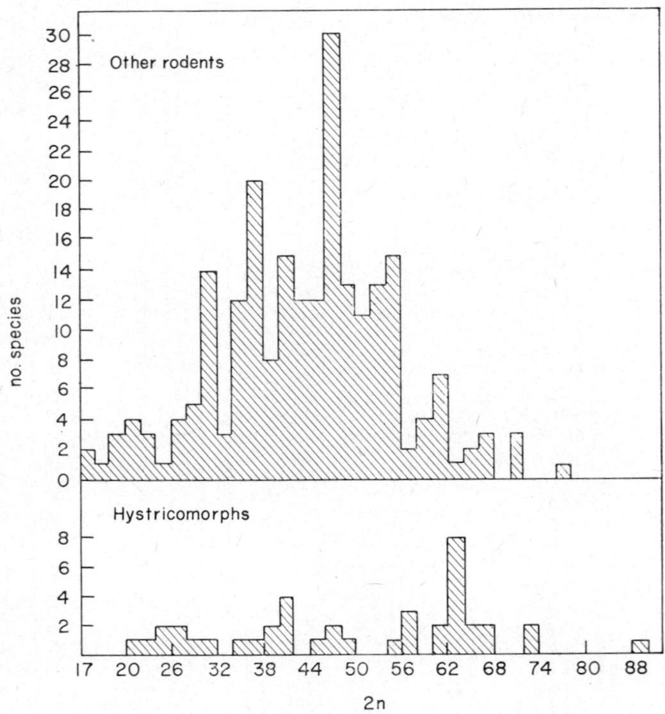

Fig. 1. Histograms of rodent diploid numbers.

The distribution pattern of the *nombre fondamental*, NF (Matthey, 1945), calculated for hystricomorphs is more widely spread than that of other rodents. It varies from 44 in some species of *Proechimys* and *Ctenomys occultus* to 136 in *Galea musteloides* and *Geocapromys brownii* with a peak at 120 to 129. The NF is considered to be the equivalent of the number of chromosome arms but authors calculate the NF in different ways (see George & Weir, 1972b). We have allocated two arms to subacrocentric chromosomes which have a long arm up to seven times as long as the short arm (see Levan *et al.*, 1964), and one arm to acrocentric chromosomes with proportionately longer long arms. This method gives a higher NF than the value found by some other authors, but even if all the subacrocentrics are considered to be one-armed, the NF calculations still give a flat skewed curve with a range of 40 to 130 and a peak at 110–119. For mammals, an NF above 80 is considered high although there are some high values in other rodents such as *Anotomys leander* (98 or 100: Gardner, 1971) and *Dipodomys merriami* (104: Hsu & Benirschke, 1970), in *Rhinoceros unicornis* (84–104: Wurster & Benirschke, 1968b) and in some equids (up to 104: Benirschke & Malouf, 1967). The NFs of *Galea musteloides* at 130 and most of the chinchillids at 126, on the most conservative calculation, must be considered high by any criteria.

The high NFs of hystricomorphs can be accounted for to some extent by the high proportion of metacentric and submetacentric chromosomes in the karyotype. Several species have more than 75% metacentrics or submetacentrics, most have more than 50%, and only *Cuniculus paca, Coendou rothschildi* and a few species of *Ctenomys* have more acrocentrics than metacentrics. Study of the NF and the proportion of metacentric and submetacentric chromosomes does not suggest immediately any clear relationship between genera. An inverse relationship between the NF and the number of metacentrics and submetacentrics would be expected if genera were derivable one from another by simple fusions or fissions. The chinchillid, dasyproctid, caviid and octodontid species can be grouped together on this system and an approximate inverse relationship can be fitted on a regression line for *Geocapromys brownii, Hydrochoerus hydrochaeris, Myocastor coypus, Erethizon dorsatum* and *Hystrix cristata*, with some ctenomyid species and *Proechimys guayannensis* in between.

Thus, it can be said that hystricomorphs (that is, the caviomorph rodents and *Hystrix cristata*) have high numbers of chromosomes, most of which are metacentrics.

Another common feature of hystricomorph karyotypes is the occurrence of chromosomes with secondary constrictions—satellite or marker

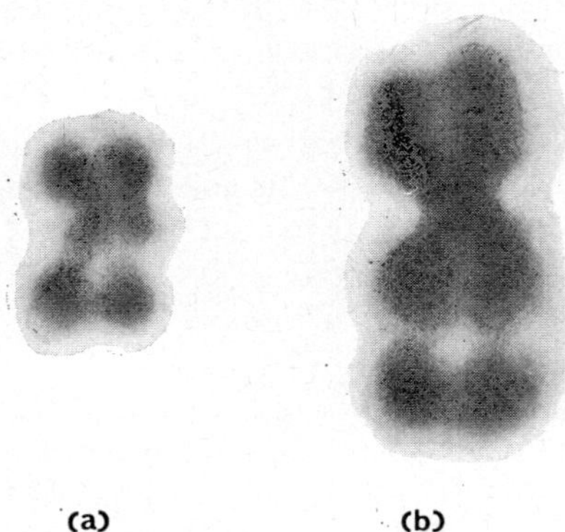

FIG. 2. The satellite chromosomes of (a) *Proechimys guairae* (♀) and (b) *Lagostomus maximus* (♀) to show octodontid and chinchillid marker chromosomes. Both ×9,000.

chromosomes (Fig. 2). A pair of satellite chromosomes is a feature of some carnivore families (Wurster & Benirschke, 1968a; Wurster, 1969; Fredga, 1972) in which it is known as the 'carnivore chromosome', and in a few insectivores (Borgaonkar, 1969; Gropp, 1969), but it is not found as a regular feature in rodents. Among the Hystricomorpha, a pair of satellite chromosomes is present in most known species of the families Hydrochoeridae, Chinchillidae, Capromyidae, Octodontidae, Ctenomyidae and Echimyidae. They do not occur as a constant feature of the karyotype in the Caviidae, Dasyproctidae or Erethizontidae. Renzoni (1967) reported satellite chromosomes in *Hystrix cristata*, but it is difficult to see these in the published karyotype and D. H. Wurster (personal communication) was unable to find satellites in her specimen of this species.

Characteristically, the caviomorph marker chromosome has satellites on the short arm but the position of the centromere varies. The ratio of the length of the short arm to the satellites and the rank of the satellite chromosome is also variable. In spite of these variables, two types of marker chromosome are apparent. In one type, called the chinchillid type (Fig. 2b), the marker is one of the biggest chromosomes in the karyotype, usually chromosome 2, is metacentric and has a short arm to satellite ratio greater than 1·25. All the chinchillids, *Hydrochoerus* and *Cuniculus* have this marker. The other type, the octodontid satellite chromosome (Fig. 2a), is never higher in rank than the eighth longest pair of chromosomes and is usually lower; it varies from sub-metacentric to acrocentric and the ratio of the short arm length to that of the satellite is 1 or less than 1. Such satellites are found in the Octodontidae, Ctenomyidae, Echimyidae and Capromyidae.

It is difficult to be certain whether or not these differences in the satellites are valid and, if so, whether the marker chromosomes should be considered as diagnostic. Wurster & Benirschke (1968a) and Fredga (1972) found a constancy in the appearance of the 'carnivore chromosome' in different families, and Wurster (1969) concluded that the satellite chromosome is a remarkably stable characteristic of the karyotype.

The possession of satellite chromosomes can, therefore, be added to the high diploid number of metacentrics to define the karyotype of an hystricomorph rodent.

The size and shape of the X–chromosome has sometimes been considered useful for separating related species (Matthey, 1966). Matthey (1967) suggested that the original sex chromosomes were a pair of acrocentrics and Ohno *et al.* (1964) calculated that the ancestral type of X–chromosomes in placental mammals constituted about 5% of the total haploid set of the female. But X–chromosomes can vary in length and in position of the centromere from species to species within a genus, as shown by Matthey (1952) in *Microtus*. In general, hystricomorph rodents have the 'normal' 5–6% sized X–chromosome which is metacentric or submetacentric, but there is a marked tendency for the X–chromosome to be bigger in some families. The Y–chromosome is usually small, forming less than 2% of the set, and may be exceptionally small as in *Myoprocta acouchy*. In *Erethizon dorsatum* the X–(12%) and the Y–chromosome (6%) are exceptionally big (Benirschke, 1968).

When the diploid number, the NF, the proportion of metacentrics, the form of the satellite chromosomes and the sex chromosomes are considered, hystricomorphs can be separated into five groups (see Table II).

Group 1

This contains two species of the Erethizontidae (*Erethizon dorsatum* and *Coendou rothschildi*). Their NF values fall well within the 'normal' range for mammals but their diploid numbers are a 'normal' 42 for *Erethizon* and a high 74 for *Coendou*. The X–chromosome is a long metacentric in both species but the Y–chromosome of *E. dorsatum* is larger than is usual for hystricomorphs and there is a pairing region between the X– and Y–chromosomes (Benirschke, 1968). There are no satellite chromosomes.

Group 2

This group, the 'cavyprocts', includes seven species of the Caviidae and Dasyproctidae, and is more homogenous than any of the other

TABLE II

The division of hystricomorph karyotypes into groups according to their characteristics

Group	$2n$	NF	% $m + sm$	X		Y		Marker rank
				size %	rank	size %	rank	
1. Erethizontids	42–74	78–86	8–81	7–12	first	1–5·6	first	none
2. 'Cavyprocts'	62–68	116–136	>75	5–8·5	first	0·8–2	last	none
3. 'Octocaps'	22–88	44–136	≃50	6–8	variable	1·5–3	variable	9–18
4. 'Lagunchoerids'	56–74	80–128	16–100	5·5–8·5	first	2–4	last	1 or 2
5. *Hystrix*	60–66	98–118	≃50	5·5–7·5	first	?2	?	?

groups, with high NFs, high diploid numbers and a high proportion of metacentrics. Typically there are no satellite chromosomes.

Group 3

The 24 'octocap' species from the Octodontidae, Ctenomyidae, Echimyidae and Capromyidae have various diploid numbers and NFs. The proportion of meta- and submetacentric chromosomes is often, but not always, more than 50% although the karyotypes are less symmetrical in the Stebbins (1971) sense (see p. 94) than those of the 'cavyprocts'. The main feature is the octodontid marker chromosome which is small and bears a satellite at the end of the short arm.

Group 4

The 'lagunchoerids' comprise four chinchillid species and *Cuniculus paca* and *C. taczanowski*, and *Hydrochoerus hydrochaeris*. The consistent features of their karyotypes are the metacentric X–chromosome which is usually the longest of the series, the small Y–chromosome which is usually the last of the series and there is always a pair of chromosomes with satellites which is either chromosome 1 or chromosome 2.

Group 5

Hystrix cristata is alone in this group because there is some doubt about its karyotype. Renzoni (1967) found $2n = 60$ in one male, but D. H. Wurster (personal communication) has found $2n = 66$ in a female. Satellite chromosomes were reported in the karyotype described by Renzoni, but none in the animal studied by Wurster. It may be that the samples were from incorrectly identified animals, or that *Hystrix* is a variable genus. It is unfortunate that a comparison of *Hystrix* with the other caviomorphs cannot be made until this problem is settled. Renzoni and Wurster agree that *Hystrix* has a moderately high NF and a large X–chromosome which is probably the first of a series of about 50% meta- and submetacentric chromosomes.

KARYOTYPE RELATIONSHIPS

On the basis of the karyotype patterns an attempt was made to arrange the families within the groups and to derive one group from another.

We tested two hypotheses. First, a Robertsonian theory of centric fusion in which actual fusion occurs, or the same effect is achieved by translocation of chromosome arms broken in heterochromatic regions near the centromeres. In both types, the immediate genetic effect is minimal except for alteration of linkage groups. The second theory was an adaptation of the Stebbins' (1971) hypothesis of chromosome evolution which postulates a progression from symmetrical to asymmetrical karyotypes. Both schemes were used to suggest phylogenetic relationships within the Hystricomorpha without any consideration of characters usually applied in taxonomic determinations. Then, conventional phylogenies were considered to determine the chromosomal changes that would be necessary to support them.

Robertsonian theory

According to Robertson (1916), karyotypes can be derived one from another by a process of chromosome fusion or fission. It was this theory which led Matthey (1945) to develop the concept of the *nombre fondamental* or total number of chromosome arms with which Robertsonian transformations could be made (see Fig. 3a). Subsequent knowledge of karyotypes has made it obvious that karyotype evolution is not that simple. Among the hystricomorph rodents there is no consistency in the NF values: the peak at 120–129 is not pronounced and the distribution curve is not a normal one. Too few hystricomorphs have enough acrocentric chromosomes to make a straight Robertsonian transformation likely.

Most authors have added tandem fusions (Fig. 3b), translocations (Fig. 3c,d,e), pericentric inversions (Fig. 3f), duplications and deletions to the simple Robertsonian transformations (see Meylan, 1970). They have also taken the view that one type of change is likely to be found consistently in the evolution of related species (White, 1959; Matthey, 1964; Hsu & Arrighi, 1966). Thus, when attempting to derive one karyotype from another it should be possible to postulate a combination of, say, fusions and translocations or fusions and inversions, but not both. Regardless of whether it is permissible to mix translocations with inversions, it is usual to adopt either fusion or fission as the basic method of change in karyotype derivations, although Nadler (1969) and Hoffman & Nadler (1968) disregarded this convention and suggested fusion and fission to account for changes in the Sciuridae. Centromeric fission has not yet been demonstrated but has been favoured for plants (Marks, 1957) and eutherian mammals (Todd, 1967). The suggestion of Matthey (1958, 1963) that change in myomorph rodents was achieved mainly by fusion is supported by the work of Hsu & Mead (1969). Centric fusion has been clearly demonstrated in *Sigmodon minimus* (Hsu & Mead, 1969) and by reciprocal translocation to give the same

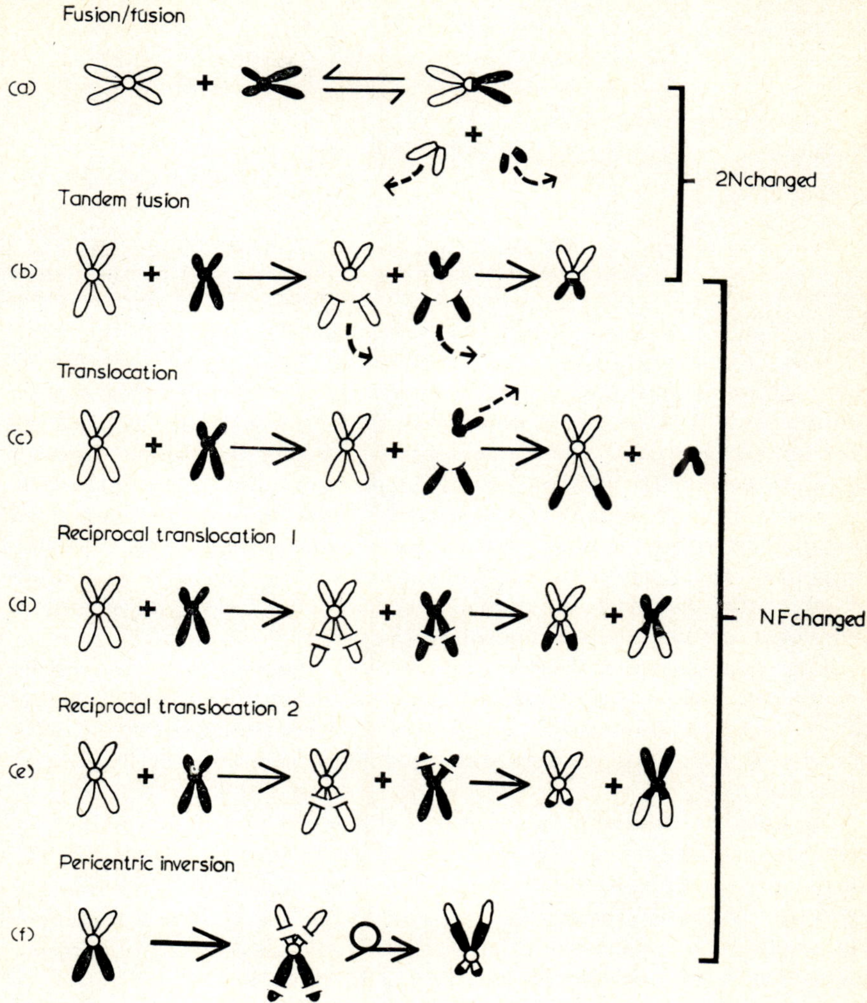

Fig. 3. Diagrams to show some possible chromosomal changes which alter the diploid number (2n), the *nombre fondamental* (NF), or both.

result has been observed in man (Hamerton *et al.*, 1963) and shrews (Meylan, 1968).

The choice for investigating karyotype evolution in hystricomorphs by Robertsonian methods was, therefore, chromosome fusion. A schematic representation of the karyotype was used, as shown in Fig. 4. Logically, the ancestral hystricomorph should have had a full comple-

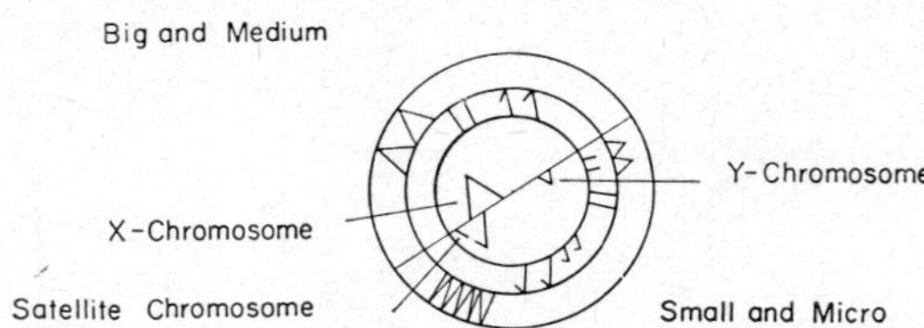

Fig. 4. Schematic representation of haploid autosomes and the sex chromosomes. Outer circle contains metacentric and submetacentric chromosomes (∧); inner circle contains subacrocentric chromosomes (1) and acrocentric chromosomes (I). Small and microchromosomes are distinguished, but there is no distinction between big and medium chromosomes because there are few big (>9% of total haploid set) chromosomes in the hystricomorph karyotypes so far examined.

ment of small acrocentric chromosomes. This would entail a diploid number of 136. This seems so unlikely that such a condition has been rejected. If, however, the NF were fixed at 136, the diploid number could be reduced to 98, comprising 18 pairs of small metacentrics, 30 pairs of small acrocentrics, a metacentric X–chromosome of the normal mammalian size (5%) and a fairly big Y–chromosome with a pairing region with the X–chromosome. Thus, species with a high diploid number of acrocentrics would be more primitive than ones having a low diploid number of metacentric chromosomes (Fig. 5).

Our hypothetical ancestor, 'Chechecdoc' (derived from the initial letters of hystricomorph families), gives rise directly to the line leading to the erethizontids (Group 1, p. 85), either by centromeric fusions, with translocations to adjust sizes and centromere positions, or by centromere loss.

In another direction, karyotypes are derived with a smaller Y–chromosome. One line leads, by many centric fusions, to 'Cavyproct', the ancestor of Group 2 (p. 85). A second line leads to the Group 3 ancestor 'Octocap' by centric fusions. A translocation or acrocentric attachment of a small chromosome would give rise to the type of satellite chromosome that is the octodontid marker. The third line is indicated by satellisation of a big chromosome to give the chinchillid marker chromosome and leads to 'Lagunchoerus' (Group 4, p. 87).

Group 1

Early along the erethizontid line, *Coendou rothschildi* becomes a relict form with a reduced NF but still a high number of acrocentrics. Centric fusions with further translocations, particularly to the X–chromosome, are all that is necessary to convert the *Coendou* karyotype to that of *Erethizon dorsatum*.

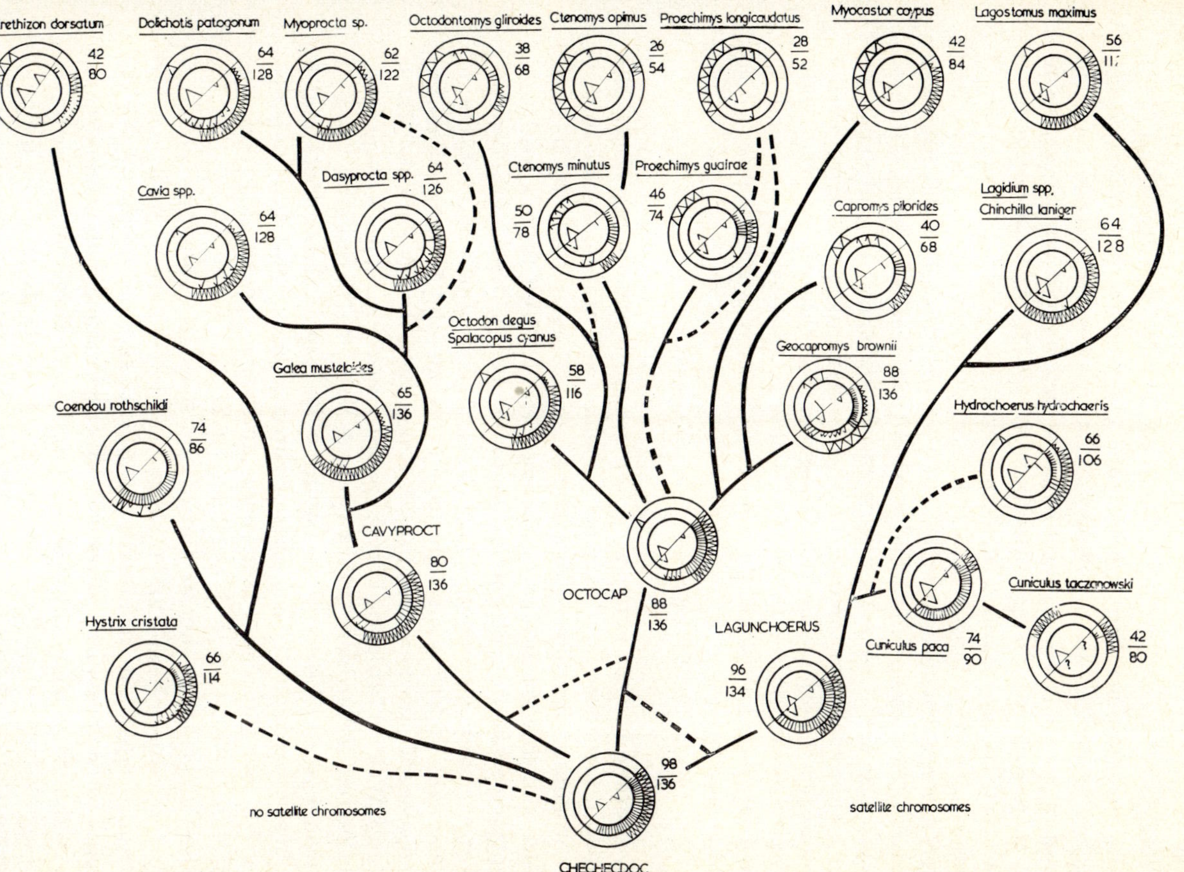

FIG. 5. Scheme showing possible phylogenetic relationships according to a theory of Robertsonian fusions. The karyotypes are expressed as in Fig. 4. Fractions represent diploid numbers, *2n* above and *nombre fondamental* (NF) below. Broken lines are possible but less favoured pathways.

Group 2

'*Cavyproct*' has more meta- and submetacentric and fewer acrocentric chromosomes than '*Chechecdoc*'. The modern cavy *Galea musteloides* can be derived from '*Cavyproct*' with only a few changes, but the derivation of the *Cavia* species requires fusions and translocations or even more complex changes (George *et al.*, 1972). At least two pairs of subacrocentrics are retained throughout the 'cavyprocts'. Chromosome 1 becomes noticeably longer than the rest by accumulating translocated material from other chromosomes. The Dasyproctidae probably branched off soon after chromosome 1 had started to increase in length. All three dasyproctids and *Dolichotis* have a big X–chromosome and small Y–chromosome such that the ratio of X to Y distinguishes them from other caviomorphs. Chromosome 1 of *Myoprocta* resembles that of *Dolichotis*. *Dolichotis* seems nearer to the *Cavia* line in that all the acrocentrics have disappeared although the big subacrocentric chromosome of *Cavia* is not represented.

Group 3

Evolution in the 'octocap' group requires fusions to reduce the number of acrocentric chromosomes, fusions and translocations to give the octodontid satellite chromosome and translocations to increase the size and linkage groups of chromosome 1. More fusions lead to a karyotype without acrocentrics as found in the octodonts *Octodon*, *Spalacopus* and *Aconaemys*. The development of the karyotypes of *Octodontomys* and *Ctenomys* species by fusions and translocations has been discussed elsewhere (Kiblisky & Reig, 1966; Reig & Kiblisky, 1969; George & Weir, 1972b). The karyotypes of *Proechimys* species (Reig *et al.*, 1970; Patton & Gardner, 1972; George & Weir, 1973) resemble those of *Octodontomys* and *Ctenomys* in having a few small acrocentrics and a very big chromosome 1. The sizes of chromosomes 2 and 3 may also be increased, like those in *Octodontomys* and some species of *Proechimys* (Reig *et al.*, 1970; Patton & Gardner, 1972).

The Capromyidae (hutias and coypu) may also be in the 'octocap' group. *Capromys pilorides* and *Geocapromys brownii* are island forms and *Myocastor coypus* has an aquatic habitat so the karyotypes might be expected to be specialized. However, all three capromyids studied have the octodontid marker chromosomes and have three to four pairs of very long chromosomes, like some of the octodonts. It would be possible to derive the capromyids from '*Octocap*', leaving *Geocapromys* as an early offshoot with a high NF, equal to that of '*Octocap*' (136), a high diploid number of 30 pairs of acro- and subacrocentrics and only 13 pairs of meta- and submetacentric chromosomes. This karyotype of

Geocapromys could be achieved by translocations to the first pairs of chromosomes, but pericentric inversions would also be needed to change metacentrics to acro- or subacrocentrics, a method suggested for *Leggada* and *Mastomys* (Matthey, 1964, 1966) and *Eutamias* (Nadler, 1964). This is the first time it has been necessary to depart from the simple hypothesis of fusions and translocations only. There is some similarity of pattern between the karyotype of *Capromys* and that of *Geocapromys* but, from a near-*Geocapromys* type, fusions, translocations and inversions would be needed to derive *Capromys* and *Myocastor*. These last two karyotypes are not obviously connected although there are three pairs of big chromosomes, reasonably similar X– and Y–chromosomes, the octodontid marker and a similar diploid number of 40 and 42 respectively. The relationships between these three capromyid genera are difficult to establish but there seems to be no good reason for separating them. None of the karyotypes resembles that of any other hystricomorph more closely.

Group 4

The 'lagunchoerids' all have the big chinchillid marker chromosome and a relatively big X–chromosome. The Chinchillidae have a high NF and a high diploid number of mainly meta- and submetacentric chromosomes, derivable from 'Chechecdoc' by fusions and translocations. *Lagostomus* differs from the other two genera in having fewer chromosomes. *Lagidium* and *Chinchilla* are more similar than either is to *Lagostomus* (George & Weir, in preparation).

The position of *Hydrochoerus* and the two *Cuniculus* species (*C. paca* and *C. taczanowski*) is difficult to determine. The only reason for putting them near the chinchillids is because they have the chinchillid marker chromosome and it is possible to derive them by Robertsonian methods from a common ancestor with the chinchillids. The karyotype patterns of *Hydrochoerus* and *Cuniculus* show some similarity, although this may be due mainly to their high numbers of small acrocentric chromosomes. *Cuniculus taczanowski* can be derived directly from *C. paca* by centric fusions and translocations.

Group 5

For the reasons already given (p. 87) *Hystrix* is difficult to fit into the scheme. If *Hystrix* does not have a marker chromosome then it could be derived directly from 'Chechecdoc' or from an ancestor on the erethizontid line or from the 'cavyprocts'.

The relationships of the hystricomorph rodent karyotypes shown in Fig. 5 can be derived by Robertsonian methods: centric fusions, with translocations. The species of Capromyidae do not conform because pericentric inversions and deletions are necessary to derive their karyotypes. The rest of the 'octocaps' and the 'cavyprocts' are reasonably homogenous. *Coendou rothschildi, Geocapromys brownii, Cuniculus paca, Hydrochoerus hydrochaeris* and, possibly, *Hystrix cristata* have 'primitive' karyotypes and can be regarded as relics.

The main objections to the scheme are (1) that the original ancestor would have had a high diploid number of chromosomes, outside the range of that of any known rodent, although there are some that are nearly as high (Patton & Gardner, 1972; George & Weir, 1972a), and (2) that the original ancestor and the intermediate progenitors would have had heterogenous karyotypes. The scheme also depends on the assumption that the total chromatin content of the genome is the same for all rodents and that the changes in karyotype pattern do not depend on extensive loss or acquisition. Bachmann (1972) has calculated the genome size, that is the DNA content of the nucleus, of several rodents and obtained values ranging from 157 ($\equiv$7·5 pg) in *Peromyscus flavicola* to 228 ($\equiv$10·9 pg) in *Sciurus carolinensis*. The only two caviomorph species measured were *Cavia porcellus* and *Chinchilla laniger* and the results were 194 ($\equiv$9·2 pg) and 173 ($\equiv$8·2 pg) respectively in spite of their identical diploid numbers and NF.

An alternative Robertsonian interpretation is to work by fission from small numbers of metacentric chromosomes to higher numbers with a large proportion of acrocentrics. To maintain this system a great many complicated inversions, fusions and losses are involved.

Stebbins' hypothesis

Stebbins (1971) formulated a more sophisticated variation of the Robertsonian method which he applied to evolution in higher plants. The theory suggests that the primitive members of a family have symmetrical karyotypes with high numbers of chromosomes (Fig. 6a,c) and that the unspecialized forms gave rise to more specialized species with an asymmetrical pattern of few chromosomes (Fig. 6b,d). The application of this theory to animals has been found useful to interpret relationships within the Caviidae (George *et al.*, 1972), Octodontidae (George & Weir, 1972b) and Chinchillidae (George & Weir, in preparation), and in other rodent groups (Wahrman *et al.*, 1969; Berry & Baker, 1971).

The classification of karyotypes as symmetrical or asymmetrical with high or low numbers of chromosomes makes no difference to the groups recognized in Table II, but the Stebbins' theory changes the specific relationships (Fig. 7). In Fig. 7 only the symmetry or asymmetry

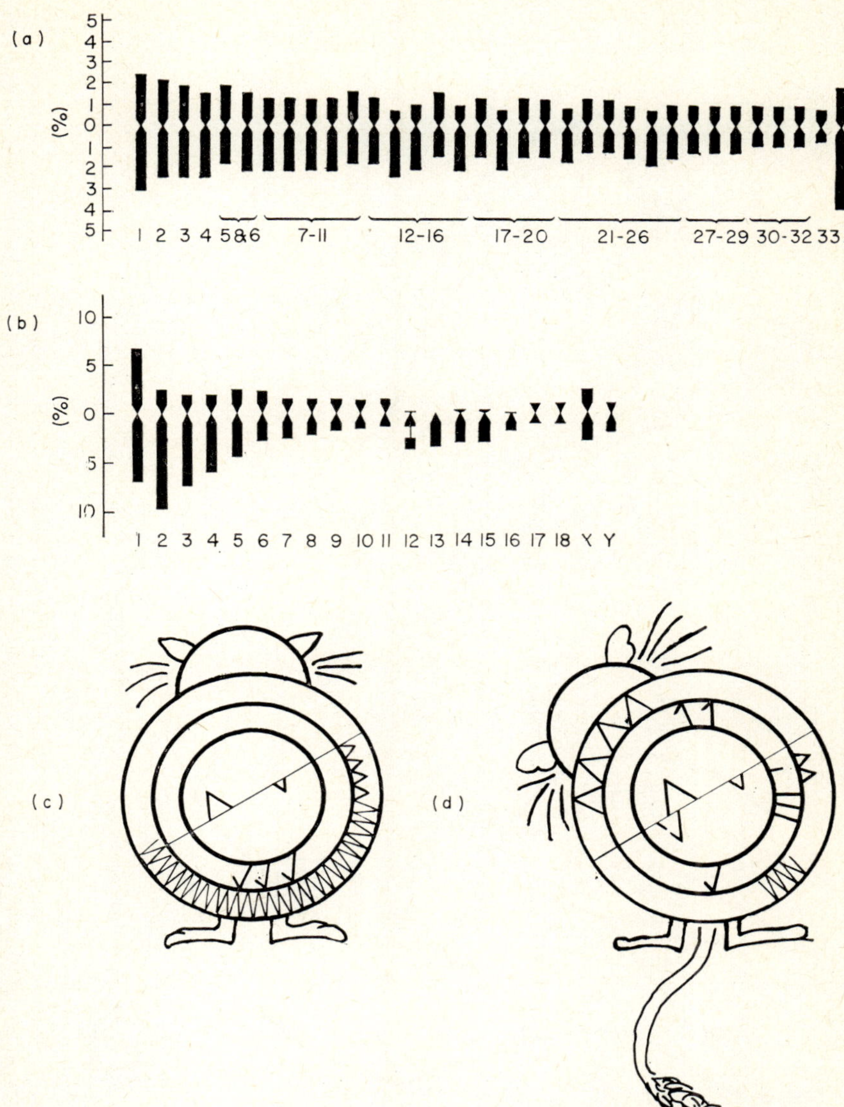

FIG. 6. Representation of symmetrical (a, c) and asymmetrical (b, d) karyotypes: (a) idiogram of *Galea musteloides* showing the high proportion of meta- and submetacentric chromosomes (from George *et al.*, 1972); (b) idiogram of *Octodontomys gliroides* which has approximately equal numbers of metacentric and acrocentric chromosomes, giving the karyotype an heterogeneous appearance (from George & Weir, 1972b); (c) diagram to represent the symmetrical karyotype of *Galea musteloides* (chromosomes as in Fig. 4); (d) diagram to represent the asymmetrical karyotype of *Octodontomys gliroides* (chromosomes as in Fig. 4).

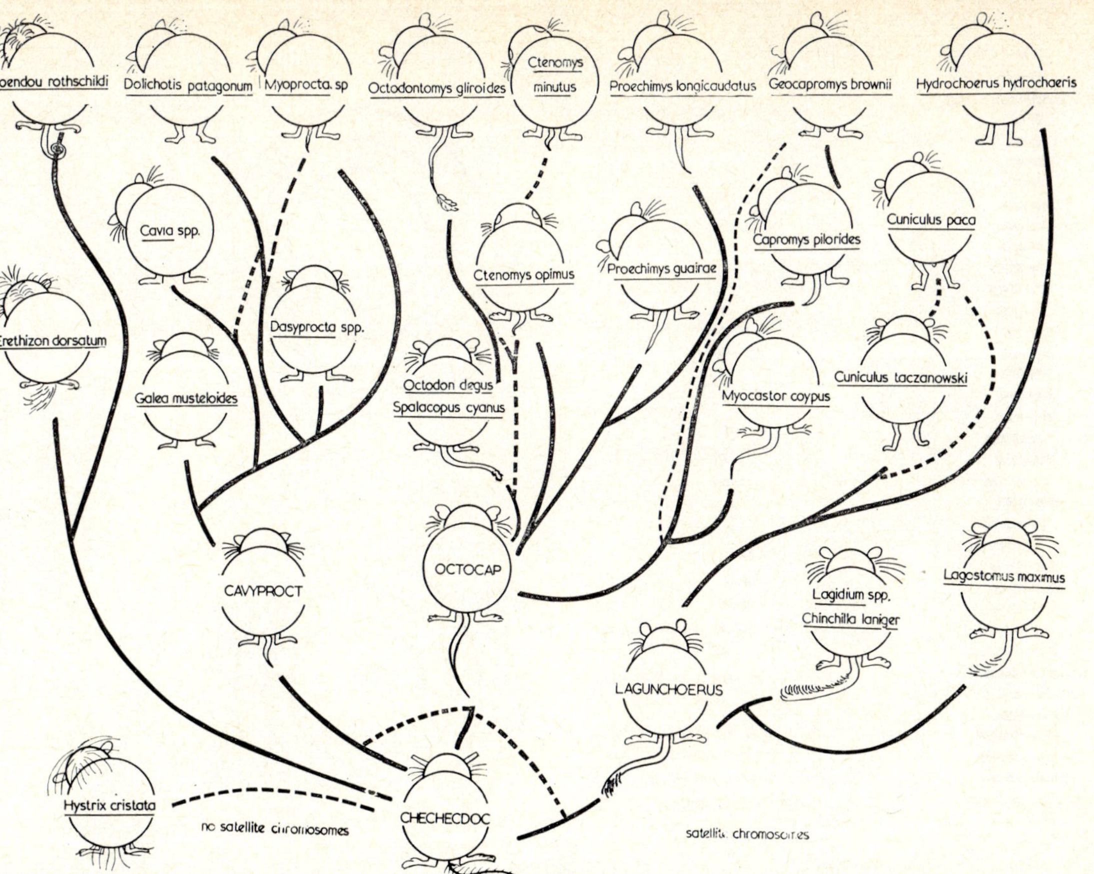

Fig. 7. Scheme showing possible phylogenetic relationships according to Stebbins' theory of evolution from symmetrical (straight head) to asymmetrical (angled head) karyotypes. Broken lines are possible but less favoured pathways.

of the karyotype is represented; the chromosomes are not depicted as only those of the ancestors are different from those shown in Fig. 5.

The original ancestor '*Chechecdoc*' has a symmetrical karyotype. The NF is 136 (as used for the Robertsonian scheme, p. 91) and there is a full complement of meta- and submetacentric chromosomes. The diploid number would then be 68, which is well within the normal range for mammals and close to that commonly found in hystricomorphs. The Y–chromosome might have been relatively big, with a pairing segment with the X–chromosome.

Group 1

The erethizontids are derived in a different way from that proposed for the Robertsonian method. The karyotype of *Erethizon dorsatum* can be derived from that of '*Chechecdoc*' with only a slight increase in asymmetry by complex translocations to the X–chromosome and some of the autosomes and by tandem fusions to reduce the diploid number and NF and to produce two pairs of minute acrocentrics. *Coendou rothschildi* could come directly from an *Erethizon*, or from a near ancestor, since the NFs are similar, but most of the meta- and submetacentric chromosomes would have to be converted to acrocentrics. This could be achieved to some extent by pericentric inversion but the need to increase the number of centromeres to 74 from 42, if from *Erethizon*, or 68, from '*Chechecdoc*', involves centromeric fissions or duplications. Centromeric fissions have been suggested for the origin of the prairie dogs (*Cyonomys*) from *Spermophilus* relatives (Nadler & Harris, 1967; Nadler, 1969; Nadler *et al.*, 1971). On this hypothesis, *Coendou rothschildi* could be a recent island inhabitant and not the relict form suggested by the Robertsonian scheme.

Group 2

The 'cavyprocts' could also be derived from the same symmetrical metacentric ancestor, '*Chechecdoc*'. *Galea musteloides* would retain the symmetry, diploid number and NF of its ancestors while changing the position of the centromeres of one pair of chromosomes, by pericentric inversion or translocations to chromosome 1, and the relative sizes of chromosomes by translocation. The other 'cavyprocts' are more specialized; they have slightly asymmetrical karyotypes and reduced diploid numbers, all of which could be achieved by translocations and inversions, with loss or fusions to reduce the number of centromeres.

Group 3

'*Chechecdoc*' is also the ancestor of the 'octocaps', but the changes are complicated. Firstly, satellisation of a small chromosome must occur to give the octodontid marker chromosome and, secondly, there must be translocations to increase the size of chromosome 1. Along the *Octodon* line, the NF and 2*n* are reduced without loss of symmetry in *Octodon degus* and its similar relatives, *Spalacopus* and *Aconaemys*, presumably by tandem fusions. Thereafter, specialization gives rise to asymmetrical karyotypes and reduced 2*n* and NF. In *Octodontomys* and several species of *Ctenomys* and *Proechimys* there are four pairs of very big chromosomes. The Capromyidae are again difficult to associate with the other 'octocaps', or with each other. No capromyid has a symmetrical karyotype and this supports the theory that specialized species are asymmetric. The ancestors of *Myocastor coypus* and *Capromys pilorides* could have had reduced diploid numbers and NFs by translocations, tandem fusions and inversions. *Myocastor* would be nearer the common ancestor than would *Capromys*. *Geocapromys brownii* could be a modified polyploid version of *Capromys*, although such a possibility is contrary to our earlier opinion (George & Weir, 1972b) and to that of most authors (for example, Matthey, 1953) who do not accept polyploidy in animals (but see Comings, 1972, on tetraploidy in mammals). Alternatively, an ancestral form may have given rise to *Geocapromys* by centric fissions before the NF had been reduced on the line to the other capromyids.

Group 4

Within the 'lagunchoerids', the chinchillids retain the primitive symmetrical karyotype with a high number of chromosomes and a high NF. Only *Lagostomus maximus* has shown any deviation although the karyotype is still essentially symmetrical and metacentric. All members of the Chinchillidae have acquired the chinchillid marker chromosome, presumably by translocation. *Hydrochoerus* and *Cuniculus* also have this form of satellite chromosome. *C. taczanowski* is symmetrical and could be ancestral to *C. paca*, derived by centric fissions and pericentric inversions. *C. paca* and *H. hydrochaeris* have asymmetrical karyotypes and a reduced NF but not a reduced diploid number. Numerous pericentric inversions and some translocations are needed to reach these karyotypes. With *Coendou rothschildi*, they conform less to Stebbins' view than do any other species because of the association of high numbers of chromosomes with asymmetry in their karyotypes.

Group 5

Hystrix cristata has a slightly asymmetrical karyotype, almost the same diploid number as the postulated ancestor 'Chechecdoc', but a reduced NF because of the number of acrocentric chromosomes. It could be derived from the early ancestral line leading to the caviomorphs, but further speculation is impossible until the karyotype is determined unequivocally.

Discussion

Comparison of the relationships derived from the application of the Robertsonian and Stebbins' theories indicates that the Stebbins' theory can be used for animals and, in fact, is the more satisfactory hypothesis for establishing relationships within the caviomorph rodents. It has the advantage that 'Chechecdoc' has a realistic karyotype and that all the families can be derived from this ancestor in fewer moves than are needed by Robertsonian methods. It does, however, require inconsistencies of chromosome changes since polyploidy or fission and simultaneous translocations and inversions are needed.

The main differences between the two evolutionary schemes are that on the Stebbins' theory, *Erethizon dorsatum* is more primitive, or generalized, than *Coendou rothschildi*; the two *Cuniculus* species, the *Ctenomys* species and the capromyid species change places; and the chinchillids become generalized and primitive.

It seems reasonable to associate an asymmetrical karyotype with specialization. The asymmetrical pattern is usually correlated with a reduced chromosome number and a karyotype made up of contrasting long and short chromosomes. The long chromosomes represent big linkage groups. Big chromosomes and a small number of chromosomes will lower the variability, or recombination index, of a species, but the limited evidence available suggests that the number, rather than the length, of chromosomes is the important factor in determining this index (Ford, 1969; George & Weir, 1972b, 1973; George *et al.*, 1972). Low chromosome numbers, therefore, should be associated with specialized forms adapted to predictable environments. But caviomorph species with low chromosome numbers cannot always be associated with such habitats. For example, *Octodontomys gliroides* and *Cuniculus taczanowski* live in specialized habitats, but so do *Lagidium boxi* and *Lagostomus maximus* and yet these have high numbers of chromosomes.

Different karyotype groupings can, of course, be made. It could be argued, for example, that satellisation occurred once only in the caviomorph stock and that the octodontid marker chromosome is a modified version of the chinchillid marker chromosome or *vice versa*. If this were so, the 'lagunchoerids' would be related more closely to the 'octocaps' than either is to the 'cavyprocts', and the Hydrochoeridae and Cuniculinae need have no greater affinity with the chinchillids than with the 'octocaps'.

Whichever method of karyotype evolution is used, the suggested relationships, based exclusively on karyotypes, give rise to the following conclusions:

1. the Erethizontidae are isolated from other caviomorphs;

2. the Caviidae and Dasyproctidae are closely related;

3. the Octodontidae, Ctenomyidae and Echimyidae are related and the Capromyidae may also be associated with these families;

4. the Chinchillidae may or may not be related to the 'octocaps';

5. *Hydrochoerus hydrochaeris* and the *Cuniculus* species are difficult to place but may have some relationship to the chinchillids;

6. it is not yet possible to determine the position of *Hystrix* in relation to the New World species;

7. two unrelated 'island' forms, *Coendou rothschildi* and *Geocapromys brownii*, are remarkable for their large number of acrocentric chromosomes and high diploid number.

COMPARISON WITH OTHER CLASSIFICATIONS

It is interesting to compare these conclusions on caviomorph relationships with those reached by other authors who have used features such as anatomy, palaeontology and ecology, and have relied on a greater number of individuals and species. Taxonomy being what it is, a precedent can be found in the literature for practically any suggestion. We have compared five of the most recent and detailed classifications with that based on karyotype patterns.

Simpson (1945) grouped the suborder Hystricomorpha into five superfamilies, the Hystricoidea, Erethizontoidea, Cavioidea, Chinchilloidea and Octodontoidea. The Cavioidea contained the Caviidae and Dasyproctidae, our 'cavyprocts', as well as the Dinomyidae (for which we have no karyotypes), and the Hydrochoeridae; but *Cuniculus* was placed with the Dasyproctidae. This grouping is difficult to accept if there is any validity for assuming the chinchillid marker chromosomes to be diagnostic, since *Cuniculus* and *Hydrochoerus* have this marker and the other cavies do not. A satellite chromosome may appear erratically in some individuals. For example, one *Galea musteloides* had a distinct satellite on one chromosome of a pair (George *et al.*, 1972), and some cells of one *Lagidium peruanum* have a similar appearance (George & Weir, in preparation). However, these satellites are not regular features of the karyotypes; their significance is not understood but they do not seem to be the same as the regular satellites on both chromosomes of a pair which we have called the octodontid and chinchillid marker chromosomes and which Wurster & Benirschke (1968a) have called the carnivore chromosome. If *Cuniculus* is to be accepted in a subfamily of the Dasyproctidae, it would need to form a satellite chromosome in parallel with the chinchillids and in contrast to the rest of the Cavioidea, except for *Hydrochoerus*. And if *Hydrochoerus* is in its own family of the Cavioidea then it must either be related to the cuniculine branch of the dasyproctids, or have invented a chinchillid marker chromosome in its own right. The final possibility is that all the Cavioidea had a marker chromosome which was lost in some lines only (Fig. 8).

The Octodontoidea of Simpson (1945) consists of the Capromyidae, Octodontidae, Ctenomyidae and Echimyidae, all equivalent to our

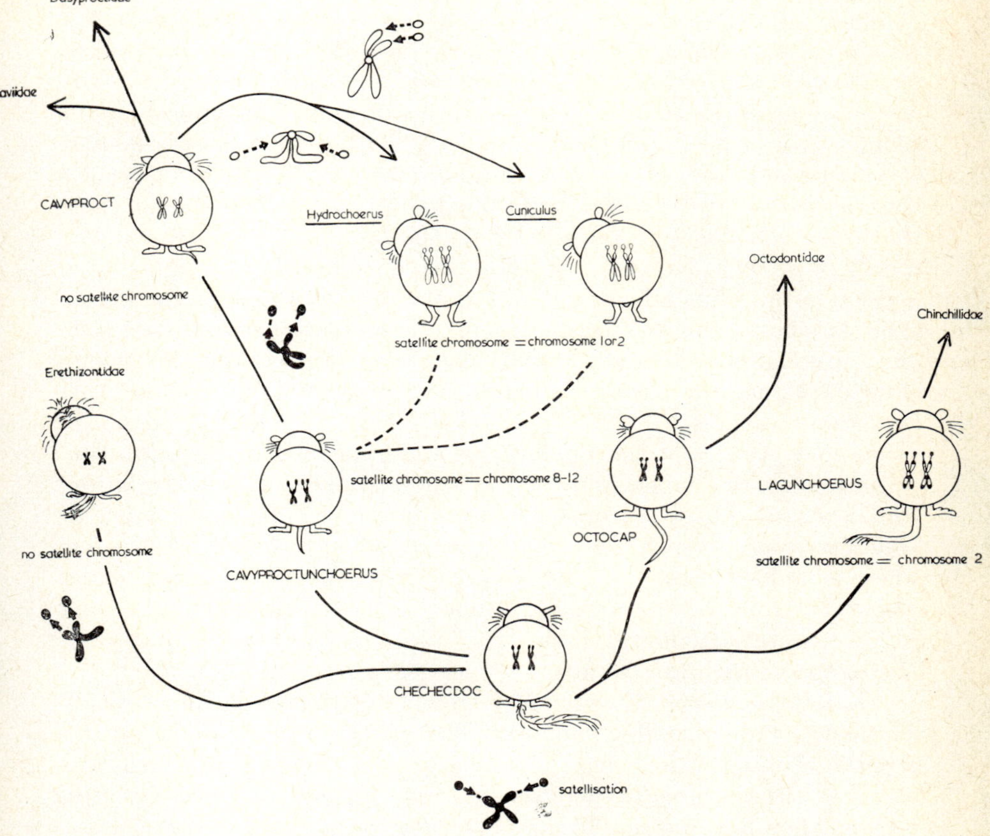

FIG. 8. Diagrams to show satellite loss or addition in caviomorphs if it is assumed that their common ancestor had a pair of satellited chromosomes and that *Hydrochoerus* and *Cuniculus* are cavioids with affinities with the Caviidae and Dasyproctidae respectively.

'octocap' group, together with the Abrocomidae, Thryonomyidae and Petromyidae for which there are no karyotypic data. Simpson's Chinchilloidea contains only the Chinchillidae. The Hystricoidea, of course, contains *Hystrix* but since its karyotype relationships are still uncertain no decision can be reached on its ancestry with respect to *'Chechecdoc'* or to the other caviomorph lines.

Thus, apart from *Hydrochoerus* and *Cuniculus*, our karyotype classification fits the more orthodox one of Simpson (1945) quite well.

Wood (1955) proposed a classification for rodents in which a new suborder, the Caviomorpha, was made for the New World hystricomorphs. The Caviomorpha included the Octodontoidea (Octodontidae, Ctenomyidae, Echimyidae, and Abrocomidae), Chinchilloidea (Chinchillidae and Capromyidae), Cavioidea (Caviidae, Hydrochoeridae, Dasyproctidae, Dinomyidae and Cuniculidae) and Erethizontoidea. The Octodontoidea correspond to our *'Octocap'* derivatives with a satellite chromosome at the 10 or 12 position, specialization by asymmetry and a reduction in the chromosome number. The Cavioidea again include *Cuniculus* and *Hydrochoerus* which do not tally with the karyotypes for the reason already discussed, but at least this classification has the merit that *Cuniculus* has been removed from the Dasyproctidae and put in a family of its own. However, Wood unites the Chinchillidae and Capromyidae in the Chinchilloidea. This has the disadvantage for the karyotype classification of making the two marker chromosomes indistinguishable, although it would not be difficult to derive the capromyids from a chinchillid ancestor rather than from an octodontid ancestor on other karyotype features.

Landry (1957) put the Echimyidae and Capromyidae with the Erethizontidae in one superfamily. There is no justification for this relationship in the karyotype patterns. The Dasyproctidae, including *Cuniculus*, and *Myocastor*, given its own family, were put into the Octodontoidea. There is no karyotypic evidence to support this degree of separation of *Myocastor* from the capromyids, although it is possible that *Myocastor* is a separate octodontid line. Like all other authors, Landry left *Hydrochoerus* with the caviids.

On fossil evidence, Wood & Patterson (1959) included the Cuniculidae in the Chinchilloidea, but also moved the dasyproctids to this superfamily since they were closely related to *Cuniculus*. Obviously, we agree with the move of *Cuniculus* but not with that of the dasyproctids because they have no satellite chromosome.

Patterson & Pascual (1968) suggested yet other groupings. Their Octodontoidea included the Ctenomyidae, Abrocomidae, Octodontidae, Echimyidae and Capromyidae. *Myocastor* was considered to be more closely related to the Echimyidae than to the Capromyidae. The karyotype of *Myocastor* has little resemblance to those of *Proechimys* and there seems no reason to remove *Myocastor* from the Capromyidae. Their Chinchilloidea still includes *Cuniculus* and the Dasyproctidae with the Chinchillidae. The Cavioidea contains the Caviidae and *Dolichotis* and *Hydrochoerus*. The karyotype of *Dolichotis* supports relationship with the caviids but does not require separation from the Dasyproctidae. The inclusion of *Hydrochoerus* presents the problem already discussed.

Clearly, there is some agreement between these five classifications and the karyotype arrangements. The Erethizontidae is agreed by everyone (except Landry, 1957) to represent a single evolutionary line: its differences from other hystricomorphs are confirmed by the karyotypes. There seems to be general agreement that the Caviinae and Dolichotinae are related and a majority are in favour of uniting them with the Dasyproctidae.

There is no doubt that the Octodontidae and the Ctenomyidae are closely related. The Echimyidae are mostly placed with these families and the karyotype may suggest a closer relationship of the Echimyidae with the octodontids than some authors have assumed.

The Chinchillidae form a discrete group but their associates in the superfamily Chinchilloidea have varied. The Capromyidae, usually with *Myocastor*, have been placed close to the Echimyidae (Patterson & Pascual, 1968), with the Chinchillidae (Wood, 1955) and then in the Octodontoidea (Wood & Patterson, 1959), while Simpson put them somewhere near the octodontid–chinchillid divergence. Karyotype analysis gives them a better fit in the Octodontoidea than with the Chinchillidae. However, if the satellite chromosomes of the octodontids and chinchillids are not distinguishable, then the groups are closely related and the position of the capromyids becomes academic. There is no justification on karyotypic evidence for separating *Myocastor* from the other capromyids.

Hydrochoerus is always placed with the Caviidae, indeed Ellerman (1940) contrasted these two taxa with all the other hystricomorphs. To justify this karyotypically it would have to be assumed that satellite chromosomes are not meaningful in evolutionary terms or that they occur regularly as parallel adaptations to something in the karyotype. It would then be possible to derive *Hydrochoerus* from a cavioid line in which the X–chromosome and chromosome 1 were already becoming long by translocation and the rest could be achieved by pericentric inversions and satellisation (see Fig. 8).

Cuniculus is put among the Cavioidea by Simpson (1945) and Wood (1955) but Wood & Patterson (1959) allied it with the Dasyproctidae and with the Chinchillidae. There is little resemblance between the karyotypes of *Cuniculus* and the dasyproctids. *Cuniculus* could be put in the Cavioidea only on the assumption that it is related to *Hydrochoerus*, particularly by virtue of the marker chromosome.

CONCLUSIONS

There is a good measure of agreement between our interpretation, based on karyological evidence, and those reached by other, well established, methods on the relationships of families in the Caviomorpha. Thus, cytotaxonomy confirms the generally accepted association of the Dasyproctidae and Caviidae, and the homogenous nature of the Octodontoidea. Conventional taxonomy which groups the Octodontoidea and Chinchilloidea, with the Capromyidae changing places within the group, should perhaps be taken as an indication that the satellisation of a chromosome in the caviomorph rodent karyotype was a once only event and that the suggestion of separate origins for the octodontid and chinchillid marker chromosomes is mistaken.

Even with this revision, the karyotype scheme agrees with Patterson & Pascual (1968) that *Cuniculus* should be placed with the chinchillids. If this is correct, then *Hydrochoerus* should be looked at again. As this is probably a monotypic genus there is little hope that more information can be gained from further karyotypes, unless chromosome banding patterns reveal any definite affinities (Seabright, 1972).

It is possible that satellisation had occurred in the early ancestor and that it is a characteristic feature of caviomorph rodents which has subsequently been lost by the Erethizontidae, Caviidae and Dasyproctidae. Wurster & Benirschke (1968a) have suggested that pairs of minute chromosomes that occur in some karyotypes are alternatives to satellites. *Erethizon* and *Coendou* have such a pair of chromosomes but the caviids do not, and *Myocastor* has these very small chromosomes as well as a satellite pair.

Satellisation in the primitive ancestor and subsequent loss may be found to be the explanation of the *Hystrix* karyotype. Unfortunately none of the other Old World hystricomorphs has been karyotyped. Of the families placed *incertae sedis* with the Hystricomorpha by Simpson (1945), only the gundis (F. Ctenodactylidae) have been karyotyped (W. George, unpublished). They have satellite chromosomes like, but not identical with, the octodontid marker. This is interesting in view of the affinities which have been suggested by many workers between the Octodontoidea and some of the African hystricomorphs.

Acknowledgments

We are grateful to Dr D. H. Wurster and Dr O. A. Reig for information on unpublished karyotypes, and to Dr I. W. Rowlands for his encouragement. The laboratory work of B. J. Weir was supported by the Ford Foundation.

References

Bachmann, K. (1972). Genome size in mammals. *Chromosoma* **37**: 85–93.

Benirschke, K. (1968). The chromosome complement and meiosis of the North American porcupine. *J. Hered.* **59**: 71–76.

Benirschke, K. & Malouf, N. (1967). Chromosome studies of Equidae. In *Equus*: 253–284. Dathe, H. (ed.). Berlin: Tierpark.

Berry, D. L. & Baker, R. J. (1971). Apparent convergence of karyotype in two species of pocket gophers of the genus *Thomomys* (Mammalia, Rodentia). *Cytogenetics* **10**: 1–10.

Borgaonkar, D. S. (1969). Insectivora cytogenetics. In *Comparative mammalian cytogenetics*: 218–246. Benirschke, K. (ed.). Berlin: Springer-Verlag.

Cohen, M. M. & Pinsky, L. (1966). Autosomal polymorphism via a translocation in the guinea pig *Cavia porcellus* L. *Cytogenetics* **5**: 120–122.

Comings, D. E. (1972). Evidence for ancient tetraploidy and conservation of linkage groups in mammalian chromosomes. *Nature, Lond.* **238**: 455–457.

Egozcue, J. (1969). Primates. In *Comparative mammalian cytogenetics*: 357–389. Benirschke, K. (ed.). Berlin: Springer-Verlag.

Ellerman, J. R. (1940). *The families and genera of living rodents.* **I**: 1–689. London: British Museum (Natural History).

Fernández, D. R. (1968). El cariotipo del *Octodon degus* (*Rodentia-Octodontidae*) (Molina, 1782). *Arch. Biol. Med. exp.* **5**: 33–37.

Ford, C. E. (1969). Meiosis in mammals. In *Comparative mammalian cytogenetics*: 91–106. Benirschke, K. (ed.). Berlin: Springer-Verlag.

Fredga, K. (1966). Chromosome studies in five species of South American rodents (Suborder Hystricomorpha). *Mammalian Chromosomes Newsl.* **20**: 45–46.

Fredga, K. (1972). Comparative chromosome studies in mongooses (Carnivora Viverridae). I. Idiograms of 12 species and karyotype evolution in Herpestinae. *Hereditas* **71**: 1–74.

Gardner, A. L. (1971). Karyotypes of two rodents from Peru, with a description of the highest diploid number recorded for a mammal. *Experientia* **26**: 1088–1089.

George, W. & Weir, B. J. (1972a). Record chromosome number in a mammal? *Nature, New Biol.* **236**: 205–206.

George, W. & Weir, B. J. (1972b). The chromosomes of some octodontids with special reference to *Octodontomys* (*Rodentia*: *Hystricomorpha*). *Chromosoma* **37**: 53–62.

George, W. & Weir, B. J. (1973). A note on the karyotype of *Proechimys guairae* (Rodentia: Hystricomorpha). *Mammalia* **37**: 330–332.

George, W., Weir, B. J. & Bedford, J. (1972). Chromosome studies in some members of the family Caviidae (Mammalia: Rodentia). *J. Zool., Lond.* **168**: 81–89.

Gropp, A. (1969). Cytologic mechanisms of karyotype evolution in Insectivores. In *Comparative mammalian cytogenetics*: 247–266. Benirschke, K. (ed). Berlin: Springer-Verlag.

Hamerton, J. L., Gianelli, F. & Carter, C. O. (1963). A family showing transmission of a D/D reciprocal translocation and a case of regular Down's syndrome. *Cytogenetics* **2**: 194.

Hoffman, R. S. & Nadler, C. F. (1968). Chromosomes and cytogenetics of some North American species of the genus *Marmota* (Rodentia, Sciuridae). *Experientia* **24**: 740–742.

Hsu, T. C. & Arrighi, F. E. (1966). Chromosomal evolution in the genus *Peromyscus* (Cricetidae, Rodentia). *Cytogenetics* **5**: 353–359.

Hsu, T. C. & Benirschke, K. (1970). *An atlas of mammalian chromosomes*. Dipodomys merriami. Folio 163. Berlin: Springer-Verlag.

Hsu, T. C. & Benirschke, K. (1971). *An atlas of mammalian chromosomes*. Capromys pilorides. Folio 282. Berlin: Springer-Verlag.

Hsu, T. C. & Mead, R. A. (1969). Mechanisms of chromosomal changes in mammalian speciation. In *Comparative mammalian cytogenetics*: 8–17. Benirschke, K. (ed.). Berlin: Springer-Verlag.

Hungerford, D. A. & Snyder, R. L. (1964). Karyotypes of two more mammals. *Am. Nat.* **98**: 125–127.

Kiblisky, P. & Reig, O. A. (1966). Variation in chromosome number within the genus *Ctenomys* and description of the male karyotype of *Ctenomys talarum talarum* Thomas. *Nature, Lond.* **212**: 436–438.

Landry, S. A. (1957). The interrelationships of the New and Old World hystricomorph rodents. *Univ. Calif. Publs Zool.* **56**: 1–118.

Levan, A., Fredga, K. & Sandberg, A. A. (1964). Nomenclature for centromeric position of chromosomes. *Hereditas* **52**: 201–220.

Marks, G. E. (1957). Telocentric chromosomes. *Am. Nat.* **91**: 223–232.

Matthey, R. (1945). L'évolution de la formule chromsomale chez les vertébrés. *Experientia* **1**: 50–56.

Matthey, R. (1952). Chromosomes de Muridae (Microtinae et Cricetidae). *Chromosoma* **5**: 113–138.

Matthey, R. (1953). A propos de la polyploidie animale. *Revue suisse Zool.* **60**: 466–471.

Matthey, R. (1956). Cytologie chromosomique comparée et systématique des Muridae. *Mammalia* **20**: 93–123.

Matthey, R. (1958). Les chromosomes des mammifères euthériens. Liste critique et essai sur l'évolution chromosomique. *Arch. Julius Klaus-Stift.* **33**: 253–297.

Matthey, R. (1963). Polymorphisme chromosomique intraspécifique chez un mammifère *Leggada minutoides* Smith. (Rodentia-Muridae). *Revue suisse Zool.* **70**: 173–190.

Matthey, R. (1964). La signification des mutations chromosomiques dans les processus de spéciation. Etude cytogénétique du sous-genre *Leggada*. *Archs Biol.* **75**: 169–206.

Matthey, R. (1966). Le polymorphisme chromosomique des *Mus* africains du sous-genre *Leggada*. Révision générale portant sur l'analyse de 213 individus. *Revue suisse Zool.* **73**: 585–607.

Matthey, R. (1967). Étude des deux femelles hétérozygoté pour une délétion partielle portant sur un bras du chromosome X chez *Mus* (*Leggada*) *minutoides musculoides* Temm. *Cytogenetics* **6**: 168–177.

Matthey, R. (1969). Les chromosomes et l'évolution chromosomique des mammi-
    fères. In *Traité de zoologie* **16** (VI): 855–909, 999–1004. Grassé, P.-P. (ed.).
    Paris: Masson.
Meylan, A. (1968). Formules chromosomiques de quelques petits mammifères
    nord-americains. *Revue suisse Zool.* **65**: 691–696.
Meylan, A. (1970). Chromosomal polymorphism in mammals. *Symp. zool. Soc.
    Lond.* No. 26: 211–222.
Nadler, C. F. (1964). Contributions of chromosomal analysis to the systematics of
    North American chipmunks. *Am. Midl. Nat.* **72**: 298–312.
Nadler, C. F. (1966a). Chromosomes of *Spermophilus franklini* and taxonomy of
    the ground squirrel genus *Spermophilus*. *Syst. Zool.* **15**: 198–206.
Nadler, C. F. (1966b). Chromosomes and systematics of American ground squirrels
    of the subgenus *Spermophilus*. *J. Mammal.* **47**: 579–596.
Nadler, C. F. (1969). Chromosomal evolution in rodents. In *Comparative mammalian
    cytogenetics*: 277–309. Benirschke, K. (ed.). Berlin: Springer-Verlag.
Nadler, C. F. & Harris, K. E. (1967). Chromosomes of the North American prairie
    dog *Cynomys ludovicianus*. *Experientia* **23**: 41–42.
Nadler, C. F., Hoffman, R. S. & Pizzimenti, J. J. (1971). Chromosomes and serum
    proteins of prairie dogs and a model of *Cynomys* evolution. *J. Mammal.* **52**:
    545–555.
Nadler, C. F. & Hughes, C.E. (1966). Chromosomes and taxonomy of the ground
    squirrel subgenus *Ictidomys*. *J. Mammal.* **47**: 46–53.
Nes, N. (1963). The chromosomes of *Chinchilla laniger*. *Acta vet. scand.* **4**: 128–135.
Ohno, S., Beçak, W. & Beçak, M. L. (1964). X–autosome ratio and the behaviour
    pattern of individual X-chromosomes in placental mammals. *Chromosoma*
    15: 14–30.
Patterson, B. & Pascual, R. (1968). The fossil mammal fauna of South America.
    Q. Rev. Biol. **43**: 409–451.
Patton, J. L. & Gardner, A. L. (1972). Notes on the systematics of *Proechimys*
    (Rodentia: Echimyidae), with emphasis on Peruvian forms. *Occ. Pap., Mus.
    Zool. Louisiana St. Univ.* No. 44: 1–30.
Reig, O. & Kiblisky, P. (1969). Chromosome multiformity in the genus *Ctenomys*
    (*Rodentia, Octodontidae*). *Chromosoma* **28**: 211–244.
Reig, O. A., Kiblisky, P. & Löbig, I. S. (1970). Isomorphic sex-chromosomes in
    two Venezuelan populations of the spiny rat genus *Proechimys* (*Rodentia,
    Caviomorpha*). *Experientia* **26**: 201–202.
Reig, O. A., Spotorno, O. A. & Fernández, D. R. (1972). A preliminary survey of
    chromosomes in populations of the Chilean burrowing octodont rodent
    Spalacopus cyanus Molina (Caviomorpha, Octodontidae). *Biol. J. Linn. Soc.*
    4: 29–38.
Renzoni, A. (1967). Chromosome studies in two species of rodents. *Mammalian
    Chromosomes Newsl.* **8**: 11–112.
Robertson, W. R. B. (1916). Chromosome studies: 1. Taxonomic relationships shown
    in the chromosomes of Tettigidae and Acrididae. V-shaped chromosomes and
    their significance in Acrididae, Locustidae, and Gryllidae: chromosomes and
    variation. *J. Morph.* **27**: 179–331.
Seabright, M. (1972). The use of proteolytic enzymes for the mapping of structural
    rearrangements in the chromsomes of man. *Chromosoma* **36**: 204–210.
Simpson, G. G. (1945). The principles of classification and a classification of mam-
    mals. *Bull. Am. Mus. nat. Hist.* **85**: 1–350.

Stebbins, G. L. (1971). *Chromosomal evolution in higher plants*. London: Edward Arnold.

Todd, N. B. (1967). A theory of karyotypic fissioning, genetic potentiation and eutherian evolution. *Mammalian Chromosomes Newsl.* **8**: 268–279.

Wahrman, J., Goiteiv, R. & Nevo, E. (1969). Geographic variation of chromosome forms in *Spalax*, a subterranean mammal of restricted habit. In *Comparative mammalian cytogenetics*: 30–48. Benirschke, K. (ed.). Berlin: Springer-Verlag.

White, M. J. D. (1959). Speciation in animals. *Aust. J. Sci.* **22**: 32–39.

Wood, A. E. (1955). A revised classification of the rodents. *J. Mammal.* **36**: 165–187.

Wood, A. E. & Patterson, B. (1959). The rodents of the Deseadan Oligocene of Patagonia and the beginnings of the South American rodent evolution. *Bull. Mus. comp. Zool. Harv.* **120**: 281–428.

Wurster, D. H. (1969). Cytogenetic and phylogenetic studies in carnivores. In *Comparative mammalian cytogenetics*: 310–329. Benirschke, K. (ed.). Berlin: Springer-Verlag.

Wurster, D. H. & Benirschke, K. (1968a). Comparative cytogenetic studies in the order Carnivora. *Chromosoma* **24**: 336–382.

Wurster, D. H. & Benirschke, K. (1968b). The chromosomes of the Great Indian Rhinoceros (*Rhinoceros unicornis* L.). *Experientia* **24**: 511.

Wurster, D. H., Snapper, J. R. & Benirschke, K. (1971). Unusually large sex chromosomes: new methods of measuring and description of karyotypes of six rodents (Myomorpha and Hystricomorpha) and one lagomorph (Ochotonidae). *Cytogenetics* **10**: 153–176.

Symp. zool. Soc. Lond. (1974). No. 34, 109–111

ECOLOGY AND BEHAVIOUR
CHAIRMAN'S INTRODUCTION:

OLIVER P. PEARSON

*Museum of Vertebrate Zoology,
University of California, Berkeley, U.S.A.*

Fellow admirers of the hystricomorphs:

I am looking forward this afternoon to becoming re-acquainted with some of my South American hystricomorph friends and to meeting new ones from Africa. Innumerable colleagues have gone off to Africa to study and have come home with gorgeous pictures of mammals—none of them smaller than a London omnibus. I have always suspected that smaller mammals must be lurking among the hooves of rhino and wildebeest, for it is no wonder the vermin underfoot are eclipsed when the big game is so spectacular. In South America the visibility problem is different because in many areas the hystricomorphs *are* the big game. In any event it is cheering to know that the smaller mammals of Africa in general and the hystricomorphs in particular are being studied and that new insights are being drawn from the South American hystricomorphs.

I should like at this time to advance an idea that I should not dare to expose to a more hostile audience—or even to an impartial audience. My thesis is simply that one of our beloved hystricomorphs was responsible for planting the seed of origin-of-species-through-natural-selection in the fertile mind of Charles Darwin. In the published journal that traces his thoughts and his travels on H.M.S. Beagle he is provoked on several occasions by the lowly tuco-tuco (*Ctenomys*) into asking penetrating questions about animal distribution and what we have come to know as evolution. Indeed, tuco-tucos are mentioned more often in his journal of researches than almost any other wild animal.

In Argentina, Darwin was impressed by finding fossil tuco-tucos lying side by side with the fossil remains of extinct, huge, bizarre mammals, and this led to lengthy thoughts about the extinction of tuco-tucos, the survival of modified tuco-tucos, and the extinction of animal species in general. He pondered also the occurrence of tuco-tucos on both sides of the Straits of Magellan. He suspected that the correct explanation of how an animal "so delicate and helpless as the tucu tuco" could have achieved this distribution was to assume that the land on the two sides of the Straits was once joined. This provided him with a proper geological and temporal framework within which to speculate on the distribution of animals.

Most appropriate of all, however, is the passage inspired by an incident in Uruguay when he was brought a blind tuco-tuco:

"The Tucutuco (*Ctenomys brasiliensis*) is a curious small animal, which may be briefly described as a Gnawer, with the habits of a mole. It is extremely numerous in some parts of the country, but is difficult to be procured, and never, I believe, comes out of the ground. ... The tucutucos appear, to a certain degree, to be gregarious: the man who procured the specimens for me had caught six together, and he said this was a common occurrence. They are nocturnal in their habits; and their principal food is the roots of plants, which are the object of their extensive and superficial burrows. ...

"The man who caught them asserted that very many are invariably found blind. A specimen which I preserved in spirits was in this state; Mr Reid considers it to be the effect of inflammation in the nictitating membrane. When the animal was alive I placed my finger within half an inch of its head, and not the slightest notice was taken: it made its way, however, about the room nearly as well as the others. Considering the strictly subterranean habits of the tucutuco, the blindness, though so common, cannot be a very serious evil; yet it appears strange that any animal should possess an organ frequently subject to be injured. Lamarck would have been delighted with this fact, had he known it, when speculating (probably with more truth than usual with him) on the gradually–*acquired* blindness of the Aspalax, a Gnawer living under ground, and of the *Proteus*, a reptile living in dark caverns filled with water; in both of which animals the eye is in an almost rudimentary state, and is covered by a tendinous membrane and skin. In the common mole the eye is extraordinarily small but perfect, though many anatomists doubt whether it is connected with the true optic nerve; its vision must certainly be imperfect, though probably useful to the animal when it leaves its burrow. In the tucutuco, which I believe never comes to the surface of the ground, the eye is rather larger, but often rendered blind and useless, though without apparently causing any inconvenience to the animal: no doubt Lamarck would have said that the tucutuco is now passing into the state of the *Aspalax* and *Proteus*." (Darwin, 1890: 52–54).

Who can vouch that on this day in 1833 Darwin's synapses and neural pathways were not cocked and then held patiently in readiness until triggered by his reading of Malthus (1798) in 1838, whereupon all of the scattered pieces of evidence condensed into the grand synthesis that we now honour as Darwin's theory.

So, in this Festspiel for the hystricomorphs, we pause to honour the tuco-tuco and Charles Darwin. The greater honour is due Darwin, however. He was told that a great many tuco-tucos are found blind. I have never found a blind tuco-tuco in the wild, B. J. Weir found none, and other biologists do not mention them. Furthermore, Darwin believed, mistakenly, that tuco-tucos never venture above ground and consequently never need to use their eyes. Who but Darwin could have arranged the available truths, half-truths, and untruths such as these into so durable a theory?

I trust that the speakers (this afternoon) will strive to convey accurate information and not rely upon the audience and serendipity to repeat the miracle of the construction of a mighty theory using misinformation.

REFERENCES

Darwin, C. (1890). *Journal of researches into the natural history and geology of the countries visited during the voyage round the world of H.M.S. 'Beagle'.* London: Murray.

Malthus, T. R. (1798). *Essay on the principle of population.*

Symp. zool. Soc. Lond. (1974). No. 34, 113–130

THE TUCO–TUCO AND PLAINS VISCACHA

BARBARA J. WEIR

*Wellcome Institute of Comparative Physiology,
Zoological Society of London,
Regent's Park, London, England*

INTRODUCTION

The tuco-tuco (*Ctenomys talarum*) and the viscacha (*Lagostomus maximus*) are among the typical mammals of the pampas in Argentina. There are over 50 named forms of tuco-tuco which are widely distributed and a single species of *Lagostomus* (Fig. 1). The ecology of these burrowers was studied in 1967 and 1970 during expeditions to capture live

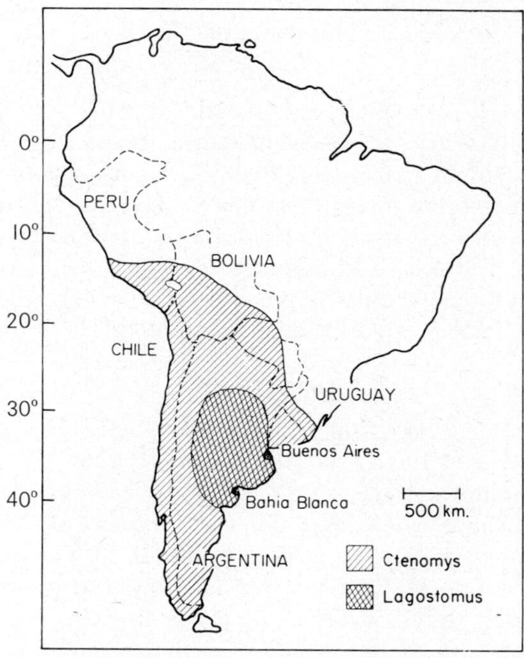

FIG. 1. Approximate distributions of tuco-tuco (*Ctenomys*) and the plains viscacha (*Lagostomus maximus*).

specimens to found laboratory colonies. Before the first expedition, I had not seen a live tuco-tuco or a live plains viscacha, and had to rely on the relatively meagre reports in the literature for these species and adaptations of methods used for other species for suggestions as to how best to catch them.

TUCO–TUCO

The first reference to the ecology of *Ctenomys* seems to be that of King (1835) who described it as a "little animal, very timid; feeds upon grass; and is eaten by the Patagonian Indians. It inhabits holes which it burrows in the ground, and from the number of holes it would seem to be very numerous". Darwin (1839) also comments on the burrowing activity of tuco-tucos (see Pearson, 1974, p. 110). Details of tuco-tuco ecology are given by Eisentraut (1933), Talice & Momigliano (1954a), Pearson (1959), Pearson *et al.* (1968) and Weir (1971), but only the accounts of Pearson (1959) and Weir (1971) are in English. Pearson's paper on Peruvian species of tuco-tuco is extensive and so only general comments on tuco-tucos and *C. talarum* in particular are needed.

Distribution

The genus *Ctenomys* is found from sea level on the pampas up to 4,000 m on the altiplano although its distribution is not continuous. Tuco-tucos can live only in areas where the soil is suitable for burrowing and the colonisation of such areas (leaving aside the problem of how the intervening inhospitable territories were crossed) has led to the recognition of many species (Rusconi, 1928; Cabrera, 1957–1961). Closer study using modern karyological and serological techniques suggests that the number of species that should be recognized is less than that of named forms (Reig & Kiblisky, 1969; Roig & Reig, 1969).

Description

Tuco-tucos (Fig. 2) are gopher-like animals which range from 170 to 250 mm in length and 100 to 700 g in weight (Walker, 1964). *C. talarum* is one of the smallest species and *C. tucumanus*, which is probably the largest, is about the size of a small domestic guinea-pig. Males are larger than females and have a more robust body. In all species the body is compact and there is no distinct neck. The limbs are short but the feet are broad, having five toes with strong claws on each. The hind feet are fringed with hairs and there is a pad of stiffer bristles on the hind toes, particularly the third toe (hence the name from *ktenos* and *mus*, comb-mouse). The pinnae are small and barely noticeable but the eyes

FIG. 2. Frontal view of a newly captured tuco-tuco eating a pellet of laboratory rat diet. Note the protuberant eyes, the vibrissae and the nearly naked, fleshy tail (arrowed).

are larger than would be expected for a burrower. The vibrissae are present but not long. The large size and orange colour of the incisors are obvious features of *C. talarum* but they are not as protuberant as in some other species. Another conspicuous feature is the rounded and fleshy tail which is about one third to one half the length of the head and body, and is sparsely covered with hair along the dorsal and ventral mid-lines and at the tip. It is suggested that the tail is used as a sensory organ during backward locomotion. Pearson (1959) did not consider it likely that tuco-tucos moved backwards within the tunnels, but observations on *C. talarum* in captivity indicated that these animals could run easily in both directions. Tuco-tucos dislike being picked up by the tail and may succumb rapidly to infections resulting from injuries of this organ. The pelage does not take the form of the dense plush that occurs in moles and some other fossorial mammals (Ellerman, 1956), and in some species of tuco-tucos it could almost be described as long. However, the skin seems to be very loose around the body and must facilitate movement in the burrow. The colour ranges from a light reddish-orange colour to almost a grey above and white to light brown below, according to species and locality.

Burrows

Tuco-tucos are extensive burrowers. They loosen the earth with the fore feet and excavate the soil with long sweeps of the hind feet. The incisors are rarely, if ever, used primarily for digging although any roots protruding into the tunnel will be cleared away by these means. Pearson's (1959) description of the methods of digging of *C. opimus* can be applied to those used by *C. torquatus* in Uruguay (Talice & Momigliano, 1954a) and *C. talarum*. The tunnel is long and meandering, being extended mainly as a foraging highway. It may be 60 cm below the surface (Cabrera & Yepes, 1960) and in most species only one tuco-tuco is in occupation; several female *C. peruanus* have been found in the same tunnel but males were solitary (Pearson, 1959). Various small chambers may be found along the tunnel for nests or food storage and they may be sealed off by grass plugs. Such plugs may also be found along the main tunnels of *C. talarum* and *C. torquatus* where they act as air filters. The temperature in a tuco-tuco burrow rarely varies by more than a few degrees in spite of much greater changes in air temperature above ground (Talice & Momigliano, 1954a; Pearson, 1959). The mean temperature in the tunnels of *C. torquatus* and *C. talarum* is 20–22°C. Humidity is also fairly constant and remains high (Talice & Momigliano, 1954a). Most burrow entrances face away from the prevailing winds and are on the sunny aspect. These entrances are plugged with earth and are distinct from the holes which may be found under the excavated soil heaps. Holes which are not plugged do not usually lead to an occupied part of the burrow. Most digging takes place during the day, but *C. talarum*, at least, can be active on warm nights. Tuco-tucos are rarely seen far from their holes when they do venture above ground

When disturbed by subsidence or trapping attempts tuco-tucos will block off a tunnel, often for distances up to 50 cm, and the soil may be so well packed that it is impossible to distinguish the tunnel from the surrounding earth. When a tunnel is broken into below ground it has a distinct edge, with a diameter usually just larger than its inhabitant. The walls are firm and well packed with soil. Captive *C. talarum* often tried to pack sawdust on the walls of their cages, thus indicating the method probably used in the burrows. The animal stands on its front feet and walks the hind feet up the wall, then it urinates and packs the moistened soil with the hind feet as they are lowered. These movements were often performed by caged tuco-tucos, mainly when they were disturbed, but urination was not invariable. Although tuco-tucos are said to be found only in light-textured soils, they can cope with heavier soils.

Vocalization

Darwin (1839) described the noise that *C. brasiliensis* makes and indicated that other tuco-tucos did not necessarily make the same noise. The best description of the vocalization of a tuco-tuco is that by Hudson (1892: 14): "not seen, but heard; for all day long and all night sounds its voice, resonant and loud, like a succession of blows from a hammer, as if a company of gnomes were toiling far down underfoot, beating on their anvils, first with strong measured strokes, then with lighter and faster, and with a swing and a rhythm as if the little men were beating in time to some rude chant unheard above the surface." It is not clear whether Hudson was referring to *C. magellanica* or to *Ctenomys* in general, but his description well fits the noise made by *C. talarum*. Not every species of tuco-tuco is vocal; *C. peruanus* is the only Peruvian species which calls using a "musical bubbling sound … reminiscent of that made by liquid poured from a bottle" (Pearson, 1959). The noise seems to be made only by *C. talarum* males (Pearson *et al.*, 1968; B. J. Weir, unpublished), although females may also call in more muted and shorter tones.

Senses

Tuco-tucos have large auditory bullae and are probably sensitive to ground vibrations. The large eyes are set well towards the top of the head; the main lines of vision are upwards and sideways, suggesting a need for alertness to aerial predators. Pearson (1959) considered that *C. opimus* could see a man at 50 m and that *C. peruanus* could distinguish between hawks and gulls flying overhead. The main predator of *C. talarum* is the burrowing owl, *Speotyto*, although it probably only takes young tuco-tucos. The sense of smell is probably the most important (Talice & Momigliano, 1954b). Tuco-tucos seem to react to strange urine in their tunnels (see Trapping, p. 118) and their frequent marking by anal glands in captivity is presumably also performed in the wild. Such marking by anal glands and urine would distinguish a tunnel enough for another animal to avoid it; tuco-tuco urine has a characteristic odour which is more pervasive than that of many other caviomorphs.

Feeding

Tuco-tucos seem to be entirely herbivorous, eating roots, stems, leaves and seeds. Most of these are obtained during burrowing but some foraging presumably takes place above ground. The animals are rarely seen during the day and it seems likely that such foraging occurs at night, probably at dawn. Feeding at this time would be advantageous because the dew would serve as a source of water; the tuco-tuco has not been seen to drink water in captivity.

Food is picked up by the teeth and then held in the fore feet while being rotated and moved between the mouth and the nostril. Blades of grass or pieces of hay are eaten in a characteristic manner. Initially such food is held in both fore-feet about 3 cm apart and passed rapidly sideways, between the incisors, several times. It is then bitten into two; one piece is dropped and the other is inserted end-on into the mouth. The amount of grass between the severed end and the fore-foot holding the grass, say the right, is chewed off. The left fore-foot replaces the right which is then positioned 3 cm away. After passing this bit of grass between the teeth, it is chewed from the end to the right foot again, and this is continued until the grass blade is consumed. The whole process is then repeated on a new piece of grass or on the other half of the original piece. The grass is not always eaten during this activity; in captivity all hay provided for bedding is chopped into 2–3 cm lengths, and the filters found in the tunnels of wild *C. talarum* and *C. torquatus* are composed of short lengths of dry grass. In captivity, tuco-tucos will eat most things they are offered and may become obese. A diabetic syndrome developed in the captive colonies of *C. talarum* (Wise *et al.*, 1968, 1972; Weir *et al.*, 1969; B. J. Weir, unpublished) which was not necessarily related to the obesity. It was caused by overeating and insufficient exercise. The most obvious symptom was the development of cataracts. Cataracts have not been found in wild tuco-tucos by me (Wise *et al.*, 1972) or other authors (Talice & Momigliano, 1954b; O. P. Pearson, personal communication) although Darwin (1839) reported that they were common. When attempts were made to prevent the development of diabetes mellitus in the captive colony by providing activity wheels, it was found that tuco-tucos 'ran' about 6·5 km a day (B. J. Weir, unpublished data). Most deaths in the tuco-tuco colonies were connected with the diabetic condition (Wise *et al.*, 1972). Some deaths, however, occurred from infection of wounds inflicted by bites during fighting. Such aggression has been reported in other species (Talice & Momigliano, 1954c).

Trapping

Tuco-tucos are reported to be difficult animals to trap alive (Talice & Momigliano, 1954a; Pearson *et al.*, 1968). Apart from shooting the larger species (Pearson, 1959), tuco-tucos have been trapped by Talice & Momigliano (1954a), Contreras & Reig (1965), Pearson (1959) and Pearson *et al.* (1968) but these methods have been unsatisfactory because the animals were injured. *C. talarum* was successfully caught unharmed in Longworth small mammal traps (Weir, 1971) and a similar trap could be effective for the larger species of tuco-tuco. The secret of success for trapping seemed to be the selection of the best holes (those plugged but

Fig. 3. A Longworth trap which has been filled with earth by a tuco-tuco while set in its tunnel.

not associated with mounds of earth), burying the trap in the earth to prevent obvious temperature changes in the burrow, and frequent checking with resetting of traps.

Using 48 of these traps and observing the above requirements, 100 tuco-tucos were caught in four days, mostly between 10.00 and 18.00 hours. The traps were unbaited and were more likely to be effective if they had previously held a tuco-tuco. The animals were probably attracted to the smell of another tuco-tuco's urine. Though many tuco-tucos did not enter the traps after first setting, they had obviously investigated them because the traps were packed with earth when lifted 2–4 h later (Fig. 3) and several centimetres of tunnel would also be blocked. If the trap was emptied and reset in the re-opened tunnel it often contained an animal when next lifted. All the tuco-tucos caught by this method (Weir, 1971) survived the air journey to London.

PLAINS VISCACHA

Little has been published on the habits and ecology of the plains viscacha (Fig. 4) although at one time "a man could ride for 500 miles

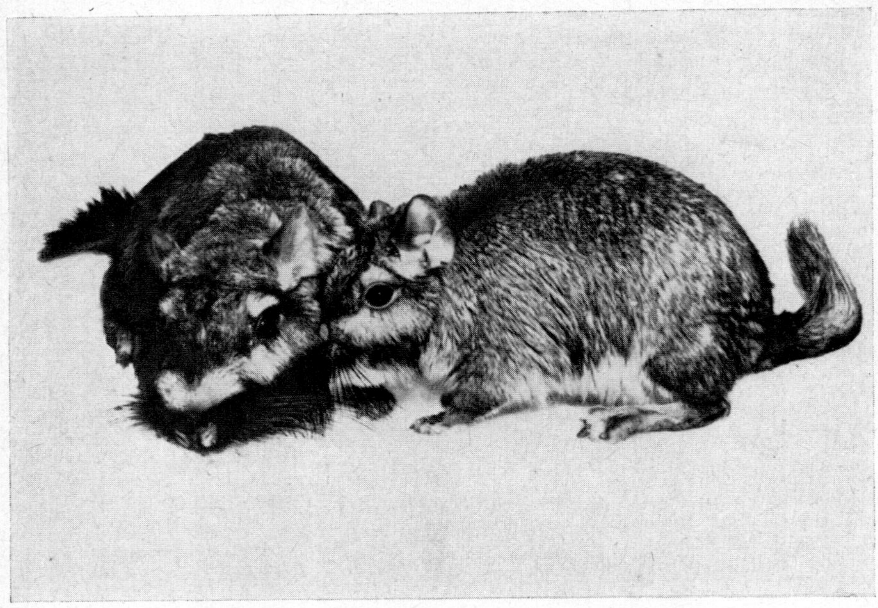

Fig. 4. A pair of plains viscacha; male on the left, female on the right. Note the size difference, the pronounced moustache of the male, and the strong claw on the hind feet.

without losing sight of viscacheras" (Hudson, 1872). Darwin (1839) comments on the hoarding behaviour of viscacha but the account by Hudson, used almost word for word in his book of 1892, is the only one in English. The only other extensive work on viscacha is that by Llanos & Crespo (1952) in Spanish. There are a few minor references to viscacha biology but they are mostly in South American papers that are hard to obtain, and/or are diatribes against the animals because of the damage they do (Gibson, 1877; Ambrosetti, 1893; Fernandez, 1949).

The first viscacha brought to Europe was exhibited in a travelling menagerie (de Blainville, 1820) and its anatomy was later described by Brookes (1828) who changed the trivial name from *maximus* to *trichodactylus* because of the erroneous use of the superlative degree: inadmissable, however, by application of the Law of Priority (I.C.Z.N., 1964). In spite of this early record of their hardiness and the survival of those kept on oats and water for six months by Camus & Gley (1922), viscacha have not often been exhibited in zoological collections, and those that have, have not received the attention that other hystricomorphs such as the acouchi, mara and pacarana have been given.

Distribution

Viscacha may be found throughout the pampas region of Argentina but not that of Uruguay (but see Barlow, 1969). Although viscacha can swim quite well and have crossed the Rio Paraná, they do not appear to have crossed the Rio Uruguay which is no greater a barrier. In the north and west of their range (Fig. 1) viscacha may have secondarily invaded scrub areas, called monte, probably because of persecution pressures in their favoured grassland habitat. Viscacha were once densely spread on the pampas but they have recently been reduced in numbers and territory because they are considered to be pests. They may do considerable damage by foraging in cultivated crops as well as compete with domestic animals for grazing; ten viscacha eat as much as one sheep. The burrowing proclivities of viscacha are also feared because cattle, horses, sheep and even men can fall down the large entrance holes of the burrows or through a collapsed tunnel and break a leg. A third reason for classifying viscacha as pests is that their urine, probably acidic like that of many other caviomorphs such as the guinea-pig, devalues the soil for many years. The organised control of viscacha for agricultural reasons has been in operation for nearly a decade. The value of the flesh and fur has provided a further incentive for killing these animals and although isolated or individually protected colonies will survive unmolested for some time, it is probable that viscacha will be nearly extinct in about 20 years.

Description

Plains viscacha are large rodents, weighing up to 9 kg. Males are larger (5–8 kg) than females (2–4·5 kg) and more heavily built with a massive head and neck. Both sexes have the typical black and white face markings (Fig. 4) but the stiff-whiskered black moustache of the male is pronounced. Apart from the white or cream eye patches, the upper pelage varies from grey to a light brown colour. The coloration varies according to the locality; in sandy areas the fur is lighter than it is on viscacha from areas with darker soils. The underfur is white to yellow. The tail is 'feathered' with stiff, often chestnut-coloured hairs; it looks sturdy but there is an autotomy plane which can fracture to leave only about four vertebrae outside the body. The underside of the tail is usually bare and horny at the base as it is used as a third 'leg' when the animal sits erect on the haunches for grooming or coprophagy. The hind limbs are long and the length of the hind foot is about 10 cm. The fore limbs are short but not unduly so. There are four digits on the fore feet and three on the hind. The claws on the latter are very strong

                          B. J. WEIR

but are not the primary digging agents. As in the other chinchillids, there
is a pad of very stiff bristles on the middle toe of the hind foot and this
comb would be useful for cleaning the fur of soil and seeds. The range of
the hind feet is limited, however, to the head and front of the body and
most of the grooming is done by the incisors. The incisor enamel is not
coloured as in many other hystricomorphs. The ears and eyes are larger
than would be expected for a burrower and the vibrissae are notable.
The intricate nostrils are finely furred and can be completely closed.

Burrows

Viscacha are colonial and creatures of habit. A colony is initiated
by a male who digs a simple burrow having only two or three entrances
which he and his followers extend as the years pass (Fig. 5). The tunnel

FIG. 5. A viscachera in Buenos Aires Province, Argentina, just north of the Rio
Colorado. This is a 'young' colony as the vegetation beyond the limit of the tunnel system
has not yet been denuded and there is little evidence of the collecting habits of viscacha.

system is known as the viscachera and frequently remains in occupation for several decades. The particular viscachera from which most of our founding colony were derived covered an area of 20 by 300 m and contained too many entrances to count. It was known to have been in use for at least 70 years.

The entrance holes lead to sloping tunnels at an angle of 13° to 30°, but usually about 21° (Llanos & Crespo, 1952). They ramify and in places coalesce to become chambers (Hudson, 1872).

The most striking feature of the viscachera is the material which is collected by the animals and placed above the entrances. Thus, above the ground the viscachera appears to be a repository for anything that can be carried but not eaten (sticks, stones, bones and dung) from the surrounding area. There is no obvious reason for this behaviour; it may be a way of clearing the ground around the colony to improve visibility but this seems unlikely as the only enemies other than man are the pampas fox (*Dusicyon*) and the viscachera boa (*Constrictor*), who eat only the young viscacha. It is possible that, as more soil is excavated from the burrows and thrown over the objects on top of the viscachera, the height of the openings would become raised above the general pampas level and thus act as a safeguard against flooding. The carrying and hoarding behaviour of viscacha was described by Darwin (1839) and it is characteristic of the species in captivity.

Viscacha dig mainly with the fore feet. They may then kick the loosened soil away with long sweeps of the back feet, but mostly they push the soil with the nose. The speed of digging is not known but Llanos & Crespo (1952) estimated that 80 m³ of earth had been moved to elaborate a viscachera about 11 m by 9 m with five surface holes. The pattern of digging a tunnel and the arrangement of soil around the hole is well described by Hudson (1782).

Vocalization

"The language of the viscacha is wonderful for its variety . . . and I doubt if there is any other four-footed thing so loquacious, or with a dialect so extensive." (Hudson, 1872). There is so much variety in the noises made by viscacha that they are difficult to describe. The commonest sound is that made when a viscacha is investigating . . . a sort of "uh-huh?" with the last syllable being higher than the first. Annoyance is expressed by a shorter note like a truncated bark. Males announce their displeasure or aggressive intentions by a fearsome crescendo of notes culminating in a high whine and accompanied by foot stamping and tail wagging. When this performance is enacted on a frosty, moonlit night, the noise is magnified eerily and the viscacha looks and sounds a formidable foe.

The display is usually directed at other males and all adult males seem to bear scars received during fights. Some females may also be lacerated. In captivity these injuries are mostly sustained when a strange animal is introduced to a group. This is probably a cause of fighting in the wild although Hudson (1872) indicates that viscacha are so tolerant of each other they will go and visit other viscacheras. When an attack is made, an animal which does not want to retaliate turns and presents its back to the aggressor and the tail is held stiffly out with the hairs erect. The attack may then proceed with a bite and the victim will either run or turn round prepared to fight. If the latter, the animals will threaten with the teeth. This degree of fighting is not usually dangerous; it is commonest in males who stand on the hind feet and hold the antagonist, usually another male, by the shoulders. The most dangerous attacks are those made without warning; any part of the victim may be grasped by the teeth and pulled over. The aggressor (always a male) also flings himself onto his side and then kicks at his adversary with the powerful hind feet. Fur is pulled out, bites and scratches are inflicted and the animal usually dies within 24 hours, probably from shock rather than its injuries.

Senses

Viscacha probably have good vision; they appeared to be watching movements about 200 m away, but it is possible that they were hearing sounds generated by the movements (B. J. Weir, unpublished). Their hearing is acute as is possibly their sense of smell. Most viscacha object to being touched along the back.

Feeding

Viscacha seem to eat any plants they encounter, although their basic diet is grasses and seeds. In captivity they relish carrots and hard items such as dog biscuits. "Viscachas are fond of exercising their teeth on hard substances ..." (Hudson, 1872), and will even tackle things such as wood which have no food value. They emerge to feed at dusk. A male is usually the first to emerge and then gradually the rest of the colony join him. At long-established viscacheras, the feeding grounds are situated at a distance because the intervening area has already been cropped. The viscacha reach the feeding grounds by well worn pathways. After their initial hunger is assuaged individuals retreat to the colony. They may stay above ground and even gambol or they may retire into the burrows. Feeding bouts occur during the night and there is concentrated activity just before dawn. At any alarm, real or imaginary, a sharp call is uttered, taken up by the rest of the group and every viscacha almost instantaneously disappears into the ground. If an animal is disturbed away from the viscachera it will race back to it in preference to looking for another hole nearby. They can run very quickly in spite of Hudson's (1872) assertion that even "the slowest dog can overtake them".

Viscacha may sit up on the haunches to feed, holding the stem in the fore feet, but usually they hold the food on the ground while they eat.

Trapping

Upon arrival at our main catching area (at Pedro Luro, about 100 km south of Bahia Blanca, Province of Buenos Aires) we made contact with the Sanidad Vegetal, the Department responsible for pest control, who were organising a viscacha extermination campaign in the area. They were pleased to try and help us catch live viscacha and several methods were tried on different viscacheras, all of which were known to be inhabited.

Digging

The most obvious method was to try and dig the viscacha out. This did not seem too difficult in an area of light, sandy soil. However, after several attempts using hand and mechanical diggers it was agreed that the viscacha must have retreated to lower tunnels that could not be reached by our efforts.

Suffocation

This was attempted at a relatively small viscachera using smoke. Material from the top of the burrow was gathered and placed on the windward side by an entrance and ignited. After about 10 min the smoke began to pour out of the other holes; and although it was maintained for 90 min, no viscacha emerged. Wisps of smoke were seen coming out of what appeared to be old and unused exits several metres away. It was concluded that the viscacha had retreated or dug their way into an older part of a more extensive system than was apparent at the surface. A variation using exhaust fumes was equally unsuccessful. The extermination method favoured by the Sanidad Vegetal at that time was carbon disulphide which was poured down the burrows to suffocate the animals and then ignited to enable the area to be ploughed.

Fear

In the traps placed in the viscacha burrows we caught three large mustelids (*Galictis furax*). These hurónes were so vicious that we decided to use them as ferrets (Ambrosetti, 1893). The holes of the viscachera were blocked or guarded and the hurón was released into the burrows. Nothing was heard and no viscacha were seen at the surface for three days. By that time this method was considered to be as useless as the others so far investigated. A shot gun was also discharged down the main tunnel of another viscachera just before dusk, but there was no rush to the surface; the animals were seen emerging normally although several hours late.

Flooding

This was said to be an effective method, and so it may have been farther north in the more clayey soils of Entre Rios Province where most of the viscacha for the Buenos Aires market are caught. But at Pedro Luro water had no effect because, in spite of the local water cart which held 16,000 litres being emptied three times, the water just soaked into the sand and the viscacha stayed where they were. Eventually, six viscacha were caught by this method by one of the peons of the Sanidad Vegetal who refused to be beaten and managed to divert an irrigation channel into a viscachera. The viscacha came up through the water and were very cold and practically dead by the time they were brought to us. After being rubbed dry and warmed up, the viscacha recovered uneventfully.

By hand

It was very tempting to try and catch viscacha by hand at night since they would let one approach quite closely before making a sudden dash for their tunnel. They were more wary if two people tried to catch them and the best hope lay in trying to net them as they ran from the feeding grounds to the viscacheras. In spite of commandeering two tennis nets for this purpose the method was ineffective.

Up to this time I had not handled a live viscacha and the prospect of capturing a very large (8·5 kg) and irate male in a garden in the centre of Buenos Aires was daunting. This viscacha had obviously escaped from one of the depots where viscacha were hoarded before being sold to the restaurants. However, after some minutes of wary observation on both sides, the viscacha was grasped quickly by the tail and put into a box. In retrospect, this was the only viscacha of our collection that I could have handled like this and I did not then know how the animals can part from their tails in these circumstances.

Trapping

This was the method that brought consistently good results (Gibson, 1877). Tubular traps, as described by Llanos & Crespo (1952), did not work but success was achieved with collapsible, wire-mesh, treadle traps (National Live Traps). A meticulous routine was followed of setting the unbaited traps at any active hole at dawn and dusk and blocking as many other holes as possible in the vicinity. The traps were checked at dawn and dusk and any which had not been effective after two checks were moved to another hole. In this way we caught an average of two unharmed viscacha a day and no doubt the capture rate would have been higher with more than six traps. The viscacha seemed to enter the traps accidentally rather than because of lack of any other exit since many holes were re-opened during the night. Traps placed in the regular runways from the feeding grounds to the viscacheras were not as successful as those at the holes but three viscacha were caught in this way.

ACKNOWLEDGMENTS

The two expeditions were financed by the Wellcome Trust, the Royal Society (from a Parliamentary Grant in Aid), the Medical Research Council and the Zoological Society of London. The assistance of Messrs M. Rumboll and D. MacIver and Dr I. W. Rowlands is gratefully acknowledged.

REFERENCES

Ambrosetti, J. B. (1893). Notas biológicas. III. Sobre la vizcachas (*Lagostomus trichodactylus* Brookes). *Revta Jard. zool. B. Aires* **1**: 43–46.

Barlow, J. C. (1969). Observations on the biology of rodents in Uruguay. *R. Ont. Mus. Life Sci. Contr.* No 75: 1–59.

Brookes, J. (1828). On the new genus of the order Rodentia. *Trans. Linn. Soc. Lond.* **16**: 95–104.

Cabrera, A. (1957–1961). Catálogo de los mamíferos de America del Sur. *Revta Inst. nac. Invest. Cienc. nat. Mus. argent. Cienc. nat. Bernadino Rivadavia* **4**: 1-724.

Cabrera, A. & Yepes, J. (1960). *Mamíferos Sud Americanos*. 2nd edn. Buenos Aires: Ediar.

Camus, L. & Gley, E. (1922). Action coagulante du liquide prostatique de la viscache sur le contenu des vésicules séminales. *C. r. Seánc. Soc. Biol.* **87**: 207–209.

Contreras, J. R. & Reig, O. A. (1965). Datos sobre la distribucíon de género *Ctenomys* (Rodentia, Octodontidae) en la zona costera de la Provincia de Buenos Aires comprendida entre Necochea y Bahia Blanca. *Physis* **25**: 169–186.

Darwin, C. (1839). *Narrative of the surveying voyages of His Majesty's ships Adventure and Beagle, between the years 1826 and 1836, describing their examination of the southern shores of South America, and the Beagle's circumnavigation of the globe.* **3**. *Journal and remarks.* 1832–1836. London: Colburn.

de Blainville, H. D. (1820). Memoire sur une nouvelle espèce de Gerboise, et illustration de ce genre. *Dictionnaire des Sciences Naturelles* (1816–1830) **18**: 471–472.

Eisentraut, M. (1933). Biologisches studien im bolivianischen Chaco. III. Beitrage zur biologie der säugetierfaune. *Z. Säugetierk.* **8**: 47–69.

Ellerman, J. R. (1956). The subterranean mammals of the world. *Trans. R. Soc. S. Afr.* **35**: 11–20.

Fernandez, M. (1949). Sobre la vizcacha (*Lagostomus trichodactylus* Brookes) sus viviendas y su proteción. *Boln Acad. nac. Cienc. Córdoba* **38**: 347–378.

Gibson, E. (1877). On the Biscacha (*Lagostomus trichodactylus*) a South American rodent. *Proc. nat. Hist. soc. Glasgow* **3**: 136–140.

Hudson, W. H. (1872). On the habits of the viscacha (*Lagostomus trichodactylus*). *Proc. zool. Soc. Lond.* **1872**: 822–833.

Hudson, W. H. (1892). *The naturalist in La Plata.* London: Chapman & Hall.

King, P. P. (1835). Notes on several rodents collected during a survey in the Straits of Magallaens. *Proc. zool. Soc. Lond.* **1835**: 189–191.

Llanos, A. C. & Crespo, J. A. (1952). Ecologia de la vizcacha ("*Lagostomus maximus*" Blainv.) en el nordeste de la provincia de Entre Ríos. *Revta Invest. agríc., B. Aires* **6**: 289–378.

Pearson, O. P. (1959). Biology of the subterranean rodents *Ctenomys* in Peru. *Mems Mus. Hist. nat. 'Javier Prado'* **9**: 1–56.

Pearson, O. P. (1974). Chairman's introduction. *Symp. zool. Soc. Lond.* No. 34: 109–111.

Pearson, O. P., Binzstein, N., Boiry, L., Busch, C., di Pace, M., Gallopin, G., Penchaszadeh, P. & Piantanida, M. (1968). Estructura social, distribucíon espacial y composición por edades de una población de Tuco-tucos (*Ctenomys talarum*). *Investnes zool. chil.* **13**: 47–80.

Reig, O. A. & Kiblisky, P. (1969). Chromosome multiformity in the genus *Ctenomys* (Rodentia, Octodontidae). A progress report. *Chromosoma* **28**: 211.

Roig, V. & Reig, O. A. (1969). Precipitin test relationships among Argentinian species of the genus *Ctenomys* (Rodentia, Octodontidae). *Comp. Biochem. Physiol.* **30**: 665–672.

Rusconi, C. (1928). Dispersión geografica de los tuco-tucos vivientes (*Ctenomys*) en la región neotropical. *An. Soc. argent: Estud. Geogr. 'GAEA'* **3**: 235–250.

Talice, R. V. & Momigliano, E. (1954a). Investigaciones sobre la biología de los roedores del género *Ctenomys*. *Revta Fac. Human. Cienc. Montev.* **13**: 5–21.

Talice, R. V. & Momigliano, E. (1954b). La vision del roedor *Ctenomys torquatus*. *Archos Soc. biol. Montev.* **21**: 173–178.

Talice, R. V. & Momigliano, E. (1954c). Captura, cautividad y manipulacion de *Ctenomys torquatus*. *Archos Soc. Biol. Montev.* **21**: 179–185.

Walker, E. P. (1964). *Mammals of the world.* 1st edn. Baltimore: Johns Hopkins.

Weir, B. J. (1971). A trapping technique for tuco-tucos, *Ctenomys talarum*. *J. Mammal.* **52**: 836–839.

Weir, B. J., Wise, P. H., Hime, J. M. & Forrest, E. (1969). Hyperglycaemia and cataract in the tuco-tuco. *J. Endocr.* **43**: vii-viii.

Wise, P. H., Weir, B. J., Hime, J. M. & Forrest, E. (1968). Implications of hyperglycaemia and cataract in a colony of tuco-tucos (*Ctenomys talarum*). *Nature, Lond.* **219**: 1374–1376.

Wise, P. H., Weir, B. J., Hime, J. M. & Forrest, E. (1972). The diabetic syndrome in the tuco-tuco (*Ctenomys talarum*). *Diabetologia* **8**: 165–172.

DISCUSSION

DEANESLY (*Cambridge*): Is the noise made by the tuco-tucos made by themselves; there is no banging on anything else?

WEIR: No, the noises made by the tuco-tucos are their own. There was some banging [on the tape recording] in the background by cross viscacha, but the 'tucing' noise was completely tuco-tuco.

PEARSON (*Berkeley*): I should welcome comments from the audience as to the function of this sound. It is made by a solitary animal, in my experience only by the males. It does not seem to be territorial, because the females are territorial without the benefit of making a sound like this.

SALE (*Nairobi*): I am still not quite clear; is this a vocalization or is it literally a thumping with the feet?

WEIR: We have watched tuco-tucos often in the laboratory and it is clearly just a vocalization; there is no thumping with the feet. The 'tucing' reverberates much more when they are doing it in their own tunnels, of course, and then it really is a deep down, bubbling noise.

SALE: It is made just by the males?

WEIR: Yes, in this species anyway. The female tuco-tuco can make a very scaled down version of the noise but it never develops into the full-blooded 'tuc'. I don't believe the sex of the performer has been determined in any other species.

PEARSON: *C. peruanus* in Peru makes a similar sound but I have not checked whether it was males or females. My checking of the sex of the *C. talarum* performers was made only a mile or two from this place you have shown us.

SALE: Is there any information on the social organization of these beasts; are they territorial, do they live singly or in colonies?

WEIR: As Professor Pearson has mentioned, he and his colleagues have studied *C. talarum* extensively and each animal of this species lives in its own tunnel. This was one of the reasons why my tuco-tucos did not breed very well even in their large tank of earth—it was too small an area. Obviously they kept wandering into other tunnels because when I dug them up they all had fight wounds round their necks. But there are some species of tuco-tucos, such as *C. peruanus*, which are more colonial; at least, Pearson found two or three animals in a burrow so one assumes they were there by choice.

PETTER (*Paris*): Do you know anything about the frequency of this noise?

WEIR: We have not recorded tuco-tucos on a sonograph, but individual animals do have recognizably different voices. Males 'tuc' mainly when they are disturbed. In the wild, we listened to see if there was a time of day when 'tucing' was most frequent but there didn't appear to be. It partly depends, I think, on the weather and the degree of burrowing. They seem to 'tuc' when they are actively burrowing as if to warn other tuco-tucos that they are there and do not want their burrows disrupted. But I do not know how a female warns off other tuco-tucos.

Lavocat (*Montpellier*): I should like to mention that there are a lot of morphological similarities between the skulls of recent *Lagostomus* and *Diamantomys* of the Miocene of East Africa, particularly in the middle ear structure. There are many other extraordinarily similar characters, and I am beginning to think that the ecology of *Diamantomys* could have been similar to that of *Lagostomus*.

Symp. zool. Soc. Lond. (1974) No. 34, 131–141

MOUNTAIN VISCACHA

I. W. ROWLANDS

Wellcome Institute of Comparative Physiology,
Zoological Society of London,
Regent's Park, London, England

SYNOPSIS

Details of the taxonomy, distribution, habitats and behaviour of mountain viscacha are drawn and described from the literature and personal observations.

INTRODUCTION

The genus *Lagidium* (mountain viscacha, Fig. 1) is one of three which compose the family Chinchillidae; the others are *Chinchilla* (chinchilla) and *Lagostomus* (plains viscacha) both of which are monotypic. Mountain viscacha and chinchilla have many more features in common than either have with the plains viscacha. The latter genus is fossorial and restricted to the alluvial pampas of Argentina (see Weir, 1974), but the others are rock dwellers, usually in very remote areas and often at high altitudes.

The name viscacha was a phonetic rendering by the Spanish invaders of Peru of the Quechuan name *uiskacha* (Cieca, 1533) but the reason for calling the pampean genus a viscacha has never been clarified.

The chinchilla (*C. laniger*) survives only as a domestic species bred for its luxurious fur. Weir (1974) believes that the future existence of *Lagostomus maximus* in the wild is threatened. The status of *Lagidium* in the wild is unknown as few travellers in South America have seen these animals and even fewer have studied them.

TAXONOMY

'Mountain viscacha' were originally placed in the genus *Viscaccia* by Oken (1816) but the use of the name *Lagidium* applied by Meyen (1833) has been validated by suspension of the International Rules of Zoological Nomenclature. Although d'Azara (1802) distinguished the 'Vizcacha de la Sierra' and the 'Vizcacha de la Pampa', other zoologists and travellers in South America were not so careful and the nomenclature of all three chinchillid genera, and particularly that of the mountain viscacha, has been changeable (see Lahille, 1906; Tate, 1935).

Fig. 1. A Peruvian mountain viscacha. Note the long, strongly curled tail, the combs on the hind feet, the vibrissae and the hair-fringed ears.

The confusion over *Lagidium* is probably caused by the tremendous range of its constituent species (Fig. 2) and the inaccessibility of their habitats has made a complete study of them all by one worker a virtual impossibility.

Ellerman (1940) and Cabrera (1957–61) have both attempted specific distinctions. Ellerman working only with the skins and skulls available in the British Museum (Natural History), London, concluded that there were four species: *L. peruanum* (Fig. 1), *L. viscaccia*, *L. boxi* and *L. wolffsohni*. Cabrera's catalogue was the outcome of this popular book on the natural history of South American mammals (Cabrera & Yepes, 1940). His material was clearly derived from specimens in local collections, the literature and his own information. In the catalogue, only *L. peruanum*, *L. viscacia* [sic] and *L. wolffsohni* are recognized as full species but four and ten subspecies are given for *L. peruanum* and *L. viscacia* respectively. Many of these subspecies are raised to specific status in the second edition of the natural history (Cabrera & Yepes, 1960). Crespo (1963) refers all Argentine mountain viscacha to *L. viscacia*.

The primary characters used to distinguish the species appear to be size, colour and locality, and their value is clearly dependent upon the number and state of the specimens available.

DISTRIBUTION

Mountain viscacha are known to occur on the altiplano of Peru (10°S) and Bolivia along the chain of the Andes passing through Argentina and Chile to the Andean foothills of Patagonia (52°S), a distance of nearly 4 700 km (Fig. 2). In the north of the range the animals live at altitudes of 4 000 to 5 000 m, although Pearson (1957) found a small colony living within the fog belt near Naña (Dept. of Lima), at only 670 m. The southern viscacha are found at lower altitudes as the latitude increases; for example, *L. boxi* Thomas, 1921 occurs at

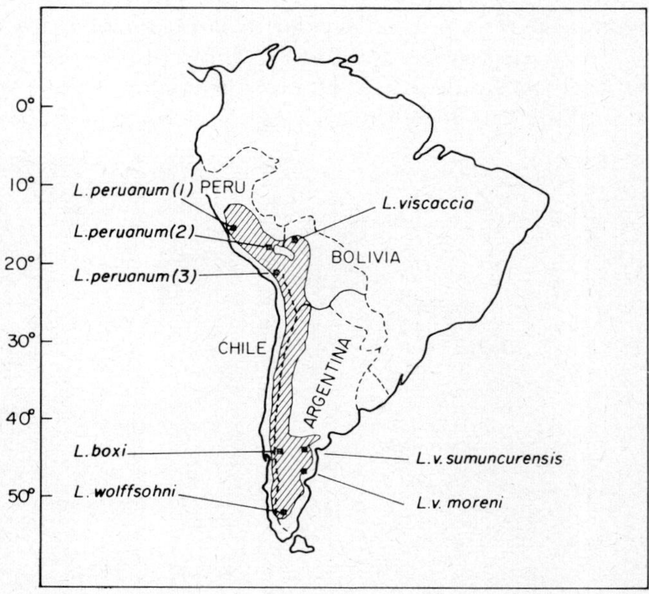

FIG. 2. Map indicating the alleged distribution of the genus *Lagidium* (shaded area) and the location of some reported species (■). *L. peruanum* (1) at Naña near Lima at 670 m (Pearson, 1957); *L. peruanum* (2) at La Raya (4 200 m), animals sent to the Wellcome Institute in London: *L. peruanum* (3) near Caccachara, at 4 850 m (Pearson, 1948, 1949). *L. viscaccia* at Ulla Ulla, Bolivia, 5 100 m (see Fig. 5); *L. boxi* near Pilcaniyeu, Patagonia, at 1 200 m and source of type specimen of Thomas (1921); *L. wolffsohni* at Ultima Esperanza, 52°S (Thomas, 1907); *L. v. sumuncurensis*, at eastern edge of Patagonian plateau at about 700 m (Crespo, 1963); *L. v. moreni* also in eastern Patagonia (Thomas, 1897).

1 200 m and *L. viscacia sumuncurensis* Crespo, 1963 lives at 700 m and spreads along the Patagonian plateau towards the east coast at 40 to 42°S (Thomas, 1897). The most important factor determining distribution is almost certainly suitability of habitat (see below).

DESCRIPTION

Mountain viscacha (Fig. 1) are similar to chinchilla in general appearance although they are larger (900–2 000 g) and have relatively smaller ears which are fringed with white hairs, smaller eyes and a longer tail. They have been likened to monkeys, rabbits and squirrels (Pearson, 1948) and in Patagonia they are called 'ardillas' or 'rock squirrels'; Bennett (1829) refers to the chinchilla as being known by the name 'ardilla' or its diminutive 'arda'. The fur is dense but the guard hairs are prominent unlike those of chinchilla. The colour ranges from all shades of grey (northern animals) to buff or orange (*L. wolffsohni* Thomas, 1907). The underparts are usually paler. A dark dorsal stripe may be a distinguishing characteristic (Cabrera & Yepes, 1960). Pearson (1948) states that there is frequently, but not always, a narrow, dark mid-dorsal line running up to the shoulder but this line is not obvious from

Fig. 3. Dorso-lateral view of *L. boxi* to show the pronounced dorsal stripe seen on all the specimens of this species observed at Pilcaniyeu.

Fig. 4. The hind foot of *L. peruanum* showing the pallipes and pad of stiff bristles on the inner toe.

his pictures of live or dead viscacha (*L. peruanum*). A similar diffuse line was present in some animals we saw in Bolivia (at Ulla Ulla, 14°S 71°W; I.W. Rowlands & B. J. Weir, unpublished), but there was only the faintest of lines along the back of six viscacha sent to the Wellcome Institute from Peru. These animals had been trapped in the vicinity of the Veterinary Institute for Tropical and High Altitude Research at La Raya (4 200 m). Crespo (1963) describes "lista negra bien marcada desde el pescuezo hasta los $\frac{3}{4}$ del dorso" for the holotype of his subspecies from Patagonia, and it is clear from the photograph of the mounted animal that the line does not extend to the base of the tail. In our specimens of *L. boxi* Thomas, 1921 the dorsal line was pronounced from the forehead to the base of the tail where it merged with the dark coloured dorsal tail hairs (Fig. 3).

The tail of all the mountain viscacha (with the possible exception of Crespo's *L. v. sumuncurensis*) appears to be at least as long as the body. It is held half-curled or completely curled up when the animals are at rest but is fully extended when they are jumping or running. Loss of the tail, however, does not seem to impede activity; Pearson (1948)

FIG. 5. The habitat of mountain viscacha (*L. viscaccia*) at Ulla Ulla, Bolivia, seen in misty conditions. The animals were active in these fallen rocks in spite of the snowstorm. Three viscacha can be seen (arrows); the pair (right) are in a typical nursing posture although the one did not appear very much smaller than the other.

mentioned several animals with damaged tails in the wild and one at the Institute with an amputated tail was just as agile as his mates. The tail of our specimens of *L. boxi* (see Weir, 1971) and those of *L. v. sumuncurensis* (see Crespo, 1963) seem to be proportionately more sturdy than those of the Peruvian or Bolivian species.

Like the other chinchillids, mountain viscacha have fleshy pads or pallipes, on the soles of the feet (Fig. 4). Such projections probably aid the grip of the feet on vertiginous or crumbling rocks. On the inner toe of the hind feet there is a pad bearing stiff bristles, similar to that of *Chinchilla* and *Lagostomus*.

HABITAT

An excellent account of the habitat and ecology of mountain viscacha in Peru has been given by Pearson (1948) at Caccachara (16°41′S, 70°4′W) at 4 850 m. The most essential feature for viscacha habitation is the availability of rocks. Viscacha may live in rocky cliffs and out-crops as long as there are "some deep crevices and narrow stony tunnels suitable as nesting sites but the favourite habitat is a vast tumble of boulders". Such a site provides numerous ledges for lookout and other purposes, and plenty of crevices and tunnels in which to escape and nest. Viscacha were seen by us at a similar type of site and altitude (5 100 m) in Bolivia (Fig. 5). At the immediate base of these rocky screes, there is often a flatter area where vegetation can be supported. According to Pearson (1948) viscacha will eat most types of vegetation that are available at these high altitudes. Above this line the vegetation is restricted to lichens.

The habitat of the *L. boxi* studied at Estancia Pilcañu (Pilcaniyeu, Province of Rio Negro, Argentina; 41°50′S, 71°W) was of a different nature (see Fig. 6). This part of the Patagonian plateau which consisted of broad flat stretches of grassland that supported Merino and Corriedale sheep was broken up by rocky outcrops of soft sandstone, some of which were eroded by the endless wind into bizarre shapes. The 25 to 30 m rock faces were usually vertical (Fig. 6a) and scored by deep clefts and chimneys (Fig. 6b). Sometimes the base of the outcrop was surrounded by large, loose, weatherworn rocks covering numerous well-worn runs connecting with the fissures of the rockface. The top of the outcrop was flattened and the patches of urine-stained rock, scattered droppings and browsed vegetation indicated the feeding areas.

Similar types of rock were seen in many parts of Patagonia as we travelled from Pilcaniyeu to the eastern coast; the reports of Thomas (1897), Crespo (1963) and residents in Chubut indicate that *Lagidium* may well be more widely spread in Patagonia than previously thought.

Fig. 6. The rocky habitat of *L. boxi* at Estancia Pilcañu, Pilcaniyeu, Argentina. (a) general view of a rocky outcrop on the top of which the viscacha feed, and (b) typical crevices in rock face in which these animals retreat and nest; the size of these rocks is indicated by the man (arrowed).

BEHAVIOUR

In the wild

Mountain viscacha are diurnal, unlike chinchilla and plains viscacha. "They quit their abodes which are amongst the loose pieces of rock, shortly after sunrise and rather before sunset, and then display the most extraordinary activity, leaping from rock to rock; in a moment as it appears, they may be seen to ascend an almost precipitous rock of 20 or 30 feet in height and, if fired at, they disappear as if by magic, having retreated to the holes and crevices. The sunshine they scrupulously avoid, and it is therefore only in shaded spots they are seen, unless the day be cloudy." (Tschudi, 1845). This routine is confirmed by Pearson's (1948) descriptions of viscacha in Peru except that sunbathing may be enjoyed in the early morning. We certainly saw fewer *L. boxi* in Patagonia on a sunny day than we did the following day which was dull, bitterly cold and with occasional snow flurries. We also saw many groups each of two to four viscacha sitting motionless on the flat tops of large rocks in conditions of swirling mist and slight snow fall at Ulla Ulla, Bolivia.

Viscacha are certainly gregarious and although they probably live in family groups within the home den (Pearson, 1948) groups will mix within the colony. In the colonies of Peruvian viscacha which Pearson studied little hostility was found between family groups and, even in the breeding season, belligerence was manifested only by a lunge, a squeak-growl and a short chase. The *L. boxi* seemed much more irascible to each other. Considerable altercation was heard as the animals moved about the colony and what has been thought to be a warning 'whistle' or trill was used intraspecifically.

This whistle is described by Pearson (1948: 361) as a warning sound used by male and female viscacha to alert other viscacha. The high pitched whistle "which might be described as a "wheee . . ." is usually repeated three or four times at intervals of a little more than a second. Each "wheee" lasts about one second, has a slight change in pitch at the very beginning and just the faintest quaver . . .". The equivalent noise made by *L. boxi* is similar but recognizably distinct. This whistle, which has been likened to the call of a goshawk, starts with unexpected abruptness quite loudly at a high pitch and is quavered strongly at this intensity for 10 to 15 secs before fading out during another 5 to 10 secs. Some whistles are of shorter duration and truncated.

At rest, mountain viscacha will walk quadrupedally or progress by one or two hops. The powerful hind legs propel the animals at considerable speed, and the tail is spread as they jump horizontally or vertically with sudden turns and ricochets off the rocks.

In captivity

L. boxi were never heard to whistle to each other at the Institute, and only did so when they were disturbed or frightened. Our captive *L. peruanum* whistle strongly when they are alerting each other and only faintly when they are upset or frightened. Both species use the tail flick described by Pearson (1948: 362) when annoyed and the *L. peruanum* will tooth chatter violently when being approached for capture. Our *L. peruanum* will also bite, unlike those caught by Pearson. In spite of these exhibitions of protest, however, it is clear from our experience of breeding other hystricomorphs at the Institute (Weir, 1967, 1970, 1973) that mountain viscacha do not take well to captivity. They have remained throughout nervous, timid, and highly strung animals, and they accept domination by chinchilla housed in the same pen.

Fig. 7. A newborn mountain viscacha (*L. peruanum*), at the Wellcome Institute (21.12.71) after 140 days of gestation.

The *L. boxi* female gave birth in captivity but did not rear the baby (see Weir, 1971). One *L. peruanum* mated at the Institute and gave birth to a single offspring (Fig. 7), 140 days later. This is the only known conception to occur in captivity and, consequently, the only accurate indication of gestation length in this species. No further reproductive activity other than a sporadic disappearance of the vaginal closure membrane which is indicative of oestrus, has occurred within the group. It was our original intention to study the phenomenon of unilateral (right) ovarian function, first described for *L. peruanum* by Pearson (1949) and suggested for *L. boxi* by Weir (1971), but further investigations are unlikely.

TRAPPING

Mountain viscacha can be easily shot but unless killed instantaneously they will retreat into their inaccessible burrows (Pearson, 1948). Pearson believes that on open ground a man or dog could catch a viscacha but these animals rarely venture far from their protecting rocks. He also describes herding viscacha in a colony so that they entered their burrows and, afterwards, setting traps at all exits. This method could only be effective in colonies in which regular burrows were recognizable and approachable. Such a method was not possible at Pilcañu although, if a viscacha programme was to be instituted there, it might be possible to place traps on the tops of the outcrops onto which the viscacha must climb to reach the feeding areas. We caught three *L. boxi* at Pilcañu with the help of an Indian boy living on the estancia. Using two sheep-dogs, one at the top of the outcrop and one at the foot, we chased viscacha along the top until one entered a suitable chimney and was caught at the bottom. As Pearson found in Peru, it was not easy to get the animals to go into the right sort of crevice. Osgood (1914) reported that viscacha in Peru were hard to catch.

ENEMIES

Apart from man, the principal predator of viscacha is probably the fox, *Dusicyon culpaeus*. Pearson (1948) did not find any evidence of foxes capturing viscacha but they were seen prowling around the viscacha colonies. In the more accessible parts of their range viscacha would be preyed upon by any carnivores such as the puma (*Felis concolor*), grison (*Grison*), or semi-feral dogs. In Peru the Indians are known to catch viscacha for food. The fur is not used as it is not durable and has never been as valuable as chinchilla fur. In Bolivia, little use is made of viscacha for food and the Indians at Ulla Ulla seemed unaware of how to catch the animals. This probably accounts for the continued survival of mountain viscacha, since the reproductive rate is not high. No more than two litters can be produced by a female each year and only one young is born. The longevity is probably high, however, as the related chinchilla are known to live for at least 10 years and to be still breeding at 15 years (B. J. Weir, personal communication).

Acknowledgments

Grateful acknowledgment is made to Dr Barbara J. Weir for much help in many ways and I am indebted to Mr Terry Dennett for preparing the illustrations: some of these were copied from colour transparencies and a ciné-film taken in South America. The expedition was generously supported by the Wellcome Trust.

I also thank Dr P. D. L. Guilbride, F.A.O., Lima and Dr Manuel Moro for sending specimens of *L. peruanum* from Peru, Mr Donny MacIver for assistance during the expedition and Mr and Mrs Bill Roberts for hospitality at Estancia Pilcañu. The expedition was generously supported by the Wellcome Trust.

REFERENCES

Bennett, E. T. (1829). *The Gardens & Menagerie of the Zoological Society delineated:* The chinchilla. 1–12. Chiswick: Thomas Tegg.

Cabrera, A. (1957–61). Catalogo de los mamiferos de America del Sur. *Revta Inst. nac. Invest. Cienc. nat. Mus. argent. Cienc. nat. Bernardino Rivadavia* **4**: 1–724.

Cabrera, A. C. & Yepes, J. (1940). *Mamiferos Sud-Americanos.* 1st edn. Buenos Aires: Ediar.

Cabrera, A. C. & Yepes, J. (1960). *Mamiferos Sud-Americanos.* 2nd edn. Buenos Aires: Ediar.

Cieca, P. de (1533). *Chronica de Peru.* Seville.

Crespo, J. A. (1963). Dispersión del chinchillon, *Lagidium viscacia* (Molina) en el noreste de Patagonia y descripción de una nueva subespecia (Mammalia; Rodentia). *Neotropica* **9**: 61–63.

d'Azara, F. (1802). *Apuntamentos para la historia natural de los quadripedos del Paraguay y Rio de la Plata.* **2**. Madrid.

Ellerman, J. R. (1940). *The families and genera of living rodents,* **1**. London: British Museum (Naural History).

Lahille, F. (1906). El nombre cientifico de la vizcachas. *An. Soc. cient. argent.* **61**: 39–44.

Meyen, G. (1833). Das viscacha. *Nova Acta Acad. Caesar. Leop. Carol.* **16**: 575–596.

Oken, L. von (1816). *Lehrbuch der Zoologie* [*Viscaccia*] **4**: 835–836. Jena: August Schmid.

Osgood, W. H. (1914). Mammals of an expedition across northern Peru. *Publs Field Mus. nat. Hist. (Zool.)* **10**: 143–185.

Pearson, O. P. (1948). Life history of mountain viscachas in Peru. *J. Mammal.* **29**: 345–374.

Pearson, O. P. (1949). Reproduction of a South American rodent, the mountain viscacha. *Am. J. Anat.* **84**: 143–174.

Pearson, O. P. (1957). Additions to the mammalian fauna of Peru and notes on some other Peruvian mammals. *Breviora* No. 73: 1–7.

Tate, G. H. H. (1935). The taxonomy of the genera of neotropical hystricoid rodents. *Bull. Am. Mus. nat. Hist.* **68**: 295–447.

Thomas, O. (1897). On a new species of *Lagidium* from the eastern coast of Patagonia. *Ann. Mag. nat. Hist.* (6) **19**: 466–467.

Thomas, O. (1907). On a remarkable mountain viscacha from southern Patagonia with diagnoses of other members of the group. *Ann. Mag. nat. Hist.* (7) **19**: 439–444.

Thomas, O. (1921). A new mountain viscacha (*Lagidium*) from northwest Patagonia. *Ann. Mag. nat. Hist.* (9) **7**: 179–181.

Tschudi, J. J. von (1845). *Fauna peruana.* St. Gallen: Suheitlim und Zollikofer.

Weir, B. J. (1967). The care and management of laboratory hystricomorph rodents. *Lab. Anim.* **1**: 95–104.

Weir, B.J. (1970). The management and breeding of some more hystricomorph rodents. *Lab. Anim.* **4**: 83–97.

Weir, B. J. (1971). Some notes on reproduction in the Patagonian Mountain viscacha, *Lagidium boxi* (Mammalia: Rodentia). *J. Zool., Lond.* **164**: 463–467.

Weir, B. J. (1973). Another hystricomorph rodent: keeping casiragua (*Proechimys guairae*) in captivity. *Lab. Anim.* **7**: 125–134.

Weir, B. J. (1974). The tuco-tuco and plains viscacha. *Symp. zool. Soc. Lond.* No. 34: 113–130.

F

Symp. zool. Soc. Lond. (1974) No. 34, 143–160

NOTES ON THE ECOLOGY OF GUNDIS
(F. CTENODACTYLIDAE)

WILMA GEORGE

Lady Margaret Hall, Oxford, England

INTRODUCTION

The family Ctenodactylidae (Zittel, 1893) is confined to the African continent. There are four living genera and an undetermined number of species. The first gundi was reported by Rothman, a Swede, who wrote from Tripoli in 1774 to a friend that he had been brought "a small animal which the Moors called gundi and which was found in great numbers burrowing in among the rocks. It was a mouse but a new species of mouse". He noted its very small tail, truncated ears and four digits on fore and hind feet, and gave its size as that of a small guinea-pig (Rothman, 1776). Zimmerman (1778–1783) called it *Mus gundi*. In 1830, Gray gave the gundi the definitive name *Ctenodactylus* (*ktenos*, comb; *dactylos*, toe). In 1855, John Speke shot gundis among the coastal hills of Somalia and sent them to Blyth in Calcutta. Speke did not describe these gundis. From his journal (Speke, 1856) it must be assumed that they are included among "hares and rats, tailless and tailed" or "jackals and vermin". Blyth (1855) called them the "common rats" of the area. He described the long moustaches, the bushy tail and the stiff combs on the two inner toes of the hind feet and gave them the name *Pectinator spekei*. He suggested this genus should be in a family of its own "about equivalent to the Chinchillidae of South America to which, upon the whole, the Pectinatoridae would seem to be more nearly affined than to any other known form". When Blandford first saw *Pectinator*, near Massawa in Eritrea, he reported "a small animal, apparently a squirrel which had lost part of its tail by accident . . ." (Blandford, 1870). In 1881, Lieutenant Massoutier sent a gundi from Algeria to the French naturalist Lataste. It was assumed by Lataste to be a new species of *Ctenodactylus*, and he called it *C. mzabi* (Lataste, 1881). Later, however, he put it in a new genus, *Massoutiera*, mainly because of its grooved incisors (Lataste, 1885–1887). Lataste also found a gundi in Senegal which he named *Felovia vae* and likened it to *Massoutiera* because of its short bushy tail, but distinguished it from that genus by its smaller tympanic bulla (Lataste, 1885–1887). Apart from a few anatomical observations (e.g. de Lange, 1937–38; Ellerman, 1940; Saint Girons & Petter, 1965), there has been little detailed work on gundis. The following account is based on preliminary studies made during 1966–1969 on two Saharan species, *Massoutiera mzabi* and *Ctenodactylus vali* (Fig. 1), and of more detailed observations (lasting at

Fig. 1. A *Ctenodactylus vali* (Thomas, 1902) half-grown male weighing about 100 g.

least six consecutive weeks for each species) made in 1971–1972 on these two species and *Pectinator spekei* and *Felovia vae*. Examples of each species were caught by single-door, folding live traps without bait (*Pectinator* and *Felovia*), or by hand (*Ctenodactylus* and *Massoutiera*), and were sent to England. Skulls and skins were identified from specimens in the British Museum (Natural History).

FIELD STUDIES

Sites

All four genera are found in desert or semi-desert conditions in Africa, between latitudes 4°N and 34°N (Fig. 2), and at altitudes

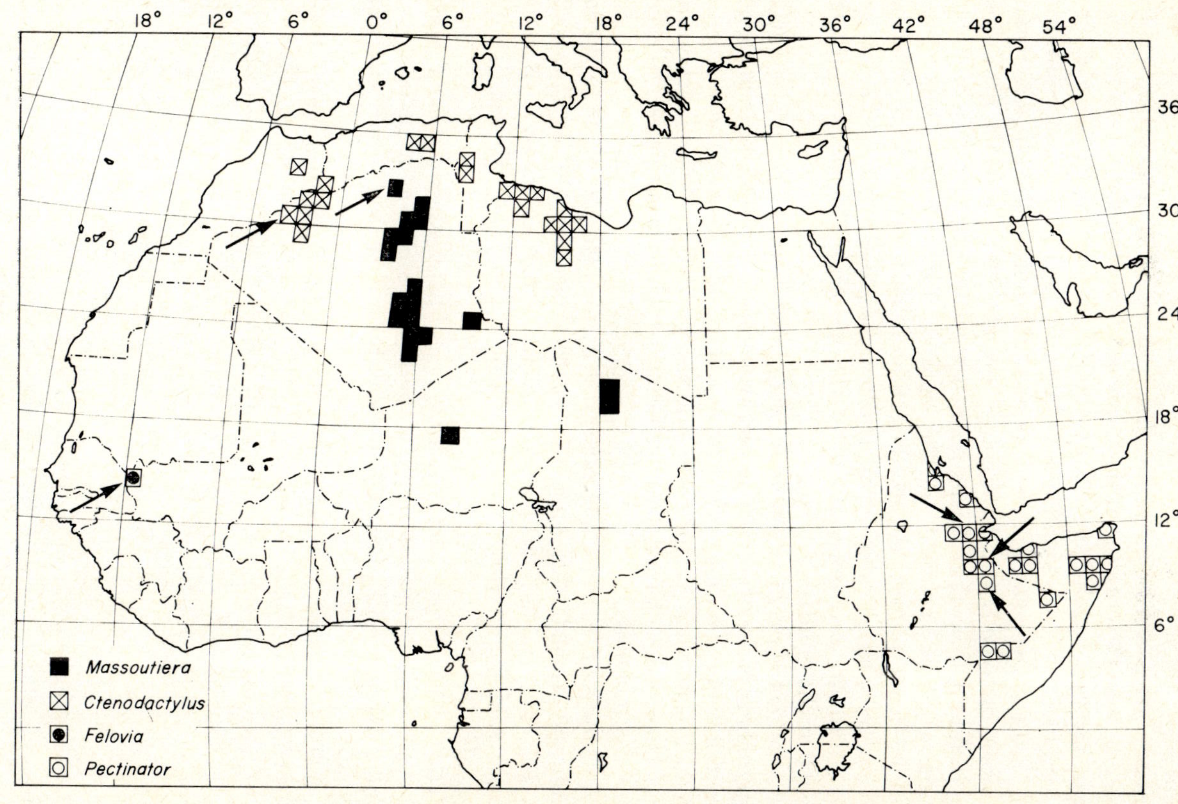

Fig. 2. Distribution of the four living genera of Ctenodactylidae (from several sources) showing study sites (arrowed).

ranging from 0 to a reported 2300 m (F. Petter, personal communication).

Study areas were chosen for their proximity to a weather station. Each site was 15 to 45 km from a source of meteorological information except for two which were about 100 km away. One of these distant sites was in the Sahara and it was known that the weather conditions were uniform over the intervening area. The other distant site was in mountainous country in Ethiopia where the climate could have been very different from that at the nearest weather station; data from this station are not included in the figures. Meteorological data were obtained from published records (Dubief, 1959, 1963; U.S. Weather Bureau, 1965–66), local airstrips and personal observations. The area of each study site was determined by the nature of the terrain and the density of the gundis.

Three colonies of *Pectinator* were observed in Ethiopia during September to November 1971 at a comparatively cool and dry time of the year, following the main rainy season (Fig. 3d). Two of these colonies were at low altitude in the desert and the third at 1 000 m in dry woodland. These sites, which had well-defined boundaries, were between 1 500 and 2 000 m².

A colony of *Felovia* was observed at an altitude of 46 m during March and April (1972) when the weather was hot and exceptionally dry (Fig. 3c). The animals covered several square kilometres with local concentrations, but three areas, each of about 1 000 m², were chosen for study.

The *Ctenodactylus* colony, at 499 m, had been studied during March and April 1966–1969 but in 1972 the site, which was a small cut of about 2 500 m², was observed during the hot season, June and July (Fig. 3b).

A *Massoutiera* site at 441 m was studied in March and April (1966–1969) when it was cold, and again in May and June (1972) at the start of the hot season (Fig. 3a). The site covered 250 000 m² because the gundis were dispersed but observation posts looked over wide areas and it was possible to follow the animals on foot without disturbing them.

Rocks

Gundis find shelter in tumbled rocks. Geologically the rocks may be of any age and any type from recent lava flows to ancient folded sandstone. Four colonies of gundis (three *Pectinator* and one *Massoutiera*) were found among the rocks of road escarpments but the ideal place is high up a shallow cut or valley.

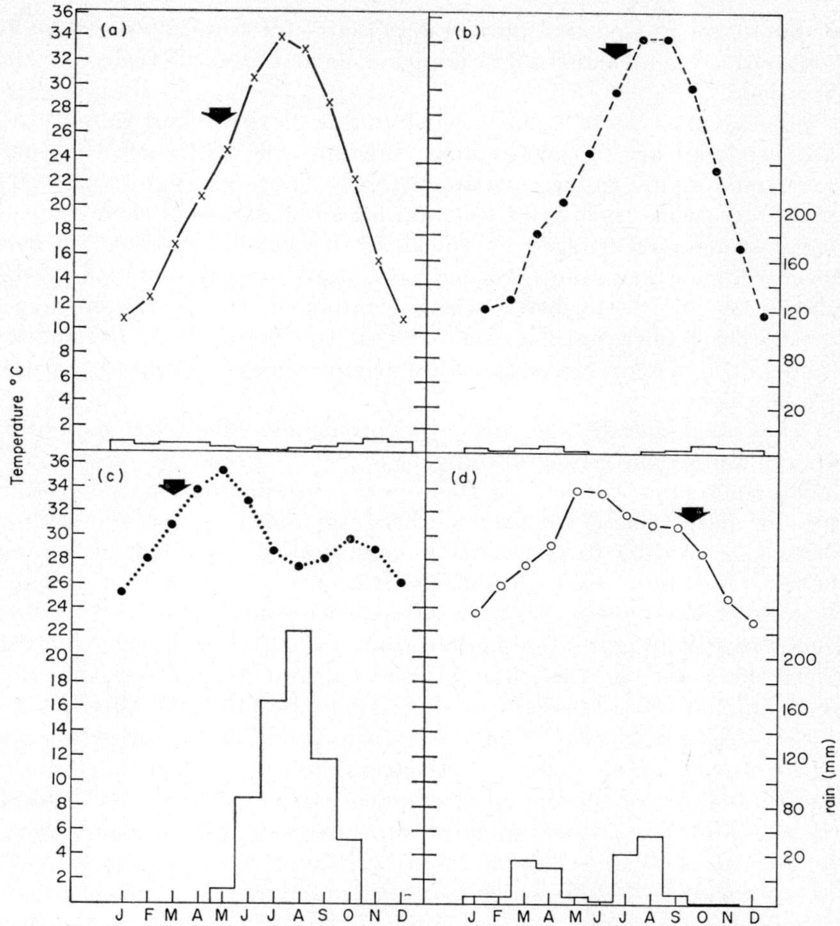

FIG. 3. Mean monthly temperature and rainfall for four ctenodactylid sites. a, *Massoutiera* (1926–1950 and 1967–1971); b, *Ctenodactylus* (1926–1950 and 1968–1971); c, *Felovia* (1941–1950 and 1967–1971); d, *Pectinator* (1964–1971). Arrows show the time of observations in 1971–1972.

The three colonies of *Pectinator* that were studied occupied three different types of rock formation. One occupied fissures in dark grey vesiculated lava, another a tumbled scree of brown lava and the third a broken fall of pink Jurassic limestone. *Felovia* was found in the long deep fissures of ancient sandstone whaleback rocks which varied from dark grey to red. *Ctenodactylus* was studied in an area of brown sandstone where recent volcanoes had come through the Ordovician and Cambrian formations. The buff-coloured *Massoutiera* colony inhabits cretaceous sandstone roughened in a pluvial period and fractured by earth movements. All colonies showed a striking colour match to the rocks in which they lived.

Gundis have never been found living in burrows and they do not excavate, contrary to some reports (Burton, 1962), although they may move small stones to improve an entrance. They can sweep away sand and pebbles with their front feet and lift small stones in their mouths. Bigger stones are dragged by the incisors. On one occasion, a *Ctenodactylus* female weighing 180 g moved a stone weighing 200 g for a distance of 30 cm. She pushed it with her nose and front feet and dragged it with her mouth until it was clear of the entrance to her shelter. Gundis do not make nests and they occupy shelters with either sandy or rocky floors.

The ideal gundi site provides permanent shelters, temporary shelters and ledges for sunbathing.

On some sites, permanent shelters are occupied all the year round and for many years. On others, there are many temporary shelters occupied according to the climatic condition. *Felovia* has permanent shelters, *Pectinator* a few alternatives, *Ctenodactylus* is more widely dispersed and *Massoutiera* is at the extreme with many temporary shelters. There are advantages in having alternative shelters and ledges in areas where there are big variations in seasonal and diurnal temperatures. New gundi sites have been made during road building and three groups of *Pectinator* are known to have moved into road sites during the last 35 years. Conversely, two *Ctenodactylus* sites have been abandoned recently because of the spread of human settlements with their associated dogs and cats. However, most sites are occupied for many years, and *Felovia* is still on the site where Lataste found it in 1885.

A permanent shelter, which may accommodate three or four gundis, has a floor space of about 25 cm^2 and a height of at least 15 cm to take a 'heap' of gundis. Sometimes they inhabit crevices or ledges within bigger shelters but this has the disadvantage of allowing access to cats and jackals. The entrance to such a shelter is, therefore, usually a narrow fissure.

Temporary shelters can be shallow crevices. They are necessary to provide escape from hawks and jackals. Many good temporary shelters extend the foraging range significantly because they provide shelter from the sun. Gundis flatten into narrow horizontal or vertical crevices when threatened, restricted only by the height of the skull because the thorax is flat and compressible. The Saharan gundis can squeeze into a space of 25 mm, and *Pectinator* and *Felovia*, even less. In such a crevice, with its back pressed against the side, a gundi can work its way up or down or sideways until it finds a safe way out or the danger has passed.

Sunbathing ledges may be the tops of big boulders, the side of a folded whaleback rock or, more typically, a ledge with a boulder behind and a shelter under it.

Where there are permanent shelters and sunbathing ledges there are distinct runways and communal defaecating places. In the more dynamic conditions of the many temporary shelters of *Massoutiera* there are runways in the sense that all the gundis in the area tend to follow much the same tracks, inspect the same rocks, and forage in the same places, but they do not necessarily use the same sunbathing ledges or shelters on consecutive days or have defaecating places.

Not all screes, tumbled rocks and fissured rocks are suitable as gundi sites. Apart from the size of the rocks, not too big but with deep cracks, the aspect has considerable importance. An escarpment or the side of a shallow valley inhabited by gundis often has an easterly aspect, exposed to the morning sun. Ideally, part of the site catches the evening sun as well and this is why gundis are often found at the head of low valleys and not on the sides of deep ones. The morning sun shines on the ledges near the permanent shelters, the evening sun on ledges and boulders not too far away. Wind is not as important as sun though it is doubtful whether a gundi colony would be found on an escarpment onto which a strong wind blew constantly. None of the sites studied faced directly into the prevailing winds; they were usually at an angle of about 45° to them. On hot days, ledges are chosen for their exposure to breezes. Depending on the combined conditions of temperature, sun and wind, a gundi may be found sunbathing in a sheltered suntrap or exposed on a high boulder with the wind ruffling its fur.

Climate

In the areas where gundis are found (Fig. 2) the mean annual temperature is high, ranging from 21°C to about 30°C (Fig. 3). This figure hides considerable differences in the characteristics of the areas. For instance, in Ethiopia and Somalia and in West Africa, where the mean annual temperature is about 30°C, the absolute minimum is never below 10°C and the mean minimum above 20°C, while the maximum reaches 46°C and the average maximum 36°C. But, in the Sahara, the temperature fluctuates from an absolute minimum as low as freezing point to a maximum of 50°C. The average minimum, however, is around 15°C and the average maximum 29°C. These temperature ranges influence the choice of sites. The southern gundis live in rocks that remain at an even temperature throughout the year. *Massoutiera* and *Ctenodactylus*, in contrast, take advantage of the sun's heat to warm their shelters and ledges in the colder seasons of the year and the cooling effect of the wind to air-condition their shelters and ledges in the summer heat.

At all sites the relative humidity is low. The mean annual value ranges from 30% at the *Ctenodactylus* site to 53% at one of the *Pectinator* sites.

Rainfall is more variable. In most areas where ctenodactylids are found the rainfall reaches a maximum of 220 mm a year, and is usually much less. At the *Ctenodactylus* site it averaged 40 mm a year. But, in some areas occupied by *Pectinator*, the average rainfall may be as high as 600 mm a year, and *Felovia* lives in a region where the annual rainfall is almost always above this figure.

The most constant climatic feature of gundi sites is the high number of hours of sunshine, $8\frac{1}{2}$ to $9\frac{1}{2}$ hours a day throughout the year (compared with 4 hours at Kew).

Vegetation

According to the values of the climatic parameters, gundis live in three of the life zones defined by Holdridge (1967). *Massoutiera* and *Ctenodactylus* are in warm-temperate and sub-tropical deserts, *Pectinator* in sub-tropical desert scrub and *Felovia* on the edge of Holdridge's "very dry tropical forest". Apart from the allocation of *Felovia* to forest, the life-zone classification holds reasonably well for gundis.

To define the plant characteristics of the gundi sites in more detail, measurements were made of the sites in accordance with Dansereau's (1957) definitions of plant formations. A grid was constructed in units of a square metre. Ten units were selected at random. In each unit the percentage area of plant coverage and the average height of the plants was measured. Each plant was identified (Ozenda, 1958; Dale & Greenway, 1961; Quézel, 1965; Hemming, 1966; Berhaut, 1967; J. Gillet, personal communication).

In 1972, the *Massoutiera* site had had 68 mm of rain in March and April which is five times the average for the two months. Therefore, when the site was studied from the end of April to the end of May there was an exceptionally high density of plants in striking contrast to the density that had been estimated on previous visits. Even so, the plant coverage was only 10% in area and the average height 15 cm. This puts the area into Dansereau's desert formation category, with Sahara–Mediterranean characteristics.

The *Ctenodactylus* site was also a desert formation. The main *Ctenodactylus* site is at the limits of the *Ctenodactylus vali* range. In June when the temperature is going up towards its maximum but soon after one of the 'wetter' periods of the year (about 10 mm during the preceding three months) the plant coverage was estimated at 2%. The plants were mainly small xerophytes averaging 20 cm in height but there were a few *Acacia raddiana* out in the broad valley. *Ctenodactylus* does not normally leave the rocks for the open sandy areas where the acacia trees grow.

The *Felovia* site was studied at one of the hotter and drier seasons of the year, five months after the last rain. The area had had exceptionally low rainfall since 1967. The plant coverage was about 9% over most of the site but the plants were taller than in the Saharan areas. *Tephrosia mossiensis* up to 300 cm, several species of tall grass and the pachy-caulus tree *Adenium obesum* predominated. Such a formation comes somewhere between Dansereau's scrub and desert categories. At the edge of the site where the flat rocks fell away there were two species of fig tree and if the leaves of this canopy are taken into consideration, the figure for plant coverage in this limited area is increased to 25%: to scrub category.

Pectinator sites were studied shortly after the rainy season and before the temperature had reached its maximum. It was a time of comparatively high density plant coverage. The sites varied from dry woodland areas to desert with only a 6–8% plant coverage. The desert sites were typically acacia-commiphora-grass associations. There were two species of acacia, *A. senegal* and *A. seyal*, and several species of grass.

Scattered clumps of grass are common to all sites at all times of the year.

Activity

All members of the Ctenodactylidae are diurnal. During a 24-h watch at one of the *Pectinator* sites, no activity was seen outside the rocks while it was dark although there was moonlight. Four hours after dark there was a brief 'whistling' of gundis from the rocks, although nothing could be seen, and then nothing was heard until just before they emerged as the sun struck the site. The diurnal activity was confirmed by trapping data. Traps were laid at night and inspected at dawn. No gundi was ever found in a trap at dawn. During the day, traps were inspected at four-hour intervals, starting one hour later each day. Most gundis were caught four hours after dawn, nothing except the occasional lizard was caught during the hottest time of the day, and only a few gundis were caught during the evening before dark.

The activity of the gundis was determined by continuous watching from dawn to dusk. Temperature and humidity were recorded every half-hour. The activity (Figs 4 and 5) is expressed as the number of sightings made at a particular hour of the day or when a particular temperature was reached during a four-day period. The four-day

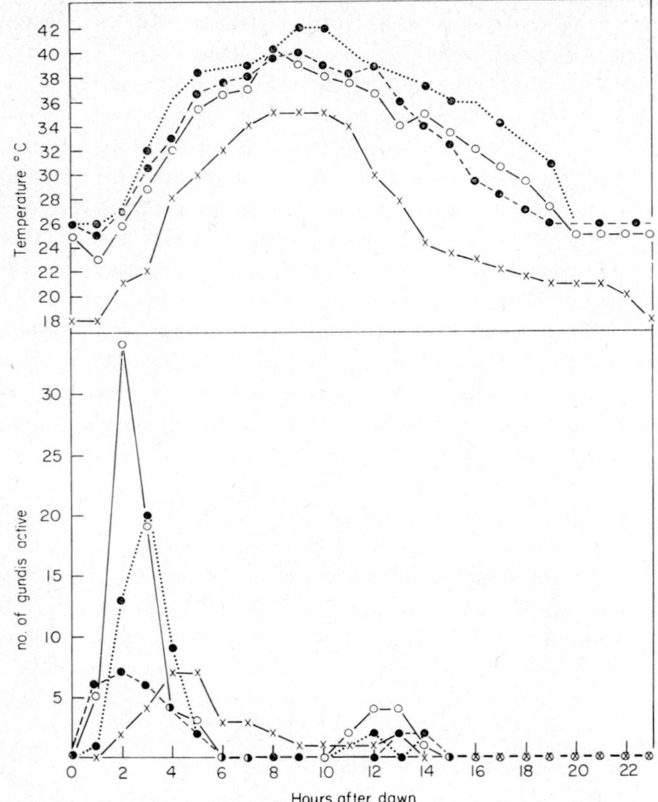

Fig. 4. Hourly temperatures (above) and activity (below) at four gundi sites, Activity recorded as the number of sightings for a four-day period. ×————×, *Massoutiera* 11–14 May 1972; ●– – – –●, *Ctenodactylus* 2–5 June 1972; ○————○, *Pectinator* 9-12 October 1971; ●············●, *Felovia* 18–22 March 1972.

periods were chosen because the general temperature pattern tended to remain stable over this length of time. The number of sightings is not expressed as a percentage of the total number of animals because the exact number of animals in the colony was not always known. The pattern of the lines is more important than the values of the points because the number of the sightings depended on the position of the observer and, to a greater extent, on the density of the gundi population. The height of the curves, therefore, gives a rough indication of density at each site.

The daily activity is dependent upon light and the ambient temperature. When the temperature is high, gundis are out at first light but they did not come out at all when the temperature was low. Unless it was cold, windy or wet, gundis emerged as soon as the first rays of the sun reached the site and they were active for the next five hours. Activity then fell off for the midday period but increased again during the two to four hours before dusk. *Massoutiera* reached its peak activity later than the others and continued, although at a reduced level, throughout the day (Fig. 4).

Although activity was closely correlated with temperature, it was modified by other environmental factors. For example, at a dawn temperature of 25°C with a high wind and flying sand, no *Ctenodactylus* emerged until evening; in March, several *Ctenodactylus* were out in a sheltered place in the sun at a dawn temperature of 10°C.

Gundi activity can be divided into three categories (Fig. 5); sunbathing, foraging and general activity (playing, chasing, exploring).

Sunbathing

In general, sunbathing does not start until the temperature is over 21°C. Therefore, although it is the first activity of *Pectinator* in September and October in Ethiopia, it only becomes a serious occupation of *Massoutiera* and *Ctenodactylus* in May and June, between or after bouts of foraging, and it is an intermittent activity for *Felovia* in March. However, there is some uniformity in the temperatures at which the four genera sunbathe. Peak sunbathing occurs at temperatures between 22°C and 29°C when rock surfaces are at about these temperatures (Fig. 6). At 31°C, when the rock temperature increases more quickly than that of the air, sunbathing stops abruptly. There is a renewal at 35°C to 38°C which represents the evening 'sunbathing', lying on a hot rock in the open rather than in direct sunlight. *Ctenodactylus* has been found 'sunbathing' on a rock as hot as 39°C and *Felovia* in the direct sun on a rock whose temperature was 38°C.

Young gundis also sunbathe within a few hours of birth. The young are born fully furred and with their eyes open after a gestation period of about 55 days. A *Massoutiera* female with two young was filmed on a ledge with a shelter behind when the young were about a week old. They shot into the shelter at a sudden movement of a plant or a shadow over the sun. If they strayed the mother retrieved them in her mouth, holding them across the body, not by the scruff of the neck as reported by Packard (1967).

Foraging

Foraging occurs over a wide range of temperature but, again, shows a drop when the temperature reaches 33°C and a slight renewal in the evening.

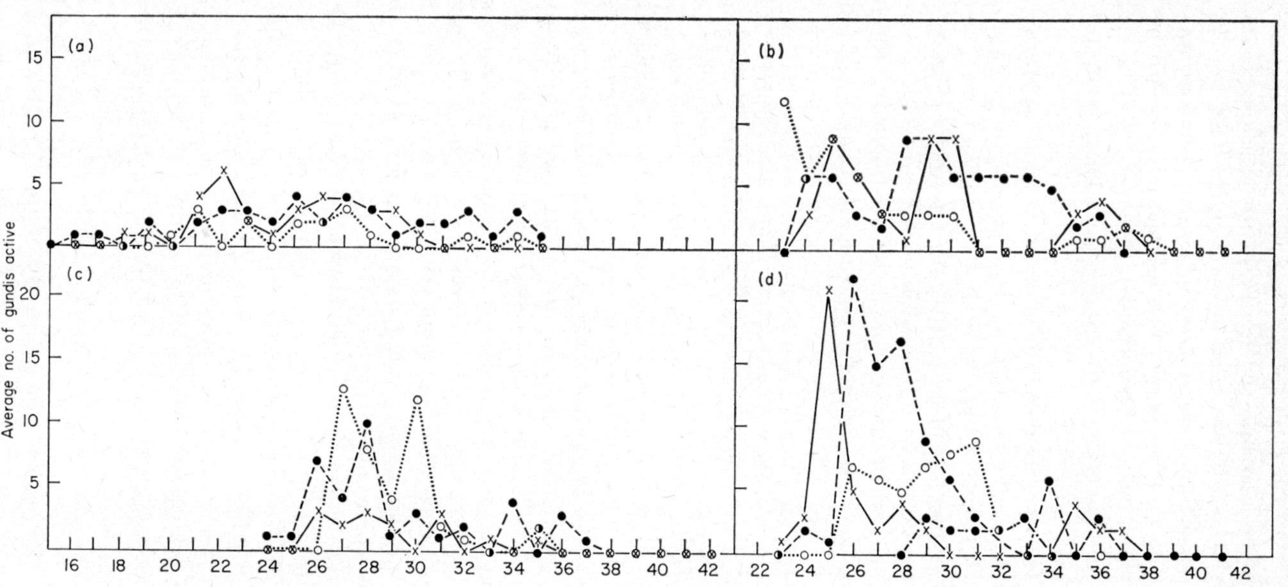

FIG. 5. Activity related to temperature; a, *Massoutiera*; b, *Ctenodactylus*; c, *Felovia*; d, *Pectinator*. ×————×, sunbathing; ○············○, foraging on ground; ●– – – –● general activity; ●–·–·–●, foraging in tree.

Gundis eat the leaves, flowers, seeds and stalks of plants. They have never been seen to eat any animal, either dead or alive, and the contents of the stomachs have always been of vegetable matter only.

Felovia and *Pectinator* experience seasons of comparative food abundance, followed by months of scarcity. *Massoutiera* and *Ctenodactylus* experience comparative abundance for a few weeks every few years and the rest of the time very low density vegetation. Gundis do not store food or lay down localized fat stores like the fat-tailed gerbil, *Pachyuromys duprasi* (Schmidt-Nielsen, 1964). As far as is known they do not aestivate or hibernate. They must, therefore, forage at all times of the year. They are economical in their eating. They eat to the end of a stalk or leaf and pick up and continue to eat a dropped leaf. *Ctenodactylus* frequently holds a leaf in its forefeet, *Pectinator* rarely, and the others never.

Gundis eat a range of plants but have preferences when there is choice. Both *Massoutiera* and *Ctenodactylus* seem to have a preference for Cruciferae, Compositae and Graminae in that order. *Massoutiera* feeds extensively on the cruciferous purple-flowered *Moricandia arvensis*, eating leaves, flowers and stalks. It nibbles at *Reseda* leaves, bites the heads of the yellow composite *Launea* and eats the whole of the grass *Stipa retorta* except for the roots. In May, 1972, after rain, all these and more were available to *Massoutiera*. In April 1969, there was hardly anything at the same site except *Stipa*, a few *Ferula* and the zygophyllous *Peganum harmala*. *Peganum* and *Ferula* are usually ignored by camels and goats, but in 1969 *Peganum* leaves were bitten off by *Massoutiera* and taken to shelters where there were young.

On the site studied, *Ctenodactylus* ate, by preference, the crucifer *Eremophyton chevallieri* and the composite *Amberboa (Centaurea) leucantha* followed by the grasses *Cymbopogon* and *Aristida*.

In the dry March of 1972 *Felovia* was eating the leaves of the leguminous *Tephrosia mossiensis* shrubs, the dropped petioles of fig trees, dry grass and seeds. On the ground *Pectinator* ate dry grass stalks and seeds and reached up for the leathery green leaves of the *Cadaba rotundifolia* bushes. When *Cadaba* bushes are prominent in a landscape it is usually a sign of overgrazing by camels and goats who ignore it. As the temperature approached 29°C, about three hours after dawn, the gundis took to the acacia trees that grew close to their rocky escarpment. Here they spent the next hour and a half in and out of the trees eating acacia leaves. They also spent some of the hottest times of the day sheltering in the shade of an acacia. *Pectinator* ate leaves from *Acacia senegal* when they were available. *A. senegal* has comparatively smooth twigs with spines only 6 mm long. But when these leaves were not available, the gundis negotiated the 65 mm sharp spines of *A. seyal*.

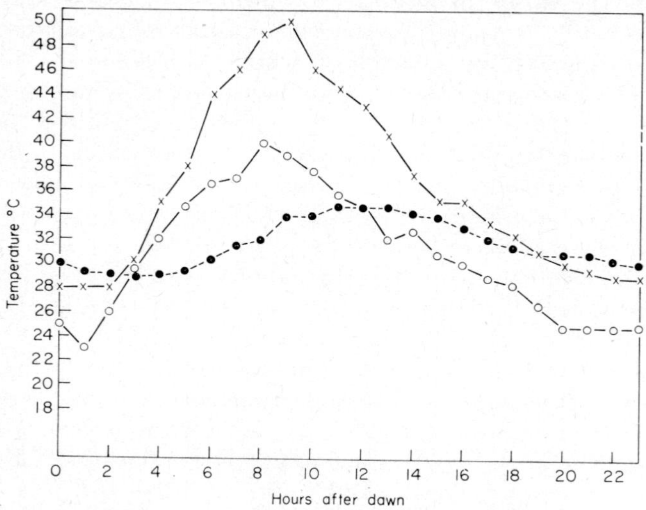

FIG. 6. Air temperature (○————○), rock surface temperature (×————×) and temperature 50 cm inside a *Pectinator* shelter (●– – – – –●), 9 October 1971.

General activity

General activity is defined here as any running about on the rocks that is not associated directly with foraging. It is an activity that is often broken by periods of cooling off in the shade. The peak of general activity for all gundis is in the temperature range 25°C to 30°C. Throughout this range gundis are out and about. After this, the activity drops to a lower level, consisting of bouts of running and longer periods of cooling off. By 35°C there is very little outside activity until the evening when it increases during the last hours of daylight, often starting at 36°C. At temperatures above 36°C and, probably, below 10°C gundis are rarely active. At the high temperatures they are mainly lying flat on cool rocks keeping to an ambient temperature below body temperature.

An observation of an individual *Massoutiera* indicated the proportion of time spent on each activity. On a day when the temperature rose from 18°C at dawn to 34·5°C by midday, the gundi spent 3 h 20 min dashing about, 60 min sunbathing, 30 min foraging and 3 h 30 min sitting in the shade. For the rest of the time it was out of sight. It was estimated that it covered about a kilometre.

The nature of the rocks in which gundis live provides shelter from the extremes of temperature outside. For example, a *Felovia* shelter

50 cm from the surface but with a wide entrance had a temperature in March varying from 28°C to 34°C during a 24-hour period when the outside temperature ranged from 25°C to 41°C. In the Sahara in March a shelter varied from 15°C to 20°C while the outside temperature changed from 9°C to 27°C over 24 hours. These temperature measurements are not meaningful at the low end of the range because desert gundis piled up together in winter in a fissure can keep their shelter at 20°C when the outside temperature drops to freezing point. The variation of shelter and ledge with outside temperatures is shown for *Pectinator* in Fig. 6.

Co-existence

In all areas, gundis are preyed on by birds, snakes and carnivorous mammals.

Tape-recordings were made of warning 'whistles' from *Pectinator* when a gabar goshawk (*Micronisus gabar*) made its daily approach to the site. The 'whistles' were followed by the rapid disappearance of all *Pectinator* into the nearest escape shelter. During filming, a wild cat jumped from a clump of grass onto a sunbathing *Felovia* but missed. Analysis of faeces from the *Ctenodactylus* site revealed the remains of a pregnant gundi in some jackal faeces.

The commonest herbivores that live with gundis are cricetids and murids. *Arvicanthus niloticus* occupied rock crevices on the *Felovia* site. *Gerbillus campestris* and *G. pyramidum* were found living in the rocks alongside *Massoutiera* and *Ctenodactylus*. These species were caught at night in traps set outside gundi shelters. Ground squirrels, *Xerus rutilus* in Ethiopia and *Euxerus erythropus* in West Africa, were common on *Pectinator* and *Felovia* sites, but they favoured the sandier and bushier areas and lived mainly on fruits, roots and bulbs. Gazelle of several species and hares are also common. They eat many of the same plants as the gundis but they have a wider foraging area. One species that might seem to be in direct competition with gundis for shelters and food is the rock hyrax *Procavia capensis* that occupied part of the *Felovia* site. At the edge of the site there are fig trees and hyrax and gundis co-exist. As the site becomes more barren and without trees the hyrax are no longer found. Where the two overlap they use the same rocks for sunbathing, the same defaecating places and feed together on the same grasses. However, they probably do not interfere seriously with one another's choice of shelter. Hyrax live in the big fissures and gundis in the little fissures. Hyrax feed in the fig trees but *Felovia* was never seen to climb a tree. They have lived together on this site for at least 86 years. One *Pectinator* site also had a small area of overlap with hyrax.

More directly competitive are the big agamid lizards *Uromastix acanthinurus* that are found on the *Massoutiera* and *Ctenodactylus* sites. Like gundis, dobs (the local Arab name) are diurnal, vegetarian and live in rock shelters. Unlike gundis, however, dobs can excavate under rocks and, therefore, have a greater range of shelters than the gundis. In the morning, dobs take time to warm up. They come out later than gundis and feed later and are still in the sun when the gundis have gone into the shade. But dobs exploit the same food supply (Dubuis *et al.*, 1971) and, since they are living in areas where food is scarce, would seem to be in competition with the gundis in this sense (Elton & Miller, 1954; Miller, 1967).

SUMMARY

Grinnell (1917) defined the ecological niche of an animal as the place where it lives. Long hours of sunshine, low humidity for most of the year, a relatively high average and high maximum temperature combined with rocky outcrops and low density vegetation describe where the ctenodactylids live. But, in 1927, Elton defined ecological niche as the animal's profession; what it does all day and, in particular, what it eats. Gundis are diurnal and their activity is restricted to the cooler times of the day and to shady places. They are adapted to the hot conditions by their behaviour which is that of a heat evader (Schmidt-Nielsen, 1964). When they get hot, they cool off by lying flat on a cold rock. They obtain water from plants most of which are eaten in the early hours of the morning when the relative humidity is comparatively high.

From an environmental point of view the four living genera can be grouped into two pairs. *Massoutiera* and *Ctenodactylus* live where there is little rain and the temperature fluctuates steeply both seasonally and daily and the vegetation is sparse. *Felovia* and *Pectinator* live in deserts where the temperature is fairly constant but where rainy seasons of comparative food abundance contrast with seasons of food scarcity.

ACKNOWLEDGMENTS

I thank George Crowther for his help with the field studies; the Directors of the Centre de Recherches sur les Zones Arides, Beni Abbès, Algeria, for their hospitality; Mr Mailak Abdelkadar for help in the Mzab; Dr Tessema Megenasa and Mr H. McKilligan for assistance and advice in Ethiopia; Mr Mainadou Soumare for help in Mali; Drs I. W.

Rowlands and B. J. Weir for looking after the gundis in England; and all the other people, too numerous to thank individually, who helped with their time and advice.

REFERENCES

Berhaut, J. (1967). *Flore du Sénégal*. 2nd ed. Dakar: Editions Clairafrique.

Blandford, W. T. (1870). *Observations on the geology and zoology of Abyssinia*. London: Macmillan.

Blyth, E. (1855). Report on a zoological collection from the Somali country. *J. Asiatic Soc. Beng.* **24**: 291 306.

Burton, M. (1962). *Systematic dictionary of mammals of the world*. London: Museum Press.

Dale, I. R. & Greenway, P. J. (1961). *Kenya trees and shrubs*. Nairobi: Buchanan's Kenya Estates Limited.

Dansereau, P. (1957). *Biogeography; an ecological perspective*. New York: Ronald Press.

de Lange, D. (1937–1938) Some remarks on the early development of *Ctenodactylus vali* Pall. *Archs néerl. zool.* **3**: Suppl. 131–147.

Dubief, J. (1959). *Le climat du Sahara* **1** Algiers. *Mém. Inst. rech. sahara Univ. Alger.*

Dubief, J. (1963). *Le climat du Sahara* **2** Algiers. *Mém. Inst. rech. sahara Univ. Alger.*

Dubuis, A., Faurel, I., Grenot, C. & Vernet, R. (1971). Sur le régime alimentaire du lézard saharien *Uromastix acanthinurus* Bell. *C. r. hebd. Séanc. Acad. Sci., Paris* **273**: 500–503.

Ellerman, J. (1940). *The families and genera of living rodents* **I**. London: British Museum (Natural History).

Elton, C. S. (1927). *Animal ecology*. London: Sidgwick and Jackson.

Elton, C. S. & Miller, R. S. (1954). The ecological survey of animal communities with a practical system of classifying habitats by structural characters. *J. Ecol.* **42**: 46–496.

Gray, J. E. (1830) *Spicilegia Zoologica* **I**. London: Treuttel, Wurtz & Co.

Grinnell, J. (1917). Field tests of theories concerning distributional control. *Am. Nat.* **51**: 115–128.

Hemming, C. F. (1966). The vegetation of the northern region of the Somali Republic. *Proc. Linn. Soc. Lond.* **177**: 173–250.

Holdridge, L. R. (1967). *Life zone ecology*. San José, Costa Rica: Tropical Research Centre.

Lataste, F. (1881). Sur un rongeur nouveau du Sahara algérien (*Ctenodactylus mzabi* n. sp.). *Bull. Soc. zool. Fr.* **6**: 214–225.

Lataste, F. (1885–1887). Sur le système dentaire du genre *Ctenodactylus*. *Naturaliste* **3**: 21; 287.

Miller, R. S. (1967). Pattern and process in competition. *Adv. Ecol. Res.* **4**: 1–74.

Ozenda, P. (1958). *Flore du Sahara septentrional et central*. Algiers: Centre National de la Recherche Scientifique.

Packard, R. L. (1967). Octodontoid, bathyergoid and ctenodactyloid rodents. In *Recent mammals of the world*; 273–290. S. Anderson & J. K. Jones (eds). New York: Ronald Press.

Quézel, P. (1965). *La végétation du Sahara*. Stuttgart: Gustav Fischer.

Rothman, D. (1776). Reise nach Garean im Gebiet von Tripoli. Im November und Dezember 1774. Ein Schrieben an den Ritter Wargentin in Stockholm. In *Briefwechsel*, Part 1, Book 339. Schlozer, A. L. (Compiler) Göttingen.

Saint Girons, M.-C. & Petter, F. (1965). Les rongeurs du Maroc. *Trav. Inst. scient. chérif.* (zool.) No. 31: 1–55.

Schmidt-Nielsen, K. (1964). *Desert animals.* Oxford: Clarendon Press.

Speke, J. (1856). Diary and observations made by Lieutenant Speke when attempting to reach the Wadi Nogal. In *First footsteps in East Africa or an Exploration of Harar*: 461–507. Burton, R. F. (main author). London: Longman, Brown, Green & Longman.

Thomas, O. (1902). On the mammals collected during the Whitaker expedition to Tripoli. *Proc. zool. Soc. Lond.* **1902**(2): 2–13.

U.S. Weather Bureau (1965–1966). *World Weather Records* 1951–1960. Washington D.C.: U.S. Weather Bureau.

Zimmermann, E. A. W. von (1778–1783). *Geographische Geschichte des Menschen,* 3 Vols., Leipzig: Weygandschen Buchandlung.

Zittel, K. A. von (1893). *Handbuch der Paleontologie Abt.* 1 *Palaeozoologie. Band* 4. *Vertebrata (Mammalia).* Munich: Oldenbourg.

Symp. zool. Soc. Lond. (1974) No. 34,161–170

THE GRASSCUTTER,
THRYONOMYS SWINDERIANUS TEMMINCK,
IN GHANA

E. O. A. ASIBEY

Department of Game and Wildlife,
P.O. Box M 239, Accra, Ghana

SYNOPSIS

The grasscutter is widely distributed throughout Ghana. Primarily, it is a savannah species but it has invaded, very successfully, the forest zone. When the savannah country becomes burnt during the dry season, the grasscutter takes shelter and feeds in reed beds.

The importance of this species as an agricultural pest and also as a source of food and income has led to it being captured in large numbers, by methods which are described. Shooting and snaring result in the capture of a very high proportion of males, and together are probably sufficient to control the population within reasonable limits. Another method, consisting of communal hunting, is an important social event for the hunters, and accounts for the capture of a much higher proportion of females. This can cause temporary local extermination, as it also involves destruction of habitat.

The food value of grasscutter meat has been shown to be high, and the meat itself is of high quality and more expensive than that of other domesticated animals. The economic value of the grasscutter to the nation as a whole requires investigation.

INTRODUCTION

Rosevear (1969), in dealing with the occurrence of hystricomorph rodents in West Africa, reported the existence of two genera of Thryonomidae—*Thryonomys* and *Choeromys*—of which only the former is present in Ghana. *Thryonomys swinderianus* (Fig. 1) has been found throughout West Africa where it is commonly called "cutting grass" or the "grasscutter"; the more formal name of "cane rat" is rarely used in West Africa. Among the West African rodents, the grasscutter is second in size only to the African porcupines of the genus *Hystrix*. It has coarse brown hair flecked with black; the under parts are white, and in the adult male the peri-genital area is stained a yellowish-brown. It has very large incisors which are grooved and yellow in colour.

The grasscutter is well-known throughout Ghana as an agricultural pest of cereals and other crops. Its flesh is a delicacy and is in demand by all levels of society in that country. Although it is extensively hunted, its market value remains very high. Both these attributes contribute to the control of its numbers which, if unchecked, would constitute a

Fig. 1. (a) A pair of grasscutters, *Thryonomys swinderianus*, in Ghana; the male is in front of the female. $\times\frac{1}{9}$.

(b) The natural habitat of the grasscutter in Northern Ghana (Mole National Park). Photographs, taken in March, 1973, of a swampy area which has been burnt during the height of the dry season.

(c) The acquired habitat of the grasscutter in the forest zone of Southern Ghana (Swedru). A photograph taken in August, 1973, of a new maize and cassava mixed farm frequently raided by grasscutters from the adjoining disused farms, where they normally lie under the thicket during the day.

serious threat to the economy of Ghana, and that of other parts of West Africa (Asibey, 1969). It is surprising, therefore, that so little is known about the biology and breeding habits of this species (Ewer, 1969). Rosevear (1969) found that even the material in the British Museum was too limited to determine whether or not there are valid races of grasscutter, and the records of their external measurements are even fewer.

The two papers on this species presented in this Symposium are intended to fill in some of the gaps in our knowledge of this African hystricomorph rodent.

GEOGRAPHICAL DISTRIBUTION

According to Booth (1959), grasscutters are abundant in most of the savannah country in Ghana—along the coast and in the cultivated and uncultivated regions of the hinterland. Throughout its range, it is commonest in dense grass, especially in thick cane-like or reedy grass such as elephant grass, *Pennisetum purpureum* (Rosevear, 1969), but it has penetrated the high forest zone including the south-west of Ghana. In the Mole National Park grasscutters are found frequently near the rivers but have never been recorded from the riverine forests of that area, as would be expected if the forests were the original habitat of this species.

In the forest zone, the grasscutter is most abundant in areas of intensive maize (*Zea mays*), cassava (*Manihot utilissima*) and sugar cane (*Saccharum* spp.) cultivations. Grasscutters also occur in areas of young cocoa (*Theobroma cacao*), coconut (*Cocos nucifera*) and oil palm (*Elaeis guineansis*) as well as in pineapple (*Ananas comosus*) and egg plant (*Solanum* spp.) cultivations. More often than not these plants are grown as a mixed crop. The farms, therefore, offer the grasscutter food all the year round, and it has settled successfully in all the major vegetational and climatic zones of the forest belt. Grasscutters are not found in undisturbed forest lands but they are known to eat *Pennisetum* spp., the dominant grass occurring in the forest zone. They should, therefore, be able to survive where there are reasonable stands of this species as well as in cultivated areas. *Pennisetum* is associated with *Panicum maximum* and both these grasses occur where the original forest has been disturbed, thus providing further evidence that the forest zone is an acquired habitat for the grasscutter.

Figure 2 indicates the natural and acquired habitats of the grasscutter. The information on its occurrence in areas where a specimen has not been collected is based on personal interviews with local residents,

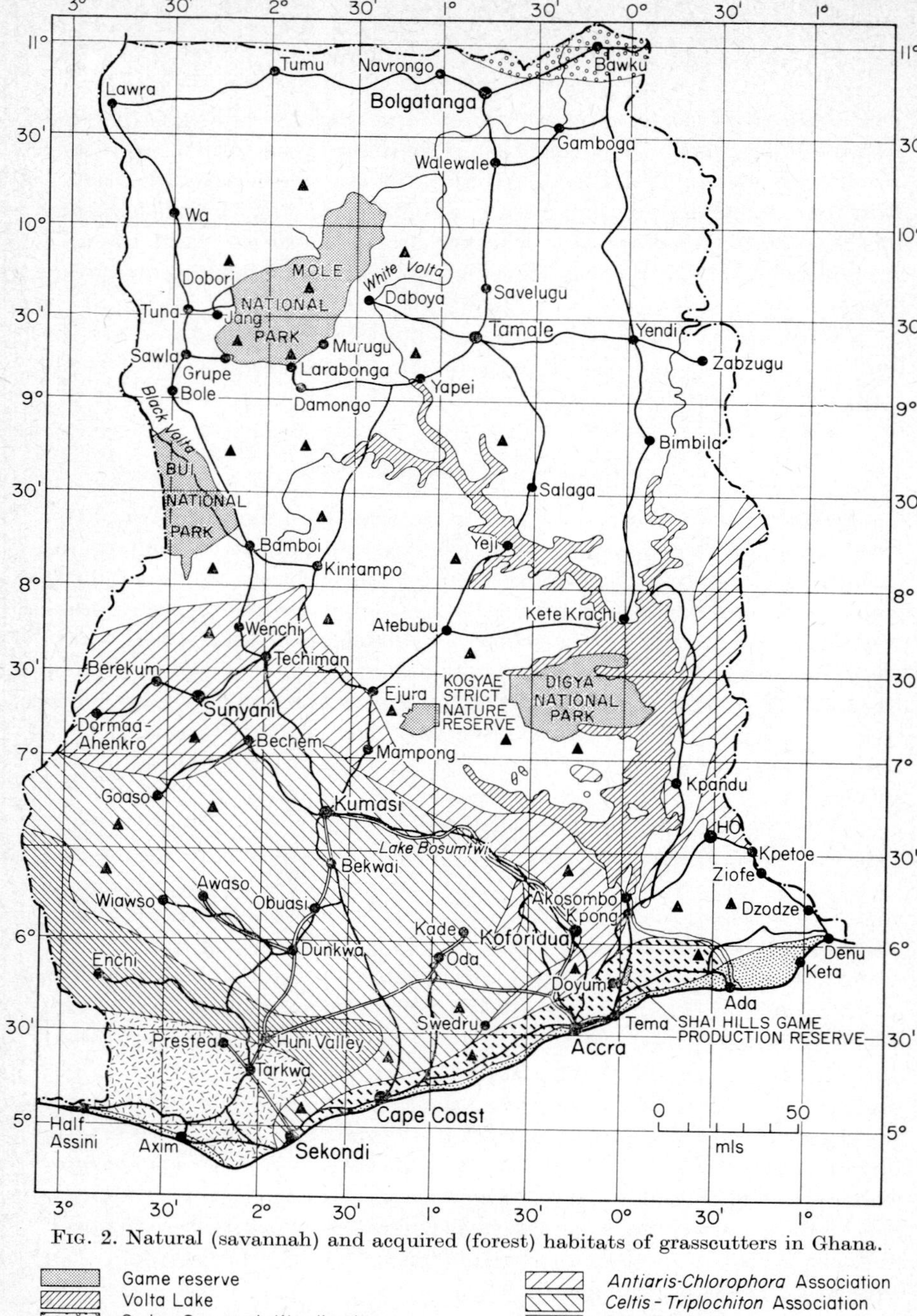

FIG. 2. Natural (savannah) and acquired (forest) habitats of grasscutters in Ghana.

traders in bushmeat, foresters, surveyors and reports of the staff of the Department of Game and Wildlife as well as on my personal field work. Detailed work on actual location by collection is required to indicate the distribution of the grasscutter in Ghana since it is feared to be locally exterminated in some of its original habitats in Tamale, Bolgatanga, Bawku and Lawra districts. In Ghana as a whole, however, the animal is not an endangered species.

LIVE AND DRESSED BODY WEIGHTS IN RELATION TO SEX AND AGE

The grasscutter matures at about five months of age when three cheek teeth have erupted. In captive stock, the heaviest animals recorded at this age were 1·0 kg (male) and 0·86 kg (female). Adult males tend to be heavier than females but no significant differences have been established in captive or wild stock throughout the country. Variation in mean dressed weight in relation to age, sex and habitat is given in Table I and the data suggest that grasscutters from northern regions

TABLE I

Mean and standard deviation of dressed weight (kg) of carcases according to age and sex of grasscutters from southern and northern Ghana

Age(months)	Southern Ghana		Northern Ghana	
	Male	Female	Male	Female
2–4 (2 cheek teeth)	0·45 ± 0·14 (4)	0·48 ± 0·17 (8)	0·63 ± 0·20 (7)	0·66 ± 0·44 (8)
5–8 (3 cheek teeth)	1·27 ± 0·63 (29)	1·29 ± 0·73 (27)	1·14 ± 0·35 (77)	0·95 ± 0·33 (86)
9 or more (4 cheek teeth)	2·87 ± 0·96 (51)	2·25 ± 0·77 (39)	2·52 ± 0·80 (102)	1·76 ± 0·44 (148)

Figures in parentheses indicate numbers of observations.

are smaller than those living in southern Ghana. The heaviest male recorded at Mole National Park was 5·4 kg and in the acquired habitat in the south, the maximum weight recorded for a male was 8·64 kg. Rosevear (1969) reported a weight of 9·09 kg but did not indicate habitat, sex or whether it was live or dressed weight.

In southern Ghana the weight of the dressed carcass of primigravid females ranged between 0·7 kg and 3·1 kg whereas in the north the range for such animals was 0·8 kg to 1·9 kg. The corresponding range of weights of parous females in their second or later pregnancy in the two locations was 0·7 kg to 4·6 kg and 0·4 kg to 2·50 kg respectively.

PEST POTENTIAL AND POPULATION CONTROL

There is no doubt that the grasscutter has been very successful in its acquired habitat where it has reached its record weight and, where it provides all-year-round hunting, particularly on plantations, and is a source of meat and income to many people. Besides man, the predators of the grasscutter are snakes, birds of prey and carnivores. There are fewer species of carnivores in the forest zone (Squires, 1962), and if the side-striped jackal (*Canis adustus* Sundevall), which is now recorded for the Mole National Park, is included this zone contains only about 36% of them. Baboons may take grasscutters but they do not occur in the forest proper. Details of grasscutter predation have not been investigated, but there is little doubt that in the absence of heavy predation by man and its natural enemies, the grasscutter could be a serious pest, and if left unmolested in the forest zone, would outstrip its food supply, and would behave in much the same way as the red deer in New Zealand.

The grasscutter is recognised as an agricultural pest in Sierra Leone (Jones, 1966). In some parts of Ghana the grasscutter could be considered as a pest, but the farmers deal very effectively with it by trapping in conjunction with fencing. An impenetrable palm-rib or stick fence is made around the maize or sugar cane farm, except that at intervals a gap is deliberately left into which a wire noose is set very low, giving the animal no chance of escape. Live trapping was unsuccessful.

The farmers also control the grasscutter by night shooting. Animals shot or trapped may be sold or eaten, as there is a ready market for the carcass.

METHODS OF CAPTURE

The two major methods of capture are by shooting (largely at night) and snaring or trapping as described above. Both methods tend to favour young animals. Normally one or other parent leads the family group when they are on the move, and it is the leader of the family that risks capture by a snare.

All the grasscutters available for sale on the Kantamanto Market of Accra were obtained by one or other of these methods, and during the period September 1970 to April 1971 they amounted to 10,375 of which 84% were males. Further analysis (see Table II) shows that both

TABLE II

Proportion of male and female grasscutters obtained by different methods of capture

Method	Male		Female	
	No.	%	No.	%
Shooting	6,958	87	1,023	13
Trapping	1,783	75	606	25
Communal hunting	32	58	23	42

methods of capture favoured the survival of females, but of the two, shooting accounts for an even greater proportion of males. As would be expected, young animals did not appear on the market; they would be less profitable than adults.

A third method of capture of the grasscutters in northern Ghana is by communal hunting parties. These parties are social events to which all able-bodied members of the village are free to join. It would be difficult to assess the total number of animals collected in this way over the country as a whole, but the method has the advantage over the others in securing a larger proportion of females, and thus having a greater effect on the growth of population (see Table II). The method of capture is to select an area of reed beds which is known to contain a colony of grasscutters. The surrounding area is burnt and allowed to dry out, after which the reeds themselves are set alight, and the escaping animals are killed by all possible means.

ECONOMIC IMPORTANCE

In the Kantamanto Market of Accra alone, an annual sale of 76·85 tons (78 073 kg) was recorded between July 1970 and June 1971, costing ₵89 595 (US $69 996). Processed (smoked) grasscutter meat is also available and, in addition, there is a direct supply of carcases to various traditional restaurants and other markets in Accra which was estimated at over 108 tons (109 714 kg) for that period and valued at ₵193 536 (US $151 200). No nationwide assessment of the consumption or production of this or any other bushmeat has been made, although sampling has been undertaken as time and opportunity have permitted (Asibey, 1966). However, in Bamboi, which is a village on the banks of the Black Volta, the 13 restaurants that cater for travellers between northern and southern Ghana annually bought about 176·54 tons (179 342 kg) of grasscutter meat, an amount which earned the hunters/trappers of the area an annual income of about ₵112 320 (US $87 750).

TABLE III

*Approximate composition (%) and mineral content (mg/100 g)
of grasscutter in relation to that of other domestic meats*

Meat	Moisture	Ash	Fat	Protein	Iron	Calcium	Phosphorus
Beef	73·8	1·0	6·6	19·6	5·1	3·9	57
Mutton	78·5	1·0	2·9	17·2	3·1	9·0	80
Pork	64·8	0·8	13·4	19·4	3·1	3·0	73
Grasscutter	72·3	0·9	4·2	22·7	2·8	83·0	111

Although the grasscutter meat used in traditional restaurants as well as most of the smoked meat is consumed largely by the low income group, my survey in Accra indicated that the large quantities of undressed grasscutter available on the market are largely bought by the high income group of people. It is the most expensive meat; in an Accra supermarket, it sold at Ø2·06/kg in comparison with beef, mutton and pork at Ø1·38/kg.

The quality of grasscutter meat, as indicated by the analysis given in Table III, shows that it compares very favourably with beef, mutton and pork available on the Ghana meat market. The size of the animal makes handling and processing of the carcass easy in a country where most people have no cold storage facility for fresh meat and the rate of decomposition is high.

It is evident that the grasscutter is an invaluable source of animal protein in Ghana in both rural and urban communities, although it is an expensive commodity, retailing at prices well above that of beef, mutton or pork. However, this meat is very popular and very acceptable to all classes of people in the country.

ACKNOWLEDGMENTS

Dr P. Grubb of the Zoology Department, University of Ghana, and Professor G. M. Dunnet of Zoology Department, University of Aberdeen, made very helpful suggestions and constructive criticisms for which I am most grateful.

Many friends and colleagues readily supplied information on the location of grasscutters. Without such co-operation I would not have been able to record the spread of these animals throughout Ghana.

I am very grateful to Messrs S. Asiedu-Gyadu and G. Awuah Mensah, of Grassland Ecology Division, Department of Animal Science, Faculty of Agriculture, Legon, Accra, for identifying the food plants of the grasscutter, and to the Chief Cocoa Officer and the Chief Survey Officer for invaluable help with map work. The Chief Survey Officer also graciously allowed me to modify his base map for this study.

I am grateful to the Ghana Government for financing this project and my participation in the Symposium, as well as for permission to publish my findings.

REFERENCES

Asibey, E. O. A. (1966). Why not bushmeat too? *Ghana Farmer* **10**: 165–170.
Asibey, E. O. A. (1969). *Grasscutter* (Thryonomys swinderianus) *as a source of bushmeat in Ghana*. Mimeographed M.S.

Booth, A. H. (1959). On the mammalian fauna of the Accra Plain. *J. West. Africa Sci. Ass.* **5**: 26–36.

Ewer, R. F. (1969). Form and function in the grass cutter, *Thryonomys swinderianus* Temm. (Rodentia, Thryonomyidae). *Ghana J. Sci.* **9**: 131–149.

Jones, T. S. (1966). Notes on the commoner Sierra-Leone mammals. *Niger Fld* **31**: 4–17.

Rosevear, D. R. (1969). *The rodents of West Africa.* London: British Museum (Natural History).

Squires, F. A. (1962). Mammalian fauna. In *Agriculture and land use in Ghana*: 170–172. J. Brain Wills (Ed.)

Symp. zool. Soc. Lond. (1974) No. 34, 171–209

PATTERNS OF BEHAVIOUR IN HYSTRICOMORPH RODENTS

DEVRA G. KLEIMAN

*National Zoological Park,
Smithsonian Institution, Washington, D.C., U.S.A.*

SYNOPSIS

The basic behaviour of the sub-order is very uniform considering the great differences that exist in the size and life style of its members. Several behavioural characteristics which are atypical of rodents have evolved, particularly in relation to the mode of their reproduction and as anti-predator devices.

Several visual signals, such as tail-up rump display and pilo-erection, are common within the suborder in courtship, agonistic and anti-predator behaviour and vary in their meaning and degree of ritualisation. 'Frisky-hop' play in juveniles and adults resembles and may be derived from anti-predator displays. Scent-marking, using urine and anal gland secretions, serves primarily to increase social cohesion. The role of odour in the stimulation of other behavioural patterns such as play is discussed.

Male courtship rituals are highly specialized. These include tail-wagging, urine spraying, body quivering and frisky hops which allow the male access to the female before the short period of sexual receptivity or oestrus. It is suggested that the brevity of copulation is related to the presence of various penile protuberances.

Parental behaviour is seen as an adaptation to the bearing of young which are mostly well developed at birth. Young may feed on solids early, but mothers may suckle and maintain close contact with dependent juveniles for many months. In many species, the male participates actively in protecting and rearing the mobile, but vulnerable young.

The majority of hystricomorphs aggregate in groups as an anti-predator strategy, but there are different levels of social cohesion in the colonial forms. Similarities between hystricomorphs and ungulate behavioural adaptations can be seen.

INTRODUCTION

The purpose of this review is to discuss some of the more unusual behavioural characteristics of hystricomorph rodents, particularly in the areas of communication, reproduction, and social behaviour. The behaviour of these animals is imperfectly known. Published accounts suggest that, as a group, they exhibit certain characteristics which may be of great interest to ethologists and psychologists and of some use to taxonomists. For example, courtship rituals are rare in mammalian species, yet male hystricomorphs display numerous complex behavioural patterns while interacting with anoestrous females (Kleiman, 1971). Detailed studies of such patterns not only might clarify taxonomic relationships, as those of Lorenz (1941) have done for the

Anatidae, but also might prove valuable in investigations of the motivation and function of social displays.

Hystricomorph rodents should be of interest to behavioural ecologists since they are found in a variety of habitat types and yet possess some basic physiological characteristics which must impose restrictions on their life style. For example, the maternal care pattern of hystricomorphs differs from that of myomorphs since the former have longer gestation periods and generally smaller litters of usually precocious young (Kleiman, 1972). Maternal care has been further modified, however, in each hystricomorph species as an adaptation to its particular ecological niche and, in some cases, paternal care has evolved as a means of increasing the survival of the young. Studies of hystricomorph social systems and their relationship to habitat utilization should prove instructive.

Throughout this review, the binomial nomenclature is used at first mention of a species and, subsequently, reference will be made only to the genus, unless several species of that genus are being discussed. I have relied frequently upon my own and the unpublished work of my colleagues, Dr J. F. Eisenberg, S. Wilson, and T. Davis (mainly on *Myoprocta, Cuniculus, Proechimys, Octodon, Octodontomys, Spalacopus, Abrocoma, Dolichotis*, and *Pediolagus*). References to this work and that of other authors (mostly quoted in the Tables) are not always included in the text as it is hoped to keep such interruptions to the continuity of thought to a minimum.

COMMUNICATION MECHANISMS

Tactile

For many genera, tactile contact is common in social interactions. The introduction of two conspecifics of *Myoprocta pratti, Octodon degus* and *Spalacopus cyanus* results initially in nose to nose contact followed by further investigatory behaviour, including sniffing and nuzzling the perineum, the rump and the neck. If a fight does not develop, tactile contact increases and nuzzling and nibbling the mouth and chin region may occur. Allogrooming is commonly seen in hystricomorphs, but varies from species to species in the length of contact and in the participants. Adult males and females groom each other most frequently; in *Octodon*, such grooming bouts can last for 10–15 minutes and a solicitation posture of presenting the throat or moving the forehead under the partner's chin may be seen. Presentation of the throat for grooming also occurs in *Thryonomys swinderianus* (Ewer, 1968),

Abrocoma cinerea, *Chinchilla laniger* and *Lagostomus maximus*, but was not observed by Rood (1972) in the cavies, *Galea musteloides*, *Cavia aperea* and *Microcavia australis*. Allogrooming between adults involves much nibbling of the head and fore-part of the body but when adults and older juveniles groom infants, licking of the perineal region predominates (*Myoprocta*: Kleiman, 1972; cavies: Rood, 1972).

The resting postures adopted by various species in captivity may involve bodily contact. Huddling is common in family groups of *Octodon* and *Chinchilla*, while *Myoprocta*, *Galea*, *Microcavia*, *Cavia*, and *Pediolagus salinicola* individuals will sit with the rumps in contact or alongside each other but facing in opposite directions. Family groups of *Pediolagus* have been seen sitting in a 'fan' or 'star' formation with heads outwards. In their natural grassland habitat, such behaviour would permit rapid visual perception of an approaching predator.

Many contactual interactions involve the exchange of olfactory information, or occur in specific contexts such as mating or nursing; these will be discussed below in the appropriate categories.

Visual

These are used predominantly in close contact interactions such as agonistic and courtship behaviour but are also important in anti-predator strategies. Many of the agonistic displays are common to other mammals but the courtship patterns seem to be unique.

Four particular visual displays which are of interest will be discussed. Each occurs in several contexts within the same and different genera and may have various functions and motivations. The derivation of each is traced below to indicate how the original pattern may have been modified and the meaning altered. Data relating to selected examples of three of these visual signals are summarized in Table I.

Pilo-erection

This phenomenon (see Fig. 1) of the autonomic erection of the rump hairs occurs in many mammals in conditions of arousal but in several hystricomorph species, in which the dorsal hair has become modified by lengthening (*Dasyprocta*, *Myoprocta*) or by conversion into quills (*Hystrix*, *Erethizon*), an exaggerated signal is produced. Gradations of the display are produced by each species in different conditions (see Table I). In *Erethizon dorsatum*, *Hystrix cristata* and *Myoprocta*, pilo-erection often precedes an attack and the likelihood of attack can be gauged by the degree of quill or hair erection. *Dasyprocta punctata*, however, exhibits rump-hair erection when mildy alarmed and often

G

TABLE I

Selected examples of three visual signals in hystricomorphs, with comments on the context of the display and its presumed function and motivation

Display	Major context	Presumed motivation	Presumed function	Genus	Remarks	References
Pilo-erection of the rump hair	(1) Before an attack, when alarmed	Aggressive threat	Withdrawal of opponent or predator	*Hystrix, Erethizon* (spines)	Combined with tail rattling and backwards movement towards predator or opponent	Ewer, 1968
				Myoprocta	Accompanied by snort vocalization and approach	Kleiman, 1972
	(2) Before or during flight	Mild fear, alarm	"Arousal indicator"	*Dasyprocta*		Smythe, 1970a
Tail movements (wag, wave, beat, vibrate)	(1) Before or during attack	Aggressive	Withdrawal of opponent	*Hystrix* (spines)	Combined with pilo-erection and foot stamping	Ewer, 1968
				Lagidium	Accompanied by growl	Pearson, 1948
				Lagostomus	Accompanied by growl	B. J. Weir, personal communication
	(2) Courting ♂	Submissive	Permits approach and contact	*Myoprocta, Octodon*	Combined with forefoot or body trembling	Kleiman, 1971, unpublished
				Octodontomys, Thryonomys	Combined with hindfoot treading	D. G. Kleiman, unpublished; Ewer, 1968

	(3) During agonistic encounter	Defensive threat? ambiva-lent?	Repulse opponent? "arousal indicator"	*Octodon, Octodontomys*	While animals separated, can precede or follow withdrawal or attack	T. Davis, D. G. Kleiman & S. Wilson, unpublished
Tail-up rump display	(1) Receptive ♀	Sexual	Invites copulation	*Erethizon, Atherurus, Octodontomys Hystrix Geocapromys*	Backwards approach to ♂	Shadle, 1946; Ewer, 1968; D. G. Kleiman, unpublished; B. J. Weir, personal communication Howe, 1971
	(2) Courted ♀	(a) Sub-missive	Permits ♂ approach and contact	*Myoprocta*	During withdrawal from ♂	Kleiman, 1971; Morris, 1962
				Octodontomys	Backwards approach to ♂	D. G. Kleiman, unpublished
		(b) Defen-sive	Repulses ♂	*Galea, Cavia, Dolichotis*	Accompanied by backwards urine-spray	Rood, 1972; Kirchshofer, 1960
	(3) Courting ♂	Submissive	Permits approach to ♀	*Abrocoma*	Followed by mounting by ♀	J. F. Eisenberg, unpublished
	(4) Agonistic ♂–♂ or ♂–♀ encounter	(a) Defen-sive threat? ambivalent	Repulses opponent?	*Octodon, Octodontomys*	May induce mounting by partner; precedes or follows withdrawal of attack	T. Davis & D. G. Kleiman, unpublished
		(b) Sub-missive	Inhibits attack	*Lagostomus, Geocapromys*	Tail hairs also erected	B. J. Weir, personal communication; Howe, 1971

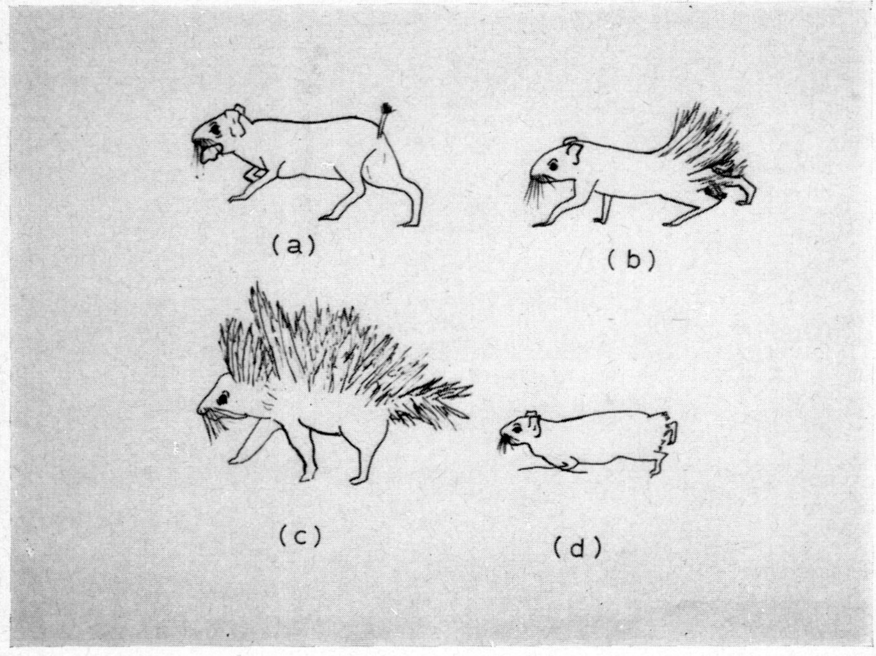

FIG. 1. Pilo-erection and the tail-up rump display in hystricomorph rodents.

(a) *Myoprocta* female tail up.
(b) *Myoprocta* rump hair erection.
(c) *Hystrix* pilo-erection (adapted from Mohr, 1965).
(d) *Microcavia* female tail up (adapted from Rood, 1972).

when fleeing from a conspecific or predator, and such lack of specificity suggests that pilo-erection is a less ritualized signal and serves mainly to indicate the level of arousal rather than a particular mood (see Eisenberg, 1974 for discussion of 'arousal indicators').

Tail-wagging

Tail movements appear to be similar to pilo-erection in their evolution. In their most primitive form, they may have been exhibited in conditions of ambivalence and high arousal, the amplitude and frequency of the tail-wagging changing at different levels of excitement. For example, highly-aroused *Octodon* males tail-wag while courting and also during agonistic encounters with other males. In *Hystrix*, however, the tail rattle, like quill erection, has taken on an aggressive intent and indicates a high probability of attack. By contrast, *Myoprocta* and

Thryonomys males appear to wag their tails only when courting females. Although the signal may still suggest ambivalent tendencies in these two genera, there is little possibility of an attack.

Tail-up rump display

Presentation of the rump with tail raised (see Fig. 1) by a female to a conspecific as an invitation to copulate is indicative of the oestrous condition in *Erethizon*, *Hystrix* and *Atherurus africanus* (see Table I). However, in *Octodontomys gliroides* this display is given by receptive and non-receptive females and may initiate sniffing and mounting by the male; it could be a submissive display to promote contact regardless of the reproductive condition of the female. In *Myoprocta*, the tail-up occurs while the female is withdrawing from a courting male; it does not initiate male mounting, but does induce continued following by the male, and could also be regarded as a submissive display. *Cavia*, *Galea* and *Dolichotis patagonum* females also exhibit the tail-up posture in this situation but often spray urine which checks the male. Clearly, the mood of the sender, the message sent and the reaction of the recipient are quite different in the cavies from those of the original pattern. The tail-up rump display has also been observed in *Abrocoma* males during heterosexual encounters and in *Octodon* males during hetero- and male homosexual encounters. In both species, mounting by the partner may occur.

These three visual signals are almost certainly more widespread amongst hystricomorphs than indicated in Table I. The tail-up rump display, originally a female pattern to solicit mating, is now used in several social contexts with different meanings whereas tail-wagging and pilo-erection have become more ritualized during their evolution. However, the contextual occurrence of tail-wagging and pilo-erection in several species suggests that visual displays with non-specific 'meanings' may be maintained in a species behavioural repertoire as indicators of the arousal level of the performer. Futher studies are needed to investigate the recipient's response to a signal so that the specificity of the message within a given context can be determined.

The 'frisky-hop' syndrome

Unlike the visual signals discussed above, this display involves several behavioural patterns in combination. The syndrome is usually considered to be play and includes vertical leaps, body twisting, head tossing, racing and pivoting, and prancing with kicking back of the hindfeet. Thus, it involves rapid locomotor and rotational body movements, approaches and withdrawals, but little physical contact between

conspecifics. The syndrome has been called 'prancing' in *Myoprocta*
(Morris, 1962), "the frenzy dance" in *Dasyprocta* and *Cuniculus paca*
(Smythe, 1970a), "locomotor-rotational movements" in *Octodon*,
Octodontomys and *Pediolagus* (Wilson & Kleiman, 1974), and "frisky
hops" in *Cavia*, *Galea* and *Microcavia* (Rood, 1972). For convenience, I
have adopted Rood's term to cover all the separate movements.

The syndrome seems to be absent in arboreal forms (*Capromys
pilorides*, *Erethizon*) and most prevalent in cursorial species (e.g.
Dasyprocta, *Myoprocta*, *Pediolagus*) or dwellers of open habitat, who do
not use burrows extensively (e.g. *Cavia*, *Galea* and *Chinchilla* exhibit
frisky hops but *Octodon* does so less often). This suggests a correlation
between frisky hops and anti-predator strategies mainly involving flight
rather than escape into burrows or climbing.

Frisky hops predominate in the juvenile play of *Cavia*, *Galea* and
Microcavia (Rood, 1972), *Pediolagus* (Wilson & Kleiman, 1974),
Myoprocta (Kleiman, 1969), *Hystrix* (Mohr, 1965), *Myocastor coypus*
(Eibl-Eibesfeldt, 1952), *Chinchilla* and *Lagostomus* (B. J. Weir, per-
sonal communication). Play interactions in *Octodon* and *Octodontomys*
include both frisky-hops and agonistic and sexual elements, and in
Capromys (Taylor, 1970) and *Erethizon* (Shadle, 1943; Shadle & Ploss,
1943), agonistic elements predominate.

Frisky hops are also reported to occur during courtship in *Dasyprocta* and *Cuniculus* (Smythe, 1970a), *Myoprocta* (Kleiman, 1971),
Cavia, *Galea* and *Microcavia* (Rood, 1972), and *Dolichotis* (D. G. Kleiman & J. F. Eisenberg, unpublished). Although it occurs primarily in
the male, it is common in females of *Dasyprocta*, *Cuniculus* and *Dolichotis* which have been sprayed with urine during an encounter although
Myoprocta females living with males do not behave in this way. Frisky
hops, therefore, may be less common in adult females than males, but are
facilitated during an encounter in captivity. Adult males of *Pediolagus*,
Myoprocta and *Octodontomys* often join with juveniles and infants in
frisky-hop play but the mother rarely does so.

The stimuli inducing frisky hops have not been enumerated, but
observations on *Myoprocta* (Kleiman, 1969), *Cuniculus* and *Dasyprocta*
(Smythe, 1970a), *Pediolagus*, *Octodon* and *Octodontomys* (Wilson &
Kleiman, 1974) suggest that olfaction plays a significant role. In *Octodon*
and *Octodontomys* juveniles, sniffing the body of a conspecific results in
head-shakes and jumps. In *Pediolagus*, juveniles and adult males often
combine frisky hops with sandbathing activity (rolling), urine marking
and sniffing conspecific odours. A typical behaviour sequence might
include approaching the site where most urine-marking and sandbathing
is localized, sniffing the substrate, performing an anogenital drag with

urination, sniffing the substrate again, rolling over in the urine, and then head tossing, leaping vertically into the air and racing to and from the site. A similar sequence is seen in *Myoprocta* (Kleiman, 1969), even though it inhabits tropical rain forests and presumably does not use sandbathing as a means of dressing the pelage. Smythe (1970a) suggests that frisky hops occur in *Cuniculus* and *Dasyprocta* females in response to the urine odour of a male on their bodies; he also noted that frisky hops occurred in *Dasyprocta* during rain, a situation in which the olfactory environment alters.

In truly dangerous situations, anti-predator displays, such as running and jumping, may be stimulated by a sudden contrast or change in the surrounding odour (as they are in play), perhaps directly from the predator. The flight may then act as a "pursuit invitation" signal (Smythe, 1970b), but since such flight patterns are contagious, their performance in the presence of conspecifics may induce other individuals to flee in an erratic manner which could distract the pursuer. During frisky hops play, there is also a lack of orientation of movement and visual contagion.

Similarities between frisky hops and adult flight behaviour are most obvious in cursorial forms and, in many respects, the flight patterns of cursorial hystricomorphs resemble those of ungulates (Walther, 1969; Smythe, 1970b), consisting of the stotting gait, prances and leaping as well as direct flight. In the ungulates, too, juvenile play is derived from patterns of fleeing, and the locomotor movements may be exaggerated (e.g. deer: Müller-Schwarze, 1968). Such convergence in play patterns implies that maintenance of the appropriate muscle tone and increase in co-ordination is important for juveniles and adults. The convergence also suggests that those hystricomorphs which show frisky hops in juvenile play but do not use flight as their main anti-predator strategy, may have been more cursorial in their past evolutionary history and only secondarily have developed other anti-predator mechanisms (e.g. *Hystrix*: Mohr, 1965).

Auditory

Hystricomorphs produce a variety of sounds, both mechanical and vocal, which communicate to conspecifics the mood and level of arousal of the sender (see Eisenberg, 1974). Warning or alarm calls are found in most genera and serve to alert conspecifics of approaching danger. Two vocalizations are closely tied to male–female interactions, the courtship whimper and post-copulatory squeaks or snorts. Mothers and infants have contact calls. Several sounds are associated with agonistic behaviour, including those produced by tooth gnashing and hindfoot thumping;

in addition, most forms have agonistic vocalizations such as snorts or whines. Eisenberg (1974), has shown that the physical characteristics of the sounds may relate to the function of the sound and the habitat in which it is produced.

Olfactory

Hystricomorphs, like most mammalian species, use chemical signals for communicating with conspecifics in certain contexts. A major source of scent are the well developed paired anal glands (Pocock, 1922), the secretions of which are typically applied to a substrate by an anogenital drag (e.g. *Cavia*: Kunkel & Kunkel, 1964; *Galea, Microcavia*: Rood, 1972; *Dasyprocta*: Roth-Kolar, 1957; Smythe, 1970a; *Geocapromys ingrahami*: Howe, in press). Most hystricomorph males mark more frequently than females (*Cavia*: Kunkel & Kunkel, 1964; *Dasyprocta*: Roth-Kolar, 1957; *Thryonomys*: Ewer, 1968). The anal glands are larger in males of the first two genera and of *Hystrix* (Mohr, 1965), suggesting that anal glands and their use may be controlled by androgens. In *Dasyprocta*, the odours may be deposited merely as the result of the animal having sat on the substrate or by similar contact during other activities, e.g. scatter-hoarding. Smythe (1970a) has remarked on the strength and long-lasting quality of the odour produced by *Dasyprocta* anal glands: he could 'smell' agouti up to 10 m away, and after handling an animal it took two days to remove the odour.

Other specialized glandular areas whose secretions are used for scent marking are not common in hystricomorphs and their appearance in the suborder is sporadic. A supracaudal gland occurs in *Cavia* (Martan, 1962) and a chin gland in *Galea* (Weir, 1971; Holt & Jones, 1973). *Erethizon* (Shadle, Smelzer & Metz, 1946) and *Dinomys branickii* (Collins & Eisenberg, 1972) rub the nose on objects, the latter species with a white secretion that arises in the eyes and drains through the nostrils. A large sebaceous gland is present on the snout of the male *Hydrochoerus hydrochaeris* (Rewell, 1949) which is rubbed on vegetation (Ojasti, 1973). Smythe (1970a) suggests that *Cuniculus* may possess glands in the folds leading to the cheek pouches and has noted the occurrence of cheek rubbing on the female during courtship (see below).

Urine is another major source of odour, and appears to be used extensively in communication. Few hystricomorphs employ specialized postures for urinating but *Spalacopus*, a fossorial form, urinates on vertical surfaces using a leg lift. During enurination, when urine is sprayed on a conspecific, specialized postures such as rearing-up on the hindlimbs are seen (see below). In hystricomorphs, the information gained from urine or glandular secretions, as in other mammals, could include an individual's racial group, or even personal identity. The marker's reproductive state may be apparent and possibly even its mood at the time the signal was deposited (see Eisenberg & Kleiman, 1972). Nothing is as yet known about the type of information that is relayed by marking in hystricomorphs or if the message is different when urine and anal gland secretions are used.

There is, however, some information on the contexts in which marking occurs. The anogenital drag and/or urination is often seen when an individual is introduced into a strange enclosure (*Myoprocta*: Morris, 1962; *Octodon*: D. G. Kleiman, unpublished) or confronted with a novel object (*Geocapromys*: Howe, in press). It is also seen during agonistic interactions when the anal glands may be everted without scent being applied to the substrate (*Cavia*: Kunkel & Kunkel, 1964; Rood, 1972; *Dasyprocta*: Smythe, 1970a). Scent marking also occurs after an animal has sniffed the mark of a strange or known conspecific (*Octodon*, *Pediolagus*: D. G. Kleiman, unpublished) or seen a conspecific mark (*Geocapromys*: Howe, in press), and the mark may be deposited on the same site. The results of tests in our laboratory with *Octodon* suggest that males and females deposit a greater percentage of urine marks over the urine of like-sexed individuals. However, there is no significant change in the basal marking frequencies when they enter an empty arena versus one containing scattered urine marks of a conspecific. Sniffing frequencies also increase when males and females encounter conspecific urine; both observations suggest that in *Octodon* conspecific odour is attractive (similar results have been found for *Cavia*: Beauchamp, 1973) and does not inhibit further marking behaviour. We have also found that urine marking (and defaecation) increases as the familiarity of a strange situation increases, and thus urination may be inversely related to fear.

Howe (in press), based on similar observations in *Geocapromys*, suggests that marking may be non-agonistic in intent and serves mainly to co-ordinate the social behaviour of this species. A similar cohesive role of odour is probably important in other hystricomorphs; in *Octodon*, *Octodontomys*, *Pediolagus* (in captivity) and *Microcavia* (in the wild), localized sites become saturated with urine and anal gland secretions, and are used preferentially for sandbathing. This ensures that members of the same group share the same scent. Group members also share scent by enurinating over one another (see below). These studies do not support the notion that scent marking, in hystricomorphs at least, is solely agonistic in intent or functions primarily to intimidate a conspecific or defend a territory (see Ralls, 1971 and Eisenberg & Kleiman, 1972).

REPRODUCTIVE BEHAVIOUR

Courtship by the male

In several hystricomorph species, complex courtship rituals have evolved which are performed almost exclusively by the male (Kleiman, 1971). The patterns of courtship in species for which descriptions are available are summarized in Table II. Differences in detail of some of these rituals exist. For instance, tail-wagging in *Myoprocta* is a rapid vibration whereas in *Thryonomys* it is a lashing movement. The body movements in *Dasyprocta*, *Myoprocta* and *Dinomys* consist of a trembling of the forequarters which may result in alternate stepping movements of the forefeet and/or quivering of each fore-limb as it is raised off the ground whereas in *Cavia* it consists of a swaying motion of the hindquarters, termed the 'rumba.' *Octodontomys* males also show treading or stepping movements of the hindfeet. Not included in Table II, since they are common to most rodent courtship patterns, are activities such as sniffing the perineum, nose to nose sniffing and driving of the female. Males may attempt to mount females (especially during staged encounters) regardless of her reproductive state, and may demonstrate pelvic thrusting. Unsuccessful mounting attempts usually include erection of the penis followed by genital grooming. While courting, males may whimper or squeak softly and repetitively (see Eisenberg, 1974).

Numerous species spray urine over the female while courting (enurination or Harnspritzen), either in a bipedal posture with the urine directed forward (e.g. *Erethizon*, *Myoprocta* and *Dolichotis*) or with one or both hindlimbs thrown over the back of the female (e.g. *Cavia* and *Octodon*) (see Fig. 2). In two genera that have not been reported to enurinate, the male approaches the female in a bipedal posture (*Dinomys*) or rears up after approach (*Galea*) with an erect penis. Presumably this pattern is closely related to that of enurination and urine spraying may possibly occur.

Marking of the female by the male using glandular secretions is uncommon. *Galea* males rub the secretions of their chin glands onto females during the 'chin-rump follow' behaviour pattern which brings the female into oestrus (Weir, 1971, 1973). *Microcavia* males also chin-rump follow but do not possess a chin-gland, (see Weir, 1973 for discussion of the possible reasons for this difference). Male *Cuniculus* also exhibit chin and throat rubbing of the female during courtship but no glands have been confirmed. The behaviour may be to effect the transfer of urine from the venter and throat of the male which have

FIG. 2. Enurination (Harnspritzen) postures in some male hystricomorph rodents.

(a) *Myoprocta*, bipedal.
(b) *Octodon*, tripedal.
(c) *Cavia*, tripedal (adapted from Rood, 1972).
(d) *Dolichotis*, bipedal (adapted from Kirchshofer, 1960).
(e) *Cuniculus*, bipedal (adapted from Smythe, 1970a).

become soaked during enurinations of the male in a partly erect, bipedal posture.

The frequency of courtship in hystricomorphs may vary with the female's oestrous cycle (as in *Myoprocta*: Kleiman, 1971), increasing as the female approaches heat. However, since oestrous cycles are long and variable in length, courtship behaviour is often displayed irregularly to females who may be neither in nor closely approaching heat. The intensity of courtship seen at any given time will depend on (a) the type of male–female relationship common to the species (solitary, pair-bonded, promiscuous), (b) the reproductive state of the animals (seasonal versus non-seasonal breeders; stage of the female's oestrous cycle) and (c) the conditions under which the observations were conducted (staged encounters versus cohabiting animals).

TABLE II

Male courtship rituals in some hystricomorph rodents

Genus	Urine-spraying on ♀	Glandular marking of ♀	Mutual groom-ing	Bipedal approach	Tail-wagging	Trembling/treading	Frisky hops	References
Hystrix	—	—	+	+	—	—	—	Mohr, 1965; B. J. Weir, personal communication
Erethizon	+ (bipedal)	? (nosing)	0	+	0	0	+ (bipedal)	Shadle *et al.*, 1946; Spalsbury, 1956
Cavia	+ (tripedal)	0	+	0	0	treading hindfeet	+	Rood, 1972
Microcavia	0	0	+	0	0	0	+	Rood, 1970, 1972
Galea	0	+ (chin rub)	+	+ (erect penis)	0	0	+	Weir, 1971; Rood, 1972
Dolichotis	+ (bipedal)	0	0	0	0	0	+	Kirchshofer, 1960; Critch, 1966; D. G. Kleiman & J. F. Eisenberg, unpublished
Dinomys	0	? (nosing)	+	+ (erect penis)	0	trembling forefeet	+ (bipedal)	Collins & Eisenberg, 1972
Dasyprocta	+	0	+	0	0	trembling forefeet	+	Roth-Kolar, 1957; Smythe, 1970a
Myoprocta	+ (bipedal)	0	+	0	+	trembling forefeet	+	Morris, 1962; Kleiman, 1971

Genus								Reference
Cuniculus	+ (bipedal)	+ (cheek, chin rub)	+	0	0	0	+	Freiheit, 1965; Smythe, 1970a; D. G. Kleiman, unpublished
Octodon	+ (tripedal)	0	+	0	+	trembling body	0	D. G. Kleiman, unpublished
Octodontomys	?	0	+	0	+	treading hindfeet	0	D. G. Kleiman & S. Wilson, unpublished
Proechimys	0	0	+	0	0	0	0	Maliniak & Eisenberg, 1971
Thryonomys	0	0	+	0	+	treading hindfeet	0	Ewer, 1968, 1969
Myocastor	+	0	—	0	0	0	0	B. J. Weir, personal communication
Chinchilla	0	0	+	+	0	0	0	B. J. Weir, personal communication
Lagostomus	0	0	+	+	+	body trembling, forefeet treading	0	B. J. Weir, personal communication
Geocapromys	0	? (chin rub)	+	0	0	0	0	Howe, 1971

+ = presence; 0 = absence; — = not known.

Role of the female

In most conditions, females who are being courted withdraw, and periodically halt and raise the rump and tail, thus exposing the perineum (*Myoprocta, Dasyprocta, Dolichotis, Galea* and *Cavia* (see p. 177)). When a male persists, the female may squirt urine over him while in the tail-up posture (*Cavia, Galea* and *Dolichotis*) or with a leg raised over the male (*Octodon*). The tail-up rump display in some cases may be a submissive signal (Kirchshofer, 1960; Critch, 1966) but when accompanied by a urine-spraying it has been termed defensive in *Galea* and *Cavia* (Rood, 1972).

If a male persists in courting and/or attempting to mount, females will often threaten or lunge at the male, e.g. *Cavia* (Rood, 1972), *Chinchilla* (Bignami & Beach, 1968), *Octodon* (D. G. Kleiman, unpublished), *Spalacopus* (D. G. Kleiman & T. Davis, unpublished), *Proechimys semispinosus* (Maliniak & Eisenberg, 1971), and *Erethizon* (Shadle *et al.*, 1946). This high level of aggression by females is common when they are reproductively active, but nonreceptive. For example, Bignami & Beach (1968) reported periodic but active aggression by female *Chinchilla* in which 'heat' was induced by oestrogen and progesterone, but *Chinchilla* can be aggressive at all times. Pearson (1948) also remarked on the increased belligerence shown by *Lagidium peruanum* at the approach of the breeding season. Recent observations on *Octodon* (during male–female encounters) in this laboratory suggest that females are more resistant to male courtship while the vaginal closure membrane is perforate than at other times.

Function and motivation of courtship

The fact that females may be aggressive to males while non-receptive to mating suggests that the major function of male courtship is to appease the female, thus permitting the male access to her. An analysis of the displays exhibited during *Myoprocta* courtship (Kleiman, 1971) indicated that several may be derived from behaviour patterns associated with fear and hesitation, and are performed while the male is in an ambivalent state since he periodically approaches but then withdraws from the female. Urine-spraying has been discussed by Kirchshofer (1960) and Kleiman (1971) and presumably serves as an odour-distributing mechanism which familiarizes each individual with the partner's odour and thus promotes increased tolerance (see p. 181) even if the enurinating animal seems defensive (e.g. *Cavia* and *Octodon*). Presumably, enurination is derived from urination autonomically induced in conditions of fear. It does not appear to be motivated by sexual tendencies in males.

Copulatory behaviour

The most complete accounts are for the domestic guinea-pig, *Cavia porcellus* (Young & Grunt, 1951), and *Chinchilla laniger* (Bignami & Beach, 1968). Table III summarizes the available information, including new observations from our laboratory on *Pediolagus*, *Octodon* and *Octodontomys*.

Based on Dewsbury's (1972) classification of mammalian copulatory behaviour, the majority of hystricomorphs conform either to Pattern 9 (no lock, thrusting during intromission, multiple intromissions, multiple ejaculations) or Pattern 10 (no lock, thrusting during intromission, multiple intromissions, single ejaculation). *Octodontomys* alone fits Dewsbury's Pattern 11 showing a single intromission per ejaculation, but multiple ejaculations. All species, including *Octodontomys*, may show multiple mounts before an intromission, with or without pelvic thrusting movements.

Thus, the basic hystricomorph pattern involves a few intromissions lasting five to ten seconds followed by a single ejaculation. By comparison with mammals that display locks, multiple intromissions and ejaculations, the copulation period is extremely short in most hystricomorphs. The three major exceptions to this are *Erethizon*, for which Shadle (1946) reports 'sexual union' as lasting one to five minutes; *Proechimys* (Maliniak & Eisenberg, 1971) in which the ejaculatory mount may exceed five minutes; and *Octodontomys* (D. G. Kleiman, unpublished) where the ejaculatory mount lasts slightly less than one minute.

Bignami & Beach (1968) suggested that matings tend to be shorter in hystricomorphs than in myomorphs because (1) oestrus is short and variable in length and (2) cervical stimulation is less essential for ovulation and the formation of the corpus luteum which provides sufficient progesterone for implantation. However, some hystricomorphs do exhibit multiple intromissions (Kleiman, 1971), and in *Galea* and *Cavia*, Rood (1972) has shown that the mating pattern is not fixed but depends on the dominance rank of the male, alpha males exhibiting several intromissions before ejaculation and subordinate males ejaculating on the first intromission.

The question as to whether hystricomorphs need or receive less cervical stimulation than do myomorphs is partly answered by considering the various styles, small spicules and the erectile ridges and horny plates on the glans penis (Pocock, 1922; Dathe, 1937; Hooper, 1961). It would be reasonable to suppose that such an array would ensure that the females at mating would receive extensive cervico-vaginal stimulation. The distribution of these accessory structures ap-

TABLE III

Copulatory patterns in hystricomorph rodents

Genus	Multiple intromissions	Multiple ejaculations	Length of intromission	Length of ejaculatory intromission	Thrusting during intromission	Length of female receptivity	References
Erethizon	+	—	1–5 min	—	+	several hours	Shadle, 1946
Cavia	+ (median = 5)	+ (1–2)	5 sec	—	+	2–3 hours	Young & Grunt, 1951; Rood, 1972; Boling *et al.*, 1939
Galea	+ (few)	0	—	—	+ (1/sec)	2–3 hours	Rood, 1972
Microcavia	+	+ (1–2)	2–6 sec	—	+ (1/sec)	several hours	Rood, 1972
Pediolagus	+ (3, 10)	0	6–7 sec	10 sec	+ (1/sec)	—	D. G. Kleiman & E. Maliniak, unpublished
Hydrochoerus	+	—	4–6 sec	—	+	—	Ojasti, 1968
Dinomys	+	0	<20 sec	—	+	—	Collins & Eisenberg, 1972
Myoprocta	+ (median = 11)	0	5–10 sec	15–20 sec	+	several hours	Kleiman, 1971
Chinchilla	+ (median = 5)	+ (1–2)	5 sec	8–10 sec	+	several hours but erratic	Bignami & Beach, 1968
Myocastor	+	—	—	—	—	—	Klapperstück, 1954
Octodon	+ (1–2)	+ (1–3)	10 sec	12–18 sec	+	several hours	D. G. Kleiman, unpublished
Octodontomys	0	+ (1–2)	40–60 sec	40–60 sec	+	—	D. G. Kleiman, unpublished
Proechimys	+	0	1–2 min	≥5 min	+ (1–5/sec)	—	Maliniak & Eisenberg, 1971

+ = presence; 0 = absence; — = not known.

pears to correlate with the known copulatory patterns. Of the species examined, only a few such as *Proechimys*, *Lagostomus*, *Dolichotis* (there is some disagreement for this genus) and *Capromys* lack these accessory structures. Although copulation has not been described for *Capromys* or *Lagostomus*, *Proechimys* is unique in having an intromission time of up to and greater than five minutes with thrusting throughout (Maliniak & Eisenberg, 1971). Shadle's (1946) reports of *Erethizon* copulation (if I have interpreted correctly his rather anecdotal account) would suggest that *Erethizon* lacked spikes or erectile plates, but spikes are reported for this genus (Pocock, 1922).

A post-copulatory vocalization has been reported for *Octodon*, *Octodontomys* and *Myoprocta*, but not for any of the caviids or *Chinchilla*. In *Octodon* and *Octodontomys*, it is a modified version of the alarm squeak while in *Myoprocta* it resembles the cough or snort vocalization heard only before an attack. The calling is prolonged; an *Octodontomys* male vocalized continuously for approximately six minutes after mating while an *Octodon* male did so for five to six minutes following first ejaculation, but up to 18 minutes following a second ejaculation. The squeaks usually were produced at the rate of one every one to three seconds. Vocalization at this time may serve to keep away all conspecifics, including other males: in *Myoprocta*, the snorting was often coupled with threats or chasing of subordinate or juvenile males (Kleiman, 1971). Rood (1972) also describes *Cavia* and *Galea* males chasing subordinates after ejaculation, and I have observed a *Pediolagus* male repeatedly chase a newborn infant after copulating at the post-partum oestrus which suggests that refractory males tolerate little physical contact. In some species, a resumption of interest in the female occurred with the termination of vocalizing; this agrees with Barfield & Geyer's (1972) suggestion that the post-ejaculatory "song" of the laboratory rat is synchronous with the refractory period.

Behaviour during oestrus

The behaviour of hystricomorph females in heat is unpredictable, as noted for *Chinchilla* by Bignami & Beach (1968). *Chinchilla* females are more receptive at the post-partum oestrus than at other times (Weir, 1967) and *Cavia*, *Galea* and *Microcavia* withdraw from and resist males at this time (Rood, 1972). *Dinomys* and *Proechimys* become aggressive before mating (Maliniak & Eisenberg, 1971; Collins & Eisenberg, 1972) and I have observed resistance or withdrawal from the male in *Octodon*, *Pediolagus*, and *Myoprocta*. Only in a few genera (e.g. *Erethizon*:

Shadle, 1946; *Atherurus*: Ewer, 1968; *Geocapromys*: Howe, 1971; *Octodontomys*: D. G. Kleiman, unpublished) do receptive females solicit a male by presenting the anogenital region or maintaining lordosis. Mounting of conspecifics by oestrous females is common in *Cavia* and *Dinomys* (Beach, 1968; Collins & Eisenberg, 1972) and possibly in many other genera; it occurs in many mammals, e.g. in cows, in which it is a reliable sign of oestrus. A stiff-legged hopping gait may be seen in receptive females of *Proechimys* (Maliniak & Eisenberg, 1971) and *Octodontomys* (D. G. Kleiman, unpublished), and may serve to attract the male.

The length of receptivity for most hystricomorph females is not accurately known because females do not exhibit lordosis for any length of time if at all and repeated testing with new males has rarely been done. On the present evidence (see Table III) it would seem that oestrus lasts not more than three hours in most genera. In *Chinchilla*, oestrus may persist for up to 48 hours and mating may occur thrice in that period (Weir, 1970). The shortness of the receptive period provides another possible explanation for the persistent courtship of the male and his increased attention to the female prior to oestrus (*Myoprocta*: Kleiman, 1971). Rood (1972) indicates that *Cavia* and *Galea* males begin consorting with pregnant females several days before parturition and in *Proechimys guairae* there is evidence of pre-partum mating (Weir, 1974b).

The erratic nature of female behaviour during oestrus and the complex structure of the penis may be two factors associated with the brevity of copulation in hystricomorphs. Short matings may also have evolved as an anti-predator mechanism, and, if so, it may be that the prolonged copulatory pattern of *Erethizon* (Shadle, 1946) is associated with the protective nature of the quills. It would be of value in this connection to observe these patterns in species which copulate in burrows and other protected sites.

Parental care

Maternal behaviour

I have recently reviewed patterns of maternal care in the South American hystricomorphs (=Caviomorpha) (Kleiman, 1972) and have discussed how certain characteristics of caviomorph maternal behaviour are related to the bearing of small litters of well developed young. For example, in cursorial species a nest may be constructed and used temporarily (*Myoprocta*) or not at all (*Cavia*), and retrieval behaviour is rare or absent. In species which typically burrow (*Octodon* and *Octodontomys*), young are less precocious at birth and both nest-building and retrieval are seen to a limited extent. Mothers and young of many species have special calls for maintaining contact. The following brief account considers only selected aspects of maternal behaviour and derives from recent observations in our laboratory and the studies of Rood (1972).

The position adopted by the young during suckling depends in part on the mother's nursing posture (see Fig. 3). In most genera, the young crouch but *Chinchilla* (Weir, 1970) and *Octodon* and *Octodontomys* (S. Wilson, unpublished) infants often lie on their backs. Nursing is undertaken in an exposed area or in a well defined nest and the length of the suckling period and the position of the mother (Fig. 3) seem to be related to the nature of the site (Table IV). Thus the caviids and

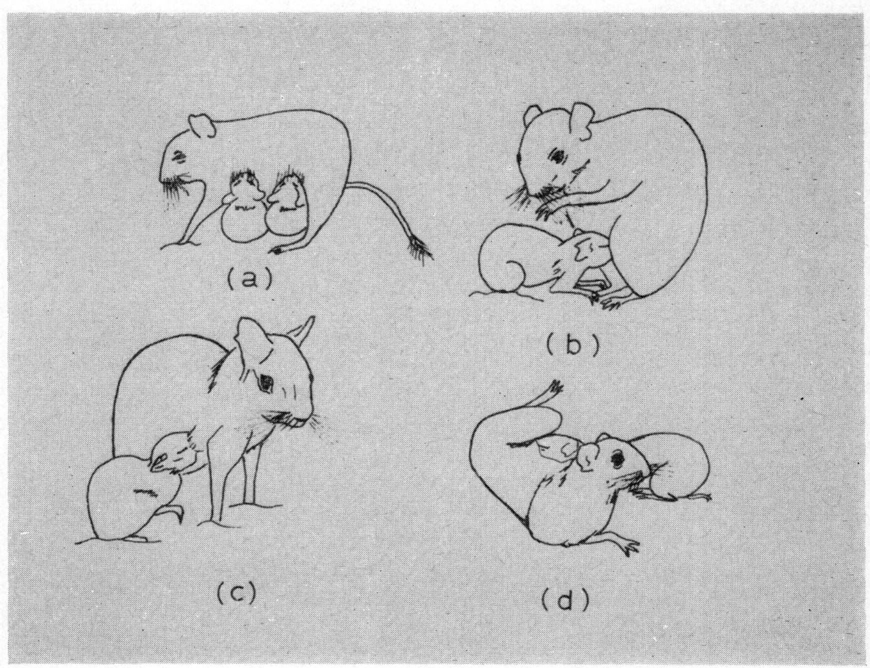

FIG. 3. Nursing postures of hystricomorph rodents.

(a) *Octodon*, huddling (adapted from S. Wilson, unpublished).
(b) *Dasyprocta*, sitting.
(c) *Pediolagus*, sitting.
(d) *Myoprocta*, lying.

TABLE IV

Variations in nursing patterns in hystricomorph rodents

Genus	Average duration of nursing bouts (min)	Nursing position	Nursing site	Teat position	Promiscuous nursing	Weaning age (weeks)	Average litter size	Reference
Hystrix	—	sitting	—	lateral	—	16	1–2	Mohr, 1965
Cavia	—	sitting, lying	open exposed area	ventral	occasionally	3–4	2·1	Rood, 1972
Microcavia	6	sitting, lying	open exposed area	ventral	yes	3–4	2·8	Rood, 1970, 1972
Galea	—	sitting, lying	open exposed area	ventral	yes	5	2·7	Rood, 1972
Dolichotis	—	sitting, lying	open exposed area	lateral	no	8–12	1–2	Mohr, 1949
Pediolagus	5	sitting	open exposed area	lateral	no	>4	2	D. G. Kleiman & S. Wilson, unpublished
Hydrochoerus	—	sitting	open area	?	no	16	4	Ojasti, 1971; Zara, 1973

Dasyprocta	2	lying, sitting	open exposed area	ventral	no	12	1	Smythe, 1970a; Roth-Kolar, 1957
Myoprocta	4	lying, sitting	open exposed area	ventral	no	8–12	2	Kleiman, 1972
Lagidium	—	—	open exposed area	lateral	—	>4 < 16	2	Pearson, 1948
Myocastor	—	sitting	open area or nest	lateral	—	8	5–6	Newson, 1966; Klapperstück, 1954
Capromys	—	sitting	open area	lateral	—	22	2	Taylor, 1970
Octodon	25	huddle	nest	ventral	yes	5–6	5	S. Wilson & D. G. Kleiman, unpublished
Octodontomys	25	huddle	nest	ventral	—	5–6	2	S. Wilson & D. G. Kleiman, unpublished
Proechimys	25	huddle	nest	ventral	—	6	2·8	Maliniak & Eisenberg, 1971, unpublished; Enders, 1935
Abrocoma	—	sitting, lying	nest	ventral	—	4	2–3	J. F. Eisenberg, unpublished

dasyproctids sit or lie in the open and suckle for two to six minutes with the young exhibiting a rapid alternation of teats. By contrast, *Octodon*, *Octodontomys* and *Proechimys* are huddled over their young in the nest for 25 to 35 minutes at a time. Active sucking is brief and alternates with periods of quiescence lasting three to five minutes with the young still attached to the nipples. Nursing is usually terminated by the mother leaving the nest or by the young leaving the mother.

Thus, the use of a protected 'nest' area permits long suckling bouts while mothers nursing during the day in the open must restrict suckling time to ensure that neither they nor the young are left vulnerable for longer than necessary. The establishment of teat ownership may also be important to prevent fighting over nipples and thus reduce the length of nursing bouts in species that suckle in exposed conditions. A 'teat order' exists in *Myoprocta* and each young uses only one area on the mother's venter until weaning. *Pediolagus* (and *Dolichotis*: Mohr, 1949) young do not change sides during a bout but may alternate teats on the side they are sitting; they may change sides between bouts. Clearly, short nursing bouts, rapid milk let-down and a teat order are not related to litter size or precocity but to the degree of protection afforded the mother and young at the chosen nursing site. The promiscuous nursing seen in some caviids may also be an anti-predator strategy to provide milk at the shortest possible notice. Similar adaptations occur in ungulates (Lent, in press) and a teat order has been described in pigs (McBride, 1963) and domestic cats (Ewer, 1960).

Many young hystricomorphs can eat solids within a day or two of birth and can survive away from the mother after two weeks (see Kleiman, 1970; Weir, 1974b), but are suckled for much longer periods (Table IV). There is a correlation between apparent weaning age and puberty (see Weir, 1974b), i.e. species that reach sexual maturity late are weaned late regardless of when they can be weaned from milk. The prolonged "nursing" must serve to maintain the mother–young bond during the juvenile period when the young are still vulnerable and are unable to assume an adult role. The protection afforded by the mother (and sometimes the family group) is in the form of access to feeding areas, a home range in which the young can shelter and move with freedom, and warning at the approach of a predator.

In hystricomorphs, independence of the mother's milk (nutritional weaning) need not be coincidental with independence from the mother (social weaning). Therefore, when discussing hystricomorph nursing, care should be taken to distinguish (i) when the young are independent of the mother's milk, (ii) when lactation ends, (iii) when suckling ceases, and (iv) when the mother–young bond is terminated.

Paternal behaviour

The father is known to be tolerant towards the young in several hystricomorph species (see Table V). One would also expect tolerance in *Hydrochoerus* and *Lagostomus* which are known to live in family groups, and intolerance or disinterest in solitary forms such as *Erethizon*, *Plagiodontia* and *Ctenomys*.

The degree of father–young interaction is partly dependent upon the female. Hystricomorph females may exhibit increased aggression after birth and drive off the male (*Dasyprocta*: Smythe, 1970a; *Myoprocta*: Kleiman, 1972). The male can interact with the young, however, in the absence of the female by waiting for the young to find him or approaching while the female is foraging. At these times, the male will follow and enurinate over the young, and groom and sleep with them. In *Myoprocta* levels of enurination over the young may be extremely high during the first two weeks (Kleiman, 1971).

In *Cuniculus*, the male rears up and enurinates extensively over the female after parturition, but urinates over the young only when the infant attempts to nuzzle him in the perineal region. This stimulates the male to lift a hind leg and walk over the young and to expel a few drops of urine (D. G. Kleiman & J. Eisenberg, unpublished). The attractiveness of the male anogenital region to the infant may be due to the conspicuous white scrotum which, in part, mimics the abdominally located white pendulous nipples of the female. *Octodontomys* males also urinate over the young and play vigorously with them. By contrast, *Octodon* males do not urinate or play with young, but do spend as much time as possible huddling over them (Wilson & Kleiman, 1974). One male in our laboratory vigorously groomed his mate while she was huddled over the litter nursing. After a short period of male grooming she would leave the nest, whereupon he would take her position over the young. Another male tried to push his mate away from the young immediately following the birth. It was also noted that females which cohabited with males during lactation only huddled over the litter while nursing, but solitary mothers often huddled when they were not nursing.

The degree of participation by the hystricomorph male in parental care appears to correlate with the amount of physical contact experienced by a male and female outside the context of reproduction, e.g. in *Pediolagus*, which is a little- or non-contact species, parental care is not well developed even though the male is tolerant. In *Octodon*, on the other hand, pairs show contact-promoting behaviour patterns such as grooming at all times. In some species, e.g. *Capromys pilorides* (Bucher,

Table V

Paternal behaviour in hystricomorph rodents

Genus	Level of tolerance	Behaviour towards young	References
Hystrix	XX	Retrieve, groom and rest with young	Mohr, 1965
Cavia aperea	X	Ignore young; chin-rump follow juvenile females beginning at 2 weeks	Rood, 1972
Microcavia	X	Chin-rump follow all juveniles from 3 weeks	Rood, 1970, 1972
Galea	X	Chin-rump follow juvenile females and males from birth, more frequent after 2 weeks	Rood, 1972; B. J. Weir, personal communication
Pediolagus	X	Rest with young; occasionally play with them	D. G. Kleiman & S. Wilson, unpublished
Dasyprocta	X	Enurinate over young, trembling mounting	Smythe, 1970a

Myoprocta	XX	Enurinate on, groom, and rest with young	Kleiman, 1972
Cuniculus	X	Urinate over, groom, and sleep with young	D. G. Kleiman & J. F. Eisenberg, unpublished
Capromys	XXX	Groom and sleep with young; aggressive to intruders	Bucher, 1937; Taylor, 1970
Octodon	XXX	Huddle over, and groom young; try to displace mother. Enurination from Day 27	S. Wilson, unpublished
Octodontomys	XX	Urinate and perineal drag over young; huddle over and groom young. Play with young	S. Wilson, unpublished
Chinchilla	XX	Sit with mother and young, may be protective	B. J. Weir, personal communication

X = tolerant but not friendly;
XX = tolerant and fairly friendly;
XXX = very friendly and protective.

1937; Taylor, 1970), the males may become as aggressive as the mother if the infants are threatened.

The initial tolerance towards neonates probably results from the similarity in odour between the mother and young. Through physical contact and scent marking, the male rapidly covers the infants with his own odour, thus decreasing the contrast in odour between himself and the litter. For example, no urination over young has been observed by *Octodon* males who huddle over offspring. *Octodontomys*, *Cuniculus* and *Myoprocta* males huddle less but do urinate over neonates. Little physical contact is seen between *Pediolagus* adults and young and neonates acquire male and group odour by rolling (sandbathing) in urination sites within a few hours of birth.

Ontogeny

Most hystricomorphs are born fully furred with eyes open and within a day are able to walk, run and perform frisky hops. Two or three days later they begin to eat solid foods, first by mouthing and nibbling and very soon afterwards ingesting small portions.

Adult behaviour patterns that are prominent in the infants at birth can often be related to the major adaptive strategy of that genus. For example, *Erethizon* young exhibit the full adult defensive posture, with quills erect, when only a few hours old (Shadle, 1954), and newborn *Pediolagus* are well co-ordinated and can sandbathe and run rapidly.

Although all hystricomorphs are well advanced at birth, the developmental state of a genus at that time depends primarily on whether the young remain in a protected nest site or leave the nest at an early age and follow the mother. *Octodon* appears to fall into the first category and is less developed at birth than *Pediolagus* whose young are probably born in the confines of a thorn bush (Smythe, 1970a) and begin following adults within a few days.

The young of many species tend to approach and follow moving objects at birth (*Dasyprocta*: Smythe, 1970a; *Cavia*: Kunkel & Kunkel, 1964; *Pediolagus*: D. G. Kleiman, unpublished). Several studies of *Cavia* have attempted to demonstrate that an imprinting-like process occurs (Shipley, 1963; Sluckin, 1968; Harper, 1970), as a result of exposure to, and following of, animate or inanimate objects. The following response has been emphasized in these studies, but little attention has been paid to whether the neonate does normally follow the mother after birth. In fact, many hystricomorph neonates appear to remain in a single location and wait for the mother to approach them for nursing (*Dasyprocta*: Smythe, 1970a; *Microcavia*, *Galea*: Rood, 1972). No species is known in which the young immediately follow the mother from birth and stay with her, as is seen commonly in some ungulates. Smythe (1970a) reports that young *Dasyprocta* are led from the birth site to a new location immediately after parturition and are then left by the mother, but this has not been recorded for other species for which there are field data, e.g. *Cavia*, *Galea* or *Microcavia* (Rood, 1972).

During the juvenile period, young apparently learn what is edible by stealing food from the mouths of adults (*Myoprocta*: Kleiman, 1969; *Capromys*: Bucher, 1937; *Proechimys*: Enders, 1935). Adults are extremely tolerant of juvenile food thefts, even when the young persistently pull food from the adults without even stopping to eat it (e.g. *Myoprocta*: Kleiman, 1969). This habit may also play an important role in the development of food preferences.

ECOLOGY AND SOCIAL ORGANIZATION

Hystricomorphs occupy a diversity of ecological niches with a corresponding variation in general habits (see Table VI). Very few detailed accounts have been given of the behaviour of species in field and captive conditions; Rood's (1972) study of *Cavia*, *Galea* and *Microcavia* is an exception. Some entries in Table VI are from anecdotal accounts which may be misleading, and conflicting data are given when they occur.

Very few species are solitary at all times except for mating. *Plagiodontia* and *Proechimys* are good examples, being nocturnal and living in a forest habitat. Such a habitat may impose a restriction on group size; *Cuniculus* lives in pairs and *Hystrix* in small family groups.

In grasslands and montane areas, hystricomorphs are often colonial but the complexity of the social structure of the colony varies from that of some species of *Ctenomys* in which each animal occupies its own burrow through family groups like those of *Lagostomus*, to the well organized social hierarchies of *Cavia*, *Galea* and *Microcavia*. In species inhabiting savannas and open country, a colonial existence may serve as an anti-predator mechanism; alarm calls are commonly found in these species (see Eisenberg, 1974). Seasonal variations in group size, reflecting environmental changes, may also occur. Recent studies on *Hydrochoerus* (Ojasti, 1971, 1973) indicate that capybaras congregate in large groups only when rivers and lakes dry up and that they disperse to their original pairs or family units after the arrival of the rains. These changes thus reflect only seasonal changes in water availability and not true differences in social structure.

In captive conditions, observed levels of social tolerance in a genus

TABLE VI

Habitat, mode of life, activity rhythm and social organization of some hystricomorph rodents

General	Habitat	Mode of life	Activity rhythm	Social system	References
Hystrix	Tropical forest, scrub savanna	T	N	F	Roberts, 1951; Shortridge, 1934; Mohr, 1965
Atherurus	Tropical rain forest	T	N	F	Rahm, 1962
Erethizon	Temperate coniferous forest	A	N	C(?)/S	Struthers, 1928; Costello, 1966; Marshall *et al.*, 1962
Coendou	Tropical evergreen rain forest	A	N	S	Enders, 1935
Cavia	Subtropical to tropical moist savanna	T	D/N	C/H	Rood, 1972
Galea	Subtropical, tropical or thornbrush savanna	T	D	C/H	Rood, 1972
Microcavia	Tropical to subtropical thornbush savanna	T	D	C/H	Rood, 1970, 1972
Dolichotis	Temperate steppe	T	D	F	Darwin, 1860; Krieg, 1929; Smythe, 1970a
Pediolagus	Temperate steppe	T	D	P/F	Smythe, 1970a
Hydrochoerus	Tropical to subtropical wet savanna; riverine	Aq	D/N	Cs/F	Ojasti, 1971, 1973
Dinomys	Lower montane tropical forest	A/T	N	S/P	Sanborn, 1931; Tate, 1931; Collins & Eisenberg, 1972
Dasyprocta	Tropical evergreen rain forest	T	D/N	S/P	Smythe, 1970a; Enders, 1935

Genus	Habitat				References
Myoprocta	Tropical evergreen rain forest	T	D	P/F	Morris, 1962; Walker, 1964; Kleiman, 1971
Cuniculus	Tropical evergreen rain forest	T	N	P	Enders, 1935; Smythe, 1970a; Mondolfi, 1972
Lagidium	Montane semi-arid	T	D	C/F	Pearson, 1948, 1951
Chinchilla	Montane, arid to semi-arid	T	N	C	Osgood, 1943; Weir, 1970
Lagostomus	Temperate to sub-tropical steppe	T/B	N	C/F	Llanos & Crespo, 1952; Hudson, 1872; Weir, 1974a
Capromys	Tropical evergreen forest	A/T	D/N	S/P/F	Bucher, 1937; Mohr, 1939
Geocapromys	Montane tropical forest, arid montane scrub	A/T	N	C/H	Clough, 1972
Plagiodontia	Tropical evergreen forest	A/T	N	S/P	Walker, 1964; J. F. Eisenberg, unpublished
Myocastor	Temperate to subtropical riverine habitat	Aq	D	C/H	Adams, 1956; Atwood, 1950; Laurie, 1946; Ehrlich, 1966
Octodon	Subtropical to temperate savanna and scrub	T/B	D	C/F	Osgood, 1943; Fulk & Schamberger, 1972
Octodontomys	Arid to semi-arid montane scrub	T	N	P(?)/F(?)/S(?)	Walker, 1964; M. L. Schamberger, personal communication; B. J. Weir & I. W. Rowlands, personal communication
Spalacopus	Temperate semi-arid coastal and montane	B	D	C/S(?)/F(?)	Reig, 1970
Aconaemys	Montane scrub	B	N	C	Osgood, 1943; Walker, 1964

continued

TABLE VI—*Continued*

Habitat, mode of life, activity rhythm and social organization of some hystricomorph rodents

Genera	Habitat	Mode of life	Activity rhythm	Social system	References
Ctenomys	Subtropical to temperate scrub-savanna; arid to semi-arid	B	D	C/S/F/	Pearson, 1951, 1959; Talice *et al.*, 1959; Pearson *et al.*, 1968; Weir, 1974a
Abrocoma	Semi-arid montane and coastal	A/T/B	N	C/H	Osgood, 1943; Cabrera & Yepes, 1960
Proechimys	Tropical evergreen forest	T	N	S	Enders, 1935; Fleming, 1971
Echimys	Tropical to subtropical evergreen forest	A/T	N	C	Tate, 1931; Walker, 1964
Carterodon	Thorn savanna	B	N		Walker, 1964
Clyomys	Thorn savanna	B		C	Walker, 1964
Hoplomys	Tropical evergreen forest	T	N		Goldman, 1920; Walker, 1964
Diplomys	Tropical evergreen forest	A	D/N		Goldman, 1920; Tesh, 1970
Thryonomys	Moist savanna	T	N	S/F	Shortridge, 1934; Ewer, 1969; Asibey, 1974
Petromus	Arid to semi-arid montane	T	D	S/P	Shortridge, 1934; Walker, 1964
Heterocephalus	Semi-arid to arid savanna	T		C/H	Hill *et al.*, 1957; Walker, 1964; Jarvis & Sale, 1971
Bathyergus	Semi-arid to arid savanna	T		C	Walker, 1964
Heliophobius	Semi-arid to arid savanna	T	D/N	C	Walker, 1964; Jarvis & Sale, 1971

A = arboreal; Aq = aquatic; B = fossorial; C = colonial; Cs = colonial (seasonally); D = diurnal; F = family; H = hierarchy; N = nocturnal; P = pair; S = solitary; T = terrestrial.

may necessitate a re-evaluation of social structure as described in the field. For example, *Octodon* and *Spalacopus* are considered to be colonial in habit as well as highly social, living in extended family groups (Reig, 1970; G. W. Fulk & M. L. Schamberger, personal communication). We have never had difficulty in pairing our *Octodon*; males and females, after an initial introduction, settle down and within a day huddle and groom each other. Our *Spalacopus*, however, will not share a nest box and pairs have fought during introductions. Thus, within the same presumed level of social organization (from field reports), different levels of social tolerance may be seen. For each genus, however social, the critical distance may differ; thus, some of the colonial, fossorial *Ctenomys* are as solitary and intolerant as terrestrial *Proechimys*. Clearly, the colonial and communal habits differ, as has been emphasized by Eisenberg (1966) and others. Individualized relationships need not exist in a colonial species while complex social relationships may be seen in genera living in pairs and small family groups. Finally, although there is little information on territoriality, there does not appear to be even one hystricomorph species whose individuals intermix freely in a herd-like association, although Hudson (1872) does describe *Lagostomus* individuals visiting neighbouring burrows with impunity.

If we discount differing levels of social tolerance, we still find that, for many hystricomorph genera, individuals are often found aggregated or in close association, i.e. they are not widely spread. Of the 38 genera listed in Table VI, 80% are colonial, live in family groups, or are pair-bonded (show contact-promoting behaviour outside the context of reproduction). These genera are found in forest, montane and open grassland habitat; they are terrestrial, fossorial, arboreal and aquatic and include both diurnal and nocturnal forms. Of course, the genera not listed may well include several of the more solitary forms, for which life history data have not been assembled. However, the percentage of colonial, communal and pair-bonded forms is probably no less than 70–75%. Thus, it would appear that there is a tendency for hystricomorphs to live in close association with conspecifics, regardless of the niche they occupy or the complexity of the social system.

This high level of colonial and/or social habit has probably evolved in conjunction with the reproductive and anti-predator mechanisms which have already been discussed. The young are mobile at birth and often move independently from the mother although they are still extremely susceptible to predation. Moreover, they are slow to develop and may not reach sexual maturity until over three months of age. For those species which have no means of defence, a colonial existence may enhance survival in the young since adults can warn young of danger and may have flight displays which are distracting. Species which do not associate with conspecifics for most of their life are often armed with anti-predator devices, e.g. spines in *Erethizon* and *Proechimys*.

CONCLUSIONS

Although it has not been repeatedly emphasized in this paper, some of the major behavioural adaptations of hystricomorphs (especially cursorial forms) are similar to those of ungulates, e.g. play, anti-predator behaviour, short copulations, prolonged mother-young bond, short nursing bouts, paternal protection, juvenile following and aggregative behaviour (see Fraser, 1968; Lent, in press). The similarities extend to morphology (Dubost, 1968), suggesting that hystricomorph rodents in Africa and South America and ungulates have been exposed to similar selective pressures in the course of their evolution. These behavioural adaptations are, in my opinion, a particularly fertile ground for further investigation.

ACKNOWLEDGMENTS

This review would not have been possible without the help of numerous individuals. I should especially like to thank J. Eisenberg, S. Wilson and T. Davis for many stimulating discussions, and access to their unpublished observations. The animal collection has been under the excellent care of Messrs E. Maliniak, J. Hough and M. Deal. Mrs W. Holden typed the manuscript through several drafts and G. Montgomery made helpful comments on its organization.

REFERENCES

Adams, W. H. (1956). The nutria in coastal Louisiana. *Proc. La Acad. Sci.* **19**: 28–41.

Asibey, E. O. A. (1974). The grasscutter, *Thryonomys swinderianus* Temminck, in Ghana. *Symp. zool. Soc. Lond.* No. 34: 161–170.

Atwood, E. L. (1950). Life history studies of nutria, or coypu, in coastal Louisiana. *J. Wildl. Mgmt* **14**: 249–265.

Barfield, R. J. & Geyer, L. A. (1972). Sexual behavior: ultrasonic post-ejaculatory song of the male rat. *Science, N.Y.* **176**: 1349–1350.

Beach, F. A. (1968). Factors involved in the control of mounting behavior by female mammals. In *Reproduction and sexual behavior*: 83–131. Diamond, M., (ed). Bloomington: Indiana University Press.

Beauchamp, G. K. (1973). Attraction of male guinea pigs to conspecific urine. *Physiol. Behav.* **10**: 589–594.

Bignami, G. & Beach, F. A. (1968). Mating behaviour in the chinchilla. *Anim. Behav.* **16**: 45–53.

Boling, J. L., Blandau, R. J., Wilson, J. G. & Young, W. C. (1939). Post-parturitional heat responses of newborn and adult guinea pigs. Data on parturition. *Proc. Soc. exp. Biol. Med.* **42**: 128–132.

Bucher, G. C. (1937). Notes on life-history and habits of *Capromys*. *Mems Soc. cub. Hist. nat. 'Filipe Poey'* **11**: 93–107.

Cabrera, A. & Yepes, J. (1960). *Mamiferos Sud-Americanos.* Buenos Aires: Historia Natural, Ediar.

Clough, G. C. (1972). Biology of the Bahaman hutia, *Geocapromys ingrahami.* *J. Mammal.* **53**: 807–823.

Collins, L. R. & Eisenberg, J. F. (1972). Notes on the behaviour and breeding of pacaranas, *Dinomys branickii,* in captivity. *Int. Zoo Yb.* **12**: 108–114.

Costello, D. F. (1966). *The world of the porcupine.* New York: J. B. Lippincott.

Critch, P. J. (1966). *The behaviour of the mara,* Dolichotis patagona (*Zimmerman,* 1785). Honour's Thesis, University of Witwatersrand.

Darwin, C. (1860). *Journal of the researches into the natural history and geology of the countries visited during the voyage round the world of H.M.S. "Beagle" under command of Captain Fitzroy, R.N.,* 2nd ed. (1901). London: John Murray.

Dathe, H. (1937). Über den Bau des männlichen Kopulations organes beim Meerschweinchen und anderen hystricomorphen Nägetieren. *Morph. Jb.* **80**: 1–65.

Dewsbury, D. A. (1972). Patterns of copulatory behavior in male mammals. *Q. Rev. Biol.* **47**: 1–33.

Dubost, G. (1968). Les niches écologiques des forêts tropicales Sud-Americaines et Africaines, sources de convergences remarquables entre rongeurs et artiodactyles. *Terre Vie* **1**: 3–28.

Ehrlich, S. (1966). Ecological aspects of reproduction in nutria, *Myocastor coypus* Mol. *Mammalia* **30**: 142–152.

Eibl-Eibesfeldt, I. (1952). Einige Beobachtungen an einer Freiheit gehalten weiblichen Biberratte (*Myocastor coypus*). *Zool. Gart., Lpz.* **19**: 277–283.

Eisenberg, J. F. (1966). The social organizations of mammals. *Handb. Zool. Berl.* **10**(7): 1–92.

Eisenberg, J. F. (1974). The function and motivational basis of hystricomorph vocalizations. *Symp. zool. Soc. Lond.* No. 34: 211–244.

Eisenberg, J. F. & Kleiman, D. G. (1972). Olfactory communication in mammals. *A. Rev. Ecol. Syst.* **3**: 1–32.

Enders, R. K. (1935). Mammalian life histories from Barro Colorado Island, Panama. *Bull. Mus. comp. Zool. Harv.* **78**: 383–502.

Ewer, R. F. (1960). Suckling behaviour in kittens. *Behaviour* **15**: 146–162.

Ewer, R. F. (1968). *Ethology of mammals.* London: Logos Press.

Ewer, R. F. (1969). Form and function in the grass cutter *Thryonomys swinderianus* Temm. (Rodentia, Thyronomyidae). *Ghana. J. Sci.* **9**: 131–141.

Fleming, T. H. (1971). Population ecology of three species of neotropical rodents. *Misc. Publs Mus. Zool. Univ. Mich.* **143**: 1–77.

Fraser, A. F. (1968). *Reproductive behaviour in ungulates.* London: Academic Press.

Freiheit, C. F. (1965). Courtship activity of the paca. *J. Mammal.* **46**: 707.

Fulk, G. W. & Schamberger, M. L. (1972). Mammals of Fray Jorge National Park, Coquimbo Province, Chile. *Am. Zool.* **12**: 657.

Goldman, E. A. (1920). Mammals of Panama. *Smithson. misc. Collns* **69**: 1–309.

H

Harper, L. V. (1970). Role of contact and sound in eliciting filial responses and development of social attachments in domestic guinea pigs. *J. comp. physiol. Psych.* **73**: 427–435.

Hill, W. C. O., Porter, A., Bloom, R. T., Seago, J. & Southwick, M. D. (1957) Field and laboratory studies on the naked mole rat, *Heterocephalus glaber*. *Proc. zool. Soc. Lond.* **128**: 455–514.

Holt, W. V. & Jones, R. C. (1973). The ultrastructure and histochemistry of the chin gland of the cuis, *Galea musteloides. J. Anat.* **114**: 221–233.

Hooper, E. T. (1961). The glans penis in *Proechimys* and other caviomorph rodents. *Occas. Pap. Mus. Zool. Univ. Mich.* No. 623: 1–18.

Howe, R. (1971). *Social behavior of the Bahaman hutia with an investigation of some aspects of marking behavior.* M. S. Thesis, University of Rhode Island.

Howe, R. (in press). Marking behaviour of the Bahaman hutia (*Geocapromys ingrahami*). *Anim. Behav.*

Hudson, W. H. (1872). On the habits of the vizcacha (*Lagostomus trichodactylus*). *Proc. zool. Soc. Lond.* **1872**: 822–833.

Jarvis, J. U. M. & Sale, J. B. (1971). Burrowing and burrow patterns of East African mole rats, *Tachyoryctes, Heliophobius*, and *Heterocephalus. J. Zool., Lond.* **163**: 451–479.

Kirchshofer, R. (1960). Über das "Harnspritzen" der Grossen Mara (*Dolichotis patagonum*). *Z. Säugetierk.* **25**: 112–127.

Klapperstück, J. (1954). Der Sumpfbiber (*Nutria*). *Neue Brehm Büch.* No. 115.

Kleiman, D. G. (1969). *The reproductive behaviour of the Green acouchi*, Myoprocta pratti. Ph.D. Thesis, University of London.

Kleiman, D. G. (1970). Reproduction in the female Green acouchi, *Myoprocta pratti* Pocock. *J. Reprod. Fert.* **23**: 55–65.

Kleiman, D. G. (1971). The courtship and copulatory behaviour of the Green acouchi, *Myoprocta pratti. Z. Tierpsychol.* **29**: 259–278.

Kleiman, D. G. (1972). Maternal behaviour of the Green acouchi (Myoprocta pratti Pocock), a South American caviomorph rodent. *Behaviour* **43**: 48–84.

Krieg, H. (1929). Biologische reisestudien in Südamerika. XV. Zur Ökologie der Grossen Nager des Gran Chaco und seiner grenzegebiete. *Z. Morph. Ökol. Tiere* **15**: 755–786.

Kunkel, P. & Kunkel, I. (1964). Beiträge zur Ethologie des Hausmeerschwein-chens *Cavia aperea f. porcellus* (L.). *Z. Tierpsychol.* **21**: 603–641.

Laurie, E. M. O. (1946). The coypu (*Myocastor coypus*) in Great Britain. *Anim. Ecol.* **15**: 22–34.

Lent, P. C. (in press). Mother-infant relationships in ungulates. In *The behaviour of ungulates and its relation to management.* Geist, V. (ed.). Morges: I.U.C.N.

Llanos, A. C. & Crespo, J. A. (1952). Ecologia de la Vizcacha (*Lagostomus maximus maximus* Blainv.) en el nordeste de la provincia de Entre Ríos. *Revta Invest. Agric.* **6**: 289–378.

Lorenz, K. (1941). Vergleichende Bewegungsstudien an Anatiden. *J. Orn., Lpz.* **89**: 194–294.

McBride, G. (1963). The "teat order" and communication in young pigs. *Anim. Behav.* **11**: 53–56.

Maliniak, E. & Eisenberg, J. F. (1971). Breeding spiny rats, *Proechimys semi-spinosus*, in captivity. *Int. Zoo Yb.* **11**: 93–98.

Marshall, W. H., Gullion, G. W. & Schwag, R. G. (1962). Early summer activities of porcupines as determined by radio-positioning techniques. *J. Wildl. Mgmt* **26**: 75–79.

Martan, J. (1962). The effect of castration and androgen replacement on the supracaudal gland of the male guinea pig. *J. Morph.* **110**: 285–293.

Mohr, E. (1939). Die Baum- und Ferkelratten-Gattungen *Capromys* (Desmarest) (sens. ampl.) und *Plagiodontia* (Cuvier). *Mitt. hamb. zool. Mus. Inst.* **48**: 48–118.

Mohr, E. (1949). Einiges vom Grossen und vom Kleinen Mara (*Dolichotis patagonum* Zimm. und *salinicola* Burm.). *Zool. Gart., Lpz.* **16**: 111–133.

Mohr, E. (1965). Altweltliche Stachelschweine. *Neue Brehm Buch.* No. 350.

Mondolfi, E. (1972). La Lapa o Paca. *Defensa Natur. Venez.* **2** (5): 4–16.

Morris, D. J. (1962). The behaviour of the Green acouchi (*Myoprocta pratti*) with special reference to scatter hoarding. *Proc. zool. Soc. Lond.* **139**: 701–732.

Müller-Schwarze, D. (1968). Play deprivation in deer. *Behaviour* **31**: 144–162.

Newson, R. M. (1966). Reproduction in the feral coypu (*Myocastor coypus*). *Symp. zool. Soc. Lond.* No. 15: 323–334.

Ojasti, J. (1968). Notes on the mating behaviour of the capybara. *J. Mammal.* **49**: 534–535.

Ojasti, J. (1971). El Chigüire. *Defensa Natur. Venez.* **1** (3): 4–14.

Ojasti, J. (1973). *Estudio biologico del chigüire o capibara.* Caracas: Fondo Nacional de Investigaciones Agropecuiarias.

Osgood, W. H. (1943). The mammals of Chile. *Publs Field Mus. nat. Hist. Zool. Ser.* **30**: 1–268.

Pearson, O. P. (1948). Life history of mountain viscachas in Peru. *J. Mammal.* **29**: 345–374.

Pearson, O. P. (1951). Mammals in the highlands of Southern Peru. *Bull. Mus. comp. Zool. Harv.* **106**: 117–174.

Pearson, O. P. (1959). Biology of the subterranean rodents, *Ctenomys*, in Peru. *Mems Mus. Hist. nat. "Javier Prado"* **9**: 1–56.

Pearson, O. P., Binsztein, N., Boiry, L., Busch, C., di Pace, M., Gallopin, G., Penchaszadeh, P. & Piantanida, M. (1968). Estructura social, distribucion espacial, y composicion por edades de una poblacion de tuco-tocos (*Ctenomys talarum*). *Investnes zool. chil.* **13**: 47–80.

Pocock, R. I. (1922). On the external characters of some hystricomorph rodents. *Proc. zool. Soc. Lond.* **1922**: 365–427.

Rahm, U. (1962). Biologie und Verbreitung des afrikanischen Quastenstachlers *Atherurus africanus* Gray (Hystricomorpha). *Revue suisse Zool.* **69**: 344–359.

Ralls, K. (1971). Mammalian scent marking. *Science, N.Y.* **171**: 443–449.

Reig, O. A. (1970). Ecological notes on the fossorial octodont *Spalacopus cyanus* (Molina). *J. Mammal.* **51**: 592–601.

Rewell, R. E. (1949). Hypertrophy of sebaceous glands on the snout as a secondary male sexual character in the capybara, *Hydrochoerus hydrochaeris*. *Proc. zool. Soc. Lond.* **119**: 817–819.

Roberts, A. (1951). *The mammals of South Africa.* Johannesburg: Central News Agency.

Rood, J. P. (1970). Ecology and social behavior of the desert cavy (*Microcavia australis*). *Am. Midl. Nat.* **83**: 415–454.

Rood, J. P. (1972). Ecological and behavioural comparisons of three genera of Argentine cavies. *Anim. Behav. Monog.* **5**: 1–83.

Roth-Kolar, H. (1957). Beiträge zu einem Aktionssystem des Aguti (*Dasyprocta aguti aguti* L.). *Z. Tierpsychol.* **14**: 363–375.

Sanborn, C. C. (1931). Notes on *Dinomys*. *Publs Field Mus. nat. Hist. Zool. Ser.* **18**: 148–163.

Shadle, A. R. (1943). The play of American porcupines (*Erethizon d. dorsatum* and *E. epixanthum*). *J. comp. Psychol.* **37**: 145–150.

Shadle, A. R. (1946). Copulation in the porcupine. *J. Wildl. Mgmt* **10**: 159–162.

Shadle, A. R. (1954). Natural parturition of a porcupine and first reactions of a porcupette. *Ohio J. Sci.* **54**: 42.

Shadle, A. R. & Ploss, W. R. (1943). An unusual porcupine parturition and development of the young. *J. Mammal.* **24**: 492–496.

Shadle, A. R., Smelzer, M. & Metz, M. (1946). The sex reactions of porcupines (*Erethizon dorsatum dorsatum*) before and after copulation. *J. Mammal.* **27**: 116–121.

Shipley, W. U. (1963). The demonstration in the guinea pig of a process resembling classical imprinting. *Anim. Behav.* **11**: 470–474.

Shortridge, G. C. (1934). *The mammals of Southwest Africa* **1**. London: Heinemann.

Sluckin, W. (1968). Imprinting in guinea pigs. *Nature, Lond.* **200**:1148.

Smythe, N. (1970a). *Ecology and behavior of the agouti* (Dasyprocta punctata) *and related species on Barro Colorado Island, Panama.* Ph.D. Thesis, University of Maryland.

Smythe, N. (1970b). On the existence of "pursuit invitation" signals in mammals. *Am. Nat.* **104**: 491–494.

Spalsbury, J. R. (1956). Unusual sex behavior of a male porcupine, *Erethizon dorsatum epixanthum*. *J. Mammal.* **37**: 452–453.

Struthers, P. H. (1928). Breeding habits of the Canadian porcupine (*Erethizon dorsatum*). *J. Mammal.* **9**: 300–308.

Talice, R. V., Tedeschi, E., Laffitte de Mosera, S. & Lagomarsino, J. C. (1959). Investigaciones sobre *Ctenomys torquatus*: un roedor autoctono del Uruguay. *An. Fac. Med. Univ. Montevideo* **44**: 452–462.

Tate, G. H. H. (1931). Random observations on habits of South American mammals. *J. Mammal.* **12**: 248–256.

Taylor, R. H. (1970). *Reproduction, development and behavior of the Cuban hutia conga*, Capromys p. pilorides, *in captivity*, M.S. Thesis, University of Puget Sound.

Tesh, R. B. (1970). Observations on the natural history of *Diplomys darlingi*. *J. Mammal.* **51**: 197–198.

Walker, E. P. (1964). *Mammals of the world* **2**. Baltimore: Johns Hopkins Press.

Walther, F. R. (1969). Flight behaviour and avoidance of predators in Thomson's gazelle (*Gazella thomsoni* Guenther 1884). *Behaviour* **34**: 184–221.

Weir, B. J. (1967). *Aspects of reproduction in some hystricomorph rodents*. Ph.D. Thesis, University of Cambridge.

Weir, B. J. (1970). Chinchilla. In *Reproduction and breeding techniques for laboratory animals*. Ch. 11: 209–223. Hafez, E. S. E. (ed.). Philadelphia: Lea & Febiger.

Weir, B. J. (1971). The evocation of oestrus in the cuis, *Galea musteloides*. *J. Reprod. Fert.* **25**: 405–408.

Weir, B. J. (1973). The role of the male in the evocation of oestrus in the cuis, *Galea musteloides* (Rodentia: Hystricomorpha). *J. Reprod. Fert.* Suppl. **19**: 421–432.

Weir, B. J. (1974a). The tuco-tuco and plains viscacha. *Symp. zool. Soc. Lond.* No. 34: 113–130.

Weir, B. J. (1974b). Reproductive characteristics of hystricomorph rodents. *Symp. zool. Soc. Lond.* No. 34: 265–301.

Wilson, S. & Kleiman, D. G. (1974). Eliciting play: a comparative study (*Octodon, Octodontomys, Pediolagus, Phoca, Choeropsis, Ailuropoda*). *Am. Zool.* **14**: 341–370.

Young, W. C. & Grunt, J. A. (1951). The pattern and measurement of sexual behavior in the male guinea pig. *J. comp. physiol. Psychol.* **44**: 492–500.

Zara, J. L. (1973). Breeding and husbandry of the capybara, *Hydrochoerus hydrochaeris*, at Evansville Zoo. *Int. Zoo Yb.* **13**: 137–139.

Symp. zool. Soc. Lond. (1974) No. 34, 211–247.

THE FUNCTION AND MOTIVATIONAL BASIS OF HYSTRICOMORPH VOCALIZATIONS

J. F. EISENBERG

*National Zoological Park, Smithsonian Institution,
Washington, D.C., U.S.A.*

SYNOPSIS

The author has studied the vocalizations of 17 species of New World hystricomorphs drawn from seven families. Special attention has been given to the vocal repertoires of: *Octodon degus, Myoprocta pratti, Dinomys branickii* and *Cavia porcellus*. The maximum number of different sounds produced by a given species ranges from 11 to 14. Generally, about nine to ten sound forms are true vocalizations; the remaining sounds are produced by tooth chattering, stamping or in some cases, quill vibrations. Vocalizations can be derived from four basic syllable types which appear to reflect at least three different fundamental moods. The fundamental syllable type can be transformed or combined in such a way as to indicate shifts in arousal. Within the vocal repertoire of all caviomorph species there appears to be a series of syllable subtypes connected by intermediate forms which can serve as 'excitation indicators'.

Natural selection has produced stereotyped calls in all species which may be rigidly tied to a given context and reflect a specific mood. A restricted function is strongly implied. Alternatively, part of the vocal repertoire appears to be unspecific with respect to context and serves as a general indicator of the sender's arousal state. The functional interpretation of calls relies heavily on studies of animals in captivity and field data. In many cases the latter are almost totally lacking for a given species.

INTRODUCTION

The systematic study of animal sounds has gained considerable impetus in the last 15 years and reviews have appeared at regular intervals (Tembrock, 1959; Lanyon & Tavolga, 1960; Busnel, 1963, 1967). Much of the initial work on mammals was on the Chiroptera (see Griffin, 1958) and Primates (Andrew, 1963; Struhsaker, 1967) but, following the discovery of ultrasonic cries in young mice (Zippelius & Schleidt, 1956), the rodents have received ever-increasing attention. Progress in the analysis of ultrasonic cries has been reviewed by Sewell (1970), who made a significant step in her syntactic analyses (Sales, née Sewell, 1972a,b). The most complete analysis of rodent vocalizations published to date is that of Brooks & Banks (1973) for the collared lemming, *Dicrostonyx*.

To date, almost all studies on vocalization and ultrasonic cries in rodents have been confined to murids. A great amount of attention has been given to the ultrasonic sounds and it is hoped that it will not mask the equal importance of rodent sounds below 18 kHz.

In 1964, the National Zoological Park set up breeding colonies to study the behaviour of some South American hystricomorph rodents (Caviomorpha) and material for this paper (see Table I) was obtained from a wide range of sounds made by 17 different species (referred to throughout by their Latin names) bred in captivity. Although some of the signals exceed 25 kHz many species produce sounds which are readily audible and easily recorded in the field with standard equipment.

MATERIALS AND METHODS

Animals

The behaviour of the following seven species in captivity has been studied in detail: *Cavia porcellus*, *Octodon degus*, *Myoprocta pratti*, *Octodontomys gliroides*, *Proechimys semispinosus*, *Dinomys branickii* and *Pediolagus salinicola* (see Kleiman, 1971, 1972; Maliniak & Eisenberg, 1971; Collins & Eisenberg, 1972). Field data collected in Panama and Mexico are available for *Cuniculus paca*, *Dasyprocta punctata* and *Proechimys semispinosus*. The vocal repertoires are most completely analysed for *Octodon*, *Myoprocta*, *Cavia* and *Dinomys*. We have not extended our studies to Old World hystricomorphs except for occasional observations on *Hystrix* in the Zoo.

Equipment

Vocalizations and sounds were recorded with a Uher Report-L 4000. When recording at tape speeds of $7\frac{1}{2}$ in./sec with the Uher microphone, a reasonably flat response can be maintained over a range of 50 to 15 000 Hz. An Electrovoice Model 460 directional microphone extended the range to 18 kHz. The sound recordings were analysed with a Missilyzer (Kay Elemetrics Company, Pine Brook, New Jersey). Tapes recorded at $7\frac{1}{2}$ in./sec can be played back at $3\frac{3}{4}$ in./sec and energy to 30 kHz can be noted, but relatively accurate analyses can be carried out only by playing back at the recorded speed and analysing to 15 kHz, which is the normal top range of the Missilyzer.

Ultrasonic pulses were monitored with an ultrasonic microphone and transducer complex which allowed the observer to hear rectified ultrasonic sounds (for details, see Eisenberg & Gould, 1970: 131). Sound spectrographs produced by the Missilyzer were used to describe the form, duration, interval and frequencies of the energy distribution. Energy distributions are imprecise when maximum temporal resolution is needed because the analysing filters can only display frequencies in 600 Hz increments at the 15 kHz setting. Accuracy to 60 Hz can be made with the narrow band setting but temporal resolution is lost.

Recordings

Recording sessions were designed to allow complete recordings of vocalizations exhibited during defined contexts. Recordings were made in the following situations: (a) during male–female encounters when the female was receptive and non-receptive; (b) during the maternal-care phase of the life cycle; (c) when fighting; (d) when young were isolated from parents and exploring a novel environment; (e) when animals were startled, frightened, and emitting warning or submissive sounds. Special manipulations were performed with *Cavia porcellus* (see p. 218).

Definitions

The following classification is a derivation (see Eisenberg, Collins & Wemmer, in press) of the system established by Andrew (1963) to analyse the vocalizations of primates, and is proposed to aid the discussion of the form of a syllable (see Table I).

1. *According to duration*. The click is a non-harmonic sound less than 20 msec in duration (Type III syllable). Short syllables are between 50 and 600 msec in duration, and long syllables are over 600 msec.

2. *According to the distribution of energy*. Type I syllables may be long or short but have in common the property of organized energy distributions into discrete frequency bands. There are several subtypes of Type I syllables depending on the presence or absence of frequency modulation. Type II syllables are mixtures of an harmonic form with the superimposition of noise. Type IV and Type III syllables are noisy. The energy itself is not organized into discrete bands.

A syllable is an uninterrupted tracing on the horizontal axis of the sonograph. Syllables may be organized into phrases. Syllables within a phrase are not separated by time intervals greater than that separating two phrases.

ANALYSIS OF THE SOUNDS OF *Cavia*

Cavia porcellus is such a vocal species (Kunkel & Kunkel, 1964) that one wonders whether domestication has altered the thresholds for vocal expression and, if so, has complicated the interpretation of the

TABLE I

Physical measurements of caviomorph sounds

Species	Name of sound	Syllable type*	Duration		Interval		Audible frequencies emphasized (kHz)	Emphasis on frequencies above 20 kHz
			n	(msec)	n	(msec)		
Spalacopus cyanus	Trill (a) syllable	I ds; r	103	Av. 57	99	15–25	0·4–1·2	—
	Trill (b) phrase		4	300–8000	2	225–2000†	0·9–1·2	—
Octodontomys gliroides	1) Protest squeak	IV s; r	5	30–32	4	112–150	1·5–2·0; 3·0–5·5	?
	2) Post-copulatory cry ♂	IV s; r	4	Av. 31	3	Av. 65	<7·0	+
Octodon degus	Adult:							
	1) Sharp squeak (warning)	II s; r	4	Av. 70	1	76	12–24	+
	2) Protest squeak	II ad	2	230–480		variable	2·0–4·0	—
	3) Protest growl	II d–IV	1	690		variable	<4·0	—
	4) Clucks	II r; s	6	17–46	4	40	2·0; 4·0	—
	5) Chuck-wee						<2·0	—
	a) Entire phrase	(see below)		400				
	b) Twi portion	I d, s	5	40–48	4	146–154	<2·0	
	c) Long portion	I ad; u; II	5	250–415		variable	<2·0	
	d) Terminal chuckles	II–III	3	19–40	2	19–40	<2·0	
	6) Post copulatory cry ♂	II s–IV s	7	Av. 143	5	200–310	<5·0	?
	Infant:							
	1) Isolation	I ad, r	4	85–165	3	200–230	1·2; 2·8	—(?)
	2) Gurgles	I as; II–III	22	42–86	15	42–135	0·9–3·4; 3·5	—

Dasyprocta punctata	Warning growl	II u–IV	3	Av. 380	2	Av. 200	<1·0	—
Cuniculus paca	Growl	II u–IV	5	393–2056	4	385–771	<1·0	—
Myoprocta pratti	Adult:							
	1) Sharp squeak	I chevron	5	45–50		variable	0·6–0·8; 1·4–1·6; 0·8–1·2; 1·8–2·2	—
	2) Tooth chatter	III	10	7–11	9	Av. 740	1·5–2·2	+
	3) Whine	II–IV	3	340–1540		variable	<2·0	—
	4) Snort	IV r, s	8	55–60	7	170–200	<3·0	—
	5) Courting squeak	I ad; or I u	9	54–410	6	130–410	2·2; or 3·0 1·8 –2·0 or 2·0	—
	6) Purr	I ad–II	4	170–400		variable	0·6–1·0	—
	Infant:							
	1) Squeak	I d	6	77–200	4	77–250	2·4–6·0 or 2·4–4·8	?
	2) Chirp	I d	4	Av. 350	2	254–310	2·0–0·7; 0·7–2·0 or 1·4–4·0	?
Proechimys semispinosus	1) Growl ♂	II u	1	560		variable	0·8–1·2	—
	2) Cluck series							
	a) Slow	II u	10	80–110	9	Av. 140	0·2–0·4; 0·8–1·2	—
	b) Whimper	II u		∼ 90		92–100	0·2–0·4; 0·8–1·2	
	3) Wail	I u or I a	4	600–1030	3	∼ 40	<3·0; 1·1; 1·8–2·2; 2·6–2·9	—

Continued

TABLE I—*Continued*

Species	Name of sound	Syllable type*	Duration		Interval		Audible frequencies emphasized (kHz)	Emphasis on frequencies above 20 kHz
			n	(msec)	n	(msec)		
Dinomys branickii	Adult:							
	1) Hiss	IV	4	Av. 612		variable	<5·0	+
	2) Tremulous whine pulsed or fused	I ad–I u	3	850–2270	3	360–3465	0·6–1·0; 0·5–0·7	
	3) 'Song' Syllables:	Whole phrase	8	500–63750	>750	<4250		
	α	II u, s	53	Av. 108			1·0–1·5–1·9 or	?
	β	II ads	96	Av. 46	All syllables 15–246		1·2–1·8	—
	γ	I ad; I d	3	Av. 303	n = 151; mode = 85		0·2–0·6; 0·8–1·2	—
	Δ	I ad L	7	Av. 785				
	Sub-adult:							
	4) Wha whee	I ad	9	Av. 290	8	Av. 412	0·6; 1·2; 1·8	—
	5) Grunt	II u	4	Av. 102	2	∼700	<0·5	
Lagidium peruanum	High twitter							
	a) Phrase			<2000		variable		
	b) Syllable	I ad, r	8	77–195		∼70–150	4·8–5·4; 3·2–4·4	+
Lagostomus maximus	1) Grunt	II u, r	3	103–900			<0·6	—
	2) Oogah			<2000		variable		—
	a) Preface squeak	I ad	2	<400		variable	0·18–1·32	
	b) Oo ah	III u–IV	2	1600		variable	<0·6 or <1·2	

Capromys *pilorides*	1) Twitter							+
	a) Phrase		2	1100–1800				
	b) Syllable	I ad, r–II	9	80–120			2·4–10·2	+
	2) Sveet	I a, s	2	330–470		variable	3–13·2–10·6	?
Plagiodontia *aedium*	1) Tweet	I u–I d	4	107–325		variable	5·0–1·6	?
	2) Oogah	II ad (compound)	10	800–2000		variable	0·2; 1·0; 1·8; 0·6; 1·2; 2·2	—
	a) Oo	I ad–II ad	4	460–720		variable	∼0·8	
	b) Gah	II ad	4	460–640		variable	∼0·3	
Cavia *porcellus*	Adult:							
	1) Tutt-tutt	II s	4	26	4	36	0·5; 1·0; 2·0	—
	2) Purr	II u, s	32	34–39	31	30–33	<0·4; <2·4	—
	3) Clucks	II ad, s	12	Av. 57	11	Av. 188	∼2·0	
	4) Chuckle	II u, s						
	5) Short squeak	I ad–II ad	4	Av. 267	4	Av. 160	0·4–1·2	?
	6) Protest squeak							
	a) Mild	I ad, s	10	Av. 112	9	Av. 70	0·6–1·2	?
	b) Strong	I ad, L	3	Av. 469	2	118–127	1·0–1·6	—
	Juvenile:							
	1) Gurgle	II ad	6	Av. 58	5	Av. 34	<2·0	?
	2) Short squeak							
	a) When huddled	II ad, s	11	Av. 79	10	Av. 88	0·8–1·4; 0·4–1·0	?
	b) When tickled	II ad, s	14	Av. 161	13	Av. 122	<2·0	?
	3) Whee-wheet							
	a) Rapid	I ad	18	Av. 193	21	Av. 116	0·5–3·5	?
	b) Prolonged	I ad, L	4	Av. 421	21	Av. 116	0·5–3·5	
Dolichotis *patagonum*	1) Wheet		2	620–925				?
	a) Preface	I u; I a	5	200–300		variable	4·8; 0·3–4·2	
	b) End	IV	2	∼300			0·6	
	2) Grunt	II r–IV r	5	115–310	6	200–320	<2·4(0·2–0·4)	—

* I, II, III, and IV = syllable types as defined in text.

† Compound call (see text). a = ascending frequency; d = descending frequency; ad = complex frequency modulation; s = short syllable; L = long syllable; r = repeated emission; u = no appreciable frequency modulation.

functional significance of its sounds. Rood's (1972) discovery of a similar complex in the vocal repertoire of the wild guinea-pig, *Cavia aperea*, makes this possibility seem unlikely but there still remains a chance that any modification in the repertoire of the domestic species (Table I) could have resulted from the selective pressures of domestication.

Correlation of stimulus input and vocal output

Simple manipulations were performed on guinea-pigs causing minimal disturbance and stress to record their vocal output. Juvenile and adult males were subjected to simple manipulations in order to stimulate vocal response, as follows:

(a) lightly 'tickling' on the throat and neck which produced short cluck-like squeaks;

(b) strong stroking of neck and shoulders to produce a lengthened syllable or inflected squeak ('wheet wheet');

(c) tickling of groin or rump which produces a long, loud series of inflected squeaks plus the introduction of noise, i.e. the harmonic system blurred; the adult males kicked violently while squeaking (see Fig. 1).

A similar progressively more intense series of vocalizations which grade into one another may be shown by a juvenile animal when displaced from the mother and siblings. If an isolated sibling is allowed to explore alone in an open arena, it will begin to vocalize with a series of clucks. Clucks change to chirps which grade to the 'wheet wheet' inflected squeal so well known to students of *Cavia*. Such a series is depicted in Fig. 1. The similarity between the two sets of sound forms and their transitions is striking. Such a graded series appears to be generated by two simple rules of transformation: (i) to shorten the interval separating syllables so that fusion occurs and results in a long call, and (ii) to lengthen the syllable while holding the interval relatively constant. A graded series of calls so generated may convey a great deal of information about the degree of arousal in the vocalizing animal.

In *C. porcellus* the basic graded series from clucks to wheets is the fundament on which selection has acted to produce specific calls associated with rather predictable stimulus situations. The steps in such selective processes have been outlined by Hinde & Tinbergen (1958), and the chief innovation which is introduced here is the recognition that selection can also act to promote the enhancement of a series of calls which may be termed 'arousal indicators'. This series is not tied to a specific form of stimulus input. Rather, as Andrew (1964) points out,

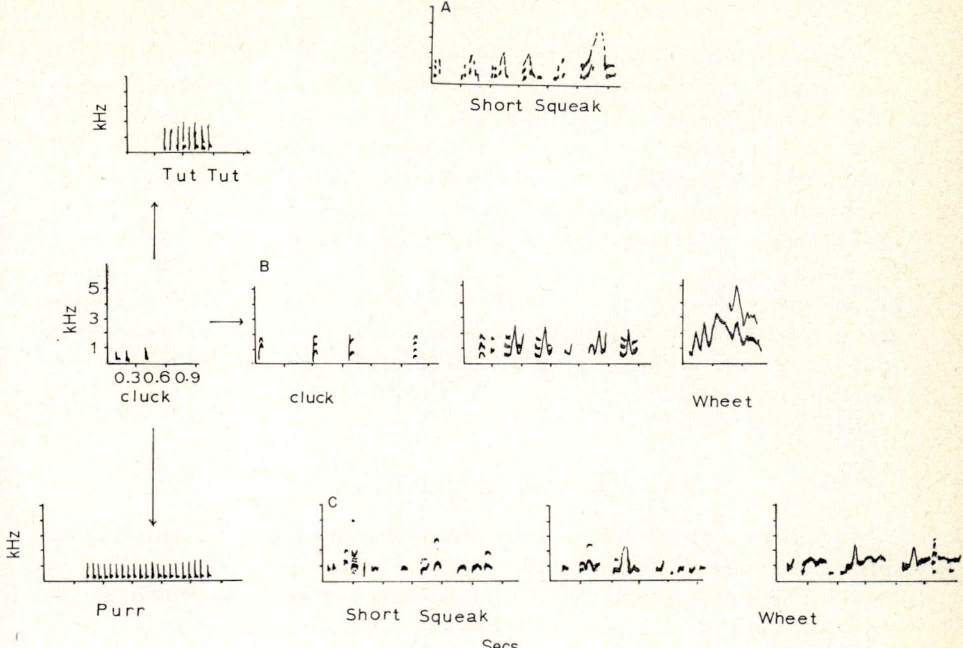

FIG. 1. Transformation of syllable types in *Cavia porcellus* (see text for discussion). The basic cluck (Type II) syllable can be uttered with a specific temporal patterning to yield the tut-tut, or purr. Longer clucks (Type I syllable) may transform to short squeaks or loud wheets. Series C was produced by scratching a male on the neck and then on the groin. Series B was produced by a two-week old juvenile while exploring an open area but separated from the mother and siblings. Series A is an inflected squeak series produced by 'grooming' the young around the neck. The ordinates are in 1 kHz increments; the abscissae are in 0·3 sec increments.

three overlapping types of stimulus contrast can be recognized which induce calling: (a) contrast with the background level or the immediately preceding stimulation; (b) contrast resulting from the discrepancy between what the recipient anticipates as input and what is actually received, and (c) contrast resulting from some specific, high-energy input.

The arousal potential of the second type of stimulus contrast, involving an 'anticipation' or a 'set' for a given input which then is not met by the environment, should not be underestimated. The application of such a theoretical orientation to an analysis of chemical communication has already been published (Eisenberg & Kleiman, 1972).

Motivation

Some vocalizations of *Cavia* are quite distinct with respect to their context and the immediate mood or tendency on the part of the sender. They should impart, in combination with olfactory, tactile, or visual cues, the intent of the sender to a conspecific recipient. These sounds include (a) the warning call and (b) courting purr, both derivable from the basic cluck; (c) the protest squeal and scream derivable from the inflected wheet; (d) the aggressive grunt derivable from the cluck and (e) the tooth chatter. These mood-specific sounds exist side by side with the graded series of 'excitation indicators' which are the cluck, chirp, squeak and inflected wheet. This latter series is not tied to a specific propensity to mate, attack or flee, but rather has the more general communicatory function of indicating the degree of arousal and/or location of the sender.

Auditory communication

In order to assess function we must also infer the presumptive information content of the signal.

(1) The purr apparently communicates the sex and mood of the sender (a male) to the receiver (the courted animal). It is produced by an aroused adult male and may be considered an indicator of sexual capacity. It is produced at relatively close range and can repel or attract other adult males when courting is in progress. It may be shown in homosexual courting bouts by both males and females. Females only produce the sound when kept together in the absence of males.

(2) The tutt-tutt-tutt is clearly a warning sound for close range communication to members of a family group. It is produced during moderate stimulus contrast and induces silence and immobility on the part of the receivers. Although a vocal emission, its rhythm and energy distribution suggest a pattern found in the foot drumming of such cursorial forms as *Dasyprocta* and *Myoprocta*.

(3) Clucks, though nonspecific with respect to the stimuli releasing them, nevertheless serve to inform a receiver of the location and the arousal level of the sender. A female will respond to the clucks of a displaced juvenile by uttering reply clucks. Juveniles moving as a group stay together by producing clucks which appear to be mutually facilitating.

(4) Chirps and wheets are not specific with respect to the stimuli which elicit them, and their function is defined by the context in which they are emitted. At best one can say that these calls appear to be free from sexual or aggressive moods. These calls are graded with respect to

form and intensity. The chirp may develop out of the cluck via the chuckle as an intermediate form. Chuckles may be given when juveniles huddle after a separation and may be given during grooming by the mother. The high intensity wheet call is loud and can be given when an anticipated stimulus (such as food) is not present.

(5) The protest squeal or squeak, which is similar to a wheet but with noise superimposed, may be given when handled, groomed by a dominant or bitten. Graded in its intensity, it functions to inhibit attack by a conspecific.

(6) Tooth chattering is graded but shows a typical intensity pattern at high levels of arousal. At low arousal the stimulus situation is variable but at the typical intensity level the chattering animal is aggressive and generally confronted with an alien conspecific. This is a warning signal which communicates readiness to attack on the part of the sender.

(7) The low grunt is generally given at close range prior to an attack, and appears to function as a signal warning the recipient of an impending attack.

After the completion of this review, an analysis of vocalization by *Cavia porcellus* was published by J. Coulon. My own classification system for *Cavia* agrees very closely including the conclusions with respect to the motivational analysis (Coulon, 1973).

SOUNDS OF SOME OTHER CAVIOMORPHS

During interactions with conspecifics, all movements, sounds, odours and tactile stimuli are considered to combine in various ways to produce significant communication between participants. Sounds are all one class of signal having presumptive communicative valence. The social function of sounds cannot be studied outside the context in which they are produced. An extensive quantitative analysis of the sounds made by caviomorphs, grouped according to family, is given in Table I and this section is an attempt to relate these sounds to a function.

Octodontidae

Spalacopus cyanus

Spalacopus produces several sound types. Tooth chattering is generally employed as a threat, a growl-like squeal is produced when attacked, and short squeaks are emitted during grooming bouts. The most characteristic and distinctive call is a musical trill (Fig. 2b) which,

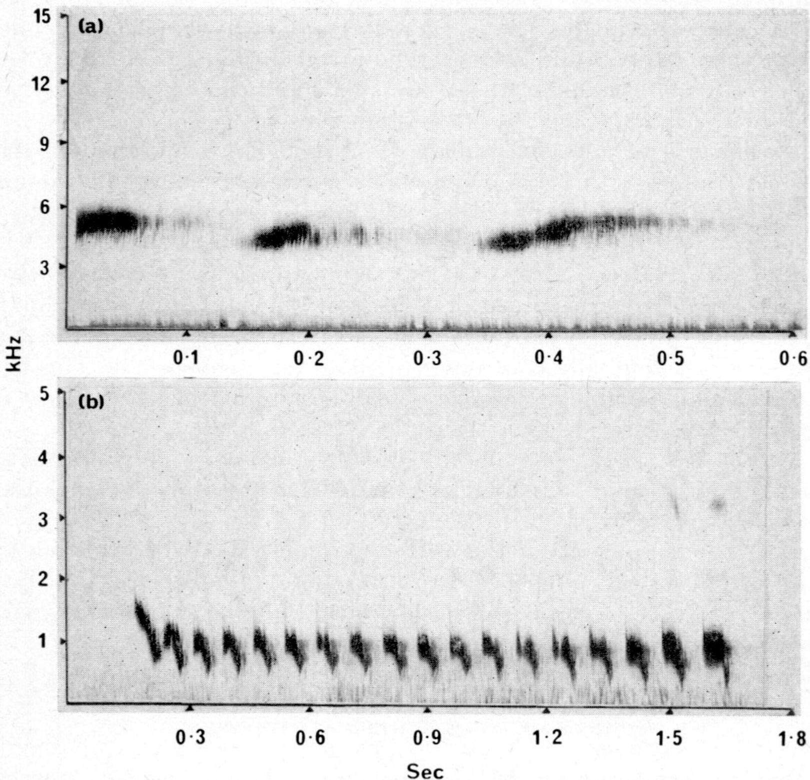

Fig. 2. Type I syllables.
(a) Portion of high twitter by *Lagidium peruanum*, Type Iu.
(b) Trill of *Spalacopus cyanus*, Type Id.

according to Reig (1972), is emitted when the animal is startled, often
when fleeing to its burrow. It is produced more commonly by females
than males. It is a warning call for the community.

Octodontomys gliroides

No field data are available for this species but, when disturbed in
the nest box, it will produce a cry composed of a series of short squeals
(Fig. 3a) and the male emits a characteristic post-copulatory cry
(Fig. 3b) which is of the same sound type, but more stereotyped in its
temporal patterning.

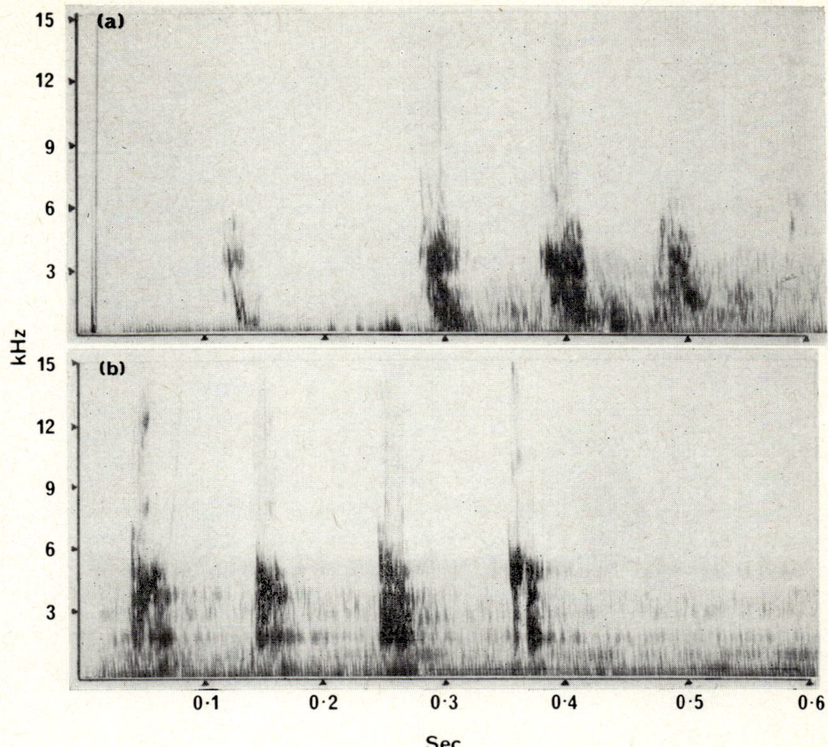

Fig. 3. Type IV syllables.
(a) Protest squeaks by *Octodontomys gliroides*, Type IV
(b) Post-copulatory call by *O. gliroides* (Type II to Type IV).

Octodon degus

Some aspects of the behaviour of *Octodon* have been examined in the field by M. Schamberger and C. Koford (personal communication). It is a communal species in which extended families occupy the same burrow system. They are diurnal and some adults frequently take up positions on rocks to alert against potential danger. When this arises they produce a sharp warning squeak which is also given in captivity when the nest box is disturbed. A similar squeak may be given by a female when approached by a male. Another but longer squeak is emitted during initial grooming which is in protest against over-vigorous treatment (Fig. 4a). All these squeaks are of similar type and quite often they grade into a very noisy sound approximating to a

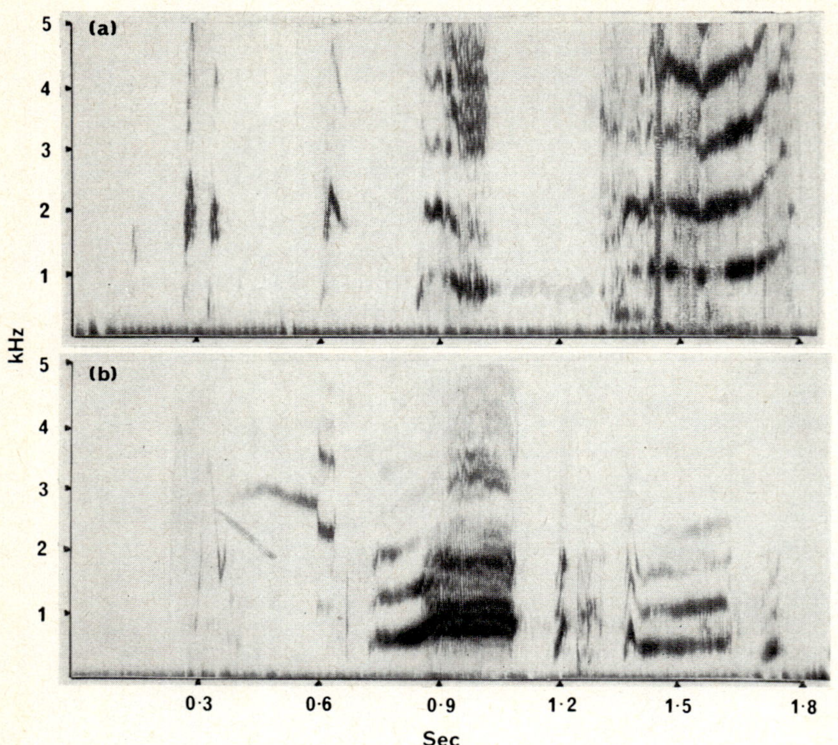

FIG. 4. Calls of *Octodon degus*.
(a) Three clucks (Type II-I sounds) followed by two protest squeaks (Type II sounds).
(b) 'Chuck-wee' by *Octodon* male. Note initial 'twi' portion followed by long syllable of Type I to buzz (Type II) with chuckles, long 'wee' and terminal chuckles.

growl. When threatening, an animal will show pilo-erection, tooth chatter and may growl before rushing at another animal. An injured animal may emit a prolonged squeal (Fig. 4a). During an encounter and allogrooming, adults utter short cluck-like sounds preceding a squeak. A special call, the 'chuck-wee', may be given when the animal is aroused but not frightened. This call (Fig. 4b) is heard in captivity when an animal sees a familiar keeper. It is a complex sound with several prefatory notes (i.e. clucks or 'twis') grading into a chuckle or sliding into a noisy buzz.

Adult males produce a post-copulatory call which is extremely uniform and repetitive (see Table I).

Infant sounds are similar to those of adults but their temporal
pattern and range are modified. When groomed, infants produce
chuckles and gurgles with very brief syllables. At seven days of age, the
young produce high-pitched squeaks which are very similar to the sharp
'warning' squeak of an adult. The young may also produce complex
calls consisting of a prefatory chirp which resembles the high intensity
anticipatory calls (the chuck-wee) of adults; these may be homologous.

Ctenomyidae

Reig (1972) noted that the warning call or trill of *Spalacopus*
differed qualitatively from that of *Ctenomys*. It is noteworthy that the
bubbling call of the latter (see Weir, 1974) although dissimilar in tonal
quality, is probably functionally identical. Pearson (1959) notes that
C. peruanus which is colonial, tending to communal, does produce the
bubbling call, unlike *C. opimus* which is solitary, and has no comparable
call.

Dasyproctidae

The vocalizations of these species that live in small family groups are
similar in many respects. Only *Dasyprocta* has been studied in the field
and captivity (Smythe, 1970a; J. F. Eisenberg, unpublished) but for
comparative purposes much can be inferred about the function of the
sounds emitted by the species studied only in captivity.

Myoprocta pratti

When held singly in a cage and startled, *Myoprocta* will often utter a
single sharp squeal, and when fleeing it occasionally emits a grunt.
Tooth chattering and pilo-erection occur when threatened. Sometimes,
in response to a stationary object or conspecific, *Myoprocta* may sit at a
distance and drum with its hind feet. At least two kinds of drumming
were described by Kleiman (1972). High intensity threat produces a
whine, and an aggressive animal, or one who has recently attacked
another, may snort, sometimes repeatedly. A submissive animal, when
approached, may emit a mewing sound (Morris, 1962) whereas an injured animal squeals loudly or screeches.

A courting male on approaching a female produces a low intensity
squeak (Fig. 5a) which is often accompanied by trembling of the forepaw. After copulation the male may produce a snorting sound similar
to that produced before an attack.

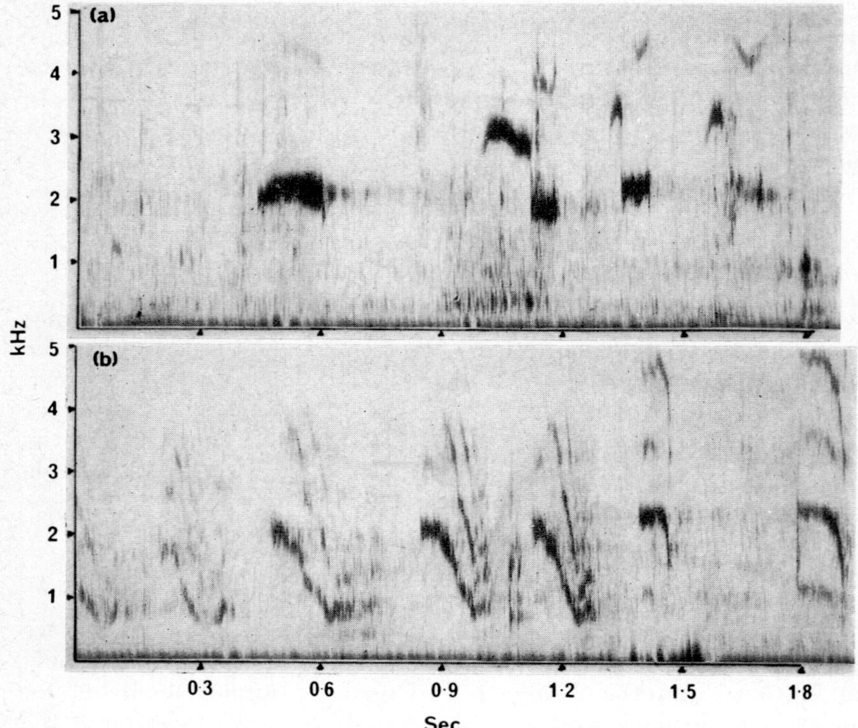

FIG. 5. Close-range calls of *Myoprocta pratti*.
(a) Courtship squeaks by male (Type I).
(b) Chirps by young as mother approaches and grooms one.

The female uses a special purring call when approaching her young
and infants may emit a chirp or squeak of extremely low intensity,
similar to the male courtship squeak but having a different form
(Fig. 5b).

Dasyprocta punctata

Dasyprocta produces sounds which are qualitatively similar to those
of *Myoprocta* in equivalent circumstances. However, when fleeing it
emits an 'alarm bark' consisting of a series of short grunts followed by a
very loud growl, which is not produced by *Myoprocta*. In response to any
stationary or sluggish predator, such as a snake, it emits a low rumbling
sound and thumps its hind feet. Threat evokes tooth chattering, thumping and pilo-erection and in response to an intimidating conspecific it

may produce a descending squeak, termed by Smythe (1970a) the "creak-squeak". During courtship, males characteristically show fore-paw trembling. When calling her young or with young a female emits a purring call. In seeking contact, the young animal produces a descending squeak form. An injured animal screams and, after winning an aggressive bout, it may growl or grunt.

Cuniculus paca

Very few recordings have been made but we have noted that, when startled, *Cuniculus* will run off emitting either a snort or a low growl. When threatening, it exhibits pilo-erection and tooth-chattering with grinding, and if provoked will growl very loudly. These sounds may be repeated at intervals. A female may call its young with a low grunt and during grooming may give off a low whine. The male does not appear to possess any courting cry.

Echimyidae

Proechimys

Some field data have been produced by Smythe (1970a) on *Proechimys semispinosus* but their nocturnal habit and solitary tendency make them difficult to observe in detail.

When threatening each other, *Proechimys* exhibit pilo-erection and tooth chatter. When submissive, and being approached by a conspecific, they may cluck. This sound may rapidly grade into a twitter series or whimper. When delivered at a fast or slow tempo, the individual pulses are of similar length but the interval between them may be decreased at the quicker rate. Aggressive males may emit a growl which may be repeated at variable intervals.

During the very long copulatory mount (see Maliniak & Eisenberg, 1971), the female produces a wailing call and she may be joined by the male thus creating a duet.

Dinomyidae

Dinomys branickii

Field data on this species are almost entirely lacking. We (Collins & Eisenberg, 1972) found that in captivity they would stamp with their forepaws when disturbed and emit hissing sounds irregularly. When provoked further by a conspecific or handling, they would intersperse these activities with tooth chattering. Full threat display involved the emission of a staccato whimper or tremulous whine of very long duration. Young animals exhibiting a following response would often produce contact grunts, frequently emitted in pairs. When being groomed they give as a mild protest a 'wha-whee' sound which appears to be a low intensity variant of the tremulous whine.

During courtship the male may utter a series of whimpering notes accompanied by forepaw trembling. These notes may be combined into an extended 'song' form (see Tables II and III). Alpha sounds are

TABLE II

Syllable patterns within a phrase from two 'songs' of Dinomys branickii*

Following syllable

		α	β	γ	Δ	$\vec{\Sigma}$
	α	12	12	15	3	42
Preceding syllable	β	17	382	8	2	409
	γ	13	7	60	10	90
	Δ	—	—	10	20	30
	$\Sigma\downarrow$	42	401	93	35	

* Based on 3 min 55 sec of calling on two separate days.

Type II syllables, short with a uniform or ascending frequency. Delta and gamma syllables are Type I sounds, highly modulated, and the two syllables differ in their average duration. As Table II indicates, within a phrase each syllable type tends to be repeated. The song is composed mainly of alternate bursts of short and long syllables.

The so-called song form composed of syllables normally employed during initial phases of courtship may be an artifact of captivity when a male is kept isolated from a female. A female may respond to the sound initially by some low intensity whimpers. The male may persist in his calling for over two minutes. A similar cry may be given by *Erethizon* during the breeding season, if the sexes are held separately but in auditory contact. In both *Erethizon* and *Dinomys*, this call may serve to attract a sexual partner.

Chinchillidae

Chinchilla laniger

Sounds of *Chinchilla* have not been analysed in any detail. When frightened or seized, an 'eek-eek' call is produced. A rather special call has been noted in isolated males who may produce at intervals a 'nyak-nyak' call. The functional significance of this call is difficult to understand at this point.

Lagostomus maximus

In 1872, Hudson described the rich vocabulary of this species. When threatening, an animal will stamp, growl, and tooth chatter. When being groomed females will often utter a clacking cry. However, the most characteristic vocalization is a braying call with several different variants, the most common being a prefatory squeak sliding into an 'oogah' sound (see Weir, 1974). This sound may be produced by a sudden disturbance in the colony, such as opening the nest box in captivity or the appearance of a dog in the wild.

Lagidium peruanum

This species has been studied in the field by Pearson (1948) who described some of its vocalizations. In captivity, the animals tooth chatter during threat and snort before an attack. One of the most characteristic calls is a high clear twitter (Fig. 2a), evidently serving as a warning call. The call made by *L. boxi* in similar circumstances is of the same type but is distinctive (Rowlands, 1974).

Capromyidae

Capromys pilorides

In captivity, *Capromys* exhibit pilo-erection when threatening and may give a 'tsit-tsit' call. Two animals starting to fight rear up and give a pulsed twitter, very high pitched and quite intense. When mildly aroused, a *Capromys* will utter a single or double note termed the "sweet" call (Taylor, 1970) which may evoke a reply from a neighbouring cage.

Plagiodontia aedium

Plagiodontia are nocturnal and semi-solitary. When threatening, they exhibit tooth chattering and pilo-erection, but a submissive animal will whimper when approached by a dominant. While following the mother, the young will exhibit a low grunt. Adults housed in separate cages will occasionally exhibit a 'twitt' sound which may be replied to by others. *Plagiodontia* also has an 'oogah' call (see Radden, 1968) similar to that of *Lagostomus* and it is produced when the animal is agitated or disturbed in the nest box. It is not generally responded to vocally by a conspecific. Typically, it has an harmonic prefatory note which may be an inflected squeak.

Geocapromys ingrahami

The calls of this species have been described by Clough (1972). He distinguishes two classes of sounds, one being a low squeak given by free-ranging animals in response to one another. The second is described as a chatter and produced when the animal is captured and held. Having kept *Geocapromys* myself, I believe that the squeak Clough describes may be homologous to the 'tweet' note of *Plagiodontia* and *Capromys*. The chatter that he has recorded is reminiscent in its form of the 'oogah' call of *Plagiodontia* but is given in a contextually different situation.

Caviidae other than Cavia

Dolichotis patagonum

Dolichotis produces calls similar to those of *Cavia*, including the inflected 'wheet' when seeking contact. A low repetitive grunt may be made when following a conspecific. When threatening, *Dolichotis* shows pilo-erection, may tooth chatter and emit low grunts. If startled, it will emit a sharp squeak.

Pediolagus salinicola

The sounds of this species are remarkably similar to those of *Dolichotis*. When following, the young also produce the 'wheet' call, and the whine, emitted when threatening a conspecific, is reminiscent of that of *Myoprocta*.

COMPARISONS OF CONTEXTUAL CATEGORIES

Table III summarizes 12 contextual categories of sound production similar to those proposed by Collias (1960). Syllable forms for five caviomorphs are represented in Figs 1–5. It can be seen that all species sounds can be grouped into an 'excitation indicator' series. An intriguing aspect involves the functional categories with signal forms, differing from species to species, which evidently show the effects of selection to produce a 'typical intensity' form that is audible in the environmental circumstances of the species concerned.

Warning sounds

These sounds are given either in response to sudden general stimulus contrast or to a specific predator. If fleeing is not induced, *Cavia* utters the 'tutt-tutt-tutt' cry which in its rhythm and sound form seems to be an analogue of drumming in other caviomorph species. In response to mild stimulus contrast, *Myoprocta* stamps or stamps rhythmically in a four or five-component pattern. *Dasyprocta* also stamps; *Cuniculus* thumps with its hind foot; *Dinomys* stamps with its forefoot; *Lagostomus* drums with its hind foot.

The cursorial forms may employ alternative warning mechanisms; *Dasyprocta* will emit an alarm bark consisting of a few short prefatory grunts followed by a loud grunting as it flees and *Myoprocta* may produce a low grunt when startled. Only in *Dasyprocta*, however, is this warning bark so conspicuous, and its function, as part of a predator distraction display, has been commented on by Smythe (1970b). Smythe also notes that in *Dolichotis*, which is cursorial in an open steppe habitat, a stotting gait is used as an anti-predator display. Such a habit lends itself well to a visual display in contradistinction to that of *Dasyprocta*, a forest dweller, for which an auditory display may be the only choice open.

Lagidium peruanum produces a high pitched loud twittering call in response to the appearance of a predator. It would appear that selection has favoured this form of warning cry in *Lagidium* since it lives among rocky crevices with excellent visibility and is certain of escaping capture even though the call is easily localized by the predator. As the cry is loud enough to be clearly heard by other members of the colony in which it lives, the signal has a distinct survival advantage; this situation is reminiscent of *Ochotona* in the Holarctic.

Capromys produces a single sharp squeak in response to disturbances. Its pitch is high and ascends rapidly. Although the call is audible to neighbours it is not easily localized by a predator.

Following or contact-promoting sounds

Such sounds may be produced by litter mates when moving out of visual contact with one another or by an infant seeking to localize the position of the mother. They may be responded to by the production of similar sound forms. In *Cavia*, the cluck-cluck is produced in these functional contexts and if contact is not achieved immediately the sound may grade into inflected squeaks culminating in the high intensity wheet. An almost identical series of sound forms is produced by *Dolichotis* and the infant *Plagiodontia* in maintaining contact with its

TABLE III

Contextual comparisons for some caviomorph vocalizations and sounds

Species (Authority)	Warning sounds		Before or after attack	Avoiding while being approached	When injured	When defeated	When calling young (♀)	When being groomed	When follow-ing	When seeking contact	When courting	Special contexts
	When startled	When threatening (to a conspecific or slow predator)										
Cavia porcellus (J. F. Eisenberg, unpublished)	tutt-tutt II s, r	tooth chatter III r	grunt, snort	low wheet I a d, s, r	sharp squeak	squeal II l	cluck II a d s	gurgle; short squeak II a d	clucks II a d, s	wheet I a d, s	purr II u s; r	inflected wheets
Cavia aperea (Rood, 1972)		tooth chatter III	grunt	bubbly squeak		squeal					rumble	
Microcavia australis (Rood, 1972)	tsit	tooth chatter III									rumble	
Galea musteloides (Rood, 1972)		tooth chatter III; drumming	stutter								churr	
Dolichotis patagonum (J. F. Eisenberg, unpublished)	sharp wheet	tooth chatter III; grunt		long wheet with grunt				short grunts	cluck II a d, s	wheet I a d, l		
Pediolagus salinicola (J. F. Eisenberg, unpublished)		whine							cluck	wheet		
Myoprocta pratti (Kleiman, 1972) (J. F. Eisenberg, unpublished)	sharp twit I u; grunt IV s	tooth chatter: drumming III r whine I to II r	snort III to IV s	mewing	squeal	screech	purr II u; r	inflected squeaks I a d; r		peeping I d; s	squeak I ad; s forepaw tremble	post-copulatory snort III–IV s
Dasyprocta punctata (Smythe, 1970a)	alarm bark	foot thumps; rumble; tooth chatter	growl; grunt	creak-squeak (infant)	scream	squeak	purr	purr		creak-squeak	forepaw tremble; squeak; grunt	

Cuniculus paca (Smythe, 1970a) (J. F. Eisenberg, unpublished)	snort; growl	tooth grinding III; thumping; growl IV l	growl-roar II u to IV l				low grunt	whine; snort			forepaw tremble	
Spalacopus cyanus (J. F. Eisenberg & D. G. Kleiman, unpublished)	musical trill I s; r to I l	tooth chatter	growl					short squeaks				
Octodon degus (D. G. Kleiman & J. F. Eisenberg, unpublished)	sharp squeak II s to I s ; r	tooth chatter	growl IV		squeal		?chuckle II s	(inflected squeak) gurgle I d, a d			short chirps I a; r	post-copulatory squeals II s; r and *chuck-wee I to II a d
Octodontomys gliroides (D. G. Kleiman & J. F. Eisenberg, unpublished		twitter IV s r										post-copulatory short squeaks IV–II s; r
Abrocoma cinerea (J. F. Eisenberg, unpublished)			grunt	squeak				gurgles				
Proechimys semispinosus (Maliniak & Eisenberg, 1971)		growl; tooth chatter		cluck; twitter	eeek						groan	whimper (duet) *during mount
Dinomys branickii (Collins & Eisenberg, 1972)	stamp hiss III; IV	tooth chatter III r; hiss IV growl II	hiss IV	grunt II r, s; staccato whimper I l				wah wee II ad low squeak	grunt II s	(see "song")	forepaw tremble; whimper I a d l	'song' ♂ (see text)
Chinchilla laniger (J. F. Eisenberg, unpublished)		snarl growl	snarl		eeek							nyak nyak ♂
Lagostomus maximus (Hudson, 1872) (J. F. Eisenberg, unpublished)	stamping III ee-uh	growl; tooth grinding						quack				post-copulatory grunt grunt ♂ and *uh eee uh (see text)

Continued

Table III—*Continued*

Species (Authority)	Warning sounds		Before or after attack	Avoiding while being approached	When injured	When defeated	When calling young (♀)	When being groomed	When follow-ing	When seeking contact	When courting	Special contexts
	When startled	When threatening (to a conspecific or slow predator)										
Lagidium peruanum (J. F. Eisenberg, unpublished) (Pearson, 1948)	clear twitter I s; r	tooth chatter	snort									
Capromys pilorides (Taylor, 1970) (J. F. Eisenberg, unpublished)	sharp squeak I s	tooth chatter	twit-twit I–II; r; s						seet I		seet I	grunt three syllable	
Plagiodontia aedium (Radden, 1968) (J. F. Eisenberg, unpublished)		oogah, tooth chatter		whimper						grunt (young)	tweet (whistle)		*oogah
Geocapromys ingrahami (Clough, 1972)	twit I d				chatter						tweet		

I, II, III, and IV = syllable types as defined in text. a = ascending frequency; d = descending frequency; ad = complex frequency modulation; s = short syllable; l = long syllable; r = repeated emission; * = compound call (see text); u = no appreciable frequency modulation.

mother. In the adult *Plagiodontia*, the grunt may develop into a whistle reminiscent of the *Cavia* call, again indicative of a higher level of excitation. At high intensities, young *Myoprocta* will produce a contact note which consists of a brief but inflected squeak or 'peeping'. In *Dasyprocta*. it is a disyllabic note, a high squeak ending in a soft grunt. *Dinomys branickii* emits a disyllabic grunt to maintain contact. The female *Cuniculus* will call to the young emitting low grunts. In *Myoprocta* and *Dasyprocta*, the adult female will call the young at close range by emitting a purr.

Grooming sounds

These apparently indicate mixed motivations. The sounds may begin as contact seeking notes gradually calming down as grooming is received. Sounds made during grooming seem to vary from low intensity grunts to more inflected squeaks depending on the nature of the grooming being received which might be rather harsh and aggressive during initial grooming between sexual partners. In *Cavia*, during mild grooming when a mother or sibling licks an infant, the sound is cluck-like. In *Dolichotis*, it consists of a series of short grunts, and in *Myoprocta* one of inflected squeaks. In *Dinomys* it consists of low inflected squeaks, as it does in *Octodon*. Such close contact notes are characterised by low intensity energy emission; the notes are brief and the degree of inflection and frequency modulation seem to be partly a function of the quality of stimuli received by the sender.

Location and aggression sounds

Tooth chattering is almost universally produced by a hostile animal. In a thwarting context, some species exhibit grunts or roars. *Dolichotis* and *Cavia* will give a grunt, *Myoprocta* produces a snort or coughing sound, *Cuniculus* typically will growl, sometimes at length, *Proechimys* growls and *Dinomys* utters a low growl of considerable amplitude.

Submission sounds

When approached by an intimidating conspecific, *Cavia* will utter a slow wheet-wheet, termed bubbling squeaks by Rood (1972), as an indication of its unwillingness to fight. Similarly, *Dolichotis* and *Pediolagus* will emit a prolonged wheet generally dropping to a terminal grunt of low intensity. In *Myoprocta*, Morris (1962) reports a mewing sound. In *Dasyprocta*, a creak-squeak is produced reminiscent of the lost call by the young (Smythe, 1970a). *Proechimys* usually utters a twittering call and *Dinomys* a staccato whimper. *Plagiodontia* females utter a whimper when being chased by males.

Excitement calls

In *Cavia*, the 'wheet-wheet' series is relatively unritualized, but in some caviomorphs a similar call form is highly ritualized and seems to announce the presence of the animal and a high intensity arousal. Such calls would include the 'oogah-oogah' of *Plagiodontia*, a similarly inflected call in *Lagostomus*, the 'nyak-nyak' call of *Chinchilla* and the 'song' of *Dinomys*. The calls of *Lagostomus* and *Plagiodontia* greatly resemble the high intensity wheet vocalizations of *Cavia* and *Dolichotis* in their temporal patterning and although they have probably evolved from the excitation indicator series they appear to have rather different functions (Fig. 6). There may be definite aggressive overtones which resemble to some extent the whine or long call of *Myoprocta* (see below).

Stereotyped mating calls

Courtship calls and sound production apparently indicate readiness on the part of the male to copulate and, at the same time, signal non-aggressiveness. The purr of the male *Cavia* is such a stereotyped example, as are the inflected squeak series of the courting *Myoprocta* male, the whimper of the male *Proechimys*, and the twitter of the male *Dinomys*. A similar function may be shown by the di- to tri-syllabic grunt sounds of the male *Capromys pilorides* (Taylor, 1970). It is noteworthy that a mechanically produced sound resulting from forepaw trembling (which is potentially a visual signal) is common during courtship by male caviomorphs. Forepaw trembling has been noted in *Myoprocta*, *Dasyprocta*, *Cuniculus* and *Dinomys*.

Sound production during mounting appears to be unique for *Proechimys semispinosus* where the duet whimper is common during the long mount lasting up to five minutes (Maliniak & Eisenberg, 1971). Post-copulatory sounds in caviomorphs range from the snort occasionally heard in *Myoprocta* to the complex sounds produced by *Octodon* and *Octodontomys*. The rhythmic post-copulatory calls resemble the protest squeaks produced during any defensive behaviour. Barfield & Geyer (1972) have reported comparable data for *Rattus*.

DISCUSSION

One might well ask the question, can the analysis and comparison of vocalizations among closely related forms be useful in resolving problems concerning affinity and relationships? In short, since we can resolve the vocalizations of caviomorphs into four fundamental syllable types, are these syllable types homologous and do their transformations obey

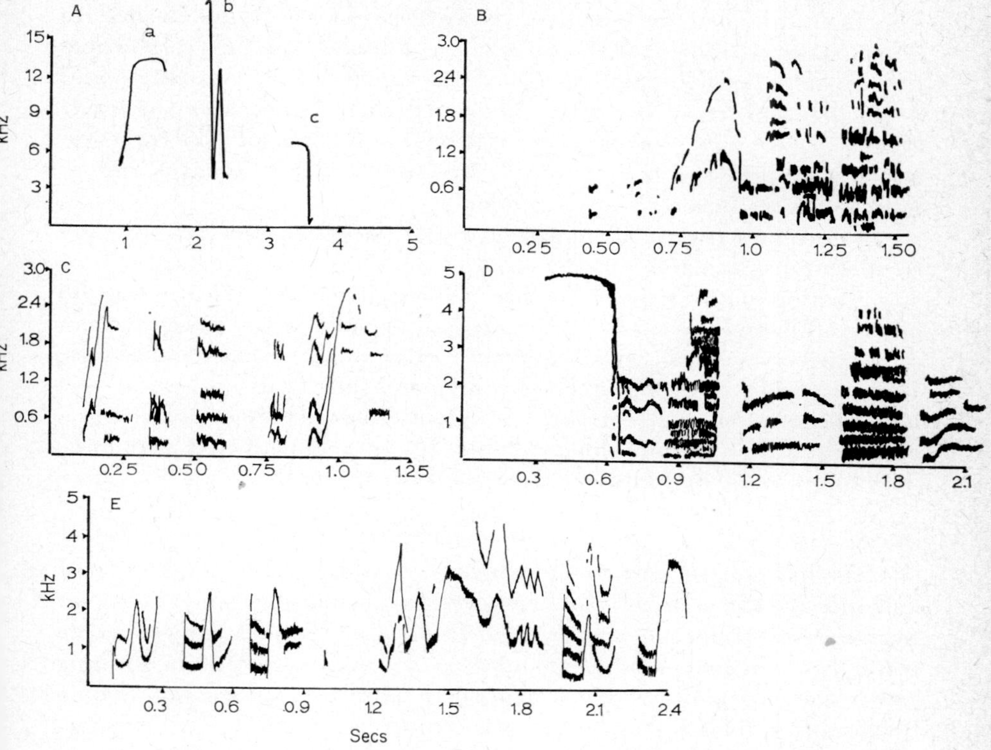

Fig. 6. Inflected squeaks of selected caviomorph rodents.
A. Selected squeaks from the Capromyidae:
   (a) 'Sveet' of *Capromys pilorides*.
   (b) One syllable of a squeak series of *Geocapromys ingrahami*.
   (c) 'Tweet' note of *Plagiodontia aedium*.
B to E. Inflected squeak series and probable homologues:
B. 'Ee oogah gah' call of *Lagostomus maximus*.
C. 'Wheet uh uh uh wheet uh' of *Dolichotis patagonum*.
D. 'Tweet oogah oogah' of *Plagiodontia aedium*.
E. 'Wheet uh wheet uh wheet uh wheet' of *Cavia porcellus*.
The examples from B to E differ in tonal quality and temporal patterning, but the juxta-position of syllables suggests an homologous sound; yet the contexts of appearance differ (see text) suggesting different selective pressures acting on a common, basic repertoire.

the same rules when a series of species are compared? An homologous sound can be identified when two species produce a sound form which is nearly identical structurally, or, if it differs, when there exist transitional sound forms which link the two homologous sound forms (Wickler, 1961). Homologues occasionally may not be perceived as the

I

same sound. When we postulate, however, that two sound forms are homologous, we infer that the same nerve–muscle structures responsible for the sound production are true homologues in an anatomical and developmental sense. But as we know very little concerning the interactions of expired and inspired air with the vocal cords and the reverberations set up in the buccal cavity, we are not able to identify nerve-muscle homologues with any confidence.

One would expect that in a wide variety of species such as we have been studying, syllable form, pitch and temporal patterning may vary as a function of lung capacity or size. Table I, however, indicates that no obvious differences exist between large and small species: larger species do not necessarily call with lower pitched sounds than do smaller ones. Both large and small species employ long and short syllables, and a very large species such as *Dinomys*, although producing short syllables, has a repertoire which includes long syllables ($> 1 \cdot 0$ sec) delivered in phrases with long ($> 1 \cdot 0$ sec) intersyllabic intervals.

Syllable homologues

Hiss-like sounds may be homologous, but highly modulated Type I syllables exhibit a variety of temporal patterning and some subtypes may very well not be homologues. One set of Type I sounds which could be homologues are the inflected squeaks with grunts forming either a compound syllable (*Cavia*) or a phrase (*Lagostomus*). Strong evidence for their homologous nature includes similarity in form and similarity in the temporal patterning of the syllable types and subtypes (see Fig. 6).

All click-like sounds are not necessarily homologous, since a click sound may be produced by a variety of non-homologous mechanisms, such as tooth chattering, stridulation of quills, or an actual glottal click produced by air expiring in short bursts. Clicks associated with tooth chattering are probably homologous, but when comparing all types of click sounds, we are surely considering a series of analogues that may have an identical function. Thus when one finds sound forms having very similar structure in the pattern of their energy distribution, they may well be functionally equivalent but not necessarily homologous.

When closely related species differ profoundly in the form and temporal patterning of their vocalizations, these differences may well be due to different selective pressures in the environment of the animal and the form of their sounds may reflect adaptations distinct to their particular mode of habitat exploitation as the following paragraphs will document.

As pointed out by Marler (1957), the warning cries of many passeriform birds have a remarkable structural similarity. They are obviously functional analogues in that they represent a sound form which allows the individual to warn conspecifics but at the same time reduces the chances of predation since the sound is localized only with great difficulty. It is also possible that the clucks and short grunts used by caviomorphs in a following context are functional analogues. The production of repetitive, low-frequency pulses with a clear onset and termination provides the receiver with the best chance of locating the sound (Marler, 1957, 1967).

Morton (1970) in his study of the song of passeriform birds in tropical rain forests stressed that the form of vocalization which an animal may employ is to some extent limited by habitat or environment. Thus, if there is background 'noise' caused by insect sounds, it may be overridden by sweeping rapidly with frequency modulation across the 4 to 8 kHz channel. This may be the explanation for the similarity in the form of the inflected squeaks produced by *Capromys*, *Plagiodontia* and *Geocapromys* (see Fig. 6). Morton also found that forest dwelling birds sing with an average lower frequency than do the forest edge or grassland species. Since lower frequency sounds travel farther regardless of the habitat type, Morton suggests that the difficulties inherent in sound propagation in the forest have set more rigid constraints on the frequency range for the advertisement song of forest adapted passeriforms.

One could analyse the sounds of caviomorph rodents along similar lines preferably by choosing a functionally analogous call across several taxa for comparative purposes. Alarm calls in response to the appearance of a potential predator seem a likely class of sounds for this sort of analysis. Such calls seem to be functional analogues and not necessarily structural homologues. They appear to be adapted for maximum efficiency of communication within a given habitat. The alarm call of *Dasyprocta* is low in pitch. Although *Dasyprocta* is solitary, a given pair occupy the same home range and their offspring are generally within hearing range. Thus, to give a warning cry is probably adaptive for the survival of the offspring. The average frequency range of the call is optimum for distance propagation within the forest itself. The alarm calls of *Lagidium* probably serve a similar purpose; they are much more highly pitched and probably related either to optimum propagation efficiency or to a relaxation of selection for a low frequency call in the rocky high altitude, non-forested habitats which they exploit (Fig. 7).

The calls made by semi-solitary rodents attempting to communicate with a potential mate are of interest but are rare. Male and female *Erethizon* call at the onset of the breeding season (Seton, 1953; J. F. Eisenberg, unpublished) with a sound which is pulsed, undulating, long

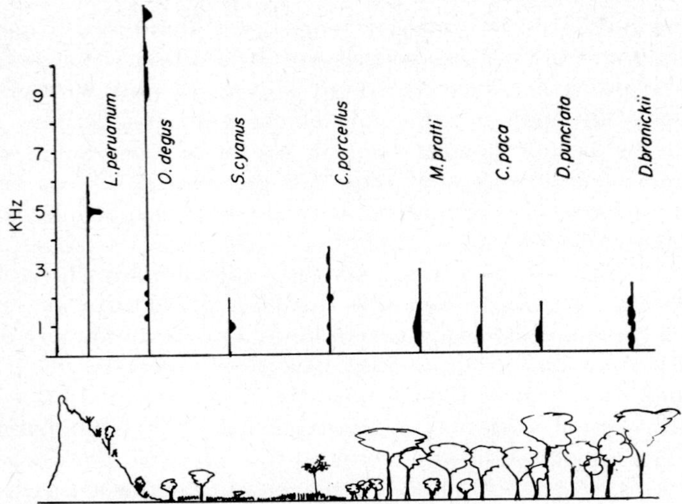

FIG. 7. Correlations between habitat and pitch.

The average pitches for long distance communication in a series of terrestrial cavio-morphs are compared. Habitats include from left to right: semi-arid, montane, *Lagidium peruanum*; semi-arid foothills, *Octodon degus*; semi-arid scrub, *Spalacopus cyanus*; wet savanna, *Cavia porcellus*; tropical, evergreen forest, *Myoprocta pratti, Cuniculus paca, Dasyprocta punctata* and *Dinomys branickii*. The calls represented include: high twitter, *L. peruanum*; warning squeak and isolation call (infant), *O. degus*; isolation 'wheet', *C. porcellus*; trill, *S. cyanus*; whine, *M. pratti*; growl-roar, *C. paca*; warning growl, *D. punctata*; courting song, *D. branickii*. The terrestrial forest forms produce calls with an average lower frequency.

and easily heard at a considerable distance in the forest. The call is similar to the 'song' recorded from an isolated male *Dinomys branickii*. The similarity may be related to identity of function, i.e. long-distance communication in the forest. *Dinomys* calls in fact with a frequency distribution (ranging from 0·9 to 1·9 kHz) optimum for propagation in forest.

Both of these calling rodents are arboreal, and it may be that all forest-adapted mammals that seem to make the most use of advertisement cries will turn out to be arboreal. Terrestrial cryptic forms seem to be the least likely to advertise their presence by means of vocalizations. There are, of course, exceptions to this generalization, but as a general rule, it appears to extend to other taxa. Such correlations have been made in the analysis of vocalizations of marsupials (Eisenberg *et al.*, in press) and small carnivores (Wemmer, 1972).

CONCLUSIONS

The young caviomorph rodents are born in a relatively well developed state. Some species, such as *Proechimys* and *Octodon*, have a nest phase and do not follow the parents for some 14 days. Other species, particularly the cursorial forms, may follow the parents after two to four days of age (see Kleiman, 1972). The young emit separation calls not easily distinguished from the high arousal calls of the adult (e.g. *Cavia* and *Octodon*). Characteristically the young produce short contact notes when following the parent. In addition, most young caviomorphs have a low, frequency-modulated call which they produce when groomed. Although ultrasonic energy is present in the separation calls of some young, they do not produce the purely ultrasonic 'abandoned cry' so characteristic of young myomorphs during the early stages of their development.

Tooth chattering (Type III sound class) with some degree of pilo-erection almost invariably accompanies threat displays. In some species the release of anal gland secretion accompanies the threat response (e.g. *Cavia*; see also Kunkel & Kunkel, 1964). Hisses are prominent in threat displays of *Dinomys*. Foot stamping often accentuates the threat posture. Before attack, grunts, barks or snorts are produced which are loud (*Myoprocta*, *Dasyprocta*). These are Type IV syllable forms with emphasis on the lower frequencies. In some cases the syllable may be prolonged and termed a growl (*Cuniculus*). When injured, seized or exhibiting intense submission, the typical sound is a squeal or scream including Type II to Type IV syllables in which the higher frequencies are emphasized.

Warning cries are highly variable when an array of species are com-pared. The call is obviously under several types of selective pressures which can be correlated with habitat and the typical social structure of the species. Special call notes have evolved to coordinate following be-tween a parent and the young. Such contact notes are generally short and easily localized.

Sounds uttered during allogrooming are highly variable but at low intensity grooming are often rhythmic, low and Type I or II syllable forms. These calls may intergrade to the sound forms produced during non-situation specific excitement, e.g. *Cavia*.

Specific calls during courtship, mating and after ejaculation have been noted for several species (*Proechimys*, *Myoprocta*, *Cavia*, *Dinomys*, *Octodon*, and *Octodontomys*). The exact significance of such calls is difficult to interpret but during courtship the call form is often variable and the syllables show affinity with either aggressive or submissive moods.

A series of highly modulated squeaks which may be considered an excitation indicator series is produced by most caviomorph rodents. In *Cavia*, *Dolichotis* and *Pediolagus*, the squeak rises and then sweeps

rapidly downwards to a terminal grunt, which may indicate a close
relationship between the three genera. *Lagostomus* and *Plagiodontia*
have a similar inflected 'wheet' as a preface to a much lower pitched
series of prolonged grunts which serves as a distance-communication
call. These inflected calls are possibly homologous. A steeply inflected
warning squeak of a frequency which is typical of *Capromys* and
Geocapromys may also be homologous.

Advertisement calls produced by *Erethizon* and *Dinomys* are of great
theoretical interest and appear to have been independently evolved.
Their functions may not be identical, but are apparently related to
either intraspecific spacing or mate attraction.

All species appear to have some three to four calls closely tied to a
given context with a specific function. Additionally, all species show a
graded series of syllable types not tied to a specific context, but serving
to communicate different states of arousal in the sender.

Whether analysed from the standpoint of function or implied motivational state, both the classification of sounds and numbers of discernible
units correspond remarkably well with the scheme developed by Brooks
& Banks (1973) for *Dicrostonyx*. What this implies then is a fundamental
unity in the sound producing repertoires of rodents. The patterns as
outlined for the caviomorphs appear to resemble closely similar sets of
vocalizations and non-vocal sounds described for carnivores (Wemmer,
1972) ungulates (Kiley, 1972), and marsupials (Eisenberg *et al.*, in
press).

ACKNOWLEDGMENTS

I gratefully acknowledge the discussions with Dr N. Smythe concerning the free ranging behaviour of *Dolichotis*, *Dasyprocta*, and
Cuniculus. Unpublished behavioural analyses of *Myoprocta* and *Octodon*
were made available to me by Dr D. G. Kleiman. Miss S. Wilson and
Mr T. Davis made helpful comments concerning the functional interpretation of several calls. Dr G. G. Montgomery and Dr D. G. Kleiman
read the manuscript and offered useful criticisms. I must express a debt
to Messrs E. Maliniak, J. Hough, and M. Deal, who did so much to
maintain our excellent breeding records with the animals. The manuscript was prepared by Mrs W. Holden.

REFERENCES

Andrew, R. (1963). The origin and evolution of the calls and facial expressions of
    primates. *Behaviour* **20**: 1–109.
Andrew, R. (1964). The displays of primates. In *Evolutionary and genetic biology
    of Primates* **2**: 227–309. Buettner-Janusch, J. (ed.). New York: Academic
    Press.

Barfield, R. J. & Geyer, L. A. (1972). Sexual behaviour: ultrasonic post ejaculatory song of the male rat. *Science, N.Y.* **176**: 1350.

Brooks, R. J. & Banks, E. M. (1973). Behavioural biology of the collared lemming (*Dicrostonyx groenlandicus*, Tràill): an analysis of acoustic communication. *Anim. Behav. Monog.* **6** (1): 1–83.

Busnel, R. G. (ed.). (1963). *Acoustic behaviour of animals*. New York: Elsevier.

Busnel, R. (ed.). (1967). *Les systems sonars animaux* (2 volumes). Jouy-en-Josas, France: Laboratoire de Physiologie Acoustique INRA-CNRZ.

Clough, G. C. (1972). Biology of the Bahaman hutia, *Geocapromys ingrahami*. *J. Mammal.* **53**: 807–823.

Collias, N. (1960). An ecological and functional classification of animal sounds. In *Animal sounds and communication*: 368–391, Lanyon, W. E. & Tavolga, W. N. (eds.). Washington, D.C.: American Institute of Biological Sciences. (Publ. No. 7).

Collins, L. R. & Eisenberg, J. F. (1972). Notes on the behaviour and breeding of pacaranas, *Dinomys branickii*, in captivity. *Int. Zoo Yb.* **12**: 108–114.

Coulon, J. (1973). Le repertoire sonore du cobaye domestique et sa signification comportementale. *Rev. Comp. Animal* **7**: 121–132.

Eisenberg, J. F., Collins, L. R. & Wemmer, C. (in press). Communication in the Tasmanian devil (*Sarcophilus harrisii*) and a survey of auditory communication in the Marsupialia. *Z. Tierpsychol.*

Eisenberg, J. F. & Gould, E. (1970). The tenrecs: a study in mammalian behaviour and evolution. *Smithson. Contrib. Zool.* **27**: 1–137.

Eisenberg, J. F. & Kleiman, D. G. (1972). Olfactory communication in mammals. *A. Rev. Ecol. System.* **3**: 1–32.

Griffin, D. (1958). *Listening in the dark*. New Haven: Yale University Press.

Hinde, R. A. & Tinbergen, N. (1958). The comparative study of species-specific behaviour. In *Behaviour and evolution*: 251–268. Roe, A. & Simpson, G. G. (eds.). New Haven: Yale University Press.

Hudson, W. H. (1872). On the habits of the vizcacha (*Lagostomus trichodactylus*). *Proc. zool. Soc. Lond.* **1872**: 822–833.

Kiley, M. (1972). The vocalisations of ungulates, their causation and function. *Z. Tierpsychol.* **31**: 171–222.

Kleiman, D. G. (1971). The courtship and copulatory behaviour of the green acouchi, *Myoprocta pratti*. *Z. Tierpsychol.* **29**: 259–278.

Kleiman, D. G. (1972). Maternal behaviour of the green acouchi (*Myoprocta pratti*, Pocock), a South American caviomorph rodent. *Behaviour* **43**: 48–84.

Kunkel, P. & Kunkel, I. (1964). Beiträge zur Ethologie des Hausmeerschweinchens *Cavia aperea f. porcellus* (L.). *Z. Tierpsychol.* **21**: 603–641.

Lanyon, W. E. & Tavolga, W. N. (eds.). (1960). *Animal sounds and communication*. Washington, D.C.: American Institute of Biological Sciences. (Publ. No. 7).

Maliniak, E. & Eisenberg, J. F. (1971). Breeding spiny rats, *Proechimys semispinosus*, in captivity. *Int. Zoo Yb.* **11**: 93–98.

Marler, P. (1957). Specific distinctiveness in the communication signals of birds. *Behaviour* **11**: 13–39.

Marler, P. (1967). Animal communication signals. *Science, N.Y.* **157**: 769–774.

Morris, D. (1962). The behaviour of the Green acouchi (*Myoprocta pratti*) with special reference to scatter hoarding. *Proc. zool. Soc. Lond.* **139**: 701–732.

Morton, E. S. (1970). *Ecological sources of selection on avian sounds*. Ph.D. Thesis Yale University, New Haven, Connecticut.

Pearson, O. P. (1948). Life history of mountain viscachas in Peru. *J. Mammal.* **29**: 345–374.

Pearson, O. P. (1959). Biology of the subterranean rodents, *Ctenomys*, in Peru. *Mems Mus. Hist. nat. "Javier Prado"* **9**: 1–56.

Radden, R. D. (1968). *The Dominican Republic hutia*, Plagiodontia aedium *hylaeum, in captivity*. Master of Science Thesis, University of Puget Sound, Tacoma, Washington.

Reig, O. A. (1972). Ecological notes on the fossorial octodont rodent, *Spalacopus cyanus* (Molina). *J. Mammal.* **51**: 592–601.

Rood, J. P. (1972). Ecological and behavioural comparisons of three genera of Argentine cavies. *Anim. Behav. Monog.* **5** (1): 1–83.

Rowlands, I. W. (1974). Mountain viscacha. *Symp. zool. Soc. Lond.* No. 34: 131–141

Sales (née Sewell), G. D. (1972a). Ultrasound and mating behaviour in rodents with some observations on other behavioural situations. *J. Zool., Lond.* **168**: 149–164.

Sales (née Sewell), G. D. (1972b). Ultrasound and aggressive behaviour in rats and other small mammals. *Anim. Behav.* **20**: 88–100.

Seton, E. T. (1953). *Lives of game animals*. 4 volumes. Boston, Massachusetts: Branford Company.

Sewell, G. D. (1970). Ultrasonic signals from rodents. *Ultrasonics* **8**: 26–30.

Smythe, N. (1970a). *Ecology and behaviour of the agouti* (Dasyprocta punctata) *and related species on Barro Colorado Island, Panama*. Ph.D. Thesis, University of Maryland, College Park, Maryland.

Smythe, N. (1970b). On the existence of "pursuit invitation" signals in mammals. *Am. Nat.* **104**: 491–494.

Struhsaker, T. (1967). Auditory communication among vervet monkeys (*Cercopithecus aethiops*). In *Social communication among Primates*: 281–324. Altmann, S. A. (ed.). Chicago: University of Chicago Press.

Taylor, R. H. (1970). *Reproduction, development and behavior of the Cuban hutia conga*, Capromys p. pilorides. Master of Science Thesis, University of Puget Sound, Tacoma, Washington.

Tembrock, G. (1959). *Tierstimmen*. Wittenberg Lutherstadt: Ziemsen Verlag.

Wemmer, C. (1972). *The ethology of* Genetta tigrina *and related genera*. Ph.D. Dissertation, University of Maryland, College Park, Maryland.

Weir, B. J. (1974). The tuco-tuco and plains viscacha. *Symp. zool. Soc. Lond.* No. 34: 113–130.

Wickler, H. (1961). Oekologie und Stammesgeschichte der Verhaltenswiesen. *Fortschr. Zool.* (N.F.) **13**: 303–365.

Zippelius, H.-M. & Schleidt, W. M. (1956). Ultraschall-Laute bei jungen Mäusen. *Naturwissen.* **43**: 502.

SALE (*Nairobi*): You mentioned vibrating in stressful situations. Have you any ideas as to its function?

KLEIMAN (*Washington*): Well, it has been suggested that the swaying, the rumba-ing, of *Cavia* and the treading movements are derived from patterns of flight. That is, they are incipient flight movements. I have seen it in *Octodontomys* but it's quite different from that of *Cavia*, it is an up-down kind of movement.

SALE: But have you no idea of its function?

KLEIMAN: It seems to be incorporated into the courtship ritual; many of the courtship displays, e.g. swaying, seem to be submissive and include a very hesitant approach, followed by withdrawal. I don't think I should ascribe a function to that particular behaviour, but it may be to reassure or appease the female.

MANLEY (*London*): You spoke about the courtship behaviour being adapted to the aggressiveness of the female, but then the question obviously arises, why are the females aggressive?

KLEIMAN: I was going to pose that question but forgot. I knew that degu females were periodically aggressive but I hadn't seen the way it changed so much until I got these latest data. It may well be that the opening of the vaginal closure membrane causes an increased arousal level in these females, much as you would find in monkeys with a sexual swelling. Now, that's just a suggestion and I'm certainly open to further suggestions, but I agree that it is worth looking into.

LAVOCAT (*Montpellier*): Having been interested in high fidelity, I was intrigued to see in your diagrams that there are a lot of transitory sounds. These sudden and abrupt inflections of sound are important for the transmission of sound because an animal can hear such transitory sounds better than it can a continuous sound. When the sound is continuous there is an adaptation of the muscles of the middle ear to the level of the sound in such a way that after a short time the level of perception is much lower so I think it is very interesting that if these sounds are long-distance advertisements they are modulated in such a way. Another thing, have you tried to determine any connection between the exact range of these sounds with normal frequencies, that is, is there any connection between the frequencies and the cavities of the middle ear?

EISENBERG (*Washington*): This business with the middle ear is an extremely interesting one to me because I did my original thesis work on kangaroo rats, *Dipodomys*, and at that time there was a man by the name of Webster, who was attempting to work out the optimum sensitivity for reception of each individual frequency in an ear with an inflated mastoid bulla. I believe there were three French workers at that time also working on a parallel problem with *Gerbillus* and *Jaculus*. This is an area of research that I would love to see someone get into but it is one that I am technically not equipped to carry out. It requires, of course, measurement of the amount of electrical activity in the tympanic nerve resulting from successive inputs at the same decibel level of every frequency that you wish to test for, and that is a bit beyond our capacity. But there is no question, however, that those people who have studied it in *Gerbillus* and in *Dipodomys* have found that an inflated mastoid bulla generally increases the sensitivity for perception of frequencies at or about the resonant frequencies of the ossicles, which was indirectly measured as I understand it, by linear measurements or weight correlations of the ossicles themselves. The bulla permits extreme sensitivity to low amplitude sounds of particular frequencies in whatever range the resonant frequency of the ossicle is and this is perhaps more of a correlation to anti-predator behaviour or predator detection than it is to the detection of long distance signals of conspecifics, even though it would seem logical that this may be the case since most desert forms live in a thin density. It appears that in those species of rodents for which predator selection is apt to produce the hypertrophied tympanic bulla with all of its advantages in perceiving a predator, selection apparently has also acted on the vocalizations to bring some of the key ones down into the area of maximum sensitivity to the ear. Thus the predator drives the ear structure to a certain configuration and then the vocalizations have to adapt to it. I don't see, at least in any of the desert rodents, i.e. those with the hypertrophied bullae, that selection has worked the other way. That is to say, I think the voice of the conspecific has to accommodate to the new tuning of the ear and that the predator is tuning the ear.

PETTER (*Paris*): First, I do not think it is the predator effect, because the predator has the same feature. Also, I think all the *Gerbillus* with big bullae make noises with their feet. I don't know about *Dipodomys*.

EISENBERG: I have recorded sounds from *Gerbillus gerbillus*, *G. nanus*, *Tatera indica*, *Meriones unguiculatus*, *M. hurrianae*, *Jaculus jaculus* and *J. orientalis*. All of these forms drum with their hind feet. *Gerbillus nanus* patters with the forefeet too. But they all produce calls and *Jaculus* especially has a courting call. In no case, however, have I found a call that I think is completely involved with long-distance communication, except for the cry of the abandoned young. In *Jaculus* there is hardly any retrieving response and the young do not cry rhythmically. In *Tatera* the call is very loud and carries a good distance, and the mother will respond to it and retrieve. This is a very good call to study. I have found that the average pitch for the call is highest in the young of *Tatera*, and lowest in *Gerbillus nanus*, and this also correlates with where the optimum hearing should be, roughly speaking. So I can get a correlation between the call of the abandoned young and the bulla size in these rodents. But I don't know if that is the primary reason why selection has promoted a large bulla in rodents. It may be that predators made it but the young have to match it.

PETTER: I just wanted to know whether *Dipodomys* has it.

Eisenberg: Yes, they do drum with the hind feet.

Petter: So the rodents that have the biggest bullae make the highest noises? They cannot make a resonant sound to call with a large bulla. Their anatomy does not permit the possession of apparatus for making loud noises and for hearing well.

Eisenberg: I think the reason Webster couldn't find a correlation between the average frequency of the call of the kangaroo rat and the optimum point for hearing was that if you average all the calls, or you look at some of the calls, this is true. But if you take a specific call like the cry of the abandoned young, made only by the young, which is highly adapted for the parent to hear at a distance, that's where you get the closest fit with the average frequency of a call and the sensitivity of the bulla. So I would never claim that there has been a matching between the frequency of the call of all notes of the animal and the optimal sensitivity for hearing.

Francis (*London*): Have you noticed any characteristic vocalizations of either the male or the female in any of these species after copulation, and if so what speculations on the function have you?

Kleiman: *Myoprocta*, *Octodon* and *Octodontomys* have post-copulatory vocalizations. In *Myoprocta*, it is not consistently heard, but it seems to resemble the threat coughing, snorting vocalization sound. In *Octodon* and *Octodontomys* it is related to the warning squeak, but in all three cases it has been modified somewhat. I suspect that it may serve to keep other animals at a distance in the same way that the vocalization of *Proechimys* acts during copulation. It may also allow the occurrence of a second ejaculation in some cases if other males are kept at a distance.

Francis: Are the animals at all aggressive towards each other after they have mated?

Kleiman: The females are not, but the males may appear aroused and may chase other animals such as juveniles and sub-adults away. We have not heard a post-copulatory vocalization in *Pediolagus* but the male chased the neonates away when they approached him. So it seems to be a keep-away-from-me-please kind of vocalization, and that includes probably everyone, females as well as young and other males.

Symp. zool. Soc. Lond. (1974) No. 34, 249–250

CHAIRMAN'S INTRODUCTION:
REPRODUCTIVE PHYSIOLOGY

J. S. PERRY

*A.R.C. Institute of Animal Physiology,
Babraham, Cambridge, England*

It is always interesting to see how far a number of animals, included within one taxonomic group on the basis of purely morphological, often skeletal, characters, share a common physiological mechanism or biochemical idiosyncrasy. In this session, and the one that follows, it will be seen that recent work in several species of South American hystricomorphs has shown that they share, in their reproductive processes, a number of features that are, as far as we yet know, exclusive to this group. We have, so far, little information about the placentation, embryology or reproductive physiology of the African hystricomorphs. As such information becomes available it will be possible to determine the extent to which these features coincide with the classification which is necessarily based on relatively few morphological characteristics. The problematic position of the African gundi (*Ctenodactylus gundi*) may be resolved in this way; the fact that it is described as having a superficial mode of implantation, and amniogenesis by folding rather than by cavitation (de Lange, 1938),* may incline one to include it with the Sciuromorpha rather than the Hystricomorpha.

All the hystricomorph species studied so far possess a 'sub-placenta'. This is a characteristic component of the allanto-chorionic placenta which does not appear in that of other rodents, or of other mammals. Some of us would like to find that it is the source of progesterone-binding protein which, as will appear, is also characteristic of hystricomorphs, and apparently restricted to them. It must be remembered, however, that little progress has been made in ascribing specific functions to morphologically distinct components of the placenta of any species or, for that matter, in explaining the striking variety to be met with among the placental structures by means of which mammals carry out essentially similar functions during gestation.

Size for size, hystricomorphs tend to have long gestation periods in comparison with other rodents. In the case of the guinea-pig, and some

---

* de Lange, D. (1938). *Archs. néerl. Zool.* **3**: 131–147.

other species, it is possible to relate the long pregnancy to the relatively advanced stage of development of the young at birth. This does not apply, however, to the degu or the tuco-tuco. It is often argued that reproductive processes are likely to have ecological significance because they are particularly susceptible to evolutionary pressure or natural selection. It may be, however, that their viviparity and homoiothermy together confer such advantages on the mammals that competition can only operate within the class, and that their origin is too recent for the effects of such selection to have become manifest.

Symp. zool. Soc. Lond. (1974) No. 34, 251–263

REPRODUCTION IN THE GRASSCUTTER (*THRYONOMYS SWINDERIANUS* TEMMINCK) IN GHANA

E. O. A. ASIBEY

Department of Game and Wildlife,
P.O. Box M. 239, Accra, Ghana

INTRODUCTION

The grasscutter, *Thryonomys swinderianus*, is a common animal and an important source of animal protein in Ghana (Asibey, 1969). It is therefore rather surprising that no systematic study of its biology has been made and even its breeding habits are not known (Ewer, 1969). Because of the high demand for the meat of this animal and its acceptability to all classes of people throughout Ghana it is regularly hunted and its market price is very high.

The grasscutter is also of economic importance as an agricultural pest in Ghana and other areas of West Africa (Asibey, 1969). Heavy hunting seems to hold the population of this species in check, otherwise it could do untold damage to cereal and other crops.

The primary aim of this study was to determine litter size and the number of litters produced per year to assess prolificacy. The study covers the period September 1970 to December 1972 inclusive.

MATERIAL AND METHODS

Sources

Material for this study was made up of dead samples obtained from traders who sell grasscutter carcases in the Kantamanto Market, Accra (Asibey, 1969), and was supplemented by dead samples from the Mole National Park in northern Ghana. Captive stock was kept in Accra and Mole to confirm data from dead samples.

Samples of the supply of animals coming daily into the market at Accra were taken twice monthly. These animals had been shot or trapped the previous night and brought in at daybreak. The number of animals dissected each month depended on the availability of carcases within a price range; relatively more females than males were chosen since they were cheaper. The majority of these carcases came from an area to the west of Accra centred approximately on Swedru and could therefore be taken as representative of the species in southern Ghana. A team of assistants caught grasscutters at random during the first ten days of every month at the Mole National Park (370 km north of Accra). A monthly sample of ten animals, regardless of sex and age, was intended but on occasion it was impossible to catch even a single specimen.

A captive stock of grasscutters was established at Mole for use by the research unit. Most of the live animals used in my investigation were provided by them, but, in addition, some members of the public maintained small breeding colonies of animals obtained elsewhere. All these animals were kept under my supervision in specially designed boxes of standard dimensions. The adults were examined, weighed and measured fortnightly. Births were recorded and the young animals were treated as were the adults but weekly.

Maintenance of live animals

Captive animals in Accra and the Mole National Park were fed entirely on grass. The diet of those kept by the public was supplemented by cassava (*Manihot utilissima*), sugar cane (*Saccharum officinarum*) and maize (*Zea mays*) fruits and stems, either alone or in combination depending upon availability.

A grasscutter can be handled easily by grasping it around the neck and the two hind legs. Gloves were unnecessary. In the Mole National Park, they were caught in this way during the dry season (January–June) when old grass could be burnt or the fresh growth was short. Attempts at live-trapping failed completely.

Collection of data

All dead animals were weighed and various linear measurements recorded for other purposes (Asibey, in preparation) as soon as they were received; they were then dissected and the number of embryos was assessed and recorded. The eyes were removed and preserved in 10% formalin for later treatment. The reproductive organs and the entire viscera were then removed. A female was considered pregnant when implantation sites were visible to the naked eye; pre-implantation stages of pregnancy were not determined, although the uterine horns were examined macroscopically for placental scars and evidence of resorption. The embryos were measured from head to rump without stretching before removal from the membranes. After removal they

were weighed and sexed by examination of the teats and measurement
of the ano-genital distance. They were then preserved in 10% formalin,
as were the ovaries and reproductive tracts of pregnant and non-pregnant females for microscopic and histological examination at a later
date.

Classification of material

The embryos were classified according to weight and development of
hair into four categories as follows:
(1) *Less than 1 g*: these range from implantation sites that are visible to
the naked eye to embryos having well developed organs but without
firm skin so that they cannot be removed from their surrounding
membranes without damage;
(2) *1–30 g*: embryos which are fully formed but are without hair or
vibrissae and can be removed from their membranes, sexed, measured
and weighed;
(3) *31–64 g*: these have some hair present, ranging from vibrissae only
to a complete coverage of hair. Embryos of this size would not survive
independently;
(4) *65–200 g*: embryos with a complete covering of hair, fully developed and capable of survival at birth.

Post-natal age was assessed by the state of tooth eruption. Animals
having up to two molar teeth were regarded as juveniles. Those having
three molars were considered to be non-parous adults in their first
pregnancy and those with a full complement of four molars were parous
animals.

RESULTS

Breeding

The grasscutter breeds throughout the year in Ghana. Analysis of the
pregnancy rate according to the month when examined indicates that, in
the southern part of the country, the proportion of females pregnant
remains at about 80% or more for 10 months and falls only as low as
64% during August and February (see Table I). In northern Ghana, the
pregnancy rate was high ($>$80%) for only about four to six months and
the minimum was twice as low (26%) as it was in southern Ghana. These
figures indicate some degree of seasonality in the breeding of the grasscutter. In the tropical range of this species the photoperiodic changes are
minimal and are not likely to be the factor responsible for differences in
breeding patterns. The pattern of rainfall in the two parts of the country

TABLE I

Proportion of pregnant grasscutters caught monthly in southern (Swedru) amd northern (Mole National Park) Ghana

Month	Southern Ghana (Swedru)				Northern Ghana (Mole N.P.)		
	Mean(*) rainfall (cm)	No. examined	No. pregnant (%)	No. with placental scars (%)	Mean(*) rainfall (cm)	No. examined	No. pregnant (%)
January	7·5	28	79	3	0·1	47	47
February	5·8	28	64	5	2·5	37	43
March	10·0	38	87	3	5·7	38	26
April	14·0	36	86	5	6·8	21	76
May	13·2	44	82	8	16·7	21	95
June	26·0	43	93	8	9·9	17	88
July	9·7	61	87	3	13·0	15	80
August	3·6	40	65	22	15·3	6	83
September	5·3	55	80	22	23·7	7	57
October	13·0	56	79	5	6·2	4	75
November	8·1	48	83	8	0·0	4	50
December	1·9	46	89	8	0·6	7	43
Total	118·1	523	—	—	100·5	224	—

(*) rainfall during 1970 to 1972 inclusive.

differs considerably and there appears to be a much closer direct relationship between its distribution and the proportion of females pregnant in northern Ghana (Table I). Table I also indicates that about 7% of the females obtained from southern Ghana had placental scars in the uterus, the period of highest incidence being August and September.

The data in Table II indicate that females in the earliest stages of pregnancy are mostly obtained in April and October. This double peak can be detected for females in more advanced stages of gestation and is present, although flattened, when the females are in late pregnancy,

TABLE II

The monthly proportion (%) of females from southern Ghana at different stages in pregnancy, assessed by weight of embryo

Month	No. pregnant	Proportion of females having fetuses of:			
		<1 g	1–30 g	31–64 g	65–200 g
January	22	32	14	18	36
February	18	50	17	17	17
March	33	45	30	3	21
April	33	71	15	3	10
May	36	39	47	8	6
June	40	33	40	15	13
July	53	23	25	30	23
August	26	27	23	19	31
September	37	52	18	5	25
October	38	59	16	16	9
November	27	48	33	4	15
December	28	51	12	15	22

suggesting that parturition is most likely in January to March and July to September, and that the gestation period of the grasscutter is about four months. The peaks in late pregnancy correspond with the times of low rainfall in southern Ghana. Unfortunately, the number of births recorded in captivity is, as yet, insufficient for comparison with the above data.

Gestation period

Estimates of the gestation period in the captive stock were based upon the interval from the date when the vaginal closure membrane (common in practically all hystricomorph rodents: Weir, 1974) was first observed to be closed to the birth of the young. When the vagina had been closed for two months it was assumed that the animal was pregnant and no further handling took place. Since observations on the vaginal closure membrane were made at fortnightly intervals, the results for the nine animals shown in Table III can be considered only suggestive

TABLE III

Gestation period in the grasscutter

Parent	Litter size	Mean birth wt and (range) in g	Gestation period (days)
621/2	3	150 (75–190)	133
818/9	4	107 (100–110)	113*
884/5	2	160 (155–165)	111*
836/7	2	152 (146–158)	111*
38Y 692/3	3	144 (118–165)	109*
870/1	3	100 (75–121)	106*
876/7	4	79 (72–98)	95*
63N 097/8	5	121 (110–135)	78
633/4	5	126 (110–139)	68

* see text below

until more precise data can be obtained from separation of the females after mating. The mean gestation period for the nine females was $102 \cdot 7 \pm 44$ days (S.D.) but for the six females indicated by an asterisk it was $107 \cdot 5 \pm 5 \cdot 3$ days. This figure seems remarkably consistent in view of the 14-day sampling error indicated above.

The specific fetal growth velocity was calculated as $0 \cdot 0614$ from the formula $\sqrt[3]{W} = a(t - t_0)$ of Huggett & Widdas (1951). The gestation length (t) was taken as 107 days (see Table III) and t_0 was estimated to be $20 \cdot 5$ days. W was the mean birth weight of the 31 young born to the nine females listed in Table III.

Resorption of embryos

Of 528 adult females, 434 (83%) were pregnant but 25 (5·8%) showed complete (in one animal) or partial (over half the litter in three animals) litter resorption. Embryos at all developmental stages were affected.

Litter size

Observations made on the uteri of 480 animals at various stages of pregnancy indicate that the modal number of embryos is four and

Table IV

Variation in litter size of grasscutters at different stages of embryonic development

Stage of embryonic development	No. of females bearing litter size of								Total no. of females
---	1	2	3	4	5	6	7	8	---
Less than 1 g	3	22	40	61	42	19	1	2	190
1–30 g	4	11	28	38	16	3	3	—	102
31–64 g	2	9	17	19	10	4	—	—	61
65–200 g	4	7	21	25	19	3	—	—	79
Placental scars	1	4	9	11	5	7	—	—	37
At birth (captive)	1	2	5	14	7	—	—	—	29

although the range is one to eight it extends from two to six in 96% of the grasscutters (see Table IV). Litter size at birth based on recent placental scars in 37 animals gave an average of four young ranging from one to seven and in 29 litters of captive animals litter size averaged 3·8, ranging from one to five. The data from these three sources are in close agreement. There was evidence of a significant relationship between litter size and state of maturity. Young adults (3 molar teeth) which are nulliparous primigravid females carry on average a smaller litter (mean size $= 3\cdot18 \pm 0\cdot14$) than do the parous adults (4 molar teeth) in which mean litter size is $3\cdot9 \pm 0\cdot06$ ($P = >0\cdot01$). There was no positive relationship between litter size and weight of the female bearing it, or between litter size and gestation length.

Sexing of embryos and the sex ratio

The female embryo has three or four pairs of teats but nipples are not visible in the male embryo. Another distinguishing feature is the shorter ano-genital distance in the female. This distance in 100 male embryos was shown to be $0\cdot80 \pm 0\cdot39$ cm whereas in a similar number of females it averaged only $0\cdot35 \pm 0\cdot20$ cm. In adults, the average ano-genital distance in 68 males was $3\cdot80 \pm 0\cdot72$ cm and in 67 females it was $1\cdot22 \pm 0\cdot20$ cm. Rudimentary teats were not visible in the adult male. The sex ratio of embryos at different stages of development was not significantly different from unity.

Birth weight

The classification of embryos described on p. 253 was applicable only to the stock of grasscutters collected in southern Ghana. Among this dead stock the weight of term embryos varied from 65 g to 190 g. In Accra, 31 young at birth (in nine litters) had a mean weight of 128·5 g ranging from 75 to 190 g. In the northern part of the country, birth weights ranged from 59 to 97 g, but, unfortunately, this sample did not contain a sufficient number of developed embryos to justify a strict comparison. Like most other hystricomorph rodents, the young at birth have their eyes open and their fur fully developed; they are able to follow the parent in less than one hour after birth.

The widest weight range in siblings of the same litter observed in animals with a gestation period estimated to be over 100 days was 115 g (75–190 g) in a litter of three; and the narrowest weight range was 10 g (100–110; 155–165 g) in litters of four and two. In litters for which the gestation periods were estimated to be below 100 days, the widest range was 29 g (110–139 g) in a litter of five and the closest was 26 g (72–98 g) in a litter of four.

Age at sexual maturity

Among the wild stock, grasscutters with three molars were found to be pregnant so that at this stage of tooth eruption they would be expected to be sexually mature. In captive stock, vaginal opening occurred at five months, but at this time none had become pregnant although the stage of dental development was similar to that in wild stock. Ewer (1969) reported that the grasscutter produced its first litter when about a year old or possibly a little earlier. My observations show that the offspring of a female born in March, 1971, as the result of a mating with a male born in February, 1971, had its first litter when 19 months old (August 1972). They were paired at the ages of seven and eight months respectively. The lowest dressed weight of 0·7 kg for a pregnant female was found for a young (3-molar) and an old (4-molar) grass-cutter. The indications to date, based on two captive females, are that two litters are born each year, but no further information is as yet available on the frequency of littering.

Breeding behaviour

Breeding in captivity has not proceeded sufficiently as yet to allow any comprehensive observations to be made. However, there is little doubt that the grasscutter is polygamous and although an adult male is rarely aggressive towards adult females he will not tolerate another male. No aggressive behaviour has been noticed in adult females. To this extent, the housing of these animals presents no insuperable difficulty although reintroduction of the male at parturition for remating could be troublesome, should the species be shown to exhibit an immediate post-partum ovulation and oestrus. Cannibalism of young by the male and the female has been observed.

Only one birth was recorded in the Mole National Park. A litter of four (including one stillborn) was delivered within 57 minutes (i.e. 07.40 to 08.37 hours) and the first born followed the mother during this period. The placenta was always eaten by the mother before the next offspring was delivered.

Weaning could be undertaken at two weeks but the best time was found to be at one month. The male was introduced to the female after the young of the previous litter had been weaned. Births have occurred five months after the introduction of a male.

DISCUSSION

The incidence of pregnancy in the grasscutters examined indicated that, in Ghana, the grasscutter breeds throughout the year and there is no defined breeding season though there is a tendency for breeding to be more frequent at certain times. In South Africa (Paradiso, 1968) and South-west Africa (Shortridge, 1934) the grasscutter is said to have a fixed breeding season. This is definitely not the case in Ghana even in the north where there is a severe dry season. Rosevear (1969) reported that for West Africa it is not known whether there is any preferred season for breeding, but juveniles have been collected from mid-October to the beginning of January. He did not indicate whether the collection was made in the region which had two peaks or one peak of rainfall. Ewer (1969) also indicated the uncertainty about breeding seasons of the grasscutter. There is an indication that breeding in the grasscutter is related to seasonality of rainfall but it is not clear whether it adapts its breeding to prevailing rainfall conditions.

Ewer (1969) reported a litter size of four to six and a sex ratio that did not differ significantly from unity. Rosevear (1969) observed a litter size of two to four and quoted a report of the birth of a litter of five in captivity while Booth (1960) indicated a litter size of six. Neither Rosevear nor Booth refer to sex ratios at birth. For South and South-west Africa, Paradiso (1968) and Shortridge (1934) report a litter size of two to four but usually three. The present study shows that in Ghana, litter size ranges from one to eight and the sex ratio does not differ significantly from unity. The average number of embryos *in utero* in primigravid grasscutters was smaller (3·1) than that (3·9) in parous animals. The estimated fetal growth velocity of 0·0614 for the grasscutter accords with values reported for other caviomorph rodents (Roberts & Perry, 1974).

Ewer (1969) reported that the gestation period was probably 70 days. The limited data in the present study suggest that it is more likely to be about 107 days, although there are some indications (from the vaginal closure membrane data and the size of the young at birth) that parturition might occur in less than 100 days from conception. It is possible that the grasscutters in northern Ghana have had to adapt their breeding to the prevailing weather conditions of drought and fire risks.

Juvenile mortality in captivity was low and if this is also true in the wild, then it is clear why the grasscutter could become a formidable pest if not used for meat (Asibey, in preparation), since about 70% of the females could be expected to produce two litters of four young each year.

SUMMARY

The grasscutter (*Thryonomys swinderianus*) breeds throughout the year. In northern Ghana, a pregnancy rate of 80% was found only from May to August, but in southern Ghana this rate of pregnancy was observed in all months except February and August. There was some suggestion of a relationship between the period of low rainfall and the time of parturition in southern Ghana.

Female grasscutters give birth to two litters of one to eight (average four) well developed young twice a year after a gestation of approximately 107 days. The weight of young at birth is variable between and within litters. The sex ratio does not differ from unity. The vaginal closure membrane first becomes perforated at five months of age but the first litter is usually born between 12 and 18 months of age.

ACKNOWLEDGMENT

Dr P. Grubb of the Zoology Department, University of Ghana, Legon, Accra, and Professor G. M. Dunnet of the Zoology Department, Aberdeen University, made very helpful suggestions and offered constructive criticism for which I am most grateful. I am also grateful to Mr Victor H. K. Awayevoo of the Mathematics Department, University of Ghana, for help with statistical analysis of data. The Director of the Meteorological Service and his staff extracted and permitted the use of the rainfall data included in this paper. This privilege and help are acknowledged with thanks.

The technicians on the Bushmeat Project in the Department of Game and Wildlife gave freely of their leisure time to assist me in compiling data reported here. Their enthusiasm, reliability and co-operation made it possible for me to carry out this study. The help of all other members of the Department who assisted me in various other ways is gratefully acknowledged.

I am most grateful to the Ghana Government for financing the project; their generous support and that of the British Council enabled me to attend the Symposium.

APPENDIX

Since this paper was presented in London, I have obtained evidence that the gestation length is probably longer than I have postulated. A female exhibited scratched flanks and this was taken as a sign of mating and the pair was separated. The female gave birth to eight young 151 days later. Using this exact gestation length, a theoretical value of $t_0 = 30$ was obtained for the Huggett & Widdas (1951) formula and the application of this to known birth-weights of 33 other litters gave a mean pregnancy length of 155 ± 9 days ranging 137–172 days. These data suggest that the shorter gestation lengths reported in this paper were, in fact, incorrectly derived from vaginal openings occurring in pregnancy.

Two females constantly with males also gave intervals of 154 and 163 days between litters and this indicates a possible post-partum oestrus.

REFERENCES

Asibey, E. O. A. (1969). *Grasscutter* (Thryonomys swinderianus) *as a source of bushmeat in Ghana*. Mimeographed M.S.

Booth, A. H. (1960). *Small mammals of West Africa*. London: Longmans Green.

Ewer, R. F. (1969). Form and function in the grass cutter *Thryonomys swinderianus* Temm. (Rodentia, Thryonomyidae). *Ghana J. Sci.* **9**: 131–149.

Huggett, A. St. G. & Widdas, W. F. (1951). The relationship between mammalian foetal weight and conception age. *J. Physiol., Lond.* **114**: 306–317.

Paradiso, J. L. (Ed.) (1968). *Walker's mammals of the world.* **2**: 1068–1069. 2nd edn. Baltimore: Johns Hopkins Press.

Roberts, C. M. & Perry, J. S. (1974). Hystricomorph embryology. *Symp. zool. Soc. Lond.* No. 34: 333–360.

Rosevear, D. R. (1969). *The rodents of West Africa*. London: British Museum (Natural History).

Shortridge, G. C. (1934). *The mammals of South West Africa.* **1**. London: Heinemann.

Weir, B. J. (1974). Reproductive characteristics of hystricomorph rodents. *Symp.zool. Soc. Lond.* No. 34: 265–301.

DISCUSSION

PERRY (*Cambridge*): When you are able to get more examples of birth weight in individual animals do you think you will find there is less variation in gestation length or have you got an animal that has a wide range of gestation periods?

Asibey (*Ghana*): I think we have got an animal that has a wide range of gestation periods because of the wide range of habitats in which it can live. There are two litters a year but, in the north, the animals cannot afford the long gestation periods that they can have in the coastal areas where the rainfall is not as severe and the weather is not so bad.

Amoroso (*Cambridge*): Have you in your dissection of these creatures looked at the ovaries?

Asibey: I have not yet looked at the ovaries but I have kept them all for future study.

Amoroso: With superficial examination, did you get the impression that there were many corpora lutea in excess of the conceptuses?

Asibey: I'm afraid I didn't look. Since I had to make every effort not to digress too far, I decided that anything that was not immediately relevant to the economics of grasscutters in Ghana could be ignored, but all the samples were kept for future examination by ourselves and other workers.

Watson (*London*): Mr Asibey, as I understand it, the minimum weight at birth for survival was 65 g, I think you said.

Asibey: No, I said that I had a high incidence of fetuses which weighed 65 g and therefore I believed that to be the minimum. When I was dealing with a large number of embryos at different stages of development, I asked myself what would have happened were the mother not killed. And I thought that any fetus that was 65 g or more in weight could have been born a day or so later.

Frazer (*London*): I think one should use a bit of care in using the Huggett & Widdas formula for hystricomorphs. As you know, the fetal growth is in two phases, initially slow and then faster and this formula normally measures the faster rate in the latter part of pregnancy. But where you have either delayed implantation or a very long first phase, as in hystricomorphs, the results tend to be too low and may be as much as 20% out.

Asibey: Thank you.

Lansdown (*London*): Essentially my question is in two parts. Firstly, to what extent do you consider that environmental conditions are contributory to the variations of gestation period? This would include not only the conditions created by climate at the time but also the availability of food before and during pregnancy. Secondly, do you find that the young born after a short gestation period are less well developed than those born after longer periods?

Asibey: The gestation period seems to be shorter in the northern animals, but there is no indication that their young are in any way premature. In fact, it was in the north that I observed young following the mother less than one hour after birth, so they could hardly be said to be premature. On the question of environmental factors—I think two things are very important, they are food and the reaction of the people. In the south there is no doubt that there is plenty of food for the animals because their acquired habitats are either farms, where, even when they have grazed initially, there is a lot left to eat, or areas that have been so disturbed that grasses have established themselves. The grass rarely dries up in the south and there is definitely good food all the time. In the north, however, at the end of the wet season, people start burning the grass, and invariably by the peak of the dry season the ground is bare. The *Thryonomys* births must occur, therefore, before December. The interval between litters in the south was about six months, but I found out that if you wean the young and pair the mother with a male, another litter will be produced about five months later, indicating that they can reduce the between-litter interval.

KLEIMAN (*Washington*): I want to make some comment on estimating gestation period from vaginal closure and opening. In the species we have looked at, you frequently get an opening of the vaginal membrane during pregnancy, so that some of those shorter gestation periods you have based on the last vaginal opening are probably already pregnant females which mated much earlier.

DEANESLY (*Cambridge*): If you can get matings in captivity, can't you isolate the female when you think there has been a mating and then check the length of the pregnancy?

ASIBEY: Yes, we are doing that now, taking as an indication of mating the hair loss and lateral scratches that females may exhibit at some time during pairing. I hope that by next month data will be available from these females.

PERRY: But that would just give you a known length of gestation in one particular animal, and I think Mr Asibey's suggestion is that these animals can vary their gestation length in different circumstances. This is a revolutionary suggestion. There are some African large animals that can bend the gestation period a little bit so that they all give birth more or less on the same day when they have all been served more or less on the same day, but the idea of this sort of variation, according to the weather or stopping when it rains or something, is really an extreme case of variation, if in fact it holds. I think we must remember that all this work Mr Asibey has done on the grass-cutter is a sort of spin-off from his main concern and we should congratulate him on what he has been able to do, and not take too much notice of his apologies for not having done more.

Symp. zool. Soc. Lond. (1974) No. 34, 265–301

REPRODUCTIVE CHARACTERISTICS OF HYSTRICOMORPH RODENTS

BARBARA J. WEIR

*Wellcome Institute of Comparative Physiology,
Zoological Society of London, Regent's Park,
London, England*

INTRODUCTION

The physiological mechanisms concerned in the reproduction of mammals are diverse and many variations have been evolved to achieve the perpetuation of the species. It seems probable that reproduction has been more subject to evolutionary pressures than any other physiological system. When similar solutions to a problem have been found, they are usually considered to be due to convergence rather than to parallel evolution determined by genetic necessity. For example, animals of two related species may have become seasonal or non-seasonal breeders because it was more 'convenient' for them and not because they shared genes from their common ancestor. It is for this reason that characteristics of reproduction have rarely been used to indicate relationships between and within taxa.

The suborder Hystricomorpha (*sensu lato*) is a relatively small group of rodents, some of doubtful affinity, in which several unusual reproductive patterns have been reported (Mossman & Judas, 1949; Pearson, 1949; Weir, 1971a,b). However, in spite of the strangeness of the reproductive patterns described by these and other authors, it has always seemed to me that hystricomorph rodents are characterized by the possession of certain common features in their reproductive processes.

In this account of some of those features, species are referred to by their common names or by the generic name since there is rarely a possibility of confusion; the binomial form associated with the commonest vernacular is given in Table I to which reference should be made when necessary.

GESTATION LENGTH

An accurate gestation length can be determined only from calculation of the interval between an observed mating (or the finding of a copulatory plug or the presence of spermatozoa in the vaginal smear—see below) and normal parturition. It is reasonable, but not always correct, to assume that copulatory behaviour represents impregnation, and it is preferable, therefore, to separate animals after mating or any other sign (such as last closure of the vaginal membrane, fur in the cage, scratches, increased or decreased aggression) believed to be indicative of such an event. Intervals between consecutive litters may also reveal a periodicity which may be a true gestation length if there is a post-partum oestrus (see below).

Whatever the source of the information, it is essential to obtain observations on many animals, if possible, to account for normal variation. Many reports of gestation lengths in the literature should be treated with reserve especially when the evidence and method of calculation is not given. The values given in Table I are those considered to be reasonably accurate because of the method used or because they have been confirmed by another author or method.

It is clear that all the hystricomorphs for which gestation lengths are known or suspected have long pregnancies for their sizes. The shortest is that of *Galea* at 52 days and the longest are those of *Erethizon* (213 days) and *Dinomys* (222–283 days). The longest known gestation lengths in myomorph rodents are those of the Australian native rat (*Mesembriomys gouldii*) at 46 days (Crichton, 1969) and the mole rat *Tachyoryctes*, at 45–50 days (Rahm, 1969), and the shortest is that of the hamster (*Mesocricetus auratus*) at 16 days (Bruce & Hindle, 1934). Gestation lengths in squirrels (see Asdell, 1964) are generally about 28–45 days, being shortest in the dipodomids (17–23 days) and longest in the beaver, *Castor fiber* (105–107 days).

Many other mammalian species are known to have very long copulation to parturition intervals in spite of their small size. But in most of these species the long 'gestation' has been caused by either delay of fertilization, as in temperate vespertilionid bats that hibernate (see Racey, 1973), or delay of implantation which is found in many unrelated species such as the roe deer (*Capreolus capreolus*), stoat (*Mustela erminea*), bear (*Ursus americanus*), seal (*Phoca vitulina*) and badger (*Meles meles*). Some notable exceptions are found among several very small tropical bats in which pregnancy is not interrupted by the above devices, e.g. in *Miniopterus australis*, weighing 4·5 to 8·0 g, gestation lasts 110 days and in three Indian species of *Hipposideros* (7 to 9 g) the gestation is 155 to 160 days (Patil, 1968). Hystricomorph rodents, with the exception of *Lagostomus* (Roberts & Weir, 1973), have not adopted either of these delaying devices but have evolved long pregnancies in which approximately the first quarter is taken up by proliferation of the placental tissues and the rest represents the very slow growth of the fetus. The fetal growth rates of hystricomorphs and those of other mammals are best compared by application of the Huggett & Widdas (1951) formula (see Roberts & Perry, 1974). The observed fetal growth velocities of hystricomorphs do not correspond with the calculated values.

Precocity and litter size

The long pregnancy of hystricomorphs is often explained by the statement that the young are well developed, or precocious, at birth. There is, of course, no absolute distinction between precocious and non-precocious young of mammals (unlike the nidifugous and nidicolous young of birds), but precocious young are usually taken to be those which are fully furred and have their eyes open at birth, as are young ungulates. This is true of most hystricomorph young, such as guinea-pigs, chinchillids, coypu and porcupines. The newborn of the tuco-tuco, however, have their eyes closed and only the guard hairs have penetrated the skin (Fig. 1a). Degu young in our colony are similarly less well furred (Fig. 1b). The eyes open in two to three days and the pelage grows rapidly but at birth these young should not be called precocious. There is some evidence that young degu of colonies in the United States are more precocious (D. G. Kleiman, personal communication). An implicit suggestion of the term precocious is that the young can fend for them-selves soon after birth, but this is not true of ungulates or of hystrico-morphs. As shown in Table II, there is a considerable difference between the age at which the neonate can survive independently (in optimum conditions) and the time at which lactation usually ceases. Young will adopt the nursing position for many weeks or months after milk secre-tion stops and Kleiman (1974) has discussed the importance of the mother–young bond at this time.

The tuco-tuco and the degu have relatively large litters (Table III) but they are no larger than those of the cuis, which has very well developed young and the shortest known pregnancy for an hystrico-morph. There is no obvious correlation of the gestation length of the species with its average litter size or with the maternal body weight. An inverse correlation of gestation length with litter size has been reported for *Cavia porcellus* (Rowlands, 1949) and *C. aperea* (Rood & Weir, 1970) but has not been found for the chinchilla, degu, cuis or casiragua (B. J. Weir, unpublished data). This is not surprising for the cuis and casiragua since the range of their gestation lengths is very small (see Table I). Only two hystricomorph species are known to have a virtually unalter-

TABLE I

Reported gestation lengths thought to be reasonably accurate in hystricomorphs

Species	Common name	Gestation length (days)		No. of observations	Method of determination	Source
		Mean	Range			
Caviidae						
Cavia porcellus	Guinea-pig	68	59–72	>2000	S, P, L	Blandau & Young 1939; Rowlands, 1949 ,
C. aperea	Native (or wild) guinea-pig	62·4	59–74	35	S, P, L	Rood & Weir, 1970
C. tschudii (= *C. cutleri*)	Wild cavy	63·3	56–59	19		Castle & Wright, 1916
C. rufescens	Brazilian cavy		62–65			Ubisch & Mello, 1940
Hybrid cavies*						
Galea musteloides	Cuis	53	49–60	>400	S, P, L	Weir, 1970b; Rood & Weir, 1970
Microcavia australis	Desert cavy	54	53–55	6	L	Rood & Weir, 1970
Dolichotis patagonum	Mara	>93		1	I	NZP
Pediolagus salinicola	Salt desert cavy	c. 77		4	C, L	D. G. Kleiman & E. Maliniak, personal communication
Hydrochoeridae						
Hydrochoerus isthmius	Capybara	104–111		1	C	Trapido, 1949
Hydrochoerus hydrochaeris	Capybara	149, 156		2	C	Zara, 1973

Chinchillidae						
Chinchilla laniger	Chinchilla	110·8	105–115	>100	S, P, L, W	Weir, 1966, 1970a; Dennler, 1940
Lagidium peruanum	Peruvian mountain viscacha	>90		Many	W	Pearson, 1949
		140		1	S, P	Weir, 1972b
Lagidium boxi	Patagonian mountain viscacha	120–140		2	W	Weir, 1971f
Lagostomus maximus	Plains viscacha	150			W	Hudson, 1872
		153·7	145–166	>100	S, P	Weir, 1971a
Dasyproctidae						
Dasyprocta sp.	Agouti	120		Several	L	Roth-Kolar, 1957
		104				Brown, 1936
		<120		1		NZP
Myoprocta pratti	Green acouchi	99		Several	S, P, L	Kleiman, 1970; Weir, 1971d
Cuniculus paca	Paca	<115		1	C	E. Maliniak, personal communication
Dinomyidae						
Dinomys branickii	Pacarana	222–283		1	C & I	Collins & Eisenberg, 1972
Octodontidae						
Octodon degus	Degu	90	87–93	>70	S, P, L	Weir, 1970b
Octodontomys gliroides	Chozchoz (pl. chozchoris)	99	99	2	S	Weir, 1972b
		104	(100–109)	1 (3)	C	D. G. Kleiman, personal communication

continued

TABLE I—*Continued*

Species	Common name	Gestation length (days)		No. of observations	Method of determination	Source
		Mean	Range			
Abrocomidae						
Abrocoma cinerea	Chinchilla rat	115–118		1	C	J. F. Eisenberg, personal communication
Ctenomyidae						
Ctenomys torquatus	Tuco-tuco	107		Several	C	Talice & Laffitte de Mosera, 1958
C. talarum	Tuco-tuco	102	93–120	8	L, C	B. J. Weir, unpublished
C. peruanus	Tuco-tuco	120			W	Pearson, 1959
Echimyidae						
Proechimys guairae	Casiragua; spiny rat	62·6	61–64	>200	S, P, L	Weir, 1973b
P. semispinosus	Casiragua	64·8	63–66	8	C	Maliniak & Eisenberg, 1971
Capromyidae						
Myocastor coypus	Coypu	128–130		>400	C	Ehrlich, 1966
		132	127–138	Several	C	Newson, 1966
Capromys melanurus	Black tailed hutia	140		1	L	Bucher, 1937
Capromys pilorides	Cuban hutia	126		1	L	Bucher, 1937
		120		1		Taylor, 1970

Geocapromys ingrahami	Bahaman hutia	110–120		1	L	Howe & Clough, 1971
Plagiodontia aedium	Hispaniolan hutia	>123–<150		2	C	J. F. Eisenberg & E. Maliniak, personal communication
Erethizontidae						
Erethizon dorsatum	North American porcupine	230		1	P	Burge, 1966
		210				Goldsmith, 1774
Erethizon epixanthum		213		2	C	Shadle, 1948
		180–252				Taylor, 1935
Hystricidae						
Hystrix cristata	African porcupine	112				Jennison, 1927
Hystrix africae-australis		112				Dekeyser, 1955
Atherurus africanus	Brush-tailed porcupine	100–110		2	C & I: L	Rahm, 1962
Thryonomyidae						
Thryonomys swinderianus	Grasscutter	155	137–172	33	C & I	Asibey, 1974

NZP = records of the National Zoological Park, Washington, by courtesy of Dr J. F. Eisenberg and Dr D. G. Kleiman.
S = Spermatozoa in vaginal smear; P = copulatory plug; L = between-litter interval; C = copulation or other signs of mating behaviour; I = isolation; W = interpretation from data collected in the field. * see Weir, 1974b, Table I.

(a)

(b)

FIG. 1. The young of two species of caviomorph rodent which are not fully precocious at birth. Compare with Fig. 7 of Rowlands (1974). (a) A 24-hour old tuco-tuco (*Ctenomys talarum*). The eyes are still closed and the posterior part of the body is poorly furred, only the guard hairs having penetrated the skin. (b) A litter of new born degu (*Octodon degus*).

able litter size; mountain viscacha (*Lagidium* spp.) are strictly monotocous because only one egg is ovulated (Pearson, 1949) and plains viscacha nearly always (95%) have twins (Table III) after selective resorption of excess embryos (see Weir, 1971a).

Fetal: maternal weight

The very large size of the newborn of two species of tree porcupine (*Coendou prehensilis* and *Sphingurus villosus*) compared with the maternal weight was noted by de Miranda-Ribeiro (1936) and the litter weight of many hystricomorphs at birth (expressed as a percentage of the maternal weight) is high compared with the maternal weight (Table IV). Leitch, Hytten & Billewicz (1959) recognized that the guinea-pig was exceptional compared with other mammals in this respect, and it is clear from Table IV that the average values for other hystricomorphs would also fall well away from the regression line of Leitch *et al.* of the relationship between maternal weight and total weight of the newborn young. This discrepancy cannot therefore be explained on the grounds of domestication.

Habitat

It can be seen from Table I that the shorter gestation lengths are characteristic of the family Caviidae and two species of Echimyidae, both in the same genus. It is clear, however, that there is no obvious correlation of gestation length with habitat (Fig. 2). The caviids are mostly plains dwellers and/or burrowers but so are other species such as plains viscacha; moreover, *Cavia* and *Galea* are widespread in South America and are found at high altitudes. Many other caviomorphs are

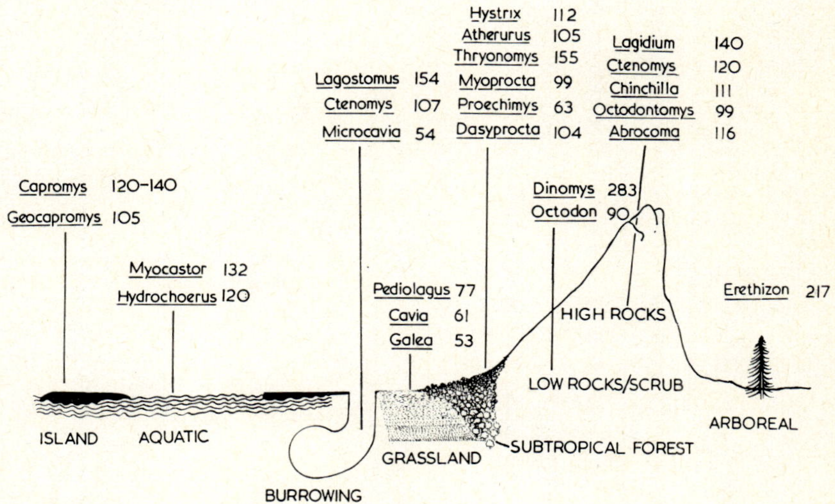

Fig. 2. A diagram indicating the habitats of some hystricomorphs and their gestation lengths in days.

TABLE II

Minimum and usual lactation periods and the age at onset of puberty of some hystricomorphs

Species	Minimum lactation for survival (days)	Normal lactation period (weeks)	Sex	Puberty* Average	Minimum
Cavia porcellus	5	3	♂	3 months	
			♀	2 months	21 days
Cavia tschudii			♀	2 months[1]	
Galea musteloides	7	3	♂	3 months	
			♀	2 months	11 days
Hydrochoerus hydrochaeris		16[2]	♀	15–24 months[2,3]	
Chinchilla laniger	21	6–8	♂	8 months[4]	$5\frac{1}{2}$ months
			♀	8 months[4]	$5\frac{1}{2}$ months
Lagostomus maximus	21		♂	15 months[5]	$5\frac{1}{2}$ months
			♀	$8\frac{1}{2}$ months[5]	$5\frac{1}{2}$ months
Lagidium peruanum		8	♂ + ♀	12 months[6]	

Proechimys guairae	14	3	♂	3 months[7]	
			♀	2 months[7]	35 days
Octodon degus	14	4	♂ + ♀	6 months	63 days
Ctenomys talarum	7[8]–10	5	♂ + ♀	8 months	6 months
Myoprocta pratti	14[9]†		♂ + ♀	8–12 months[9]	
Dasyprocta aguti		20[10]	♀	6 months[10]	
Myocastor coypus	5[11]	6–10[11]	♂ + ♀	3–4 months[11]	(born in summer)
				6–7 months[11]	(born in autumn)
Erethizon dorsatum		8[12]	♂ + ♀	15 months[13]	

[1] Castle & Wright (1916); [2] Zara (1973); [3] Ojasti (1970); [4] Weir (1970a); [5] Weir, (1971a); [6] Pearson (1949); [7] Weir (1973b); [8] Talice & Laffitte de Mosera (1958); [9] Kleiman (1970); [10] Roth-Kolar (1957); [11] Newson (1966); [12] Shadle & Ploss (1943); [13] Shadle (1952).
* Puberty = age at first conception.
† Can survive if weaned before this age but growth retarded.

TABLE III

Litter size and the number and position of nipples in hystricomorphs, grouped according to the nipple arrangement

| Species | Litter size | | Nipples | | Reference |
	Range	Mean	No. of pairs	Position	
A. *Cavia porcellus*	1–13	4·0	1	1 ing.	see Rood & Weir, 1970
C. aperea	1–5	2·3	1	1 ing.	Rood & Weir, 1970
C. rufescens		1·35		(1 ing.)	Ubisch & Mello, 1940
C. tschudii (= *C. cutleri*)	1–4	1·9		(1 ing.)	Castle & Wright, 1916
Hybrid cavies*					
Hydrochoerus hydrochaeris	1–8	5	5	ventral	Ojasti, 1970
Dolichotis patagonum	1–2		4	ventral	B. J. Weir, unpublished
Myoprocta pratti	1–3	2	4	ventral	B. J. Weir, unpublished
Dasyprocta aguti	1–3	2	4	ventral	B. J. Weir, unpublished
Erethizon dorsatum	1–2	1	2	1 th., 1 abd.	Shadle, 1950
B. *Galea musteloides*	1–7	3	2	1 ing., 1 lat. abd.	B. J. Weir, unpublished
Pediolagus salinicola	1–2		2		D. G. Kleiman, personal communication
Proechimys guairae	1–7	3	3	1 ing., 2 lat. th.	Weir, 1973b
P. semispinosus	1–5	3	3		Maliniak & Eisenberg, 1971
P. dimidiatus	1–5	3	3	(1 ing., 2 lat.)	Davis, 1947
Eurzygomatomys guaira			3	1 ing., 2 lat.	Davis, 1947
Octodon degus	1–10	5	4	1 ing., 3 lat.	B. J. Weir, unpublished
Octodontomys gliroides	1–3	2	3	1 ing., 2 lat.	B. J. Weir, unpublished; D. G. Kleiman, personal communication

Species					
Ctenomys talarum	1–7	5	3	1 ing., 2 lat.	B. J. Weir, unpublished
C. peruanus	1–5	3	3	1 ing., 2 lat.	Pearson, 1959
C. opimus	1–3	2	3	1 ing., 2 lat.	Pearson, 1959
Chinchilla laniger	1–6	2	3	1 ing., 2 lat. th.	Weir, 1970a
Abrocoma cinerea	1–2		2	2 lat. th.	Pearson, 1951
Lagostomus maximus	1–4	2	2	2 lat. th.	Weir, 1971a
Lagidium peruanum	1	1	2	2 lat. th.	Allen, 1901
L. boxi	1	1	2	2 lat. th.	B. J. Weir, unpublished
Capromys nana		1	2	2 lat. th.	Pocock, 1943
Capromys pilorides	1–4		2	2 lat. th.	Mohr, 1939
Plagiodontia aedium	1–2		3	3 lat. th.	Mohr, 1939
Dinomys branickii	1–2		4	2 lat. th., 2 lat. abd.	J. F. Eisenberg, personal communication
Atherurus africanus	1–2	1	2	2 lat. th.	Hatt, 1940
Hystrix cristata	1–4	2	2–3	2 or 3 lat. th.	Hatt, 1940; Weir. 1967b.
Myocastor coypus	1–13	5	4	4 lat.	Newson, 1966
Thryonomys swinderianus	1–7	4	2–3	2 or 3 lat.	Ansell, 1966; Asibey, 1974
Hopolomys gymnurus	1–3	2			Fleming, 1970
Geocapromys ingrahami		1			
Spalacopus cyanus	1–3				Reig, 1970
Cuniculus paca			4		Dubost, 1968
Cryptomys lechei		e	2	1 ing., 1 pectoral	Verheyen & Verschuren, 1966

ing. = inguinal; lat. = lateral; th. = thoracic; abd. = abdominal. Details in parentheses are inferred.
* See Weir, 1974b, Table I.

TABLE IV

List of hystricomorphs arranged according to increasing gestation length and demonstrating that there is no apparent correlation of gestation length or litter size with maternal weight, litter weight or maternal : litter weight ratio

Species	Maternal weight (MW) (g)	Neonatal weight (g)	Average litter size*	Litter weight (LW) (g)	$\frac{LW}{MW}$ %	Gestation length (days)
Galea musteloides	400	40	3	120	30	52
Cavia aperea	500	60	3	180	33	61
Proechimys guairae	300	22	3	66	22	63
Cavia porcellus	1000	100	6	600	60	68
Octodon degus	250	14	5	70	24	90
Octodontomys gliroides	100	15	2	30	30	99
Myoprocta pratti	1000	100	2	200	20	99
Geocapromys ingrahami	700	80	1	80	11	100
Atherurus africanus	2500	150	1	150	6	105
Chinchilla laniger	500	35	2	70	14	111
Hystrix cristata	20000	(1000)	2	2000	10	112
Ctenomys talarum	150	8	4	32	20	130
Myocastor coypus	6000	225	5	1125	20	132
Lagidium peruanum	12000	180	1	180	15	140
Lagidium boxi	1400	260	1	260	18	140
Hydrochoerus hydrochaeris	30000	1500	4	6000	20	150
Lagostomus maximus	3000	200	2	400	13	153
Thryonomys swinderianus	2000	130	4	520	25	155
Erethizon dorsatum	5000	1500	2	3000	60	217
Dinomys branickii	13000	900	2	1800	14	222–283

* See Table III for ranges.

All values given in columns 2 to 5 are approximations and therefore those in column 6 could differ greatly in extreme circumstances.

found in the same type of habitat as *Proechimys* but do not have particularly short pregnancies.

NIPPLE POSITION

The mammary primordia are developed from epidermal thickenings and are reported to be along a paired ventral (mammary) line from the axilla to the inguinum. When the nipple number is reduced, the nipples may be spaced out along the whole of the ventral line or may be restricted to the pectoral or inguinal ends (Raynaud, 1969). The number and position of the nipples is a species characteristic and can be correlated with the litter size; for example, primates usually have only one pair of nipples while the tenrec (*Tenrec ecaudatus*) has been recorded as rearing 30 of 31 young with 11 or 12 pairs of nipples (Louwman, 1973).

A similar nipple and litter size correspondence occurs in hystricomorphs (Table III) except for *Cavia porcellus* and *C. aperea* which have only a single pair of mammae but litters of more than two. This arrangement is possible because these cavies will nurse promiscuously. There is great variation in the position of the nipples among the hystricomorphs; and in some species the anterior nipples are placed along the sides of the thorax or the abdomen (Fig. 3). Only a few caviomorphs in which the

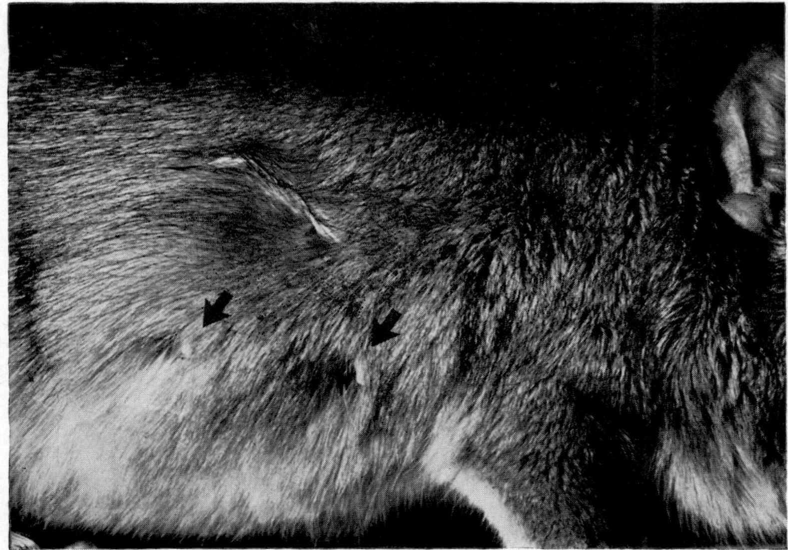

FIG. 3. A lactating plains viscacha (*Lagostomus maximus*) showing the lateral position of the mammae (arrowed).

nipple position is known have all the mammae along the ventral lines (Table III, Group A); these include the dasyproctids, *Dolichotis* and *Hydrochoerus* while *Cavia porcellus* and *C. aperea* have only an inguinal pair. Many hystricomorphs, like the chinchilla, have the inguinal pair, representing the ventral line, and the rest in a lateral position (Table III, Group B). Others appear to have lost the inguinal pair and possess only one or two lateral thoracic pairs of nipples (Table III, Group C). The coypu has three or four pairs of mammae and it was their position in this species that originally drew attention to this tendency for the nipple line to be lateral in hystricomorphs (Martin, 1835). It was thought that the high position of the nipples enabled the young coypu to suck while they and the mother were in the water (Gibson, 1877b) but this is not true. Kleiman (1974) suggests that the nipple position determines the nursing position which is itself dictated by the need for alertness at the nursing site. Thus a pampean species like *Dolichotis* will sit to nurse young and can keep a look out for predators and be ready to flee at a moment's notice, while a nest dweller such as *Octodon* will huddle over its young. This theory could be applicable to most hystricomorphs, but there are some discrepancies; for example, *Lagostomus* is a well protected nest dweller and yet the young are nursed in a sitting position from the lateral nipples. It is surprising that, if the lateral nipples confer such an advantage for predator detection and fleeing, other mammals have not evolved similarly. As far as I know, lateral nipples are not present in any other mammals except gundis (F. Ctenodactylidae). All four genera of these rodents of uncertain affinities (see George, 1974) have a pair of lateral thoracic nipples and a lateral cervical pair (W. George & B. J. Weir, unpublished observation).

VAGINAL CLOSURE MEMBRANE

The vagina of all mammals is sealed by a membrane until puberty is reached, and, thereafter, it is usually permanently patent as in ungulates, carnivores and primates. It may, however, be sealed during anoestrus in seasonal breeders, such as myomorph and sciuromorph rodents or plugged by cornified detritus, as in temperate zone bats during hibernation and delayed fertilization (Racey, 1972).

The presence in the guinea-pig of a vaginal closure membrane was first noted by le Gallois (1812) and described in detail by Stockard & Papanicolaou (1919) who related its periodic opening to the recurrence of oestrous behaviour. The cellular nature of the membrane and the changes in its histological characteristics during its dissolution and repair after oestrus have been investigated by Kelly & Papanicolaou (1927) in the guinea-pig. A similar membrane (Fig. 4), perforate at oestrus and parturition, is present in all hystricomorphs that have been examined (see Table V) except coypu. In some species, like the chinchilla, the membrane is very sturdy and its perforation and sealing is quite distinct; in others, like the degu and tuco-tuco, the membrane is more fragile, and may become partly perforate and may reseal before becoming completely open. The vaginal membrane in the capybara and African porcupine is more deeply seated than in the guinea-pig.

The duration of perforation at oestrus varies with the individual animal as well as with the species. A chinchilla may 'open' and 'close' in 12 hours but usually takes two to four days. A plains viscacha may remain 'open' for about 13 days and some degu may have a perforate vagina for weeks. The time of oestrus is not easy to determine from the appearance of the vulva. Many of the caviomorphs secrete mucus but only in the plains viscacha, tuco-tuco, casiragua and degu may vulval

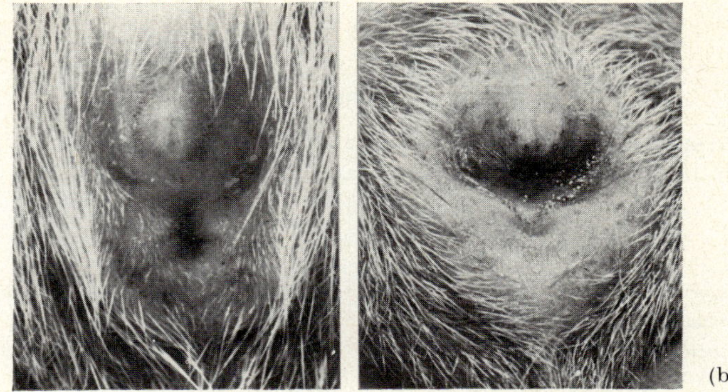

(a) (b)

Fig. 4. The vaginal closure membrane of the cuis (*Galea musteloides*). (a) Closed, as during pregnancy or when isolated from a male. The position of the vaginal opening is indicated by the slightly raised semi-circular area separating the urinary papilla (above) and the anus (below). (b) Open, as at oestrus. There is no vulval swelling in this species. The anus is much less conspicuous in this particular animal.

swelling sometimes be observed. The detection of oestrus from interpretation of vaginal smears is not as easy in other hystricomorphs as it is in *Cavia*, in which the re-appearance of leucocytes in the smear is a good indication of ovulation having occurred. In the chinchilla, for example, leucocytes are present throughout the time of perforation and mating may take place at any time in relation to the time of vaginal opening. Plains viscacha usually mate towards the end of the period of vaginal opening and cuis, degu and casiragua mate in the first day or two. Closure of the membrane usually takes place quickly once mating has occurred, but may be delayed after parturition, for several days more than is customary.

Vaginal opening can occur at times other than parturition and oestrus. A detailed study made in pregnant guinea-pigs by Ford *et al*. (1951) indicated that perforation occurred in about 80% of animals at Day 26 or 27 when, it was postulated, the placenta was gradually replacing the ovary as the main source of progesterone to maintain gestation. The vagina of the degu and tuco-tuco frequently opens sporadically throughout gestation (B. J. Weir, unpublished) and many plains viscacha open at Day 62 of the 153-day gestation (Weir, 1971a).

OESTROUS CYCLES

Oestrus may occur in mammals only once or twice a year, or at regular periods throughout a breeding season or a complete year. Cycles are perhaps artifacts produced by captivity (see Weir & Rowlands, 1973) since most wild females are pregnant, lactating or anoestrous.

TABLE V

Hystricomorph species in which a vaginal closure membrane is known to be present

Species	No. of animals	Reference
Cavia porcellus	many*	Stockard & Papanicolaou, 1919
Cavia aperea	many*	Rood & Weir, 1970
Galea musteloides	many*	Rood & Weir, 1970; Weir, 1970b
Microcavia australis	several	Rood & Weir, 1970
Hydrochoerus hydrochaeris	2	B. J. Weir, unpublished
Dolichotis patagonum	?	Pocock, 1922
Chinchilla laniger	many*	Hillemann, Tibbitts & Gaynor, 1959, Weir, 1966, 1970a
Lagidium peruanum	many*	Pearson, 1949; B. J. Weir, unpublished
Lagidium boxi	2*	Weir, 1971f
Lagostomus maximus	many*	Weir, 1971a
Dasyprocta aguti	several*	Weir, 1967a, 1971e
Myoprocta pratti	several*	Kleiman, 1970; Weir, 1971d
Dinomys branickii	2	D. G. Kleiman, personal communication
Octodon degus	many*	Weir, 1970b
Octodontomys gliroides	10	B. J. Weir, unpublished
Spalacopus cyanus	2	D. G. Kleiman, personal communication
Ctenomys talarum	many*	B. J. Weir, unpublished
C. torquatus	several	Talice & Laffitte de Mosera, 1959
C. opimus	many	Pearson, 1959
C. peruanus	many	Pearson, 1959
Proechimys guairae	many*	Weir, 1973b
P. dimidiatus	21	Davis, 1947
Hoplomys gymnurus	several	Fleming, 1970
Abrocoma cinerea	several	Pearson, 1951
Plagiodontia aedium	2	D. G. Kleiman, personal communication
Geocapromys brownii	2	B. J. Weir, unpublished
Hystrix cristata	2*	Weir, 1967a
Thryonomys swinderianus	many*	Asibey, 1974
Heterocephalus glaber	several	Hill, Porter, Bloom, Seago & Southwick, 1955

* Cyclic perforation and closure of the membrane was studied in some of these animals.

Nevertheless, the periodic recurrence of oestrus can be considered as a specific characteristic. Oestrous cycles in myomorph rodents are determined from the changing pattern of cell types in the vaginal smear (see Long & Evans, 1922) and are four to five days in rat and mouse, four days in the hamster and up to eight or nine days in *Oryzomys*, *Sigmodon* and *Meriones* (Asdell, 1964).

Most of the hystricomorphs that have been investigated are spontaneous ovulators. When isolated from a male the length of the oestrous cycle is equivalent to the interval from the first day of vaginal opening at one oestrus to the day before the first day of opening at the next oestrus providing that the species has a closure membrane (see above), a regular cycle, and the vagina does not open from any other cause. Known lengths of oestrous cycles in hystricomorphs are shown in Table VI. The caviids, and the echimyid *Proechimys*, have relatively 'short' cycles which are, however, longer than those in myomorphs; the cycles of the other hystricomorphs are rather long. The variations in cycle length in any one species may be wide, e.g. 20 to 60 days in chinchilla (Weir, 1970a), but the mean lengths of most species are in the range of 30 to 40 days, indicating a long luteal phase (see Weir & Rowlands, 1974). In general the length of the cycle is determined to a large extent by the duration of the secretory activity of the corpora lutea. The life-span of the corpora lutea of different species is very variable, but within a species it is constant, as is the extent to which this ovarian structure regresses before ovulation can recur and another cycle begin. A 'cycle' length is difficult to define for species in which ovulation is induced by copulation because the vagina often stays perforate for long periods. In these species cycle length may be determined by observing vaginal opening in females kept with an infertile (i.e. vasectomized) male to provide the necessary cervical stimulus to induce ovulation.

Two species, the coypu and the cuis, present even greater difficulties. The coypu, as indicated above, does not have a vaginal closure membrane and many workers have tried to estimate the cycle length by analysing vaginal smears. The results are variable, even within the same female, and estimates range from four to 40 days, the average being about 26 days. Asdell (1964) suggests that the variation may be because the coypu is an induced ovulator, but this possibility has not been investigated. Only 13% of isolated female cuis exhibited vaginal perforation, which more often than not is sporadic rather than cyclic. Most females do not become oestrous unless a male is present for at least 24 hours and can perform the behavioural courtship pattern, the chin-rump follow (Weir, 1971c, 1973a; Rood, 1972). Vasectomized

males will induce vaginal opening at intervals of about 20 days (Rood &
Weir, 1970) and the importance of the chin gland of the male is discussed
by Weir (1973a).

POST-PARTUM OESTRUS

Many rodents and other mammals experience oestrus very soon
after parturition; in the guinea-pig it may occur within a few hours of
the expulsion of the last placenta (Rowlands, 1949). A possible evolutionary progression to this condition from seasonal breeding through
post-lactation and mid-lactation oestrus is postulated by Weir &
Rowlands (1973). Hystricomorph representations can be found at all
these stages. Plains viscacha, for example, bear only one litter a year
in the southern part of their range (Hudson, 1872; Llanos & Crespo,
1952) and thus the next fertile oestrus is at the onset of the ensuing
breeding season. In more favourable climates further north, however,
two litters are produced (Gibson, 1877a) and evidence derived from
captive animals suggests that the post-lactation oestrus is the most
fertile and an immediate post-partum oestrus is rare (Weir, 1971a).
Chinchilla have a post-lactation and post-partum oestrus; if conception
does not occur post-partum, there is a lactation anoestrus of about 56
days (Weir, 1970b). If the young chinchilla or viscacha die at or soon
after birth, there is often a rapid return to oestrus. A lactation anoestrus
may occur also in the cuis (Rood & Weir, 1970) and acouchi (Weir,
1971d).

The post-partum oestrus is more easily recognized and, as indicated
above, the shortest repeated interval between litters is frequently
accepted as the gestation length and implies a post-partum mating. All
hystricomorphs are probably capable of mating at this time (see Weir &
Rowlands, 1974); most do so in captivity and almost certainly in the
wild (see Table VI). Captive plains viscacha rarely mate at this time,
however, and none has done so whilst suckling; acouchi, degu and tuco-
tuco may do so but not regularly; guinea-pigs, cuis, coypu and chinchilla
nearly always mate but may not always conceive. The cuis and the
casiragua are the most prolific breeders at the Wellcome Institute
because a very high proportion have a fertile post-partum oestrus and
therefore once a female has become pregnant she usually remains in this
state. The casiragua is not included in the list above because only about
50% of females experience a post-partum oestrus; the other 50% have a
fertile oestrus 24 to 48 hours before parturition (B. J. Weir, unpublished
data)—a logical extension of the progression suggested by Weir &
Rowlands (1973).

OVULATION

Female mammals have often been classified as those which ovulate spontaneously (hamster, mouse) or which ovulate in response to copulation (rabbit, ferret). It is now thought (Conaway, 1971; Weir & Rowlands, 1973) that not all species can be categorized in this way and that there is a continuum representing the most likely response. Some hystricomorph species for which information (gained from observation of ovaries of females of known reproductive history—see Weir & Rowlands, 1974) is available are indicated in Fig. 5. It is interesting that

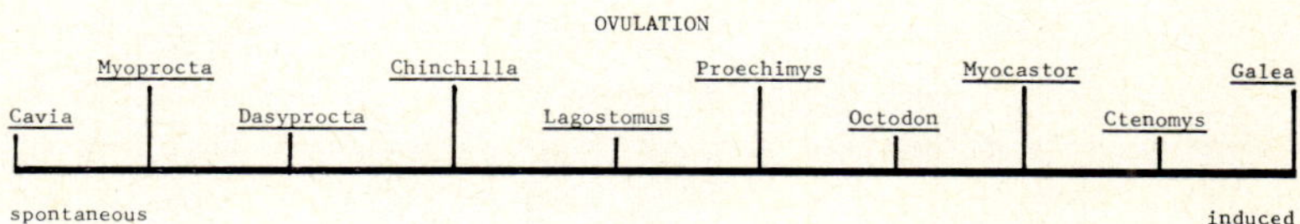

Fig. 5. The position of various caviomorph species along a continuum representing the type of ovulation.

two caviids are at the extremes of the spectrum. The evidence for *Galea* being an induced ovulator is mainly circumstantial because of the difficulty of preventing intromission in the females who have needed a male present and in contact to bring them into oestrus (see Weir, 1973a) but rests on the following data: (a) only half the females (i.e. 7% of the whole colony) in which oestrus occurs spontaneously actually ovulate, (b) females in which the vagina opens after the male has been removed rarely ovulate, and (c) only about 30% of females isolated at parturition ovulate (B. J. Weir, unpublished data). Corpora lutea (an indication of ovulation) were never found in tuco-tuco females unless they had mated; a similar picture is common for degu although some females do ovulate spontaneously (B. J. Weir, unpublished data). Casiragua probably ovulate spontaneously or reflexly with about equal facility since they come into oestrus spontaneously and will ovulate, but copulation may cause ovulation if it takes place early enough to precede the presumed LH surge. Females of low parity are less likely to ovulate spontaneously if they are isolated at parturition than are multiparous females. Chinchilla can ovulate in response to copulation (Weir, 1973c) but mostly do so spontaneously, sometimes before the vagina has opened. Some circumstantial evidence (Weir, 1971b) suggests that *Lagostomus* may be a spontaneous ovulator but that the corpora lutea are formed only if copulation has occurred.

TABLE VI

The occurrence and periodicity of oestrus in some hystricomorphs

| Species | No. of observations | Cycle length (days) | | Seasonal | | Next oestrus after parturition (days) | | Reference |
		Mean ± SE	Range	Wild	Captivity	Litter present	Litter lost	
Cavia porcellus	Many	16·5	13–25	−	−	PP	PP	Asdell, 1964
C. aperea	67	20·6 ± 0·8	12–41	±	−	PP	PP	Rood & Weir, 1970
C. rufescens			9–25					Detlefson, 1914; Ubisch & Mello, 1940
C. tschudii (=*C. cutleri*)					+			Castle & Wright, 1916
Galea musteloides	56	22·3 ± 1·4	11–48	±	−	PP; 41·8 ± 4·0	PP	Rood & Weir, 1970
Microcavia australis					+	PP; 15	PP; 15	Rood & Weir, 1970
Chinchilla laniger	323	38·1 ± 0·7	16–69	+	+	PP; 57·4 ± 2·6	PP; 57·4 ± 2·6	Weir, 1970a
			30–42					Bullard, 1953
Lagidium peruanum	6	56·7 ± 5·3	36–72	+		49(1)		B. J. Weir, unpublished

Lagostomus maximus	152	45·0 ± 1·2	16–94	+	−	55·9 ± 3·7 (41)	17·1 ± 3·6 (12)	Weir, 1971a
Dasyprocta aguti	29	34·0 ± 2·1	12–59	+	+	PP		Weir, 1971; Smythe, 1970
Myoprocta pratti	21	42·7 ± 2·5	24–62		+	PP;51(2)	PP; 16(3)	Weir, 1971d
	22	40 ± 2	16–68		+	54 ± 3(7)†	25 ± 4(11)‡	Kleiman, 1970
Octodon degus		×		+	±	PP	PP	B. J. Weir, unpublished
Octodontomys gliroides						PP		B. J. Weir, unpublished
Ctenomys talarum		×		+		PP		B. J. Weir, unpublished
Proechimys guairae	61	22·5 ± 3·4				PP**	PP**	Weir, 1973b
Myocastor coypus			4–46			PP		Gluchowski, 1954; Ehrlich, 1966; Newson, 1966
Hystrix cristata	3	35	30–37					Weir, 1967a, b
Erethizon sp.	2	29						Burge, 1966

× = do not appear to cycle; + = cycles or produces litters seasonally; − = breeds all year round; ± = breeds all year round unless weather (or some other environmental circumstance) is poor. PP = post partum. Figures in parentheses represent number of observations. * With vasectomized ♂; see Weir, 1971c, 1973a. † = data given as females housed alone. ‡ = data given as females with males. ** = 50% have a pre-partum oestrus.

All the species represented by an L in Table I (between-litter interval) are assumed to have a post-partum oestrus by implication.

Table VII

Reported sex ratios of live young at birth of some hystricomorphs

Species	Sex ratio ($\male\male/100\female\female$)	No. of animals	Reference
Cavia porcellus	115	1 393	Rowlands, 1949
Cavia aperea	83	75	Weir, 1970b
	72	117	Rood & Weir, 1970
Cavia rufescens $\male$			
$\times$ *C. porcellus $\female$*	76	37	Detlefson, 1914.
Galea musteloides	92	171	Weir, 1970b
	102	361	Rood & Weir, 1970
	97	1 175	
Microcavia australis	110	40	Rood & Weir, 1970
Chinchilla laniger	122	2 586	Hillemann, Tibbits & Gaynor, 1959
	119	10 664	Galton, 1968
Lagostomus maximus	85	119	Weir, 1971a
	79	297	
Octodon degus	110	128	Weir, 1970b
	116	389	
Ctenomys talarum	107	124	B. J. Weir, unpublished
Proechimys guairae	85	331	Weir, 1973b
	80	713	
Proechimys semispinosus	61	37	Maliniak & Eisenberg, 1971
	145	54	Tesh, 1970

The figures given without a reference are those available from the colonies at the Wellcome Institute at 31 December 1973; they include data already published for those species.

SEX RATIO

Quantitative data are available only for the few caviomorph species that have been kept in captivity (Table VII). It is interesting that the chinchilla and the plains viscacha should have such different sex ratios and that these two species and the casiragua should differ so greatly from the normal mammalian 1 : 1 ratio (see Altman & Dittmer, 1962).

PUBERTY

In spite of their apparent precocity at birth, most hystricomorphs take a long time to reach sexual maturity, even in laboratory conditions which favour early breeding (Table II). Even in captivity, chinchilla breed seasonally, their breeding season having shifted by six months in the northern hemisphere (Weir, 1970a), and puberty depends upon the time of year at which the young are born. A similar variation is found in coypu but they breed throughout the year (Newson, 1966). No such effect is found in *Lagostomus*, another seasonal breeder, or in *Galea*, a continuous breeder. Males are usually slower to mature and breed than females, in many instances because of social repression. In a group of plains viscacha, cuis or casiragua, only one adult male is usually tolerated, at least in captivity. This is not so for chinchilla, but in this species the male is usually smaller than the female and may need the additional three or four months to become physically capable of mating (Weir, 1970a). Isolated casiragua females experience first oestrus later than females living with males. The extreme condition of this arrangement is found in the cuis in which the first oestrus is also induced by the male (see above and Weir, 1973a). This is probably the reason that this species has the earliest known conception age for an hystricomorph and, indeed, for any mammal (see Vandenbergh *et al.*, 1971).

CHARACTERISTICS OF MALES

The male hystricomorph has not been as extensively studied as the female. There is no true scrotum in any hystricomorph (Pocock, 1922) and the testes are held in the inguinal canal or abdomen. In *Galea* there is an area of highly pigmented wrinkled skin around the anus and extending forward to the urinary aperture into which the testes can partly descend and *Chinchilla* has post-anal sacs (see Weir, 1972a) which can accommodate the cauda epididymidis; degu (Fischer, 1940) and chinchilla (Weir, 1967b) may be called facultative cryptorchids.

All males have a posteriorly directed penis with an S-bend which has to be straightened before erection can take place; the tip of the erect penis extends to the level of the axilla in a chinchilla or viscacha, a distance of about 11 and 23 cm respectively. The surface of the glans in most hystricomorphs is covered with spines or spicules (Pocock, 1922). The pronounced styles (see Pocock, 1922) seen in the guinea-pig, agouti (Jones, 1834), paca (Martin, 1838) and North American porcupine (Hooper, 1961) have also been observed in the acouchi and degu but they are not present in *Proechimys* (Hooper, 1961), *Hystrix*, *Chinchilla* or *Lagidium*. The filiform penis of *Lagostomus* is exceptional amongst the hystricomorphs and the absence of the sacculus urethralis, considered to be characteristic of the suborder (Dathe, 1937), in this species may be connected with the unusual shape of the penis. An os penis has been found in several species of hystricomorph rodent, and is considered a useful character to distinguish closely related species such as the echimyids (Didier, 1962).

The distance between the urinary and anal apertures varies with the

species. The penis in *Proechimys, Octodon, Lagostomus* and *Chinchilla*
is separated from the anus by an expanse of bare skin. In *Dasyprocta,
Myoprocta, Galea, Myocastor* and *Ctenomys* there is a shorter separation
but the penis is distinct. The two sexes of *Hystrix* were known to be
difficult to distinguish in 1702 (see Royal Academy of Sciences, Paris)
and a similar problem was noted by Rewell (1950) for *Hydrochoerus*
which has a pseudo-cloaca. The most difficult of caviomorphs to sex,
however, is *Cavia* in which the tip of the penis projects only slightly out
of a cloacal-type arrangement (see Paterson, 1972). Many female
hystricomorphs have a large clitoris which may be mistaken for a penis
by the inexperienced.

The testes are proportionally larger in somes species (chinchilla,
casiragua and cuis) than in others (plains viscacha). No hystricomorph
male has been reported to be a seasonal breeder, even if the female of
the species is, and there is no periodic atrophy of the gonad such as
occurs in seasonally breeding myomorphs and sciuromorphs (e.g.
Citellus). The accessory glands (the vesicular, prostate and Cowper's
glands) are probably present in all hystricomorphs, although Angulo &
Alvarez (1948) state that *Capromys pilorides* has no prostate, and an
extra paired gland between the prostate and the base of the vesicular
glands was reported in *Erethizon* (Mirand & Shadle, 1953). The prostate
is histologically divisible into two or four parts but the division is
detected macroscopically only with difficulty. In *Galea*, however, the
division of the prostate into dorsal and ventral lobes is accentuated by
the dark red colour of the former portion. A well developed prostate
gland is prominent in *Lagostomus* females but has not been found in
Chinchilla, Ctenomys, Galea, Cavia or *Proechimys* (I. W. Rowlands,
personal communication). The shape of the vesicular glands appears to
vary according to the species (see Weir, 1967b). They are large blind
sacs with no, a few or several diverticula branching from one or both
sides of a main stem. The vesicular gland provides the bulk of the
accessory gland secretions and the fluid gels when mixed with prostatic
secretions. This mixture, which varies in consistency from species to
species, forms the copulatory plug (Fig. 6) which may be considered a
useful indicator of mating (*Cavia*: Lawlah, 1930; *Lagostomus*: Camus &
Gley, 1922; Weir, 1971a; *Lagidium*: Pearson, 1949; *Chinchilla, Myoprocta*: Weir, 1967b; *Capromys*: Owen, 1832; *Octodon, Octodontomys,
Ctenomys, Proechimys, Galea*: B. J. Weir, unpublished observations).
Hart & Greenstein (1968) have suggested that the failure of the
secretion from Cowper's gland of the guinea-pig to coagulate the
vesicular gland material supports the separation of the Hystricomorpha
from the Myomorpha.

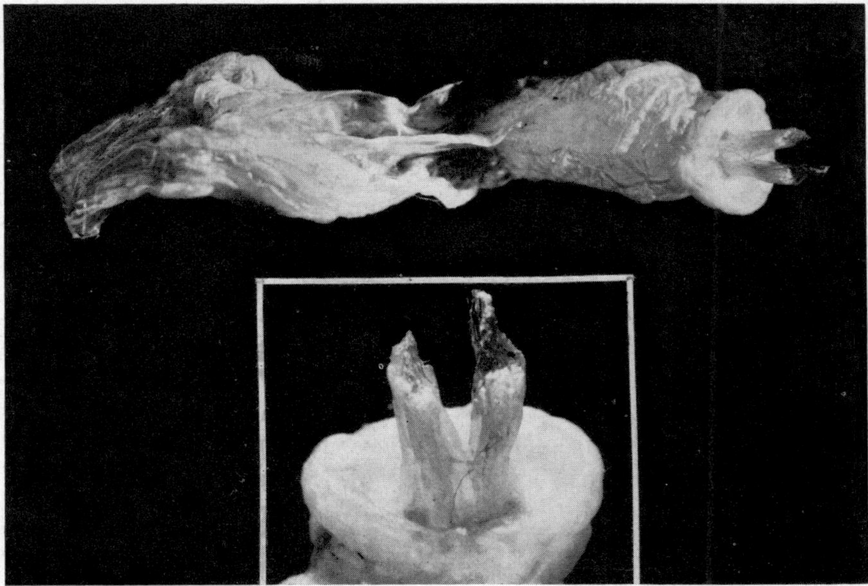

Fig. 6. A copulatory plug from a mated chinchilla female ($\times$ 1·5). The outer surface of the plug is the sheath of cornified cells derived from vaginal epithelium and the external ora cervices. The opaque parts of the plug are those containing the emissions of the male. Inset: 'cast' of the ora cervices ($\times$ 3).

The similarity of the male reproductive tract of *Thryonomys* to that of *Lagostomus* suggested relationship to Garrod (1873).

DISCUSSION

It is clear from the foregoing account that members of the suborder Hystricomorpha (*sensu lato*) share a number of characteristics of their reproductive physiology. The most outstanding features are:

    (a) a long gestation period;

    (b) a long oestrous cycle;

    (c) a vaginal closure membrane;

    (d) lateral position of the nipples;

    (e) non-scrotal (i.e. abdominal or inguinal) testes;

    (f) a sacculus urethralis;

    (g) penile styles and/or spicules on the glans.

This paper has been concerned mostly with the female characteristics (a–d) and all females of the 40 (22%) of the 180 or so species of the Hystricomorpha which have been studied have a combination of two or more of these features. Other mammals may exhibit one or even two of these characteristics but they are confined to small taxa. For example, in some microtines a membrane seals the vagina during anoestrus but these rodents do not have lateral nipples, a long pregnancy or a long oestrous cycle.

The length of the oestrous cycle of hystricomorphs is hard to explain; it is associated with a functional corpus luteum (see Weir & Rowlands, 1974) but the delay (about 40 days) before another pregnancy can be initiated after a non-fertile oestrus is much longer than that in other polyoestrous mammals (e.g. 21 days in pigs and cows).

Since the long pregnancy is characteristic of the Old World as well as the New World hystricomorphs and there is no apparent correlation of gestation length with any feature such as maternal weight, litter size, precocity of young or habitat, it is likely to have a genetic basis. Therefore hystricomorphs must have evolved through selection pressure towards long pregnancies or to have been prevented from shortening them. The latter seems the most likely (see Weir & Rowlands, 1973).

The most 'efficient' caviomorphs are obviously the cavies and echimyids (if other echimyids are found to be similar to *Proechimys*); they have the shortest pregnancies, oestrous cycles and ages to puberty, and large litters. Compared with the reproductive rate of a mountain viscacha, for example, cavies and casiragua are prolific, but less so than some cricetine rodents (e.g. hamster), or the rabbit; the selection pressures on these species have almost certainly been mainly those of predation. For this reason, the caviomorphs are now endangered by the activities of man and face competition from the more rapidly breeding cricetines and marsupials. It is feared that some caviomorphs may be exterminated, as has been the chinchilla; *Dolichotis*, *Pediolagus* and *Lagostomus* are certainly threatened with extinction, *Lagostomus* by man (Weir, 1974a), and the other two by direct competition with the introduced European hare (*Lepus europaeus*). It is comforting to note, however, that the Old World hystricomorphs, the porcupines and *Thryonomys*, have held their own in spite of the populous conditions of man and other animals in Africa and Asia. The long life-span of hystricomorphs (five years for cuis, 15 years for chinchilla—B. J. Weir, unpublished data) must be a decisive factor in their prolificacy and hence their survival.

Species with a long life-span, and therefore a long reproductive period, have a large brain at birth and in adulthood (Sacher, 1959), and Sacher & Staffeldt (in press) have suggested that gestation length is determined by the brain weight and stage of development at birth. This could be true for hystricomorphs; the nine species studied by Sacher fit this hypothesis except for the factor for brain advancement of the coypu which seems rather low. Owen (1839) commented on the very small size of the brain of the plains viscacha for its body weight (1 : 416). Sacher & Staffeldt (in press) and Kihlström (1972) suggest that the placental type may be correlated with the gestation length but their arguments seem doubtful.

The form of the placenta and the fetal membranes have been considered to be useful characters to determine taxonomic relationships since they are presumed to be unaffected by selective pressures of the external environment (Mossman, 1937; Fischer & Mossman, 1969). Apart from the basic division of the Mammalia into Prototheria, Metatheria and Eutheria according to their egg-laying, pouch incubation and extended intrauterine gestation, and the occurrence of menstruation in the Primates, characteristics of reproductive physiology are rarely used for taxonomic purposes within this class of animals.

I believe that the characteristics of the Hystricomorpha which have been surveyed in this paper should be considered as diagnostic of the group and as further evidence that the Old World hystricomorphs and the New World caviomorphs are related through a common ancestor whose genetic complement is not perpetuated in any other line of rodents.

ACKNOWLEDGMENTS

Many of the observations reported in this paper would not have been possible without the help and encouragement of Dr I. W. Rowlands, the attention, care and interest of Mrs Barbara J. Murrill and her team of animal technicians, the meticulous work of Mrs S. McLintock and Mr A. Hilton in the laboratory, and the unfailing efforts of the staff of the Library of the Zoological Society of London.

The degu were the gift of Dr Iglesias, Chile, and the casiragua were caught and sent from Venezuela by Dr O. A. Reig. Chinchilla have been donated by many commercial breeders. The plains viscacha, tuco-tuco, cuis and chozchoris were caught in South America with the financial assistance of the Wellcome Trust, the Royal Society, and the Medical Research Council.

The laboratory work was supported by the Ford Foundation.

To all these I am most grateful.

REFERENCES

Allen, J. A. (1901). On a further collection of mammals from Southern Peru collected by Mr H. H. Kedys, with descriptions of new species. *Bull. Am. Mus. nat. Hist.* **14**: 41–46.

Altman, P. L. & Dittmer, D. S. (eds.) (1962). *Growth*. Sex ratios and life spans. Washington D.C.: Fedn Am. Socs exp. Biol.

Angulo, J. J. & Alvarez, M. T. (1948). The genital tract of the male Conga Hutia, *Capromys pilorides* (Say). *J. Mammal.* **29**: 277–285.

Ansell, W. F. H. (1966). *Thryonomys gregorianus* and *Thryonomys swinderianus* in Zambia. *Puku* **4**: 1–16.

Asdell, S. A. (1964). *Patterns of mammalian reproduction*. 2nd edn. Ithaca: Cornell Univ. Press.

Asibey, E. O. A. (1974). Reproduction in the grasscutter (*Thryonomys swinderianus* Temminck) in Ghana. *Symp. zool. Soc. Lond.* No. 34: 251–263.

Blandau, R. J. & Young, W. C. (1939). The effects of delayed fertilization on the development of the guinea-pig ovum. *Am. J. Anat.* **64**: 303–330.

Brown, E. (1936). Rearing wild animals in captivity and gestation periods. *J. Mammal.* **17**: 10–13.

Bruce, H. M. & Hindle, E. (1934). The Golden hamster, *Cricetus* (*Mesocricetus*) *auratus*, Waterhouse. Notes on its breeding and growth. *Proc. zool. Soc. Lond.* **1934**: 361–366.

Bucher, G. C. (1937). Notes on life history and habits of *Capromys*. *Mems Soc. cub. Hist. nat. 'Felipe Poey'* **11**: 93–107.

Bullard, R. W. (1953). The use of vaginal smears in a study of the oestrous cycle of the female chinchilla. *Anat. Rec.* **117**: 598.

Burge, B. L. (1966). Vaginal casts passed by a captive porcupine. *J. Mammal.* **47**: 713–714.

Camus, L. & Gley, E. (1922). Action coagulante du liquide prostatique de la viscache sur le contenu des vésicules seminales. *C. r. Séanc. Soc. Biol.* **87**: 207–209.

Castle, W. E. & Wright, S. (1916). Studies of inheritance in guinea-pigs and rats. *Publs Carnegie Instn* No. 241: 1–129.

Collins, L. R. & Eisenberg, J. F. (1972). Notes on the behaviour and breeding of pacaranas, *Dinomys branickii*, in captivity. *Int. Zoo Yb.* **12**: 108–114.

Conaway, C. A. (1971). Ecological adaptation and mammalian reproduction. *Biol. Reprod.* **4**: 239–247.

Crichton, E. (1969). Reproduction in the pseudomyine rodent *Mesembriomys gouldii* (Gray) (Muridae). *Aust. J. Zool.* **17**: 785–797.

Dathe, H. (1937). Ueber den Bau des männlichen Kopulations-organes beim Meerschweinchen und anderen hystricomorphen Nagetieren. *Morph. Jb.* **80**: 1–65.

Davis, D. E. (1947). Notes on the life histories of some Brazilian mammals. *Bolm Mus. nac. Rio de J.* **76**: 1–8.

Dekeyser, P. L. (1955). Les mammifères de l'Afrique Noire Française. 2nd edn. *Init. afr.* **1**.

de Miranda-Ribeiro, A. (1936). The newborn of the Brazilian tree porcupine (*Coendou prehensilis* Linn.) and of the Hairy tree porcupine (*Sphingurus villosus* F. Cuv.). *Proc. zool. Soc. Lond.* **1936**: 971–974.

Dennler, J. (1940). Contribuciones al estudio de la *Chinchilla*. Las epocas del celo y de las pariciones. *An. Soc. cient. argent.* **130**: 129–136.

Detlefson, J. A. (1914). Genetic studies on a cavy species cross. *Publs Carnegie Instn* No. 205: 1–134.

Didier, R. (1962). Note sur l'os pénien de quelques rongeurs de l'Amérique du Sud. *Mammalia* **26**: 408–430.

Dubost, G. (1968). Les niches écologiques des forêts tropicales sud-américaines et africaines, sources de convergences remarquables entre rongeurs et artiodactyles. *Terre Vie* **22**: 3–28.

Ehrlich, S. (1966). Ecological aspects of reproduction in nutria *Myocastor coypus* Mol. *Mammalia* **30**: 142–152.

Fischer, G. M. (1940). Contribucion a la anatomia de los Octidontidos. *Boln Mus. nac. Hist. nat. Chile* **18**: 103–124.

Fischer, T. V. & Mossman, H. W. (1969). The fetal membranes of *Pedetes capensis* and their taxonomic significance. *Am. J. Anat.* **124**: 89–116.

Fleming, T. H. (1970). Notes on the rodent faunas of the Panamanian forests. *J. Mammal.* **51**: 473–490.

Ford, D. H., Webster, R. L. & Young, W. C. (1951). Rupture of the vaginal closure membrane during pregnancy in the guinea-pig. *Anat. Rec.* **109**: 707–714.

Galton, M. (1968). Chinchilla sex ratio. *J. Reprod. Fert.* **16**: 211–216.

Garrod, A. H. (1873). On the visceral anatomy of the ground-rat (*Aulacodus swinderianus*). *Proc. zool. Soc. Lond.* **1873**: 786–789.

George, W. (1974). Notes on the ecology of gundis (F. Ctenodactylidae). *Symp. zool. Soc. Lond.* No. 34: 143–160.

Gibson, E. (1877a). On the Biscacha (*Lagostomus trichodactylus*) a South American rodent. *Proc. nat. Hist. Soc. Glasg.* **3**: 136–140.

Gibson, E. (1877b). On the nutria (*Myopotamus coypu*), a South American species of rodent. *Proc. nat. Hist. Soc. Glasg.* **3**: 344–348.

Gluchowski, W. (1954). Studies on fertility factors in nutria. Part I. Preliminary experiments on the vaginal cycle. *Annls Univ. Mariae Curie-Sklodowska* E. **9**: 41–49.

Goldsmith, O. (1774). *A history of the earth and animated nature.* **2**. London.

Hart, R. G. & Greenstein, J. S. (1968). A newly discovered role for Cowper's gland secretion in rodent semen coagulation. *J. Reprod. Fert.* **17**: 87–94.

Hatt, R. (1940). Lagomorpha and Rodentia other than Sciuridae, Anomaluridae and Idiuridae, collected by the American Museum Congo expedition. *Bull. Am. Mus. nat. Hist.* **76**: 457–604.

Hill, W. C. O., Porter, A., Bloom, R. T., Seago, J. & Southwick, M. D. (1955). Field and laboratory studies on the Naked mole rat *Heterocephalus glaber*. *Proc. zool. Soc. Lond.* **128**: 455–514.

Hillemann, H. H., Tibbitts, F. D. & Gaynor, A. I. (1959). *Reproductive biology in chinchilla*. Natn. Chinchilla Breeders of America, Inc.

Hooper, E. T. (1961). The glans penis of *Proechimys* and other caviomorph rodents. *Occ. Pap. Mus. zool. Univ. Mich.* No. 623: 1–18.

Howe, R. & Clough, G. C. (1971). The Bahaman hutia, *Geocapromys ingrahami*, in captivity. *Int. Zoo Yb.* **11**: 89–93.

Hudson, W. H. (1872). On the habits of the viscacha (*Lagostomus trichodactylus*). *Proc. zool. Soc. Lond.* **1872**: 822–833.

Huggett, A. St. G. & Widdas, W. F. (1951). The relationship between mammalian fetal weight and conception age. *J. Physiol., Lond.* **114**: 306–319.

Jennison, G. (1927). *Table of gestation periods and number of young.* London: Black Ltd.

Jones, R. (1834). Some notes on the dissection of an *Agouti* (*Dasyprocta Aguti*, Ill.). *Proc. zool. Soc. Lond.* **1834**: 82–84.

296 B. J. WEIR

Kelly, G. L. & Papanicolaou, G. N. (1927). The mechanism of the periodical opening and closing of the vaginal orifice of the guinea-pig. *Am. J. Anat.* **40**: 387–411.

Kihlström, J. E. (1972). Period of gestation and body weight in some placental mammals. *Comp. Biochem. Physiol.* **43A**: 673–679.

Kleiman, D. G. (1970). Reproduction in the female green acouchi, *Myoprocta pratti* Pocock. *J. Reprod. Fert.* **23**: 55–65.

Kleiman, D. G. (1974). Patterns of behaviour in hystricomorph rodents. *Symp. zool. Soc. Lond.* No. 34: 171–208.

le Gallois, L. J. J. (1812) *Expériences sur le principe de la vie.* (English Translation by N. C. and J. G. Nancrede published in 1813.) Philadephia: M. Thomas.

Lawlah, J. W. (1930). Studies on the physiology of the accessory glands of reproduction of the male guinea-pig. *Anat. Rec.* **45**: 163–175.

Leitch, I., Hytten, F. E. & Billewicz, W. Z. (1959). The maternal and neonatal weights of some mammalia. *Proc. zool. Soc. Lond.* **133**: 11–28.

Llanos, A. C. & Crespo, J. A. (1952). Ecologia de la vizcacha (*Lagostomus maximus maximus* Blainv.) en el nordeste de la provincia de Entre Rios. *Revta Invest. agric., B. Aires* **6**: 289–378.

Long, J. A. & Evans, H. M. (1922). The oestrous cycle in the rat and its associated phenomena. *Mem. Univ. Calif.* **6**: 1–148.

Louwman, J. W. W. (1973). Breeding the Tailless tenrec, *Tenrec ecaudatus*, at Wassenaar Zoo. *Int. Zoo Yb.* **13**: 125–126.

Maliniak, E. & Eisenberg, J. F. (1971). Breeding spiny rats, *Proechimys semispinosus*, in captivity. *Int. Zoo Yb.* **11**: 93–98.

Martin, W. (1835). Visceral and osteological anatomy of the coypus (*Myopotamus coypus* Conn.). *Proc. zool. Soc. Lond.* **1838**: 52–55.

Martin, W. (1838). On the visceral anatomy of the spotted cavy (*Coelogenys subniger* Cuv.). *Proc. zool. Soc. Lond.* **1838**: 52–55.

Mirand, E. & Shadle, A. (1953). Gross anatomy of the male reproductive system of the porcupine (*E. dorsatum*). *J. Mammal.* **34**: 210–220.

Mohr, E. (1939). Die Baum- und Ferkelratten-Gattingen *Capromys* (Desmarest) (sens. ampl.) und *Plagiodontia* (Cuvier). *Mitt. hamb. zool. Mus. Inst.* **48**: 48–118.

Mossman, H. W. (1937). Comparative morphogenesis of the fetal membranes and accessory uterine structures. *Contr. Embryol.* **26**: 126–246.

Mossman, H. W. & Judas, I. (1949). Accessory corpora lutea, lutein cell origin and the ovarian cycle in the Canadian porcupine. *Am. J. Anat.* **85**: 1–39.

Newson, R. M. (1966). Reproduction in the feral coypu (*Myocastor coypus*). *Symp. zool. Soc. Lond.* No. 15: 323–334.

Ojasti, J. (1970). Datos sobre la reproduction del chiguire (*Hydrochoerus hydrochaeris*). *Acta cienc. venez.* **21**, Suppl. **1**: 27.

Owen, R. (1832). Dissection of a *Capromys fournieri. Proc. zool. Soc. Lond.* **1832**: 68–69.

Owen, R. (1839). Notes on the anatomy of the Biscacha (*Lagostomus trichodactylus* Brookes). *Proc. zool. Soc. Lond.* **1839**: 175–177.

Paterson, J. S. (1972). The guinea-pig or cavy. In *The UFAW handbook on the care and management of laboratory animals*: 223–241. 4th edn. Edinburgh: Churchill Livingstone.

Patil, D. R. (1968). *Reproduction in the Indian leaf-nosed bat* Hipposideros fulvus (*Gray*). Ph.D. thesis, Nagpur University.

Pearson, O. P. (1949). Reproduction of a South American rodent, the mountain viscacha. *Am. J. Anat.* **84**: 143–174.

Pearson, O. P. (1951). Mammals in the highlands of southern Peru. *Bull. Mus. comp. Zool. Harv.* **106**: 117–174.

Pearson, O. P. (1959). Biology of the subterranean rodents, *Ctenomys*, in Peru. *Mems Mus. Hist. nat. 'Javier Prado'* **9**: 1–56.

Pocock, R. I. (1922). On the external characters of some hystricomorph rodents. *Proc. zool. Soc. Soc. Lond.* **1922**: 365–427.

Pocock, R. I. (1943). The external characters of an adult female of the rare Cuban hutia (*Capromys nana*). *Proc. zool. Soc. Lond.* **113**: 198–200.

Racey, P. A. (1973). *Aspects of reproduction in some heterothermic bats.* Ph.D. thesis, University of London.

Racey, P. A. (1973). Environmental factors affecting the length of gestation in heterothermic bats. *J. Reprod. Fert.*, Suppl. **19**: 175–189.

Rahm, U. H. (1962). L'élevage et la reproduction en captivité de l'*Atherurus africanus. Mammalia* **26**: 1–9.

Rahm, U. H. (1969). Gestation period and litter size of the mole rat, *Tachyoryctes ruandae. J. Mammal.* **50**: 383–384.

Raynaud, A. (1969). Mamelles. In *Traité de Zoologie* **16** (6): 1–147. Grassé, P.-P. (ed.) Paris: Masson et Cie.

Reig, O. A. (1970). Ecological notes on the fossorial octodont rodent *Spalacopus cyanus* (Molina). *J. Mammal.* **51**: 592–600.

Rewell, R. E. (1950). Hypertrophy of sebaceous glands on the snout as a secondary male sexual character in the Capybara, *Hydrochoerus hydrochaeris. Proc. zool. Soc. Lond.* **119**: 817–819.

Roberts, C. M. & Weir, B. J. (1973). Implantation in the plains viscacha, *Lagostomus maximus. J. Reprod. Fert.* **33**: 299–307.

Roberts, C. M. & Perry, J. S. (1974). Hystricomorph embryology. *Symp. zool. Soc. Lond.* No. 34: 333–360.

Rood, J. P. (1972). Ecological and behavioural comparisons of three genera of Argentine cavies. *Anim. Behav. Monogr.* **5**: 1–83.

Rood, J. P. & Weir, B. J. (1970). Reproduction in female wild guinea-pigs. *J. Reprod. Fert.* **23**: 393–409.

Roth-Kolar, H. (1957). Beiträge zu einem Aktionssytem des Aguti (*Dasyprocta aguti aguti* L.). *Z. Tierpsychol.* **14**: 362–375.

Rowlands, I. W. (1949). Post-partum breeding in the guinea-pig. *J. Hyg., Camb.* **47**: 281–287.

Rowlands, I. W. (1974). Mountain viscacha. *Symp. zool. Soc. Lond.* No. 34: 131–141.

Royal Academy of Sciences, Paris (1702). *The natural history of animals containing the anatomical description of several creatures dissected by the Royal Academy of Sciences at Paris*: 147–156. The anatomical description of six porcupines and two hedge-hogs. Translated by Alexander Pitfield. London: Royal Society.

Sacher, G. A. (1959). Relation of lifespan to brain weight and body weight in mammals. In *Ciba Fndn Colloq., Ageing,* **5**: 115–133. *The lifespan of animals.* Wolstenholme, G. E. W. & O'Connor, M. (eds.). London: Churchill.

Sacher, G. A. & Staffeldt, E. F. (in press). Relation of gestation time to brain weight for placental mammals. Implications for the theory of vertebrate growth. *Am. Nat.*

Shadle, A. R. (1948). Gestation period in the porcupine *Erithizon dorsatum dorsatum. J. Mammal.* **29**: 162–164.

Shadle, A. R. (1950). The North American porcupine up-to-date. *Ward's nat. Sci. Bull.* **24**: 4–11.

Shadle, A. R. (1952). Sexual maturity and first recorded copulation of a 16-month male porcupine, *Erithizon dorsatum dorsatum. J. Mammal.* **33**: 239–241.

Shadle, A. R. & Ploss, W. R. (1943). An unusual porcupine parturition and development of the young. *J. Mammal.* **24**: 492–496.

Smythe, N. (1970). Relationships between fruiting seasons and seed dispersal methods in a neotropical forest. *Am. Nat.* **104**: 25–35.

Stockard, C. & Papanicolaou, G. N. (1919). The vaginal closure membrane, copulation and the vaginal plug in the guinea-pig with further considerations of the oestrous rhythm. *Biol. Bull. mar. biol. Lab., Woods, Hole* **37**: 222–245.

Talice, R. V. & Laffitte de Mosera, S. (1958). Parto, comportamiento maternal y comportamiento filial en *Ctenomys torquatus* ("tucu-tucu"). *Revta Fac. Hum. Cienc. Univ. Repúb. Urug.* **16**: 5–10.

Talice, R. V. & Lafitte de Mosera, S. (1959). El fenómeno de la abtura y del cierre de la vagina en *Ctenomys torquatus. Revta Fac. Hum. Cienc. Univ. Repúb. Urug.* **17**: 6–13.

Taylor, R. H. (1970). *Reproduction, development and behaviour of the Cuban hutia conga*, Capromys p. pilorides, *in captivity*. M.S. Thesis. University of Puget Sound.

Taylor, W. P. (1935). Ecology and life-history of the porcupine (*Erethizon epixanthum*) as related to the forests of Arizona and the south-western United States. *Biol. Sci. Bull. Univ. Ariz.* No. 3: 1–177.

Tesh, R. B. (1970). Notes on the reproduction, growth and development of echimyid rodents in Panama. *J. Mammal.* **51**: 199–202.

Trapido, H. (1949). Gestation period, young and maximum weight of the Isthmian capybara, *Hydrochoeris isthmius* (Goldman). *J. Mammal.* **30**: 433.

Ubisch, G. & Mello, R. F. (1940). Genetic studies on a cavy species cross (*Cavia rufescens* Lund and *Cavia porcellus* Linné). *J. Hered.* **31**: 389–398.

Vandenbergh, J. G., Drickamer, L. C. & Colby, D. R. (1971). Social and dietary factors in the sexual maturation of female mice. *J. Reprod. Fert.* **28**: 397–405.

Verheyen, W. & Verschuren, J. (1966). Breeding in Congolese rodents. *Explor. Parc natn. Garamba Miss H. de Saeger* No. 50: 1–71.

Weir, B. J. (1966). Aspects of reproduction in chinchilla. *J. Reprod. Fert.* **12**: 410–411.

Weir, B. J. (1967a). The care and management of laboratory hystricomorph rodents. *Lab. Anim.* **1**: 95–104.

Weir, B. J. (1967b). *Aspects of reproduction in some hystricomorph rodents*. Ph.D. thesis, University of Cambridge.

Weir, B. J. (1970a). Chinchilla. In *Reproduction and breeding techniques for laboratory animals*, Ch. 11: 209–223. Hafez, E. S. E. (ed.). Philadelphia: Lea & Febiger.

Weir, B. J. (1970b). The management and breeding of some more hystricomorph rodents. *Lab. Anim.* **4**: 83–97.

Weir, B. J. (1971a). The reproductive physiology of the plains viscacha, *Lagostomus maximus. J. Reprod. Fert.* **25**: 355–363.

Weir, B. J. (1971b). The reproductive organs of the plains viscacha, *Lagostomus maximus. J. Reprod. Fert.* **25**: 365–373.

Weir, B. J. (1971c). The evocation of oestrus in the cuis, *Galea musteloides. J. Reprod. Fert.* **26**: 405–408.

Weir, B. J. (1971d). Some observations on reproduction in the female green acouchi, *Myoprocta pratti*. *J. Reprod. Fert.* **24**: 193–201.

Weir, B. J. (1971e). Some observations on reproduction in the female agouti, *Dasyprocta aguti*. *J. Reprod. Fert.* **24**: 203–211.

Weir, B. J. (1971f). Some notes on reproduction in the Patagonian Mountain viscacha, *Lagidium boxi* (Mammalia: Rodentia). *J. Zool., Lond.* **164**: 463–467.

Weir, B. J. (1972a). The Chinchilla. In *The UFAW handbook on the care and management of laboratory animals*: 269–277. 4th edn. Edinburgh: Churchill Livingstone.

Weir, B. J. (1972b). In Rowlands, I. W. Scientific Report of the Zoological Society of London, 1969–71. *J. Zool., Lond.* **166**: 562–579.

Weir, B. J. (1973a). The rôle of the male in the evocation of oestrus in the cuis, *Galea musteloides*. *J. Reprod. Fert.*, Suppl. **19**: 421–432.

Weir, B. J. (1973b). Another hystricomorph rodent: keeping casiragua (*Proechimys guairae*) in captivity. *Lab. Anim.* **7**: 125–134.

Weir, B. J. (1973c). The induction of ovulation and oestrus in the chinchilla. *J. Reprod. Fert.* **33**: 61–68.

Weir, B. J. (1974a). The tuco-tuco and plains viscacha. *Symp. zool. Soc. Lond.* No. 34: 130–131.

Weir, B. J. (1974b). Notes on the origin of the domestic guinea-pig. *Symp. zool. Soc. Lond.* No. 34: 437–446.

Weir, B. J. & Rowlands, I. W. (1973). Reproductive strategies of mammals. *A. Rev. Ecol. System.* **4**: 139–163.

Weir, B. J. & Rowlands, I. W. (1974). Functional anatomy of the hystricomorph ovary. *Symp. zool. Soc. Lond.* No. 34: 303–332.

Zara, J. L. (1973). Breeding and husbandry of the capybara, *Hydrochoerus hydrochaeris*, at Evansville Zoo. *Int. Zoo Yb.* **13**: 137–139.

DISCUSSION

DEANESLY (*Cambridge*): Would you say that the pregnancy was mostly associated with elongation of the first part of pregnancy rather than the more rapid growth part?

WEIR (*London*): I think Dr Roberts will have something to say on this later, but the answer is, no. They have got a certain minimum slow growth part but the last part is also very slow, even if you plot that bit on the Huggett & Widdas formula. This long pregnancy seems such an odd thing to have developed that I can't help feeling that hystricomorphs were stuck with it when they first branched out from their phiomyid or reithroparamyid ancestor.

BRAMBELL (*London*): You said rather categorically that if you lived in a cold climate you had well developed young. This may be a general rule of rodents, but it certainly doesn't stand up outside the rodents. The polar bear is a marvellous example of an animal that lives in a very cold climate, and has a very poorly developed young.

WEIR: I'm sorry, I didn't intend to be dogmatic. What I was referring to specifically was *Chinchilla* and *Lagidium*, who because they live at high altitude have a good excuse for having furred young.

FRANCIS (*London*): Is there any chance that the black scrotal sac might be some sort of signalling device in the mating pattern?

WEIR: Not that I know of. The only time that it is obvious is when these cavies roll. They do a lot of rolling in the dust and then of course the scrotum is available for everyone to see. During courtship the female urinates backwards at the male who sometimes rears up and urinates forwards so I doubt if the female, who is at that point running away, would notice what he had got.

LAVOCAT (*Montpellier*): Why did you take *Hystrix* not *Atherurus*? You must try to get some because they are in many respects more primitive than *Hystrix*. *Atherurus* has a skull which is much nearer to the normal Hystricidae and Thryonomyidae.

WEIR: We have some information on *Atherurus* that Rahm (1962) produced; they have a long pregnancy of approximately 110 days, they have lateral nipples and a closure membrane. So all we need to know now is the oestrous cycle. *Hystrix* has a long oestrous cycle, which is very suggestive of relationship with the caviomorphs.

PEARSON (*Berkeley*): Is not 13 days some sort of world record for sexual maturity?

WEIR: I believe so, yes.

PEARSON: From a population point of view this is rather important, I believe, in that the age of sexual maturity is of enormous importance for building up a population rapidly and could well compensate for the handicap of a relatively long gestation period.

WEIR: Well it seems to me that *Galea* has really got it all worked out, it has got the shortest known puberty, the shortest known pregnancy (within the Hystricomorpha), an almost certain post-partum oestrus, a reasonable litter size and a good sex ratio. It just never stops. I have females who are in their 18th pregnancy and they are still going strong.

PEARSON: You are correct when you suggest the *Galea* is essentially a walking hamburger from the predation point of view in South America.

AMOROSO (*Cambridge*): Is this precocious puberty alike in males and females?

WEIR: No, the males have to wait for at least two months before they mature. But usually in the wild they don't have the chance even then because there is a hierarchical system and the dominant alpha male fights off everybody else. If a little male does manage to sneak in he has then got to cope with the oestrous female who is probably five times the size.

WOOD (*London*): Have you tested for a pheromonal effect in the acceleration of puberty?

WEIR: Yes, there is a pheromonal effect. *Galea* females are removed from the breeding pens at two weeks of age and can be kept for any length of time until convenient. When a male is put in there is a response within two days. To get the best effect in this induction of puberty it must be a male with a chin gland. If you pair them with a castrated male or an ovariectomized or intact female they are not really interested. But it is clearly not an olfactory pheromonal system because I can keep animals side by side in a mesh cage and nothing happens until the male actually goes in; all the breeding stock and the test males are also in the same room as the experimental females.

Symp. zool. Soc. Lond. (1974) No. 34, 303–332

FUNCTIONAL ANATOMY OF THE HYSTRICOMORPH OVARY

BARBARA J. WEIR and I. W. ROWLANDS

*Wellcome Institute of Comparative Physiology,
Zoological Society of London,
Regent's Park, London, England*

SYNOPSIS

A survey of the general appearance of the ovaries of hystricomorphs and their component tissues indicates a surprising degree of diversity.

The ovary of the cuis is unusual in that there is an ovulation fossa, lined by germinal epithelium, around which the fimbria is apposed. A peculiar condition, possibly pathological and known as lipophanerosis, characterizes the ovaries of older, heavier females of high parity.

The plains viscacha has the most bizarre ovary known for a mammal: it is liver-shaped and arranged in strands such that the 200 to 800 eggs which are ovulated have to find their way to the oviduct along numerous clefts. The mature follicle is only 300μm in diameter and apart from the ovulatory loss, eggs are also lost by the formation of accessory corpora lutea (CL) from unovulated follicles.

This form of atresia is common in the chinchilla, agouti and acouchi in which large numbers, often more than 100 in a pair of ovaries, of accessory CL are formed during pregnancy. Fewer accessory CL are formed by the mountain viscacha or Canadian porcupine during pregnancy and by the degu and coypu at the end of gestation. Accessory CL are rarely formed in the guinea-pig, cuis or casiragua.

Interstitial tissue is prominent in chinchilla, plains viscacha, mountain viscacha, agouti, degu, tuco-tuco, casiragua and African porcupine, but is sparse in the guinea-pig, cuis, acouchi and coypu. There is no correlation between the abundance of interstitial tissue and the type of ovulation usually experienced by the animals. In the degu, the interstitial tissue appears to luteinize in the later stages of pregnancy.

Embryonic remnants are common in the ovaries of cuis, casiragua and the dasyproctids, but rete cysts are unusual in species other than the guinea-pig.

It is suggested that the complexity of the ovary is characteristic of the caviomorphs. The only detailed information available for an Old World species is on the African porcupine and since it also seems to have a complex ovary with excess luteal tissue, it is probable that this may be another feature common to the New and Old World hystricomorph rodents.

INTRODUCTION

The ovary of the adult mammal during reproductive life is constantly in a state of flux because of the fluctuating nature of its two main functions. The first is its gametogenic function—the production of a determinate number of mature gametes at regular intervals during adult life. The second is its complex endocrine function—the secretion of various steroids which act on target organs such as the uterus, vagina, mammary gland, pituitary, nervous system and, directly or indirectly, on other components of the ovary itself to ensure the female's periodic capacity to receive the male gametes for fertilization. These two functions are closely related as can be understood from a brief description of the events of an ovarian cycle.

The female mammal, unlike the male, is born with a full complement of germ cells on which she depends for her reproductive capacity throughout life. In all but primates the supply lasts to the end of her life. At an early stage, the oocytes become invested with a layer of epithelial cells which multiply and differentiate to form a complex trilaminar structure, the follicle, which later expands by the formation of an antrum and the accumulation of fluid to become the Graafian follicle. The outermost layer of the follicle, the theca externa, consists of fibrous connective tissue cells and serves to contain a middle layer of glandular cells, the theca interna, which with the innermost layer, the membrana granulosa in which the oocyte is suspended, is believed to secrete oestrogen. Oestrogen affects the vagina, stimulates sexual behaviour (oestrus), and permits mating and the passage of spermatozoa through the uterus. The two inner layers of the follicles are separated by a clearly defined membrana propria. Ovulation may occur spontaneously or in response to copulation or other form of cervical stimulation. The spent follicle becomes converted into a corpus luteum (CL) which secretes progesterone that induces progestational changes in the endometrium to permit implantation and the establishment of pregnancy.

This highly complex cycle of events can be divided into follicular and luteal phases separated by ovulation but their anatomical and endocrinological details vary widely amongst the relatively few but different mammals that have been studied. Basic information on these details may be obtained from the standard reference works of Parkes (1956–1966), Young (1961), Zuckerman (1962) and Mossman & Duke (1973). There is much miscellaneous information on the anatomy of the ovary in different mammals, but it is only in recent years that biochemical techniques have permitted studies and conclusions to suggest the functions of the various ovarian tissues. Of necessity much of this work has been performed on the domestic and common laboratory species and it is an interesting and essential exercise to consider these conclusions (thought by most workers to be generally applicable) against the known anatomy of some other mammals.

The normal female in the wild state does not experience regular ovarian activity comparable to that imposed upon it by isolation of the sexes when in captivity. The wild female should be pregnant, lactating (sometimes simultaneously) or in anoestrus (Weir & Rowlands, 1973; Weir, 1974) and a luteal phase involving pregnancy is therefore the primary basic condition. Rarely is she found in a non-pregnant state, which to a large extent is an artifact of captivity or domestication. The ovarian cycle of the unmated guinea-pig, which up until 1949 was the only hystricomorph species to have been studied, resembles in greater detail that of the ungulates than that of the more closely related squirrels and myomorph rodents. The principal difference between the guinea-pig and the other rodents is that the CL of the guinea-pig is an actively secreting gland which persists for 15 to 17 days. Thus, the luteal phase in the non-pregnant (un-mated) guinea-pig is equivalent to that produced after infertile mating (pseudopregnancy) in myomorph rodents.

In the guinea-pig, as in other laboratory myomorphs and domesticated ungulates, the only ovarian structures involved in the regulation of the cycle and the maintenance of pregnancy are the Graafian follicles which ovulate, and the CL that develops from each follicle. There is no evidence to suggest that any other ovarian structure participates in the process of reproduction in these animals, and the guinea-pig possesses what Mossman (1966) has called a 'prosaic' ovary. It is only the temporal relationships of the follicular and luteal phases which distinguish the guinea-pig from other species.

This description of the ovary cannot be applied to that of some other hystricomorphs, as was shown by Mossman & Judas (1949) for the Canadian porcupine, *Erethizon dorsatum*, and by Pearson (1949) for the mountain viscacha, *Lagidium peruanum*. The results of these two studies are summarized diagrammatically in Fig. 1. Both species ovulate only one egg and the primary CL persists throughout pregnancy although terminally some involution is apparent. Furthermore, additional luteal tissue is produced during gestation in the form of accessory CL which are derived from unovulated follicles; these remain smaller than the primary CL but are histologically identical with it. In *Lagidium* the accessory CL are found only in the ovary containing the primary CL (which is the right ovary since ovulation always takes place only from this side). In *Erethizon* they occur in both ovaries, but those in the ovary containing the primary CL (which may be the right or the left) follow the life span of the primary CL while those in the contralateral ovary regress and disappear in mid-pregnancy. The explanation for this would appear to be a luteolytic influence, perhaps of the uterus, on the one side and a supporting action by the primary corpus luteum on the other, but Mossman (1966) found that no experimental procedures could alter the dimorphism of the two ovaries of *Erethizon* in pregnancy. Pearson (1949) found that if the right ovary of *Lagidium* was excised

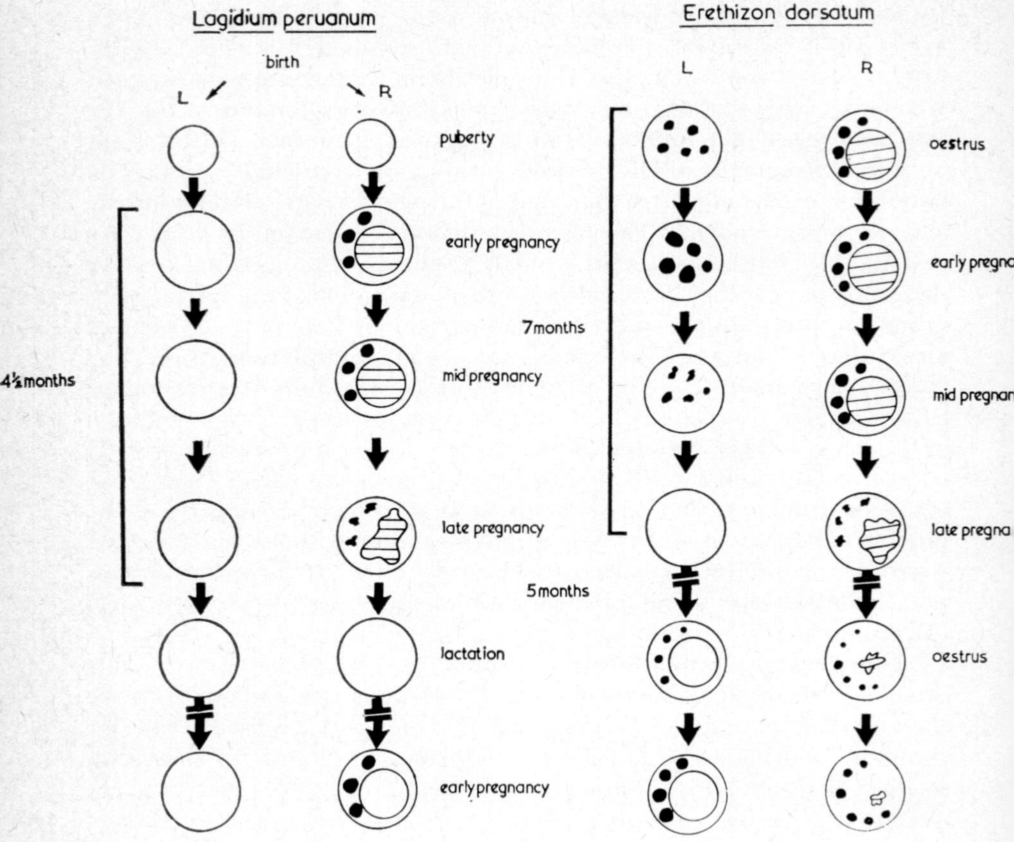

FIG. 1. Diagrammatic representation of the events in the ovaries of *Lagidium peruanum* (data from Pearson, 1949) and *Erethizon dorsatum* (data from Mossman & Judas, 1949) during pregnancy. The mountain viscacha (*Lagidium*) ovulates only from the right ovary and true and accessory CL are formed only on this side. In the North American porcupine (*Erethizon*) ovulation can take place from either ovary and accessory CL are formed in both but regress only in the ovary on the non-pregnant side.

before or after puberty the left side could and did become functional in the same way as the right would have done. It has not been possible to examine this species of mountain viscacha any further but the occurrence of a similar one-sidedness is suggested in the ovaries of one pregnant Patagonian mountain viscacha (*Lagidium boxi*) (Weir, 1971a). It is impossible to suggest a physiological explanation for this phenomenon since both ovaries are exposed to the same circulating hormones and the cells of the left are clearly capable of responding. The absence of any luteal cells in even the smallest of atretic follicles is very surprising in view of this general tendency among hystricomorphs (see below).

In spite of this promise of exciting differences in the ovaries of hystricomorphs, they have been little studied. Isolated and generally incomplete comments on the ovaries of some species have appeared from time to time (Schaeffer, 1911; Llanos & Crespo, 1952; Konieczna, 1956; Hillemann & Tibbitts, 1956; Stanley & Hillemann, 1960; Lagomarsino & Momigliano, 1961; Contreras, 1965; Kayanja & Jarvis, 1971) but detailed studies throughout the reproductive cycle have been possible only in those hystricomorphs which have been bred in laboratory conditions (Weir, 1967a, 1970a, 1972, 1973a), except for the coypu that were plentiful in Britain during attempted extermination a decade ago (Newson, 1966; Rowlands & Heap, 1966). The following account of our present knowledge of hystricomorph ovaries is thus derived mainly from our own studies, some of which are unpublished; references will be given only to published work. We have arranged the paper according to cell types rather than to species to give more emphasis to what are general features and what are specific differences. Our information is mainly anatomical because anatomy is always the first study to be made in the investigation of a new laboratory species. When experimental studies have been performed in hystricomorphs we have discussed the results in the light of the anatomy, but otherwise we have been forced to extrapolate from interpretations of function of the ovarian anatomy of other mammalian species.

Since there is little if any likelihood of confusion, the species are referred to by their common names only after the first citation.

MACROSCOPIC APPEARANCE

The ovary of the adult hystricomorph is generally ovoid and positioned near the posterior pole of the kidney. As in all mammals, the ovary is suspended from the dorsal wall of the abdomen by the mesovarium and the Fallopian tube by the mesosalpinx. This latter mesentery may partly invest the ovary forming a very thin and tenuous partial bursa ovarica, e.g., in guinea-pig, chinchilla (*Chinchilla laniger*) and casiragua (*Proechimys guairae*). In none of the hystricomorphs studied does the fimbria completely invest the ovary; it is more prominent in the capybara (*Hydrochoerus hydrochaeris*) and mara (*Dolichotis patagonum*) than in the chinchilla or casiragua.

A cranial loop of the Fallopian tube is present in all the species examined but its extent and the degree of convolution of the rest of the tube varies. The loop is prominent in the agouti (*Dasyprocta aguti*) and

chinchilla but the last third of the tube is virtually straight in the former and tightly twisted in the latter before straightening out at the
utero-tubal junction which leads to one horn of the duplex uterus.

Blood vessels and nerves enter the ovary along the cranial part of
the mesovarium and the hilum may well be defined and short as in casiragua or more extensive as in the chinchilla. Accumulations of fat in the
mesovarium can occur in all the hystricomorphs but are particularly
noticeable in chinchilla, cuis (*Galea musteloides*) and tuco-tuco (*Ctenomys
talarum*).

The surface of the hystricomorph ovary is characteristically smooth.
Follicles may appear as translucent areas but they rarely project
domes on the surface. Nor do CL protrude, even in the polytocous
species, to give the appearance of 'bunches of grapes' so characteristic
of myomorph ovaries. Stigmata, if formed, are not prominent and it is
very difficult to detect ovulation in a chinchilla, for example, a few
days later. The most obvious external feature of the chinchilla ovary is
the appearance of the patches of interstitial tissue (Fig. 2). Only in the
casiragua are the CL prominent since they become about seven times
larger in pregnancy than during the cycle (see Fig. 11c) but they are not
pedunculate as in sciuromorphs. Active CL of casiragua are a deep orange
but the CL of chinchilla, agouti, cuis and acouchi (*Myoprocta pratti*) are
not deeply coloured when active but may become more yellow with age.

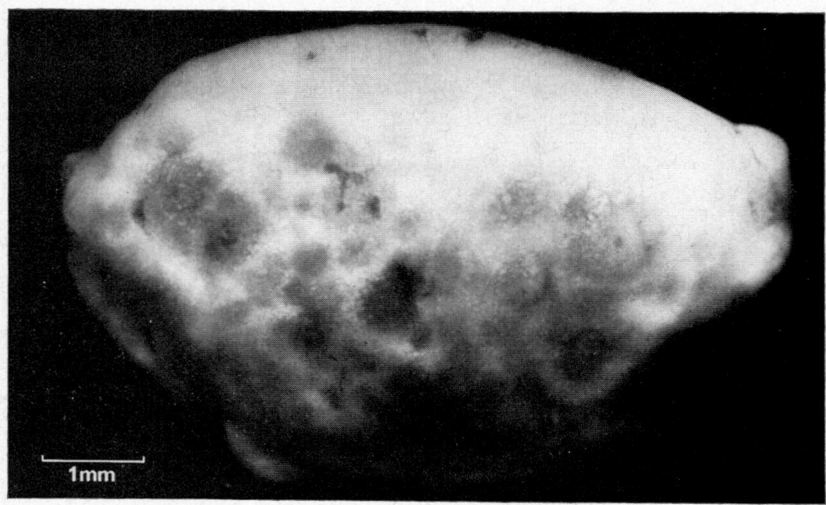

FIG. 2. Surface appearance of chinchilla (*Chinchilla laniger*) ovary midway through
pregnancy, showing the characteristic patches of interstitial tissue; CL are not obvious
because they are deeply seated.

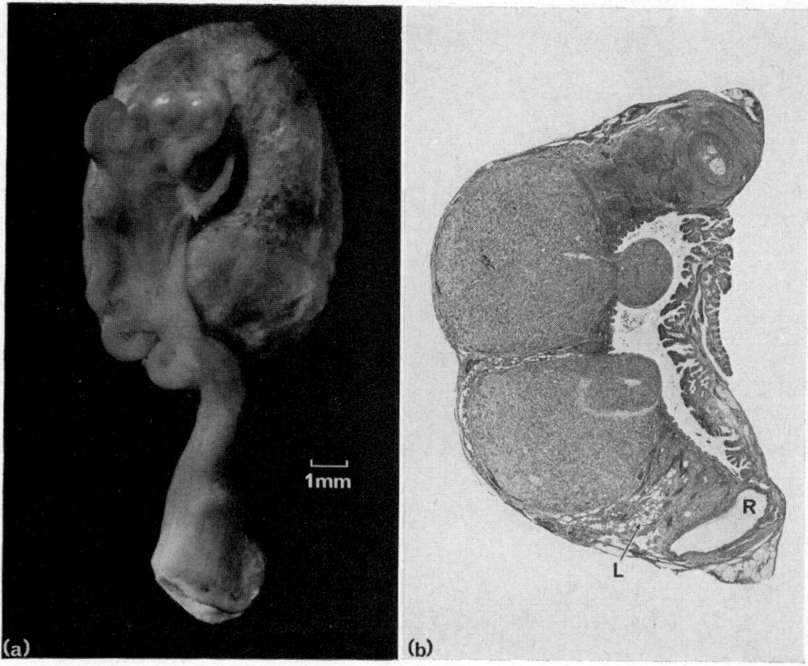

FIG. 3. The ovary of the cuis (*Galea musteloides*). (a) View of the concave surface which is developed into an ovulation fossa against which the fimbria of the oviduct is applied. Protuberances forming CL are visible only from the convex side of the ovary. (b) Longitudinal section of the ovary showing the position of the fimbria and the stigmata of two CL in the ovulation fossa. Note the lack of germinal epithelium on the convex surface and the patch of early lipophanerosis (L) near a small rete cyst (R). × 14.

There are two species which are exceptions to the above generalities —the cuis and the plains viscacha (*Lagostomus maximus*). The cuis ovary (Fig. 3a and b) resembles that of the horse in that the so-called germinal epithelium extends only over the pronounced concave (median) surface. The fimbria is attached to the ovary at the edges of this cleft and ovulation occurs only at this fossa. The rest of the ovary is covered by what appear to be several mesenteries and the limit of the organ is difficult to determine, especially when fat has been deposited and the ovary has become 'lipophanerotic' (see below).

The plains viscacha probably has the most bizarre ovary of any mammal (see Weir, 1971c). The organ is flat and liver-shaped rather than reniform (Fig. 4). The surface is greatly increased by many indentations which divide the ovary into raphes rather than lobules.

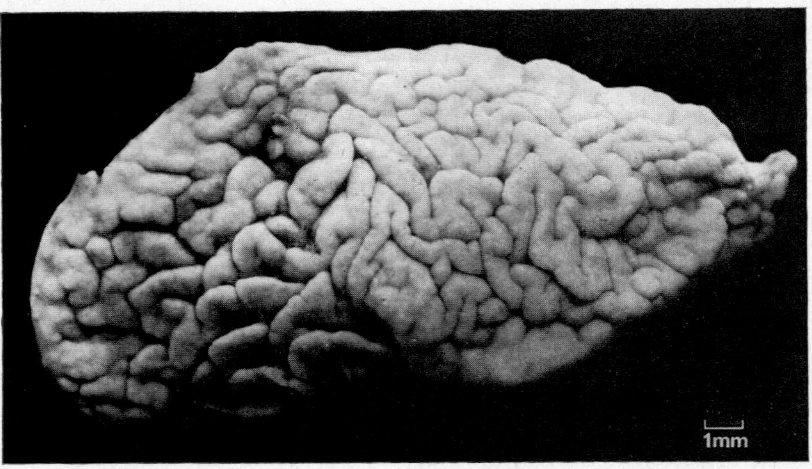

Fig. 4. Surface view of the ovary of an adult plains viscacha (*Lagostomus maximus*) showing its convoluted appearance. The follicles and corpora lutea are too small to be visible.

The fimbria is insignificant and would be unable to cover one surface of the ovary even if it were extended at oestrus. The oviduct is closely apposed to the ovary along the hilar surface. No follicular or luteal features ever mar the surface patterning because they are too small to be seen (see below).

MICROSCOPIC FEATURES

Oocytes

The oocytes of hystricomorphs, like those of other mammals, lie beneath the tunica albuginea and are more numerous at the poles of the ovary than elsewhere. At birth, many of the oocytes may be collected in nests, and their incomplete separation may account for the occurrence of polyovular follicles. Such follicles have been reported in the guinea-pig (Collins & Kent, 1964) and are particularly frequent in chinchilla (61% of animals had 0·4–19% polyovular follicles—Weir, 1967b), degu and plains viscacha. Parthenogenesis of oocytes has been described for the guinea-pig (Branca, 1925; Loeb, 1932; Harman & Kirgis, 1938) and is frequently found in the casiragua (Fig. 5). This condition of the oocytes, referred to as degenerative fragmentation, was noted by many earlier workers. Among these, Loeb (1932) showed that occasionally when the 'blastomeres' or fragmented portions of the

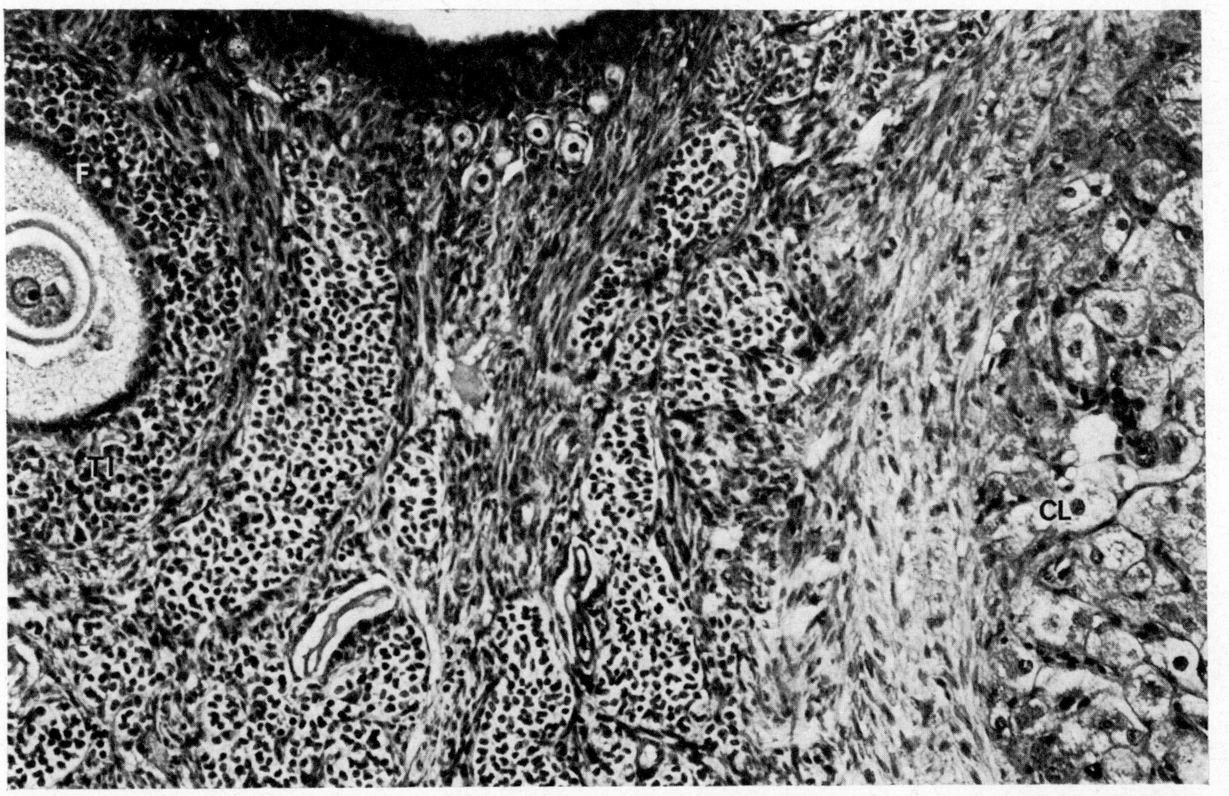

Fig. 5. Section of the ovary of a casiragua (*Proechimys guairae*) showing an atretic follicle (F) containing a parthenogenetic egg. The transformation of the theca interna (TI) to interstitial tissue can be clearly seen and many zona pellucida remnants are visible. The edge of a corpus luteum (CL) is visible on the right. $\times$ 225.

follicular eggs came into contact with the surrounding ovarian structure they developed into a tissue bearing a striking resemblance to trophoblastic rather than embryonic tissue. Many of these cells enlarged to become giant cells and migrated towards blood vessels and often replaced part of their endothelial lining. Another common feature was the presence of small haemorrhages at the junction of these tumours and the adjacent ovarian tissue. The Swiss workers, Ponse, Weihs, Libert & Dovaz (1954), confirmed the above observations but considered the trophoblastoma to be a rare type of ovarian tumour. Similar structures have been observed in chinchilla ovaries, in which they were assumed to be true trophoblastic cells which had arrived through deportation (Billington & Weir, 1967), and in plains viscacha.

Follicles

Growth

As indicated above, the follicle is the organ which surrounds and nurtures the growing oocyte; both are growing simultaneously for certain periods of time, but the growth of the oocyte is completed quite early in the developmental period of the follicle. Formulae expressing the relative growth rates of these two structures have been constructed for many species and the biphasic relationship was discussed by Brambell (1956). The same general relationship pertains in the guinea-pig and probably all other hystricomorphs except the plains viscacha in which the follicle ceases to grow to any significant extent after the formation of the antral cleft in the membrana granulosa. At this time the follicle is only about 300 μm in diameter, that is, only about four times that of the oocyte itself.

The size of the mature Graafian follicle of mammals is roughly proportional to body weight (e.g. follicle diameter of a 20-g mouse is 400 μm, of a 600-g guinea-pig about 1 mm, of a 500-kg cow about 2–3 cm and a 750-kg horse about 5–6 cm). Here again the plains viscacha, having a body weight of 2–3 kg and a mature follicle of 300 μm only, is a misfit.

Follicular layers

As in other mammals, the hystricomorph follicle contains the characteristic three layers mentioned earlier, although there are numerous specific appearances of a minor nature.

There is nothing distinctive about the granulosa cells of hystricomorphs: the nuclei of the basal layer of cells are noticeably aligned (Fig. 6a) in casiragua, cuis and African porcupine (*Hystrix cristata*).

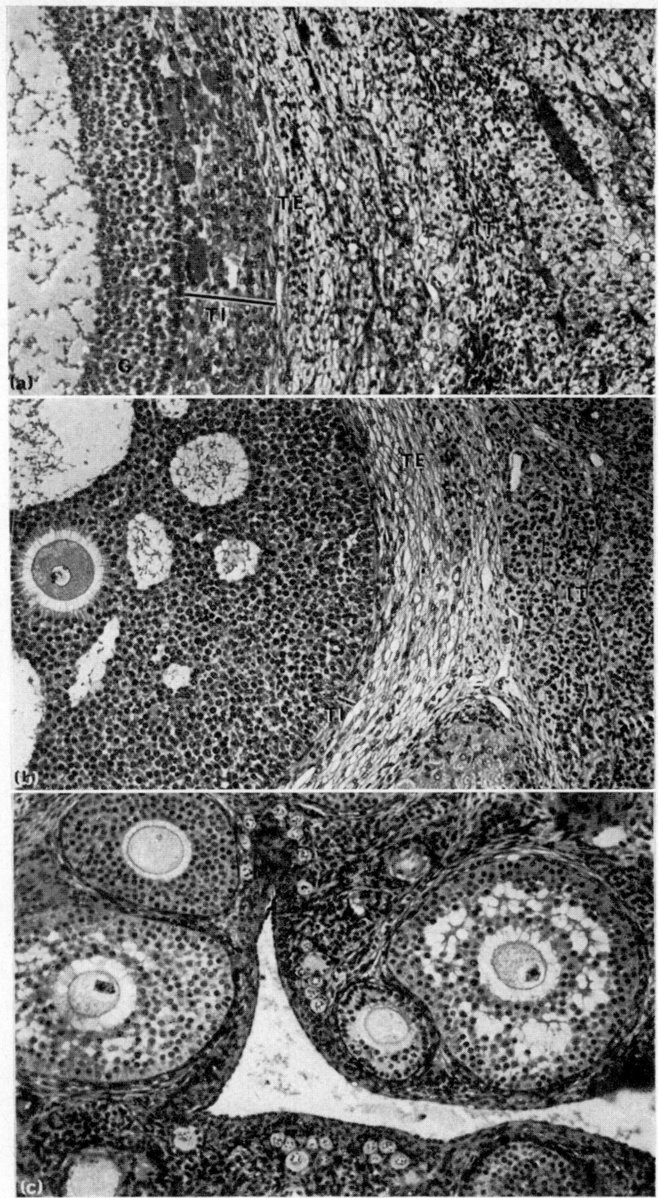

FIG. 6. Examples of the follicular layers of some hystricomorphs. (a) The edge of a follicle of the African porcupine (*Hystrix cristata*) showing, from left to right, the antrum; membrana granulosa (G) with a distinctly pallisaded layer adjacent to the membrana propria; the theca interna (TI); the theca externa (TE); interstitial tissue (IT). (b) The edge of a follicle of an agouti (*Dasyprocta aguti*) showing the attenuated theca interna (TI) and the very thick theca externa (TE) and the abundant interstitial tissue (IT). (c) Section of the ovary of a plains viscacha (*Lagostomus maximus*) showing the small size of the mature follicle and the narrow thecal layers. Note the oocytes under the extensive germinal epithelium and the patches of interstitial tissue. × 120.

The membrana granulosa of the capybara (*Hydrochoerus hydrochaeris*) is folded. The membrana propria of casiragua follicles is very prominent whether stained by haematoxylin and eosin or with a trichrome stain.

The theca interna varies in thickness with the diameter of the follicle and with the species. It becomes distinct and may be quite thick (Fig. 6a) when the follicle develops an antrum. Thinning occurs as the follicle expands but although mitoses are rarely seen, cell division is presumably frequent to enable the theca interna of the mature follicle to retain a thickness of four to five cells, as in the African porcupine or chinchilla. The theca interna of some species, e.g. agouti and plains viscacha, is barely visible and is represented by only a few cells (Fig. 6b and c), while that of coypu follicles hypertrophies and luteinizes during late gestation. A similar, but less developed, thecal gland has been described in the guinea-pig (Stafford, Collins & Mossman, 1942).

The theca externa may be a zone of cells that are stretched by the pressures of the growing follicle (Perry, 1971), but this seems to us unlikely because its thickness and definition in some hystricomorphs indicate that it is a true tissue type, often distinguishable by the fibres which are apparent after staining with a trichrome stain such as Azan. The mature follicles of a particular species are associated with a theca externa of what may be said to be a characteristic thickness. Thus the agouti has a thick theca externa (Fig. 6b), the cuis a medium one and the chinchilla a thin one.

Occurrence

Primordial follicles are present in the ovaries of most hystricomorphs at birth and the occurrence of one or more waves of abortive follicular (non-antral) growth in the pre-natal guinea-pig was described by Bookhout (1945). In some species at birth (e.g. plains viscacha and degu) follicles with more than one surrounding layer may be present. Follicles of various sizes and degrees of atresia are present in the ovaries of prepubertal animals, but in the spontaneously ovulating species the first crops of even quite large and apparently mature follicles do not ovulate and are not associated with oestrus until these processes become integrated at or shortly after the normal time of puberty. One very interesting example of this presumed imbalance in the secretion and release of gonadotrophic hormones is found in the degu (*Octodon degus*) in which puberty is reached at six months of age (Weir, 1974) but practically all females have at least one large (500 μm) antral follicle in the ovaries at two weeks of age. These follicles do not ovulate, however, but become atretic (see below). The ovaries of cuis may contain mature follicles from a very early age and, as in other induced ovulators, new crops of follicles mature about every three days, but not all crops may become physiologically mature within that short interval. As in other mammals all remaining medium-sized antral follicles disappear or become atretic immediately after ovulation in all hystricomorphs except the chinchilla (Weir, 1967b, 1969) and possibly the African porcupine. Follicles of all sizes are present in the ovaries of hystricomorphs throughout gestation. Those present in the pregnant guinea-pig may be induced to ovulate by exogenous gonadotrophins but normally they do not ovulate until after parturition (Rowlands, 1956). The follicles in the chinchilla ovary during pregnancy are refractory to this treatment (Weir, 1973b) but ovulation may be effected during anoestrus (Weir, 1969, 1973b).

Ovulation

With the exception of the plains viscacha which ovulates 200 to 800 eggs at each oestrus (Weir, 1971c), the ovulation number for each species of hystricomorph approximates to the litter size (see Weir, 1974). In polyovular species such as the guinea-pig, cuis and coypu, as well as the plains viscacha, more follicles reach ovulable size than actually ovulate. The impetus for final maturation of some follicles rather than others is unknown, though the mechanisms probably differ in induced and spontaneously ovulating animals. Some sort of physiological priming appears to be necessary before the chinchilla will ovulate (Weir, 1969, 1973b). In this species, and probably also others that are primarily spontaneous ovulators, ovulation can be induced by the male if coitus occurs early enough to cause a premature LH surge (Weir, 1973b). Cuis nearly always ovulate in response to coitus; other hystricomorphs for which information is available are discussed by Weir (1974). The normal occurrence of ovulation during gestation is rare amongst mammals. Its occurrence during early pregnancy in equids is associated with the secretion of serum gonadotrophin (PMSG) (see Rowlands, 1964) and, in the viscacha, eggs are commonplace in the ovarian clefts throughout gestation (Weir, 1971c). Many casiragua mate 24 or 48 hours before parturition (Weir, 1974) and these females are therefore ovulating during pregnancy.

The processes of ovulation are probably similar to those described for other mammals (see Blandau, 1967, 1970) but the plains viscacha again may be an exception because of the small size of the ovulating follicle. There is little preovulatory swelling and luteinization of the granulosa cells in hystricomorph follicles; it may occur in some follicles of some species but cannot be considered as characteristic of any one species. Similarly, although a prominent stigma may be formed at the point of follicular rupture its occurrence is not a regular feature of hystricomorph ovaries, except in the plains viscacha in which stigmata, and even completely everted CL, are commonly observed.

Atresia

A follicle which does not survive to maturation and ovulation may degenerate or become atretic. Such is the fate of the majority of follicles. Atresia, the filling up of a lumen, is common amongst the hystricomorphs and may take one, or sometimes both, of two courses. The commonest course, which occurs also in many other mammals to a greater or lesser extent, is the occlusion of the antrum by pyknotic granulosa cells (before or after death of the egg) and the transformation of the theca interna cells to interstitial tissue (see below and Fig. 5). The second course involves the normal transformation of granulosa cells into lutein cells but with the retention of the egg, thus converting the follicle into an accessory CL (Fig. 7). Interstitial tissue formation can only occur if the follicle has developed sufficiently to possess a theca interna (but see below), while accessory CL may be formed from follicles of any size. Very occasionally, e.g. in African porcupines and chinchilla,

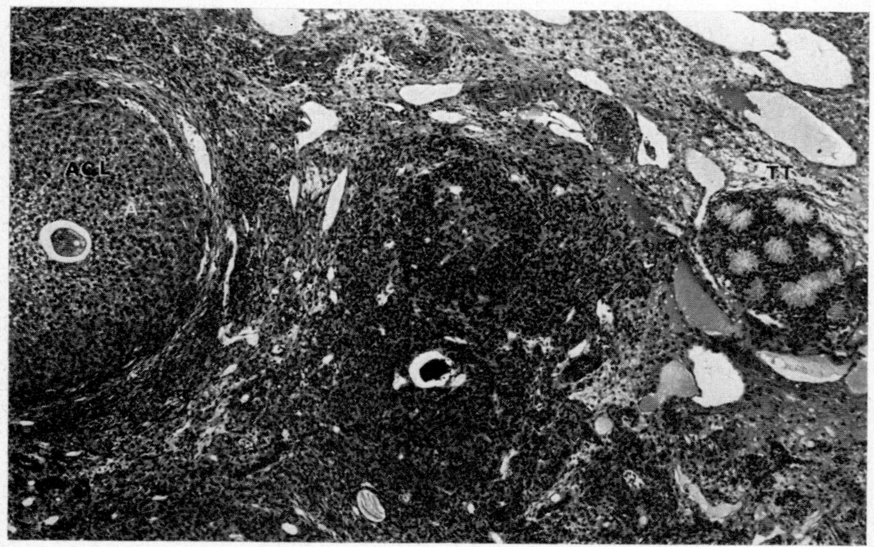

Fig. 7. Section of the ovary of an acouchi (*Myoprocta pratti*) showing a typical accessory CL with a retained egg (ACL), a zona pellucida remnant in the stroma, and a patch of 'immature testis tubules' (TT). × 62.

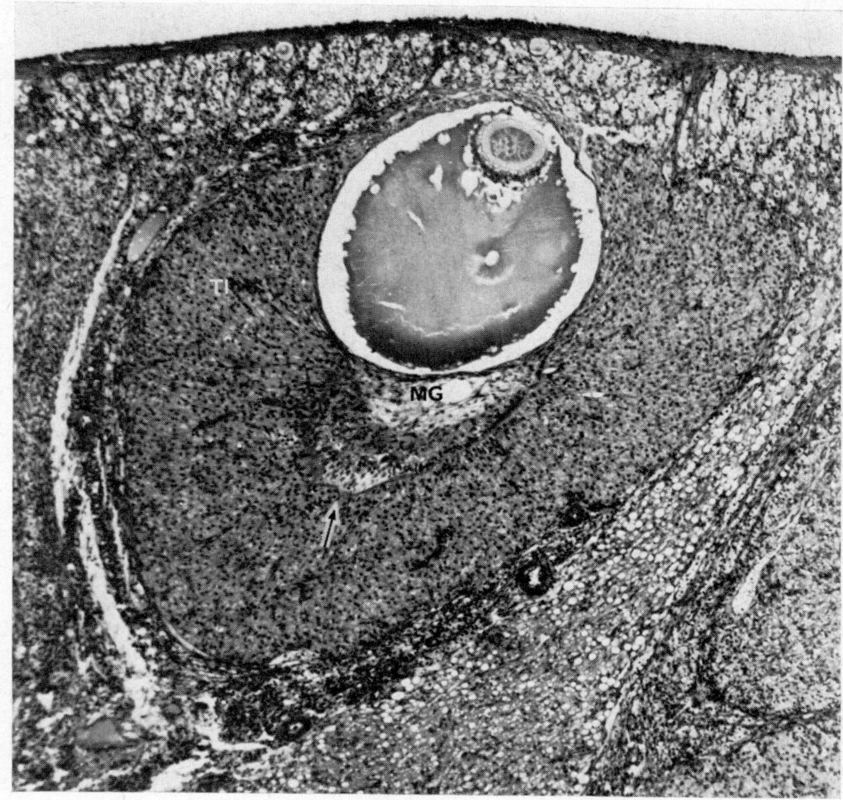

FIG. 8. Section of the ovary of an African porcupine (*Hystrix cristata*) showing a follicle undergoing atresia by luteinization of the theca interna (TI). The degenerating membrana granulosa is still bounded by the membrana propria (arrowed). The follicle is surrounded by interstitial tissue (above) and an old CL below. × 50.

and frequently in coypu during late pregnancy, accessory CL may be formed from luteinization of the theca interna cells (Figs. 8 and 9); the membrana propria remains distinct and the granulosa cells become pyknotic.

Interstitial tissue

Interstitial cells in the ovaries of mammals are claimed to be derived from various sources (Brambell, 1956). Primary interstitial tissue is common in neonatal ovaries, and in the rat (Rennels, 1951) and ferret (Deanesly, 1970) is thought to have originated from epithelial cells of cortical ingrowths (i.e. Pflüger's tubules). Secondary interstitial cells

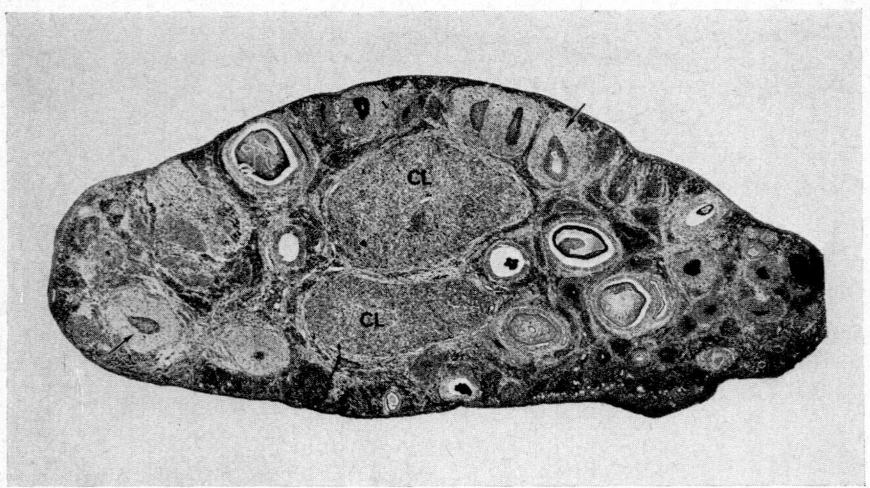

FIG. 9. Section of the ovary of a coypu (*Myocastor coypus*) in late pregnancy showing two CL of the pregnancy (CL), some small follicles and numerous follicles having a luteinized theca interna (arrowed), surrounding the remains of the membrana granulosa. × 11.

formed later in development are generally considered to be associated with medullary cords and are transformed theca interna cells of atretic follicles (see above). Some species, such as the guinea-pig or acouchi, may have very little if any secondary interstitial tissue while in others, like the agouti and degu, any ovarian component which is not a follicle or CL appears to be interstitial tissue (Fig. 10). The uneven distribution of interstitial cells renders their quantitative estimation difficult. The amount and appearance of secondary interstitial tissue varies from species to species but there does not seem to be any variation according to the stage of the reproductive cycle (see Stafford & Mossman, 1945) except in the degu (see below). The interstitial cells, especially in the degu and African porcupine, are vacuolated which suggests that either they are actively secreting or are unable to release their secretion— probably cholesterol or some other steroid precursor. In chinchilla, the cells stain moderately with fat stains but do not appear to contain Δ^5-hydroxysteroid dehydrogenase. The unusual eosinophilia that exists in the interstitial cells of degu during pregnancy suggests that these 'luteinized' cells may be involved in some steroid synthesis.

The amount and conformation of the interstitial tissue in species such as the African porcupine, agouti, degu and casiragua suggest that it may not all be derived from the theca interna cells and that recruitment from the stroma may occur. There is some evidence that in the agouti interstitial cells may be derived from the transformation of luteinized granulosa cells of atretic follicles (Weir, 1971d). If interstitial cells of a degu can become luteinized and if the theca cells themselves can luteinize, there seems to be no reason why the reverse transformation should not take place. The pluripotential character of the theca interna cells in this respect was noted by Mossman & Judas (1949) for the North American porcupine.

Corpora lutea

As indicated above, the luteal tissue of hystricomorphs may comprise true corpora lutea (CL) or accessory CL and both forms may be present in the same ovary.

True corpora lutea

The morphology and growth of these vary from species to species as for other mammals. The CL of pregnancy are much larger than those of the cycle in the casiragua (Fig. 11c) and cuis (see Weir, 1974), are about twice as big in the guinea-pig (Rowlands, 1956) and chinchilla (Weir, 1967b), and are similar in the plains viscacha. After reaching a peak the CL may remain at this size for some time or may immediately regress. The stages of regression are similar to those seen in other mammals (vacuolation and increased fibrosis) but only in the casiragua is a corpus albicans sometimes left.

In the cycle. In the cycle of the spontaneously ovulating species the luteal phase is active whether or not mating has taken place. In the guinea-pig the cyclic CL begins to decline after Day 7, as indicated by progesterone concentration (Rowlands & Short, 1959) and a reduction in size occurs between Days 10 and 12 (Rowlands, 1956) of the 15- to 17-day cycle. In chinchilla, the CL of the cycle do not regress until ovulation has recurred (Weir, 1967b); the regressing CL may disappear quickly or may still be apparent several cycles later (each of 41 days—see Weir, 1974). A similar situation seems to obtain in the casiragua which may ovulate spontaneously or in response to coitus (see Weir, 1974).

Ovulation is nearly always induced by coitus in the cuis and, since oestrus is also induced by the male (Weir, 1971e, 1973c), females in a breeding colony are nearly always pregnant. When vasectomized males were used to reveal the inherent periodicity of oestrus (Rood & Weir, 1970) it was found that the CL were significantly larger in females stimulated by a male with an intact chin gland than one whose chin gland had been excised.

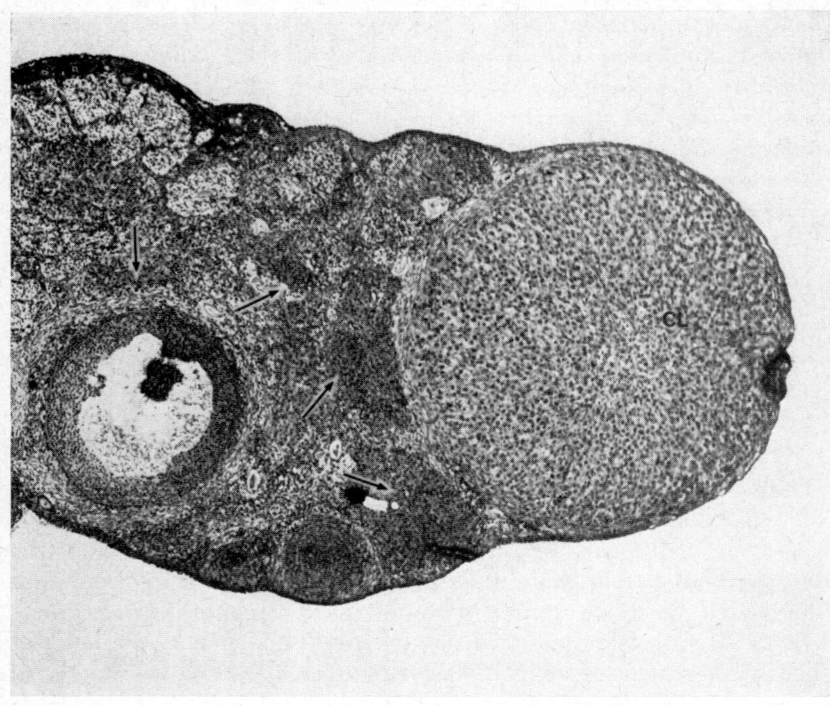

FIG. 10. Part of the ovary of a pregnant degu (*Octodon degus*) showing a corpus luteum (CL), a follicle, and blocks of secondary interstitial tissue, some of which appears to be luteinizing (arrows). × 50.

In pregnancy. The primary CL may persist until the end of pregnancy (see Weir, 1974) as in the casiragua, degu and North American porcupine or it may start to regress before the end of pregnancy, as in the guinea-pig. It is not possible to determine the life span of the primary CL of the chinchilla and plains viscacha during pregnancy as they become unrecognizable amongst the numerous accessory CL which are formed. Only a few accessory CL are sometimes formed in the coypu (Rowlands & Heap, 1966) and the acouchi (Rowlands, Tam & Kleiman, 1970) in pregnancy. In the degu, primary CL begin to regress before the end of pregnancy but may sometimes be difficult to define because of the 'luteinized' interstitial tissue (Fig. 10).

Accessory corpora lutea

By definition the accessory CL should have a retained ovum. The ovum rarely persists in follicles becoming atretic by luteinization but often does in a block of interstitial tissue. Retention of the zona pellucida remnant in accessory CL is common in the acouchi (Fig. 7) but may only be apparent in the accessory CL of other species in the earliest stages of development.

Accessory CL are usually formed during pregnancy but occur at an infertile oestrus in some chinchilla and African porcupine and in all plains viscacha. In the plains viscacha, as in the mountain viscacha and North American porcupine, the accessory CL are formed simultaneously with the primary CL and are histologically identical. Accessory CL in the chinchilla become very numerous and may be formed at any stage during pregnancy; unlike those described in other hystricomorphs they may become larger than the primary CL. It is uncertain whether the increase is due to hyperplasia or hypertrophy of the luteal cells and a contribution from thecal cells cannot be excluded. Apart from the exceptional plains viscacha, enormous numbers of accessory CL are found in chinchilla (Weir, 1966), agouti (Weir, 1971d) and acouchi (Weir, 1971f).

In the degu, several small accessory CL (Fig. 11) are formed at about Days 75 to 80 of the 90-day pregnancy and they persist for 10 to 20 days beyond parturition. Accessory CL are also formed with great regularity (not less than 80%) from the large follicles in the ovaries of two-week-old degu. They are often haemorrhagic and many are also polyovular.

Other cell types

The cell types described below constitute recognizable structures when they are present in an ovary.

Rete ovarii

These remnants of the mesonephric tubules occur in the ovaries of all hystricomorphs that have been studied. The rete ovarii is commonly extensive in the guinea-pig, casiragua, chinchilla and the dasyproctids, and rete cysts are found in 20% of guinea-pig females although they are very rare in the other species.

Epoophoron tubules

These Wolffian duct remnants may be occasionally found near the hilum in the mesovarium of any mammal but they have frequently been encountered associated with degu and cuis ovaries.

Extra-adrenal tissue

Although such tissue is common in and near the ovaries of some sciurids (e.g. *Citellus*: Chester Jones & Henderson, 1963), in the hystricomorphs it has been found only infrequently near the ovary of the cuis.

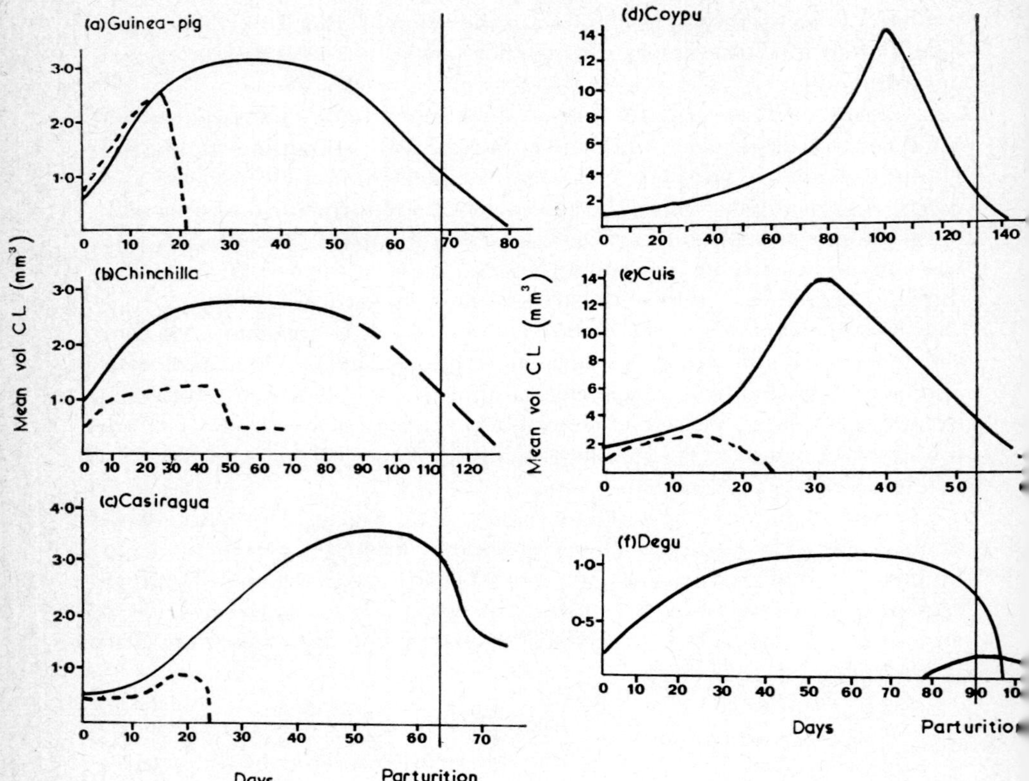

Fig. 11. Diagram of the volume changes of the CL throughout pregnancy (————) and the cycle (- - - - -) of six species of hystricomorphs. (a) Data from Rowlands (1956) transformed from the D^3 values. (b) Data from Weir (1967b). The dashed line at the end of pregnancy represents the uncertainty of identification of the primary CL because of the large numbers of accessory CL that are formed. (c) Data of Weir (unpublished). (d) Data from Rowlands & Heap (1966), who did not study non-pregnant females. (e) Cuis females rarely exhibit an oestrous cycle and the data for the cycle shown here are derived from experimental females (see Weir, 1973c). (f) The degu is primarily an induced ovulator and no cyclic CL are formed. Small accessory CL are produced in late pregnancy.

'Immature testis tubules'

These occur in acouchis and agoutis (Weir, 1971d, f), casiragua and cuis. The origin of the tubules in 50% of dasyproctids is unknown; sometimes large groups of them are found near the rete ovarii but they may also be scattered throughout the other tissues in small clumps (Fig. 7). When such clumps are separated by interstitial tissue the resemblance to the tubules in the immature testis is striking (see Weir, 1971d: Figs 11 and 12).

In casiragua the tubules are prominent at puberty in some females but the ovaries of all females retain a medullary component of the indifferent gonad until about three weeks after birth. The medulla is associated with the rete but its tubular nature is not obvious and it is doubtful whether the differentiation of the gonad could be altered by androgens, at least after birth.

Extensively scattered tubules are frequently found in the ovaries of young cuis, especially those which have been isolated for induction of puberty experiments (see Weir, 1973c).

'Lipophanerosis'

This term has been applied to a peculiar condition which develops in the ovaries of cuis. The cellular change appears to start in the stroma, perhaps as a consequence of non-dissolution of the Call-Exner bodies in the membrana granulosa of small follicles, and then spreads to the CL which become more and more infiltrated as they age (Fig. 12). The final result may have the appearance of a lace mat and only the normal follicular apparatus remains unaffected. We are not certain whether this condition occurs in the ovaries of wild cuis or is an artifact of captivity because none of the cuis we caught and killed in Argentina was mature. There is some correlation of the development of 'lipophanerosis' with age, weight and parity but the association is not complete. That the term 'lipophanerosis' may be a misnomer is indicated by the fact that the tissue does not stain fully with the normal fat stains, but we have no indication of its true nature.

FUNCTIONAL ACTIVITY

It is clear from the foregoing account that the most striking feature of the hystricomorph ovary is the development of extraordinary amounts of luteal tissue. As has already been stated, the usual condition for any mammalian female is that of pregnancy and therefore it is important to consider the relevance of this luteal tissue in the maintenance of gestation. Experiments involving bilateral ovariectomies during gestation have shown that although the guinea-pig (Artunkal & Cologne, 1949) and the cuis can carry live young to term when the ovaries are removed, the chinchilla, viscacha and casiragua cannot. In the chinchilla, progesterone replacement therapy in the form of subcutaneous implants or daily injections can prevent the threatened abortion and indicates that these hystricomorphs are dependent upon progesterone for pregnancy maintenance as are other species. It is not yet known whether or not progestins are secreted by the cuis placenta as they are by that of the

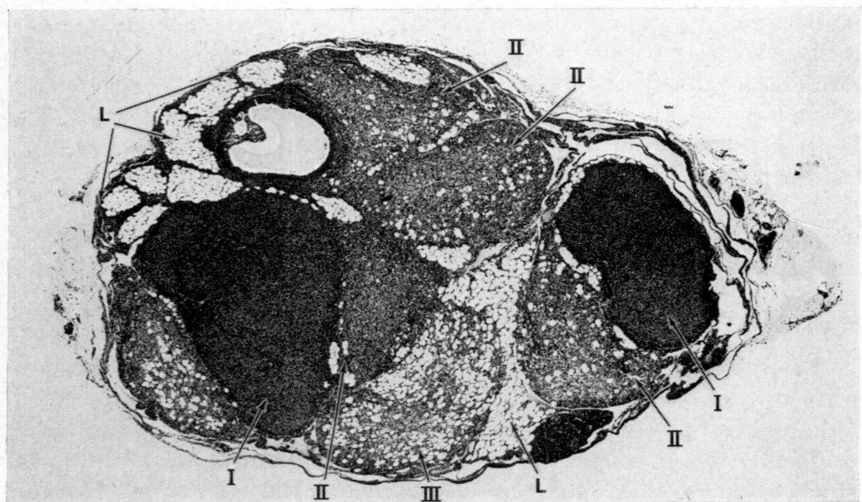

Fig. 12. Section of the ovary of a cuis (*Galea musteloides*) showing three sets of CL (I = youngest CL; III = oldest CL) and extensive patches of lipophanerosis (L). × 17.

guinea-pig (see Illingworth & Challis, 1973), but it seems a likely possibility in view of the strange morphological condition of many cuis CL in late pregnancy, and the decrease in progesterone concentration in arterial plasma after Day 30 (Tam, 1973).

Only in the guinea-pig, cuis and casiragua are accessory CL rarely if ever formed but, since the casiragua does appear to require the presence of the ovaries throughout gestation, the development of accessory luteal tissue cannot be entirely associated with failure of the placenta to take over the pregnancy-maintaining role. It is true, however, that all three of these species have 'short' gestations (see Weir, 1974) and it is possible that extra luteal tissue has been a necessary development for the maintenance of long pregnancies.

In the degu, for example, gestation lasts for 90 days and the primary CL are extant for the complete period (Fig. 11). The development of a crop of accessory CL just at the end of pregnancy indicates a 'tiding-over' function. Further support is also perhaps given by the interstitial cells which appear to luteinize in the last half of gestation (Fig. 10). In the chinchilla, agouti, acouchi, Canadian porcupine and mountain viscacha, the primary CL appear paramount for the first part of gestation and then the ovary becomes 'cluttered' by histologically identical CL, often in enormous numbers (over 100). Only in the chinchilla do these CL appear regularly to take over the role of the primary CL in such a way that the two types are indistinguishable by size. It has been shown that the accessory CL are functionally equivalent to the true CL (Weir, 1967b; Tam, 1970, 1971), and there is little doubt that these luteal bodies do act as extra sources of progesterone (see Heap & Illingworth, 1974; Tam, 1974). Coypu have several medium and large CL which outnumber the fetuses *in utero* but the tremendous atresia by theca luteinization suggests another important source of progesterone. The plains viscacha is the only species in which the accessory and primary CL are indistinguishable soon after ovulation. Whether a process so wasteful of eggs has evolved because of a need for other sources of progesterone, or because of a breakdown in the transcription processes that determine which follicles shall mature to ovulation, is uncertain.

The hormonal backgrounds to many of these complex ovaries are going to be difficult to unravel, and it will almost certainly be found that, although each species has worked on the same 'blue-print' for survival, each has embroidered it in a different way. Thus the mountain viscacha (Fig. 1) has used the same tissue types and trends as other hystricomorphs but has somehow managed to physiologically 'separate' the two ovaries in their responses to the circulating hormone levels. The regression of the accessory CL from the ovary on the side of the empty uterine horn of the North American porcupine (Mossman, 1966) is also inexplicable, even by current concepts involving luteolysins. There is no evidence for a luteolytic influence of the uterus on the CL in chinchilla as there is in the guinea-pig (Rowlands, 1961; Bland & Donovan, 1969; Blatchley & Donovan, 1972; Poyser, 1972).

The guinea-pig system may be necessary because the influence of the CL has to wane before a follicle can respond by ovulation in the guinea-pig (Rowlands, 1956). This is not so in the chinchilla or casiragua which both ovulate when 'active' luteal tissue appears to be present in the ovary. The long life-span of the CL of the cycle in hystricomorphs is very surprising since the delay to the next ovulation if conception should fail seems excessive, especially in those species with very long oestrous cycles, e.g. 41 days in chinchilla, (Weir, 1970b) and 45 days in plains viscacha (Weir, 1971b). It again suggests that this is an hystricomorph trait and emphasizes that species which are reflex ovulators are probably at an advantage (see Weir & Rowlands, 1973). The suggestion that plains viscacha ovulate spontaneously but need a coital stimulus to ensure conversion of the granulosa cells to luteal cells is an interesting one (Weir, 1971c), and such a mechanism, reminiscent of that of the myomorph rodents, would clearly be one way of reducing the interval to the next opportunity for conception. The luteotrophic stimulus may also be influenced by coitus in the cuis, since females mating with males with chin glands have larger CL than those mating with males without chin glands (Weir, 1973c). The chin-gland secretions are believed to act as a primer pheromone that influences the quality of the oestrus induced in the female. The cuis system, which can be so easily manipulated, would seem to be very suitable for detailed studies on the process of ovulation and the development of the CL.

The cuis seems to be a well organized animal in that little is left to chance. Oestrus, ovulation and puberty are induced by the presence of a male (Weir, 1971e, 1973c), and the ovulation fossa system provides a good egg-collecting mechanism. Few, if any, eggs should get lost, but it seems possible that the very high ovulation rate of the plains viscacha may be linked to the efficiency of fimbrial collection once the eggs have reached the surface of the ovary through the various clefts. However, this seems a poor excuse for the continual wastage of eggs that become entrapped in the follicles at each oestrus and which are shed throughout pregnancy (Weir, 1971c).

The ovarian activity of two-week-old degu is reflected in the vaginal epithelial cytology but no overt activity is seen and puberty does not occur until about six months of age (Weir, 1974). Many of these CL accessoriae haemorrhagicae are also polyovular and it is possible that this is a mechanism for removing such unwanted follicles at an early stage of life. The process of atresia appears to be well developed in hystricomorphs, resulting in the production of accessory CL and/or interstitial tissue. Tam (1972) has isolated various steroid metabolites from the 'residual' tissue of chinchilla ovaries (containing follicles, and probably very small luteinized follicles as well as interstitial tissue), but there is no evidence that the interstitial tissue plays a role in the ovulatory process by secreting progestins in the way that Hilliard, Penardi & Sawyer (1967) have described for the rabbit. The interstitial tissue is prominent in some species, but there is no obvious correlation of its abundance with the type of ovulation (see Weir, 1974). The degu and tuco-tuco, both mainly induced ovulators, have a lot of interstitial tissue; but so do agoutis, African porcupines, some chinchilla, mountain viscacha, plains viscacha and casiragua. The cuis, the most completely induced ovulator, and the coypu, suspected of ovulating reflexly, have very little interstitial tissue, like the acouchi and guinea-pig.

Thus, a reason can be suggested for the development of excess luteal tissue in hystricomorphs—the necessity caused by a long gestation. If hystricomorphs are related by virtue of their long pregnancies (see Weir, 1974), then the complexity of the ovary may be considered another associated factor. It is noteworthy that the African porcupine has a complex ovary in which accessory CL are abundant and interstitial tissue is active.

Acknowledgments

We are grateful to Mrs S. McLintock and Mr A. Hilton for their help in processing many of the ovaries on which this work is based and to Mr T. D. Dennett for his excellent photographic assistance. The laboratory work was supported by the Ford Foundation.

References

Artunkal, T. & Cologne, R. A. (1949). Action de l'ovariectomie sur la gestation du Cobaye. *C.r. Séanc. Soc. Biol.* **143**: 1590–1592.

Billington, W. D. & Weir, B. J. (1967). Deportation of trophoblast in the chinchilla. *J. Reprod. Fert.* **13**: 593–595.

Bland, K. P. & Donovan, B. T. (1969). Observations on the time of action and the pathway of the uterine luteolytic effect of the guinea-pig. *J. Endocr.* **43**: 259–264.

Blandau, R. J. (1967). Oogenesis—ovulation and egg transport. In *Comparative aspects of reproductive failure*: 194–205. Benirschke, K. (ed.). Berlin: Springer-Verlag.

Blandau, R. J. (1970). Growth of the ovarian follicle and ovulation. In *Progress in gynecology*: 58–76. Sturgis, S. A. & Taymor, M. L. (eds.). New York: Grunt Stratton.

Blatchley, F. R. & Donovan, B. T. (1972). The effect of prostaglandin $F_{2\alpha}$ and prostaglandin E_2 upon luteal function and ovulation in the guinea-pig. *J. Endocr.* **53**: 493–501.

Bookhout, C. G. (1945). The development of the guinea-pig ovary from sexual differentiation to maturity. *J. Morph.* **77**: 233–252.

Brambell, F. W. R. (1956). Ovarian changes. In *Marshall's Physiology of reproduction*: **I**: 397–542. 3rd ed. Parkes, A. S. (ed.). London: Longmans Green.

Branca, A. (1925). L'ovocyte atresique et son involution. *Archs Biol., Paris* **35**: 325–440.

Chester Jones, I. & Henderson, I. W. (1963). The ovary of the 13-lined ground squirrel (*Citellus tridecemlineatus* Mitchell) after adrenalectomy. *J. Endocr.* **26**: 265–272.

Collins, D. C. & Kent, H. A. (1964). Polynuclear ova and polyovular follicles in the ovaries of young guinea-pigs. *Anat. Rec.* **148**: 115–119.

Contreras, J. (1965). Datos a cerca de la actividad reproductora del genero *Ctenomys* (Rodentia: Octodontidae). *Physis, B. Aires* **25**: 169–186.

Deanesly, R. (1970). Oogenesis and the development of the ovarian interstitial tissue in the ferret. *J. Anat.* **107**: 165–178.

Harman, M. & Kirgis, H. D. (1938). The development and atresia of the Graafian follicle and the division of intra-ovarian ova in the guinea-pig. *Am. J. Anat.* **63**: 79–100.

Heap, R. B. & Illingworth, D. V. (1974). The maintenance of gestation in the guinea-pig and other hystricomorph rodents: changes in the dynamics of progesterone metabolism and the occurrence of progesterone-binding globulin (PBG). *Symp. zool. Soc. Lond.* No. 34: 385–415.

Hillemann, H. H. & Tibbitts, F. D. (1956). Ovarian growth and development in *Chinchilla. Northw. Sci.* **30**: 115–126.

Hilliard, J., Penardi, R. & Sawyer, C. H. (1967). A functional role for 20α-hydroxypregn-4-en-3-one in the rabbit. *Endocrinology* **80**: 901–909.

Illingworth, D. V. & Challis, J. R. G. (1973). Concentrations of oestrogens and progesterone in the plasma of ovariectomized and ovariectomized-Norgestrel-treated pregnant guinea-pigs. *J. Reprod. Fert.* **34**: 289–296.

Kayanja, F. I. B. & Jarvis, J. U. M. (1971). Histological observations on the ovary, oviduct and uterus of the naked mole rat. *Z. Saugetierk.* **36**: 114–121.

Konieczna, B. (1956). Sexual maturation and reproduction in *Myocastor coypus*. II. The ovary. *Folia biol., Kraków* **4**: 139–150.

Lagomarsino, J. G. & Momigliano, E. (1961). Estudio histologico del aparato genital femenini de *Ctenomys torquatus* ("tuco-tuco"). *Revta Fac. Hum. Cienc. Univ. Répub. Urig., Montevideo* **19**: 5–17.

Llanos, A. C. & Crespo, J. A. (1952). Ecologia de la vizcacha (*Lagostomus maximus maximus* Blainv.) en el nordeste de la provincia de Entre Ríos. *Revta Invest. agríc., B. Aires* **6**: 289–378.

Loeb, L. (1932). The parthenogenetic development of eggs in the ovary of the guinea-pig. *Anat. Rec.* **51**: 373–408.

Mossman, H. W. (1966). The rodent ovary. *Symp. zool. Soc. Lond.* No. 15: 455–470.

Mossman, H. W. & Duke, K. L. (1973). *Comparative morphology of the mammalian ovary*. Wisconsin: University Press.

Mossman, H. W. & Judas, I. (1949). Accessory corpora lutea, lutein cell origin, and the ovarian cycle in the Canadian porcupine. *Am. J. Anat.* **85**: 1–39.

Newson, R. M. (1966). Reproduction in the feral coypu (*Myocastor coypus*). *Symp. zool. Soc. Lond.* No. 15: 323–334.

Parkes, A. S. (ed.). (1956–1966). *Marshall's Physiology of reproduction*. **1, 2, 3**. London: Longmans.

Pearson, O. P. (1949). Reproduction of a South American rodent, the mountain viscacha. *Am. J. Anat.* **84**: 143–174.

Perry, J. S. (1971). *The ovarian cycle of mammals*. Edinburgh: Oliver & Boyd.

Ponse, K., Weihs, D., Libert, O. & Dovaz, R. (1954). Trophoblastèmes ovariens et leur activité endocrine chez le cobaye. *Acta endocr., Copenh.* **17**: 355–365.

Poyser, N. L. (1972). Production of prostaglandins by the guinea-pig uterus. *J. Endocr.* **54**: 147–159.

Rennels, E. G. (1951). The influence of hormones on the histochemistry of ovarian interstitial tissue in the immature rat. *Am. J. Anat.* **88**: 63.

Rood, J. P. & Weir, B. J. (1970). Reproduction in female wild guinea-pigs. *J. Reprod. Fert.* **23**: 393–409.

Rowlands, I. W. (1956). The corpus luteum of the guinea-pig. *Ciba Fdn Colloq. Ageing* **2**: 69–83.

Rowlands, I. W. (1961). Effect of hysterectomy at different stages in the life cycle of the corpus luteum in the guinea-pig. *J. Reprod. Fert.* **2**: 341–350.

Rowlands, I. W. (1964). Levels of gonadotrophins in tissues and fluids, with emphasis on domestic animals. In *Gonadotropins, their chemical and biological properties and secretory control*: 74–107. Cole, H. H. (ed.). San Francisco and London: Freeman.

Rowlands, I. W. & Heap, R. B. (1966). Histological observations on the ovary and progesterone levels in the coypu, *Myocastor coypus*. *Symp. zool. Soc. Lond.* No. 15: 335–352.

Rowlands, I. W. & Short, R. V. (1959). The progesterone content of the guinea-pig corpus luteum during the reproductive cycle and after hysterectomy. *J. Endocr.* **19**: 81–86.

Rowlands, I. W., Tam, W. H. & Kleiman, D. G. (1970). Histological and biochemical studies on the ovary and progesterone levels in the systemic blood of the green acouchi (*Myoprocta pratti*). *J. Reprod. Fert.* **22**: 533–545.

Schaeffer, A. (1911). Vergleichend histologische Untersuchungen über die interstitell Eierstocksdruse. *Arch. Gynäk.* **94**: 491–541.

Stafford, W. T., Collins, R. F. & Mossman, H. W. (1942). The thecal gland in the guinea-pig ovary. *Anat. Rec.* **83**: 193–207.

Stafford, W. T. & Mossman, H. W. (1945) The ovarian interstitial gland tissue and its relation to the pregnancy cycle in the guinea-pig. *Anat. Rec.* **93**: 97–107.

Stanley, A. P. & Hillemann, H. H. (1960). Histology of the reproductive organs of nutria, *Myocastor coypus* (Molina). *J. Morph.* **106**: 277–299.

Tam, W. H. (1970). The function of the accessory corpora lutea in the hystricomorph rodents. *J. Endocr.* **48**: liv–lv.

Tam, W. H. (1971). The production of hormonal steroids by ovarian tissues of the chinchilla (*Chinchilla laniger*). *J. Endocr.* **50**: 267–279.

Tam, W. H. (1972). Steroid metabolic pathways in the ovary of the chinchilla (*Chinchilla laniger*). *J. Endocr.* **52**: 37–50.

Tam, W. H. (1973). Progesterone levels during the oestrous cycle and pregnancy in the cuis, *Galea musteloides*. *J. Reprod. Fert.* **35**: 105–114.

Tam, W. H. (1974). The synthesis of progesterone in some hystricomorph rodents. *Symp. zool. Soc. Lond.* No. 34: 363–384.

Weir, B. J. (1966). Aspects of reproduction in chinchilla. *J. Reprod. Fert.* **12**: 410–411.

Weir, B. J. (1967a). The care and management of laboratory hystricomorph rodents. *Lab. Anim.* **1**: 95–104.

Weir, B. J. (1967b). *Aspects of reproduction in some hystricomorph rodents.* Ph.D. thesis, University of Cambridge.

Weir, B. J. (1969). The induction of ovulation in the chinchilla. *J. Endocr.* **43**: 55–60.

Weir, B. J. (1970a). The management and breeding of some more hystricomorph rodents. *Lab. Anim.* **4**: 83–97.

Weir, B. J. (1970b). Chinchilla. In *Reproduction and breeding techniques for laboratory animals*: 209–223. Hafez, E. S. E. (ed.). Philadelphia: Lea & Febiger.

Weir, B. J. (1971a). Some notes on reproduction in the Patagonian Mountain viscacha, *Lagidium boxi* (Mammalia: Rodentia). *J. Zool., Lond.* **164**: 463–467.

Weir, B. J. (1971b). The reproductive physiology of the plains viscacha, *Lagostomus maximus*. *J. Reprod. Fert.* **25**: 355–363.

Weir, B. J. (1971c). The reproductive organs of the female plains viscacha, *Lagostomus maximus*. *J. Reprod. Fert.* **25**: 365–373.

Weir, B. J. (1971d). Some observations on reproduction in the female agouti, *Dasyprocta aguti*. *J. Reprod. Fert.* **24**: 203–211.

Weir, B. J. (1971e). The evocation of oestrus in the cuis, *Galea musteloides*. *J. Reprod. Fert.* **26**: 405–408.

Weir, B. J. (1971f). Some observations on reproduction in the green acouchi, *Myoprocta pratti*. *J. Reprod. Fert.* **24**: 193–201.

Weir, B. J. (1972). Laboratory hystricomorph rodents other than the guinea-pig and chinchilla. In *The UFAW Handbook on the care and management of laboratory animals*: 278–286. 4th edn., Edinburgh: Churchill Livingstone.

Weir, B. J. (1973a). Another hystricomorph rodent: keeping casiragua (*Proechimys guairae*) in captivity. *Lab. Anim.* **7**: 125–134.

Weir, B. J. (1973b). The induction of ovulation and oestrus in the chinchilla. *J. Reprod. Fert.* **33**: 61–68.
Weir, B. J. (1973c). The rôle of the male in the evocation of oestrus in the cuis, *Galea musteloides. J. Reprod. Fert.*, Suppl. **19**: 421–432.
Weir, B. J. (1974). Reproductive characteristics of hystricomorph rodents. *Symp. zool. Soc. Lond.* No. 34: 265–301.
Weir, B. J. & Rowlands, I. W. (1973). Reproductive strategies of mammals. *A. Rev. Ecol. Syst.* **4**: 139–163.
Young, W. C. (ed.) (1961). *Sex and internal secretions.* 3rd edn. Baltimore: Williams & Wilkins.
Zuckerman, S. (ed.) (1962). *The ovary.* New York: Academic Press.

DISCUSSION

AUSTIN (*Cambridge*): Is it safe to talk about internal thecal cells giving rise to interstitial cells? After all, they are still going to be bounded by the theca externa, whereas the interstitial cells that one sees in a rabbit ovary, for example, morphologically resemble luteal cells and seem to be quite free. Aren't the theca interna cells you are thinking of essentially always going to be luteal cells?

WEIR (*London*): No, I don't think so. I have examples of cells breaking through a theca externa and joining up with the cells outside, with which they are identical (Weir, 1971)*. The theca externa is not a retaining structure in the same way that a membrana propria is; it is a loose arrangement of cells and in some species it is very difficult to find at all.

AUSTIN: Did you see germ cells in those ovarian tubules or are they all indeterminate cells?

WEIR: If one considers these tubules out of context one would almost certainly say that they are from an immature testis, which was just about to start spermatogenesis, but, of course, it would be impossible to be certain that germ cells are present until further development has taken place.

DEANESLY (*Cambridge*): From studying the development of a lot of ovaries of different species, I think one comes to the conclusion that almost all components of the ovary will respond to gonadotrophins and very likely 'luteinize' under certain conditions. As regards the development of the interstitial tissue in the rabbit and the guinea-pig, there is primary interstitial and secondary interstitial tissue, the secondary forming the bulk of the stuff in the adult ovary. In my experience, this secondary interstitial tissue is developed from thecal and mesodermal elements, because one can see the follicular epithelium

* Weir, B. J. (1971). *J. Reprod. Fert.* **24**: 203–211.

and the granulosa cells actually degenerating as the theca cells develop and luteinize. I have also seen bodies looking more like accessory corpora lutea in, say, the stoat, but I didn't have the material there to convince myself whether some of these were granulosa cells or whether, as in the rabbit, they had developed from thecal elements. I think it would be interesting to know, but I suspect that both things may happen.

WEIR: Yes, I agree that both things may happen quite frequently. People often say, how do you know it is a corpus luteum accessorium rather than a corpus luteum atreticum and it is very difficult to explain.

DEANESLY: But one can always find the remains of the egg.

WEIR: Not always. Agoutis and chinchilla, for example, don't usually show zona remnants, whereas acouchis do. They are very prominent in acouchis for long periods. Sometimes it is impossible to find eggs, and yet one instinctively feels that this is not an ovulated follicle.

DEANESLY: The secondary interstitial tissue in ordinary ovaries such as the guinea-pig and the rabbit does go through a phase of being a luteal body. More luteal bodies are formed all the time and they gradually sink into the body of the ovary.

WEIR: This is not the impression I get from the other hystricomorphs. I use the term 'luteinization' when the interstitial tissue cytoplasm becomes stainable with eosin. Normally this is not so for interstitial cells and one can see the derivation of the blocks of interstitial tissue from atretic follicles; it is only in those species which I mentioned, the tuco-tuco and degu, that the interstitial gland as a whole becomes stainable. I certainly saw no evidence of a normal progression of this at other times.

AMOROSO (*Cambridge*): Dr Weir, when speaking of *Lagostomus*, how many of these 400 or 500 eggs are fertile?

WEIR: Our assessment is that not more than 10 or 20% are fertilized, but this doesn't, of course, mean that they are not all 'fertilizable'.

AMOROSO: But from microscopic examination of those cells within the tube, what was the maximum number you found that were fertilized?

WEIR: We thought it was only about 20, but we admit that we are not accustomed to dealing with eggs and zygotes and it might be a few more.

AMOROSO: Only 20 out of 800 were fertilized?

WEIR: Yes. We flushed eggs from several viscacha, photographed the eggs in the embryo cups, fitted the photographs together to form a montage and then studied the eggs. Again we came up with only this very low number.

AMOROSO: Do other hystricomorphs try to emulate this feat? For example, you say there are one to four young in the litter of the African porcupine and yet you got corpora lutea in excess of four.

WEIR: Yes, I did, but some of these corpora lutea were obviously luteinizing follicles because the egg was still present.

AMOROSO: But you never got anything in excess of four eggs in the tubes?

WEIR: I didn't have the tubes to go with the ovaries, so I can't answer. But there were certainly more luteal bodies in the *Hystrix* ovaries than could normally be expected to be accounted for by ovulation.

AUSTIN: The logistics of collecting all the viscacha eggs is quite a factor. Did you notice anything specific about the infundibulum or the behaviour of it around the time of ovulation?

WEIR: There is no obvious swelling of the infundibulum at oestrus. It didn't appear that the fimbria could stretch all the way round the ovary and when one considers that these eggs have to travel through all the little clefts of the ovary it is surprising that they do all manage to get collected. I should think there must be quite a current going down the tube because all the eggs are found in more or less the same place in the tube.

AUSTIN: And the surface of the ovary isn't ciliated, or anything extraordinary?

WEIR: Not that I could see, but it would be interesting to get some electron micrographs of this extraordinary ovary.

Symp. zool. Soc. Lond. (1974) No. 34, 333–360

HYSTRICOMORPH EMBRYOLOGY

CHRISTINE M. ROBERTS and J. S. PERRY*

*Wellcome Institute of Comparative Physiology,
Zoological Society of London,
Regent's Park, London, England*

and

**A.R.C. Institute of Animal Physiology,
Babraham, Cambridge, England*

SYNOPSIS

Members of the rodent sub-order Hystricomorpha are embryologically interesting, especially because of their long pregnancies compared with other rodents. In the hystricomorphs, gestation periods range from 53 days in the cuis, *Galea musteloides*, to 213 days in the Canadian porcupine, *Erethizon dorsatum*. Except in one of the species studied, these long pregnancies are not due to phenomena such as delayed fertilization or delayed implantation. The mode of implantation is of the complete interstitial type. As in the domestic guinea-pig, *Cavia porcellus*, blastocysts reach the uterus by 3–3½ days *post coitum* (*p.c.*) and implant antimesometrially at six to seven days. The plains viscacha, *Lagostomus maximus*, is exceptional in that implantation does not occur until Day 18 *p.c.* This appears to be the longest 'delay' recorded for any rodent. The long gestation periods of the hystricomorphs are related to the extremely slow rate of growth of the embryos, particularly in the early stages of pregnancy when the development of the conceptus is more especially concerned with the establishment of the placenta rather than the formation of the embryo.

Early complete inversion of the germ layers is another characteristic of post-implantation development in hystricomorphs. All hystricomorph placentae studied so far are discoid, haemochorial and labyrinthine. They are further characterized by the development of a prominent sub-placenta.

The occurrence of fetal resorption has been reported in several hystricomorph species including the plains viscacha in which it is a regular feature of pregnancy. Despite the remarkable polyovulation of between 200 and 800 eggs at each oestrus in this species, there are rarely more than six to eight implantation sites and usually only the fetus situated nearest the cervix in each uterine horn survives. In this region, it appears that blastocysts implant slightly earlier than those at the other sites along the uterus. Resorption of embryos occurs progressively and is almost complete by Day 35 *p.c.* in those implanted at the ovarian extremities of the uterine horns. Thus it seems that the fate of the viscacha embryos is determined at the time of implantation.

INTRODUCTION

Most of the embryological literature on the rodent sub-order Hystricomorpha relates to the domestic guinea-pig, *Cavia porcellus*. The embryological development of hystricomorphs is of interest, however, in view of the long gestation periods found in these species (see Weir, 1974: Table I) compared with other rodents.

Studies have been made on the placental morphology of the agouti, *Dasyprocta aguti* (Strahl, 1905; Becher, 1921a,b), Canadian porcupine, *Erethizon dorsatum* (Perrotta, 1959), chinchilla, *Chinchilla laniger* (Tibbitts & Hillemann, 1959; Crooks, 1973), coypu, *Myocastor coypus* (Hillemann & Gaynor, 1961), and paca, *Cuniculus paca* (Strahl, 1905). In each of these studies, however, the animals used have been of unknown reproductive history and, consequently, only estimated ages of pregnancy have been given. Recently, Roberts (1973) has studied aspects of the embryology of the following species, from which accurately dated pregnancy stages were obtained: cuis (*Galea musteloides*), casiragua (*Proechimys guairae*), degu (*Octodon degus*), chinchilla (*Chinchilla laniger*), coypu (*Myocastor coypus*) and the plains viscacha (*Lagostomus maximus*). All the species studied, except the coypu, have a vaginal closure membrane which is perforate only at oestrus and parturition. Matings may, therefore, be detected by the presence of copulatory plugs or examination of vaginal smears for spermatozoa at these times.

All the species studied have been found to resemble the guinea-pig in many characteristics including completely interstitial implantation, extremely early inversion of the germ layers and the possession of a sub-placenta, a central region of the placental disc developed from the ectoplacental trophoblast which, in this group, is separated from the amnio-embryonic ectoderm soon after implantation (Sansom & Hill, 1931).

PRE-IMPLANTATION DEVELOPMENT

The unfertilized egg

The guinea-pig ovum has been described extensively by Squier (1932), Boyd & Hamilton (1952) and Austin (1961). Unfertilized eggs of other hystricomorphs have been described by Roberts (1973). By far the largest numbers of unfertilized eggs have been found in the viscacha which ovulates between 200 and 800 eggs at each oestrus (Weir, 1971a). The observations of Roberts (1973) on unfertilized eggs of some hystricomorphs are summarized in Table I. The diameters of the unfertilized eggs are between 65 and 75 μm, and are, therefore, small compared with those of many other mammals. In freshly flushed ova, the zona pellucida is an homogeneous membrane with some irregularities over its surface. Radial striations are particularly prominent in the zona pellucida of chinchilla eggs, and are thought to be the cytoplasmic extensions of the corona radiata cells. Cells of the cumulus oophorus are present around the newly ovulated eggs of the guinea-pig (Squier, 1932),

TABLE I

The numbers of unfertilized eggs observed in the Fallopian tube (F.t.) and uterus (Ut.) of the viscacha, and other hystricomorph rodents

Species	No. of animals	No. of eggs	Situation F.t.	Situation Ut.	Age (days *p.c.*)
Plains	1	197		s, f	7
viscacha	1	137		s	10
	1	255		s, f	$13\frac{1}{2}$
	1	43		s	15
	2	544		s, f	$18\frac{1}{2}$
Chinchilla	1	2	s		$2\frac{1}{2}$
	1	1		f	$3\frac{1}{2}$
Casiragua	1	1		f	$2\frac{1}{2}$
Degu	1	5	s		$3\frac{1}{2}$

s = sectioned; f = flushed

chinchilla (Weir, 1967), agouti (Weir, 1971c), viscacha (Weir, 1971b) and casiragua (Roberts, 1973). Newly ovulated eggs of other hystricomorphs have not been examined. As in most mammals, the first maturation division of the ovum is completed within the ovarian follicle some time before it ruptures. In hystricomorphs, it is possible that the first polar body undergoes rapid fragmentation and cytolysis within the perivitelline space, since the presence of both the first and second polar bodies has not been reported. The perivitelline space is difficult to distinguish in freshly flushed eggs but is more prominent in fixed material. The distribution of yolk granules within the ooplasm is even and there is no obvious polarity.

Fertilization and cleavage

An excellent account of the development of the guinea-pig egg, including information on its rate of cleavage and transport through the oviducts, is provided by Squier (1932). Fertilization takes place in the oviducts between three and eight hours after ovulation, although the fertilizable life of the ovum is approximately 20 h (Blandau & Young, 1939; Rowlands, 1957). On Day 3 *post coitum (p.c.)*, eight-cell zygotes are present in the uterus and blastocysts are formed by Day 5 *p.c.* (Squier, 1932; Blandau, 1949a, 1961). Detailed investigations on other hystricomorph rodents were made more difficult since the required numbers of animals were not readily available. Although the precise timing of ovulation and fertilization is unknown for most hystricomorph species, the preliminary findings of Roberts (1973) suggest that the timing of pre-implantation events in the cuis, casiragua, degu and chinchilla is essentially similar to that of the domestic guinea-pig.

Single-cell zygotes have been found in the Fallopian tubes of the chinchilla and casiragua (Roberts, 1973) at 1–2 days *p.c.* but fertilization is frequently difficult to detect since some eggs show no traces of polar bodies, spermatozoa or pronuclei. Information on those eggs in which fertilization was detected (Fig. 1) is summarized in Table II. Polyspermy has not been recorded in any of these species but supernumerary spermatozoa have been observed in the zona pellucida of viscacha eggs at 7 days *p.c.* (Roberts, 1973).

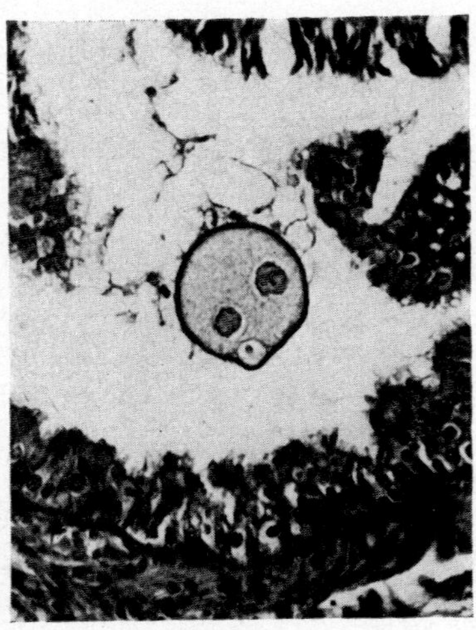

FIG. 1. Single-cell zygote of casiragua (*Proechimys guairae*) showing two pronuclei and a polar body (5 μm, H & E, × 300).

TABLE II

Observations on single-cell zygotes

Species	No. of zygotes	Age (days *p.c.*)	Observations
Chinchilla	2	$2\frac{1}{2}$	Single polar body, 2nd maturation spindle
	2	$2\frac{1}{2}$	Single polar body, spermatozoa in zona pellucida
Casiragua	2	1	2 pronuclei, 1 polar body (see Fig. 1)

The information available at present on the rate of cleavage and transport of some hystricomorph eggs suggests that they are similar to those of the guinea-pig (Roberts, 1973). In the viscacha, only about 10% of the eggs ovulated are fertilized (Weir, 1971a; Roberts & Weir, 1973). This is in contrast to findings of van der Horst & Gillman (1941) for the South African elephant shrew, *Elephantulus myurus*, in which up to 120 eggs are ovulated and the majority are fertilized (see Tripp, 1971). The rate of cleavage of eggs of the viscacha appears to be slow, since zygotes do not enter the uterus until Day 6 *p.c.* at the four-cell stage, (Weir, 1971a).

IMPLANTATION

Blastocyst formation and attachment

Although the rate of blastocyst development varies between species, hystricomorph blastocysts are all very similar in appearance (Fig. 2; and see Sansom & Hill, 1931; Blandau, 1949a) and measure about 95 μm in diameter.

Blastocysts are present in the uteri of the casiragua and degu at $4\frac{1}{2}$ days *p.c.*, but were not found until Day 18 *p.c.* in the viscacha (Table III). In this species, it seems that more blastocysts are formed than can or do implant because 17 very small blastocysts were found at the ovarian extremity of a uterine horn in which implantation was already proceeding. In hystricomorphs the blastocyst cavity is prominent and only a narrow perivitelline space exists between the trophoblast

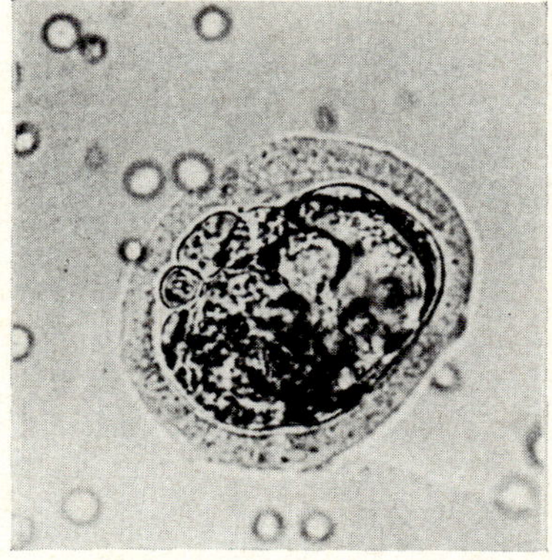

FIG. 2. Early blastocyst flushed from a casiragua (*Proechimys guairae*) uterus at $3\frac{1}{2}$ days *p.c.* The inner cell mass is enclosed by a thin wall of trophoblast, and only a narrow perivitelline space can be seen between the trophoblast and the zona pellucida ($\times$ 400).

TABLE III

Pre-implantation blastocysts

Species	No. of animals	No. of blastocysts	Sectioned (s) or flushed (f)	Age (days *p.c.*)
Casiragua	1	1	f	$3\frac{1}{2}$
	1	3	f	$4\frac{1}{2}$
	3	6	s	$4\frac{1}{2}$
Cuis	2	3	s	$5\frac{1}{4}$
Degu	2	5	s	$4\frac{1}{2}$
	1	6	s	$5\frac{1}{2}$
Chinchilla	1	1	s	$3\frac{1}{2}$
Plains	2	19	s	$18\frac{1}{2}$
viscacha	1	1	s	22

and zona pellucida. The covering trophoblast is distinguishable as a thin investment over the inner cell mass whilst the abembryonal trophoblast is slightly thickened. The diameters of blastocysts at this stage are 95–105 μm, indicating that little expansion occurs before implantation.

Shortly before implantation the blastocyst occupies a rounded bay at the antimesometrial extremity of the uterine lumen, and the abembryonal trophoblast is orientated towards the uterine epithelium (Fig. 3). The zona pellucida probably remains intact until just before implantation. Loss of the zona pellucida and attachment of the blastocyst to the uterine epithelium have been fully described only for the domestic guinea-pig (Blandau, 1949a,b) although the appearance of pseudopodial processes at the abembryonal pole of the blastocyst was first noted by von Spee (1883). Thee findings were confirmed by Blandau (1949b) who showed that as the trophoblastic processes become more numerous, they penetrate the zona pellucida, which gradually disappears in that region, and eventually form an 'attachment cone' which penetrates the uterine epithelium. Observations *in vitro* showed that the remainder was sloughed (Blandau, 1961). Such trophoblastic processes have not been observed in other hystricomorph species but this may be due to a lack of material at the critical attachment phase of implantation.

Observations on blastocysts of the cuis, casiragua, degu, chinchilla and viscacha at the time of implantation suggest that the spacing of embryos along the uterine horns is not random. Since little information is available on the precise time of entry of embryos into the uterus it is not known whether the zygotes enter singly or in groups. In the viscacha,

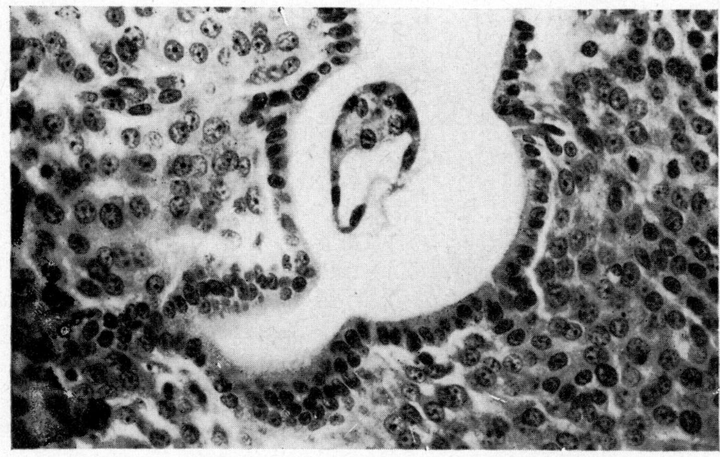

Fig. 3. Section through a casiragua (*Proechimys guairae*) blastocyst found in the uterus at 4½ days *p.c.* Traces of the zona pellucida remain around the blastocyst, which is situated in a rounded bay at the antimesometrial extremity of the uterine lumen (5 μm, H & E, ×325).

most of the eggs in sectioned uteri were found at the ovarian ends of the uterine horns, suggesting that spacing of the blastocysts occurs close to the time of implantation (Roberts & Weir, 1973). As in other rodents, decidualization following contact of the blastocyst with the uterine epithelium is considerable in the hystricomorphs. In the guinea-pig, at least partial removal of the uterine epithelium is necessary for the initiation of a decidual cell reaction (DCR) (Blandau, 1949b). In the other hystricomorphs studied there appears to be some variation in this respect (Roberts, 1973). In the casiragua and cuis, no DCR could be detected and although blastocysts appeared to be in contact with the epithelium, no damage was visible. Thus, these species resemble the guinea-pig in that some disruption of the epithelium is probably necessary to initiate the DCR. In the degu, chinchilla and viscacha, there was evidence of a DCR where blastocysts were merely in contact with the uterine epithelium and no disruption of epithelial cells was apparent. The possibility of trophoblastic processes being present between the epithelial cells, causing slight damage, cannot be ruled out, however, and electron microscope studies need to be carried out in order to clarify these findings.

Timing of implantation

The timing of implantation shown in Table IV for each of the hystricomorphs studied is the earliest stage of implantation observed in that

TABLE IV

The timing of implantation in some hystricomorph rodents

Species	Time of implantation (days *p.c.*)
Chinchilla	$5\frac{1}{2}$
Casiragua	$6–6\frac{1}{2}$
Cuis	6
Guinea-pig	6–7
Degu	$6\frac{1}{2}–7$
Plains viscacha	$18\frac{1}{2}$

species. Unfortunately, with the exception of the domestic guinea-pig, few hystricomorph blastocysts have been observed in the process of penetrating the uterine epithelium. Attached blastocysts, or blastocysts lying immediately beneath the uterine epithelium, have been described in the Canadian porcupine (Perrotta, 1959), viscacha (Roberts & Weir, 1973) and degu (Roberts, 1973). The interval between ovulation and implantation in the viscacha ($18\frac{1}{2}$ days) appears to be the longest recorded for any rodent.

Mode of implantation

Implantation in the domestic guinea-pig was the subject of some controversy in the early part of the 20th century mainly because of the misinterpretation of the structure of the newly implanted blastocyst (von Spee, 1883; McLaren, 1926). The disagreement was finally settled by a masterly communication from Sansom & Hill (1931), who concluded that implantation in the guinea-pig is of the highly specialized complete interstitial type.

The mode of implantation in the cuis, casiragua, degu, chinchilla and plains viscacha also appears to be completely interstitial (Roberts, 1973). In the Canadian porcupine, Perrotta (1959) suggested that implantation may be incompletely interstitial mainly because of the size of an attached blastocyst. In the viscacha, it appears that the blastocysts situated nearest the cervix in each uterine horn are the first to implant (Roberts & Weir, 1973). In one uterus at $18\frac{1}{2}$ days *p.c.*, the blastocysts in this situation were found lying immediately beneath the uterine epithelium (Fig. 4), those in the mid region were in the process of penetrating the epithelium and others were lying free at the ovarian extremities of the uterine lumen. The significance of these findings is discussed in relation to fetal resorption on p. 352.

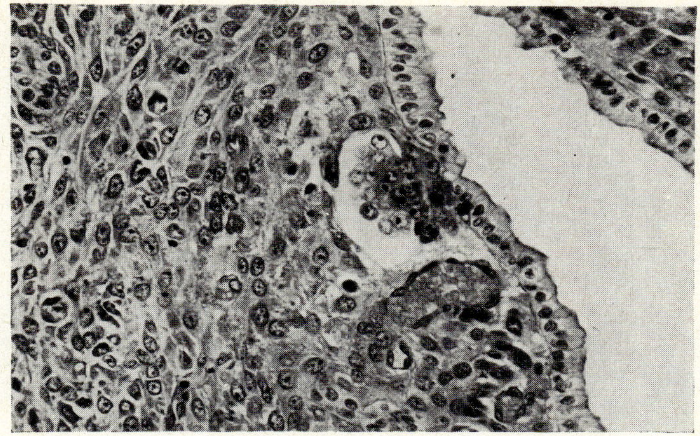

Fig. 4. Section through a plains viscacha (*Lagostomus maximus*) implantation site at 18½ days *p.c.* The blastocyst is implanted on the antimesometrial side of the uterine lumen nearest to the cervix. The central, deep-staining cells represent the amnio-embryonic mass, beneath which is a layer of endodermal cells. Above is the covering trophoblast, from which extends the parietal trophoblast enclosing the blastocyst cavity. Decidualization of the stroma is extensive (5µm, H & E, × 300).

EARLY POST-IMPLANTATION DEVELOPMENT

Accounts of the early post-implantation development of the guinea-pig are provided by Duval (1892), Sansom & Hill (1931) Harman & Prickett (1932) and Amoroso (1952). Several early embryos of the chinchilla are described by Hillemann, Tibbitts & Gaynor (1959) but only estimates of their ages are given.

The early post-implantation development of the cuis, casiragua, degu, chinchilla, coypu and viscacha has been described by Roberts (1973) from accurately dated stages of pregnancy. It appears that the basic pattern of development is similar to that of the domestic guinea-pig, but a number of interesting specializations exist.

One of the most significant and prominent features of early post-implantation development in hystricomorphs is the elongation of the blastocyst into a cylindrical structure (the 'egg cylinder') due to growth of the endoderm. At the same time, disintegration of the parietal tropho-blast is evident, so that the visceral yolk sac endoderm becomes directly apposed to the uterine tissues. These events are clearly visible in the chinchilla at 8 days *p.c.* (Fig. 5). Thus, complete inversion of the germ layers takes place at a very early stage of development in these species and, in most of the hystricomorphs studied, there is no evidence of a parietal yolk sac wall; only in the Canadian porcupine are traces of this membrane visible (Perrotta, 1959).

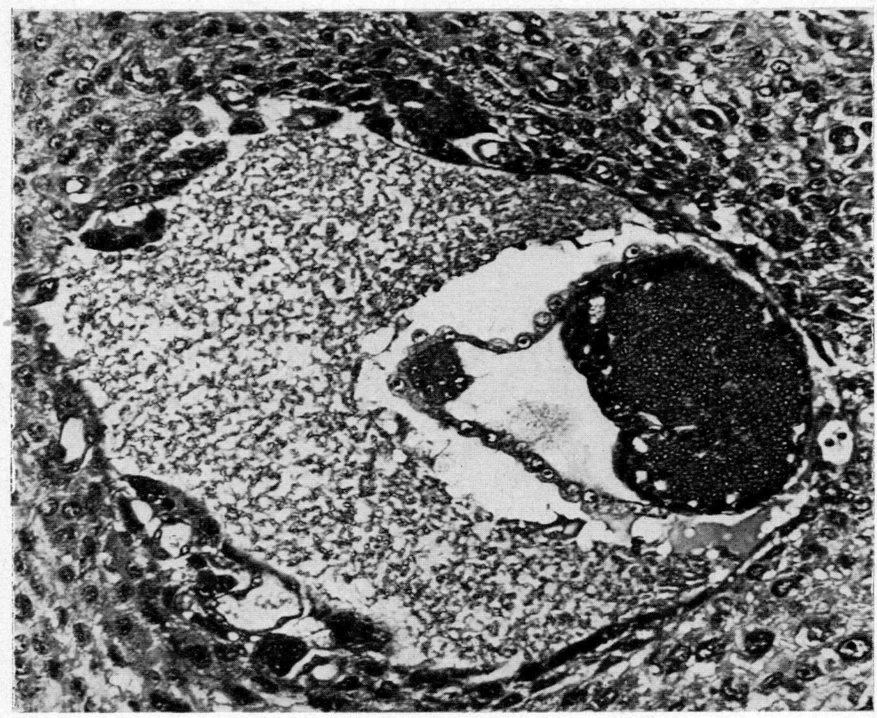

FIG. 5. Section through a chinchilla (*Chinchilla laniger*) embryo at 8 days *p.c.* The endoderm has elongated to give the blastocyst a cylindrical structure (the 'egg cylinder'). At the embryonic pole, the amnio-embryonic mass is a solid sphere of cells, whilst at the placental pole, the ectoplacental cavity is filled with extravasated maternal blood (5 μm, H & E, $\times$ 250).

The abembryonal pole of the blastocyst is composed of ectoplacental trophoblast which encloses an 'ectoplacental' cavity. In the chinchilla (see Fig. 5), coypu and viscacha, this cavity is frequently filled with extravasated maternal blood. The outer lamina of the cavity is usually only one cell thick, whilst its inner layer is thicker. In most of the hystricomorphs studied, therefore, the extensive Träger characteristic of the myomorph rodents is not developed. Continued elongation of the endoderm separates the ectoplacental trophoblast from the inner cell mass, the cavity between them being the pro-exocoelomic cavity. The decidual cavity increases considerably in size, particularly in the anti-mesometrial region where it expands and becomes dumb-bell shaped (Fig. 6). At the ectoplacental pole of the blastocyst, tubular extensions of the trophoblast invade the decidua, piercing the unilaminar endoderm wall of the blastocyst cylinder. These trophoblastic branches

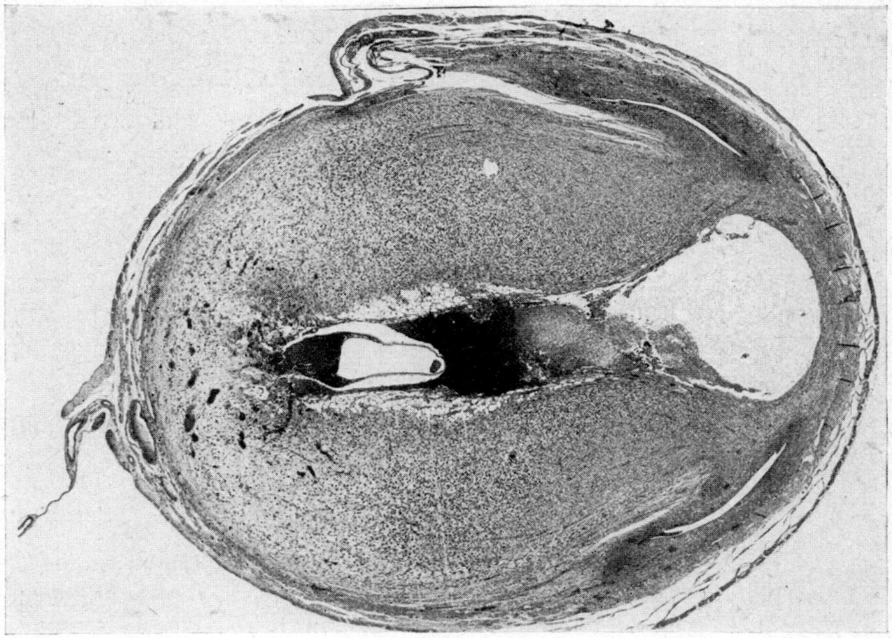

Fig. 6. Section through a chinchilla (*Chinchilla laniger*) embryo at 10 days *p.c.* The decidual cavity has increased considerably in size, particularly in the antimesometrial region where it has expanded and become dumb-bell shaped (5 μm, H & E, $\times$ 80).

anchor the blastocyst to the decidua, and establish the site of the future chorio-allantoic placenta.

In all the hystricomorphs studied so far, amniogenesis is by cavitation not by folding. At about the same time as the appearance of the amniotic cavity (Table V), the endodermal cylinder of the blastocyst becomes lined with a thin covering of mesoderm which later becomes vascularized.

The elaboration of the decidua is brought about by the formation of clefts in the endometrium establishing a new uterine lumen. The original lumen is obliterated by the expansion of the decidua, and the new lumen establishes the decidua parietalis peripherally and the decidua capsularis antimesometrially. The ectoplacental cavity gradually disappears as the chorionic ectoderm thickens and becomes more folded. The spaces between the narrow trabeculae of syncytial trophoblast which traverse the ectoplacental cone are frequently blood filled. Primary giant cells arise from the peripheral regions of the ectoplacental trophoblast at the junction of the fetal and maternal tissues. These cells appear to be

TABLE V

The time of formation of the amniotic cavity and allantois

Species	Amniotic cavity (days *p.c.*)	Allantois (days *p.c.*)
Cuis	9–12	12–15
Casiragua	10	$10\frac{1}{2}$
Guinea-pig	10–12	15
Degu	$12\frac{1}{2}$	30
Chinchilla	15	25
Coypu	15–18	20–25
Plains viscacha	28	35

phagocytic, and are especially prominent in the cuis, chinchilla and coypu.

PLACENTAL DEVELOPMENT

One of the earliest accounts of placental morphology in the guinea-pig was provided by Creighton (1878) and details of the development of its placenta and fetal membranes are included in the reviews of Mossman (1937) and Amoroso (1952).

Amoroso (1952) distinguished four main regions of the chorio-allantoic placenta of the guinea-pig.

(i) The trophoblastic area consists of labyrinthine and spongy zones. A comparative ultrastructural study of this area of several haemochorial placentae, including that of the guinea-pig, is provided by Enders (1965). This author describes the feto-maternal relationship of the guinea-pig as haemo-monochorial, in which a single layer of trophoblast cells is interposed between the two blood streams.

(ii) The sub-placenta, originating from the chorionic ectoderm of the 'central excavation' of the guinea-pig placenta, was first recognized by Duval (1892) and was described in histological and histochemical detail by Davies, Dempsey & Amoroso (1961a,b).

(iii) The junctional zone, forming a region of intimate feto-maternal contact, is composed largely of necrotic tissue derived mainly from the decidua. Its fine structure is described by Wynn (1964).

(iv) The decidua basalis is the maternal component of the placenta and reference to its ultrastructure is made by Wynn (1967).

An unusual form of yolk sac placentation is encountered in hystricomorphs; owing to the extremely early inversion of the germ layers, the ectoplacental trophoblast, instead of endoderm, as in most myomorph rodents, provides a roof to the cavity of the yolk sac at the placental pole (Sansom & Hill, 1931; Harman & Prickett, 1932; Amoroso, 1952). The yolk sac mesoderm becomes highly vascularized and a prominent sinus terminalis is formed. Thus, the vascular splanchnopleure plays an important role in the nutrition of the embryo, particularly following the disappearance of the decidua capsularis. The yolk sac then becomes directly adjacent to the uterine lumen. The nature of the placental barrier allowing physiological exchange between mother and fetus in the labyrinth of the chorio-allantoic placenta of various mammals is described by Mossman (1967) who also provides data on the thickness of the placental barrier in the guinea-pig. The histochemistry of the guinea-pig placenta has received considerable attention, with extensive contributions from Wislocki & Dempsey (1945), Dempsey & Wislocki (1947), Wislocki, Deane & Dempsey (1946), Hard (1946) and Christie (1967). These publications include information on the distribution of glycogen, lipids, alkaline phosphatase and iron in the uterus of the pregnant guinea-pig.

The pattern of placental development described for the guinea-pig is also found in the agouti (Strahl, 1905; Becher, 1921a,b), Canadian porcupine (Perrotta, 1959), chinchilla (Tibbitts & Hillemann, 1959; Crooks, 1973), coypu (Hillemann & Gaynor, 1961) and paca (Strahl, 1905). In each the chorio-allantoic placenta is formed by the growth of the allantoic mesenchyme across the exocoelomic cavity until it establishes contact with the rim of the ectoplacental cavity (Fig. 7). It spreads over the chorionic surface which thus becomes vascularized as the chorio-allantoic placenta (Table VI). Extensive branching of the fetal vessels occurs as the placenta matures. These vessels generally appear to radiate from a large central vessel, and several of these main vessels, each with its extensive branches, give rise to the lobate appearance of the definitive placenta (Fig. 8). The trophospongium forms an investing layer around the developing labyrinthine zones and from its inner surfaces thin syncytial tubules carrying maternal blood pierce the labyrinth. Peripherally, between the trophospongium and the necrotic decidual zone, a layer of chorionic giant cells is prominent in the cuis and chinchilla, but is usually discontinuous in other hystricomorphs in which only isolated nests of giant cells occur. Deportation of trophoblastic giant cells to the lungs in chinchilla has been reported by Billington & Weir (1967); it is not yet known whether this phenomenon occurs in other members of the suborder.

The necrotic area adjacent to the giant cell zone is particularly extensive in the chinchilla and viscacha, and in the coypu it contains much free maternal blood. Basally, it contains much amorphous material

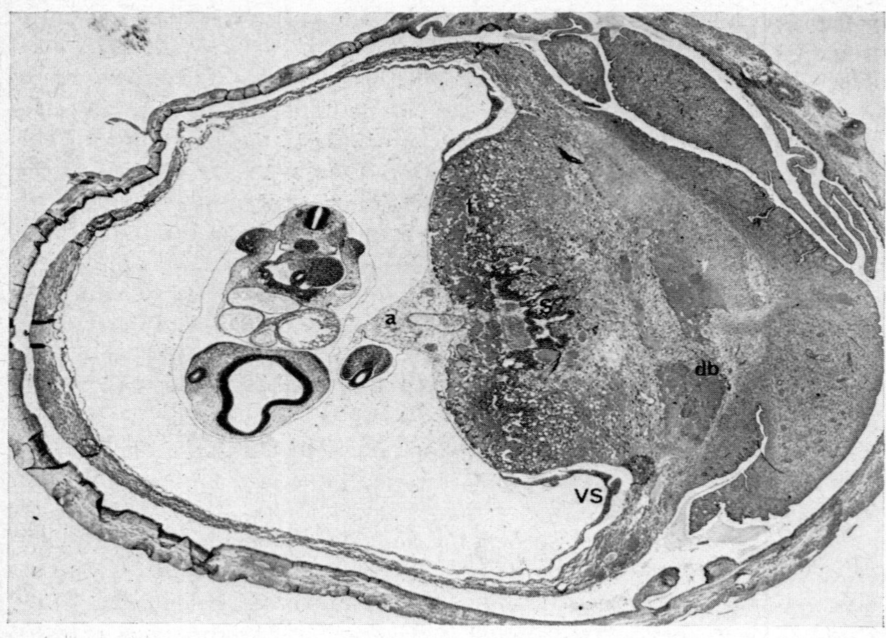

FIG. 7. Cuis (*Galea musteloides*) conceptus at 25 days *p.c.* The chorio-allantoic placenta is formed by growth of the allantoic mesenchyme (a) across the exocoelomic cavity until it establishes contact with the chorion. S = subplacenta; VS = villous portion of yolk sac; db = decidua basalis; t = trophospongium (5 μm, H & E, × 55).

TABLE VI

The time of formation of the chorio-allantoic placenta

Species	Time of formation (days *p.c.*)
Cuis	17–20
Casiragua	15–20
Guinea-pig	15–17
Degu	30–35
Chinchilla	30
Coypu	32–39
Plains viscacha	35–47

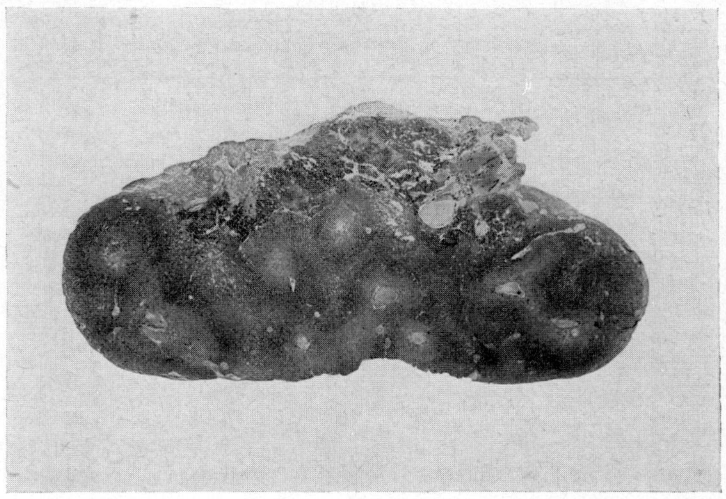

Fig. 8. Definitive placenta of a casiragua (*Proechimys guairae*) at 55 days *p.c.* showing its lobate structure. The trophospongium forms an investing layer around the labyrinthine zones and, from its inner surface, thin syncytial tubules carrying maternal blood pierce the labyrinth (5 μm, H & E, $\times$ 40).

and cellular residue which is strongly PAS positive. Extensive undercutting of the placental disc occurs in the later stages of development, and the mature disc is attached to the uterine wall only by a narrow band of tissue. Remains of this pedicle may frequently be observed in postpartum uteri.

The chorio-allantoic placenta of each of the hystricomorphs studied attains definitive size and differentiation after approximately 70–80% of gestation. Data on the average weights and diameters of the full-term placentae of some hystricomorphs are given in Table VII.

Compared with earlier stages of development the allantoic vessels and trophoblastic tubules of the labyrinthine zones of the definitive placenta follow less tortuous pathways, although there may be some more anastomoses between the fetal endothelial channels. Towards term, however, the trophoblastic tubules of the labyrinth are not readily distinguishable and the possible development of an haemo-endothelial condition cannot be ruled out.

A report by Contreras (1965) on the occurrence of two chorio-allantoic placental discs in association with each embryo in the tuco-tuco was confirmed in two animals of 80–90 days gestation by Roberts (1973). Examination of the 11 conceptuses in these specimens revealed in each the presence of one apparently normal placenta together with

TABLE VII

Average weights and diameters of full-term placentae of some hystricomorphs

Species	Placental weight (g)	Placental diameter (mm)	Body weight (kg)
Cuis	9·5	18	0·3–0·7
Guinea-pig	9·5	20	0·5–0·8
Casiragua	8·5	16	0·2–0·4
Degu	7·9	13	0·2–0·3
Chinchilla	9·0	22	0·4–0·7
Coypu	12·2	28	3·0–7·0
Plains viscacha	10·6	26	3·0–8·0
Tuco-tuco	6·7	15	0·1–0·2

another slightly smaller disc joined to it by a number of prominent blood vessels. Lack of material has prevented further study of this remarkable condition (Wise, Weir, Hime & Forrest, 1972).

The sub-placenta

The sub-placenta of the guinea-pig was studied extensively by Davies, Dempsey & Amoroso (1961a,b) and the findings of Roberts (1973) suggest that it is at least characteristic of the caviomorph rodents. Its development is marked by an increase in the thickness and convolution of the cytotrophoblastic lamellae forming the floor of the concavity that Duval termed the 'central excavation'. Extensive folding occurs at an early stage in the development of the viscacha sub-placenta, and is quite marked in later stages in the cuis, chinchilla and coypu. The adjacent syncytium gives rise to invasive outgrowths which penetrate deep into the decidua. Towards the end of pregnancy the decline of the sub-placenta becomes evident from the reduction in thickness of its cytotrophoblastic lining and the absence of mitotic figures. The sub-placenta is still evident at term and degeneration is rarely complete.

The occurrence of PAS-positive material within the sub-placenta was described by Davies *et al.* (1961a), who suggest that this structure might perform some gonadotrophic function or be implicated in the absorption of materials from the decidua. The presence of PAS-positive globules in the vacuoles of the coarse syncytium and between the trophoblastic giant cells adjacent to the sub-placenta was described in the cuis, casiragua, chinchilla and viscacha by Roberts (1973). These findings are in close agreement with those of Davies *et al.* (1961a) for the guinea-pig. In view of these results, it seems likely that the sub-placenta of the hystricomorphs is involved in the transfer of nutrients, especially those of high molecular weight such as polysaccharides and proteins, from the decidua to the fetal tissues. The invasiveness of the syncytium converting the decidua to 'uterine milk', containing much PAS-positive material, suggests that the material may be absorbed by the syncytium and carried to the chorionic giant cells. These cells are known to be phagocytic and contain considerable quantities of PAS-positive material. Because the hystricomorphs are distinguished by the presence of a sub-placenta, and by the production of a specific progesterone-binding globulin (PBG) during pregnancy (see Heap & Illingworth, 1974), it is tempting to seek the source of the PBG in the sub-placenta. In this case, however, it is difficult to explain the absence of protein from the fetal blood.

The yolk sac placenta

During the early post-implantation development of the hystricomorphs (see p. 341) a parietal yolk sac is not developed. Following disintegration of the parietal trophoblast, the visceral yolk sac endoderm becomes directly apposed to the uterine tissues. The yolk sac is then lined later with splanchnic mesoderm which becomes vascularized. Close to the junction of the yolk sac and chorion, considerable folding of the yolk sac occurs which gives it a villous appearance (see Fig. 7). These villi, however, never become fused with the placental disc although they are often closely associated with it. The sinus terminalis is situated near the umbilical stalk. The yolk sac is highly folded and extremely vascular near the margins of the placental disc.

A non-vascularized portion of the yolk sac covers the peripheral fetal aspect of the chorio-allantoic placenta, and continues mesometrially over the area of placental undercutting where it is frequently thickened and tufted. More rapid growth of the embryo in the later stages of pregnancy causes the disintegration of the decidua capsularis, and the yolk sac comes directly into contact with the uterine lumen.

Amnion

The amnion is formed by cavitation in all hystricomorphs so far studied and throughout gestation it appears to be avascular. Reports by Tibbitts & Hillemann (1959) of primordial blood cells in hollow vesicles of the somatic mesoderm of the chinchilla amnion could not be confirmed. In the later stages of pregnancy, a strong PAS-positive reaction was recorded in the amniotic ectoderm. Many white rounded pustules begin to appear on the surface of the amnion in the viscacha at about 90 days *p.c.* (Fig. 9). Histologically, they consist of a series of coiled layers of loose fibrous tissue, with small, scattered, deeply-staining nuclei, and are particularly prominent on that part of the amnion which is closely

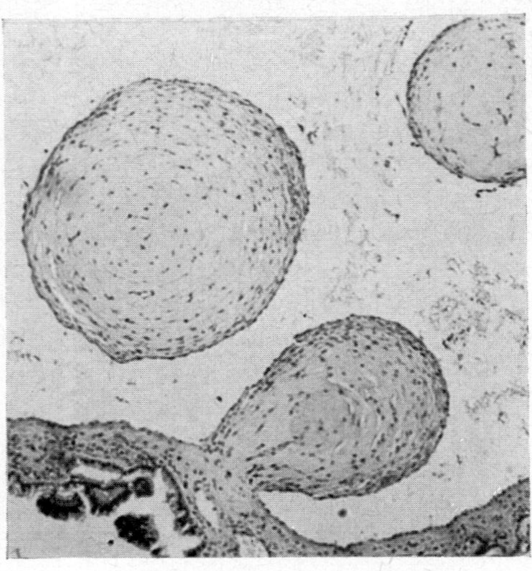

FIG. 9. Section through 'amniotic pustules' found in a plains viscacha (*Lagostomus maximus*) at 90 days *p.c.* (5 μm, H & E, $\times$ 120).

apposed to the fetal surface of the chorio-allantoic placenta. No vascular elements are present within the pustules and their function is, as yet, unknown.

FETAL GROWTH

Fetal growth rates

Fetal weight and crown rump length have been recorded for the following species at various stages of gestation; cuis, casiragua, degu, viscacha, chinchilla (Weir, 1970) and coypu (Newson, 1966). Regression lines of the cube root of fetal weight on fetal age have been plotted according to the formula of Huggett & Widdas (1951):

$$\sqrt[3]{W} = a(t - t_0)$$

where W = fetal weight, a = specific fetal growth velocity, t = gestation length in days and t_0 = the calculated intercept on the age axis. The value of t_0 is calculated using the following approximations quoted by Huggett & Widdas (1951). If $t = 50$–100 days, $t_0 = t \times 0.3$ and if $t = 100$–400 days, $t_0 = t \times 0.2$. Thus the theoretical value of a may be calculated using the cube root of birth weight and the total gestation length for t (Table VIII). The actual specific fetal growth velocities of these species are plotted in Fig. 10. It is clear that the fetal

TABLE VIII

*Comparison of theoretical and observed rates of fetal growth
in hystricomorphs*

Species	Theoretical			Observed	
	$\sqrt[3]{W_b}$	t_0	a	t_0	a
Cuis	3·405	16	0·0896	12	0·0718
Casiragua	2·466	19	0·0560	22	0·0765
Degu	2·422	27	0·0384	38	0·0483
Chinchilla	3·503	22	0·0407	30	0·0436
Plains viscacha	5·809	31	0·0472	50	0·0641
Coypu*	6·400	26	0·0603	40	0·0695
Guinea-pig†	4·500	19	0·0978	16	0·0900

* see Newson (1966);
† see Huggett & Widdas (1951).

growth rates of these hystricomorphs are extremely slow compared with
those of other mammals of similar birthweight (e.g. rat, 0·17; rabbit
0·20). The observed intercepts on the age axis agree well with the time
at which embryo development becomes more marked (Roberts, 1973).
Up to that time, histological observations indicate that much of the
development of the conceptus concerns more especially the establishment of the placenta. The formula of Huggett & Widdas (1951) is,

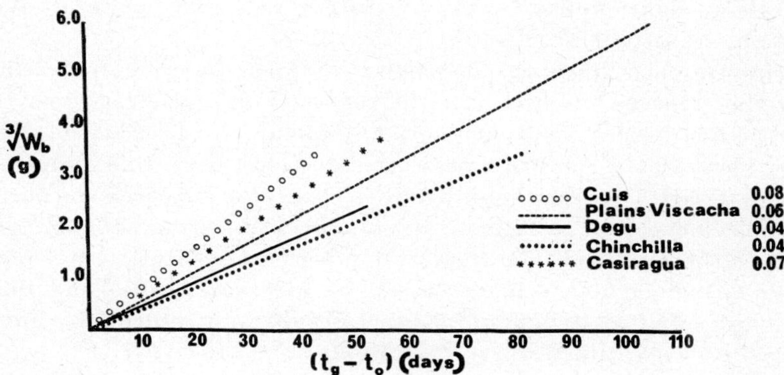

FIG. 10. Specific fetal growth rates of some hystricomorph rodents, plotted according to the method of Huggett & Widdas (1951) (see text). $\sqrt[3]{W_b}$ = cube root of fetal
weight t_g = total gestation length, t_0 = calculated intercept on the age axis.

therefore, inapplicable to hystricomorphs in so far as it makes no allowances for the slow early growth of the embryo immediately after implantation. For this reason the theoretical intercept on the age axis falls earlier than the observed intercept in each species except the cuis. The greatest deviation from the Huggett & Widdas formula is found in the viscacha in which implantation is delayed until Day 18 *p.c.* (Roberts & Weir, 1973) and gestation lasts for 154 days (Weir, 1971a). It is interesting that a fetal growth velocity of 0·0614 has been suggested for the grasscutter, *Thryonomys swinderianus*, (Asibey, 1974) and this is the first recorded estimate for an African member of this suborder.

The slow rate of growth of the hystricomorphs cannot be correlated with the precocious appearance of the young at birth in all of the species studied. Although the newborn of the chinchilla, viscacha, cuis, coypu and casiragua resemble small adults, those of the degu and tuco-tuco are not fully furred and their eyes are closed, after gestation periods of 90 and 120 days respectively (Weir, 1974). Thus it appears that the absolute fetal growth rates of the degu and tuco-tuco are even slower than those of other hystricomorphs.

Fetal resorption

Fetal resorption occurs regularly in the viscacha, frequently in the chinchilla (Weir, 1967) and coypu (Newson, 1966), and occasionally in the casiragua (Roberts, 1973).

Plains viscacha (Table IX)

Only about 10% of the viscacha eggs are fertilized of which a minority (about eight) implant and only two fetuses come to term (Weir, 1971a,b; Roberts & Weir, 1973).

Resorption is universal and begins between Days 26 and 35 *p.c.*; only the embryos situated near the cervical end of each horn survive (i.e. the first blastocyst in each horn to implant). No histological differences between the cervical and ovarian ends of the uterine horns have been detected. The blood supply to each conceptus appears to be normal and no evidence of it being more favourable to the cervical regions of the uterus has been found. The greatest number of implantation sites recorded in any single uterine horn was six, but this was exceptional. Usually there are only up to four implantation sites and resorption cannot, therefore, be explained by overcrowding.

Chinchilla

In the chinchilla, resorption may take place at any stage of pregnancy. Even in the later stages of gestation when the skeletal tissue of the fetus

TABLE IX

Distribution of live and resorbing plains viscacha embryos in utero
(*including data from Weir, 1971b*)

Stage of gestation (days)	Embryos				Total
	Resorbing		Live		
	Left	Right	Left	Right	
26			3	3	6
28			3	2	5
35	1	4	1	2	8
35	1	1	2	1	5
47	2	4	1		7
67	1	3	1	1	6
71	1	1	1	1	4
85	1			1	2
90	1	2	1	1	5
104		1	1	1	3
106			1	1	2
110			1	1	2
113			1	1	2
115			1	1	2
133			1	1	2
140				1	1
153			1	1	2

has formed, resorption rather than mummification or spontaneous abortion occurs. These sites contain only a necrotic mass in which neither placental nor fetal tissue can be recognized. There is usually a central blood-filled cavity.

DISCUSSION

Recent work on several genera of South American hystricomorphs has shown that they share a number of developmental features. There is, as yet, little information concerning the placentation, embryology, or reproductive physiology of any of the African genera and the number of 'unknown' genera far exceeds the 'known'. As more information of this kind becomes available, it will be possible to determine the extent to which these features coincide with formal classifications which have been based on various other characteristics. The problematic position of the African gundis (F. Ctenodactylidae) may perhaps be resolved in this way;

the fact that *Ctenodactylus gundi* is described as having superficial implantation and amniogenesis by folding (de Lange, 1938) suggests that it should be included with the Sciuromorpha rather than the Hystricomorpha. However, "about all that can be said is that the ctenodactylids are not sciuromorphs" (Simpson, 1945). The same argument applies with more force to the genus *Pedetes*, the African spring haas, the fetal membranes and taxonomic significance of which have been described and discussed by Fischer & Mossman (1969). The close taxonomic affinities within the South American hystricomorphs are attested by their similar modes of implantation, their extreme form of entypy and the definitive form of their placenta and its characteristic sub-placental region. The loss of the parietal trophoblast during implantation suggests that the Hystricomorpha should be regarded as a highly specialized group among the Rodentia.

Within the Rodentia, various types of implantation occur, ranging from superficial in the more primitive Sciuridae, through eccentric in the Geomyidae and Heteromyidae to interstitial in the Muridae. The extreme condition of the complete interstitial implantation is found only in the hystricomorph rodents. This is a significant specialization since it appears to have a direct bearing on other early developmental events. According to Mossman (1937), the more intimate the relationship between uterine tissues and blastocysts at implantation the more likely the cavitate form of amniogenesis—all hystricomorphs undergo amniogenesis by cavitation. The phenomenon of inversion of the germ layers is also associated with the type of implantation. In the Rodentia, there is a trend towards earlier achievement of inversion. Thus, in some sciuromorphs the bilaminar omphalopleure persists to term, whereas in the hystricomorphs it is never formed, except transiently in the Canadian porcupine (Perrotta, 1959) and the casiragua (Roberts, 1973), and the yolk sac endoderm is exposed to the decidual cavity shortly after implantation.

No trace of an endodermal allantoic vesicle has been found in any of the hystricomorphs examined so far. There appears to be a trend in the evolution of the Rodentia for the reduction of the primitive, medium-sized allantoic vesicle to a purely mesodermal allantois which becomes associated with the chorion (Mossman, 1937). The hystricomorphs, therefore, exemplify the extreme reduction. The specialized lobulate, labyrinthine placenta of the hystricomorphs results in a reduction of the amount of tissue forming the barrier between fetal and maternal tissues. All the species studied so far possess a sub-placenta and no comparable structure has been described in any other rodents, or indeed, any other mammals, although the term has been applied to a different structure elsewhere (see Davies *et al.*, 1961a). Mention has been made of functions that have been tentatively ascribed to the sub-placenta (p. 348) but it must be remembered that little progress has so far been made in defining the function of the different components of the placenta of any species, or in explaining the striking variety that exists among the placental structures by which mammals carry out the essentially similar functions involved in gestation.

A consideration of the specialized features of development in the hystricomorphs must include an attempt to correlate them with gestation length. A trend towards increased efficiency of physiological exchange between the fetal and maternal systems is indicated and a reduction of the length of gestation might be expected. The hystricomorphs, however, have extremely long gestation periods compared with other rodents (see Weir, 1974: Table I). The longest known gestation period in myomorph rodents is 43–44 days in *Mesembriomys gouldii* (Crichton, 1969) whilst in the hystricomorphs the shortest pregnancy is 52 days in the cuis (Rood & Weir, 1970). There are no obvious correlates of the long gestation of the hystricomorphs (Weir, 1974) and it is possible that physiological studies, particularly those relating to the maintenance of pregnancy, will be valuable.

Most of our knowledge of the hystricomorphs is confined to the South American species, the caviomorphs. The phylogenetic relationships of these forms to the African hystricomorphs are not clear (see Lavocat, 1974; Wood, 1974) but several taxonomists have remarked upon their similarities. The embryological details of the caviomorphs discussed in this paper indicate that they are a closely related group, but little information of this kind is available for the African species. Weir (1974) has shown that the New World and Old World species have several common reproductive characteristics, including the long gestation length. It is our belief that investigations of the embryology of Old World species will show that they also have a slow fetal growth rate, complete interstitial implantation, amniogenesis by cavitation, and a haemochorial labyrinthine discoidal placenta with a sub-placenta.

ACKNOWLEDGMENTS

We thank Dr I. W. Rowlands and Dr B. J. Weir for their encouragement and the provision of facilities and material to one of us (C.M.R.) whose laboratory work was supported by the Ford Foundation.

REFERENCES

Amoroso, E. C. (1952). Placentation. In *Marshall's Physiology of reproduction* **2**: 127–311. Parkes, A. S. (Ed.). London: Longmans Green & Co.

Asibey, E. O. A. (1974). Reproduction in the grasscutter (*Thryonomys swinderianus* Temminck) in Ghana. *Symp. zool. Soc. Lond.* No. 34: 251–263.

Austin, C. R. (1961). *The mammalian egg.* Oxford: Blackwell.

Becher, H. (1921a). Die Entwicklung des Mesoplacentariums und die Placenta bei Aguti (*Dasyprocta azarae* Schl.). *Z. Anat. EntwGesch.* **61**: 337–364.

Becher, H. (1921b). Der feinere Bau der reifen Placenta von Aguti (*Dasyprocta azarae* Schl.). *Z. Anat. EntwGesch.* **61**: 439–454.

Billington, W. D. & Weir, B. J. (1967). Deportation of trophoblast in the chinchilla. *J. Reprod. Fert.* **13**: 593–595.

Blandau, R. J. (1949a). Observations on implantation of the guinea-pig ovum. *Anat. Rec.* **103**: 19–48.

Blandau, R. J. (1949b). Embryo-endometrial relationships in the rat and guinea-pig. *Anat. Rec.* **104**: 331–359.

Blandau, R. J. (1961). Biology of eggs and implantation. In *Sex and internal secretions*: 797–882. 3rd ed. Young, W. C. (Ed.). Baltimore: Williams & Wilkins.

Blandau, R. J. & Young, W. C. (1939). The effects of delayed fertilization on the development of the guinea-pig ovum. *Am. J. Anat.* **64**: 303–329.

Boyd, J. D. & Hamilton, W. J. (1952). Cleavage, early development and implantation of the egg. In *Marshall's Physiology of reproduction* **2**: 1–126. Parkes, A. S. (Ed.). London: Longmans Green & Co.

Christie, G. A. (1967). Implantation of the rat embryo; further histochemical observations on carbohydrate, RNA and lipid metabolic pathways. *J. Reprod. Fert.* **13**: 281–296.

Contreras, J. (1965). Datos a cerca de la activadad reproductora del genero *Ctenomys* (Rodentia, Octodontidae). *Physis, B. Aires* **25**: 169–186.

Creighton, C. (1878). The formation of the placenta in the guinea-pig. *J. Anat. Physiol.* **12**: 534–590.

Crichton, E. G. (1969). Reproduction in the pseudomyine rodent *Mesembriomys gouldii* (Gray) Muridae. *Aust. J. Zool.* **17**: 785–797.

Crooks, J. H. (1973). *Aspects of reproduction in the Steppe lemming and chinchilla.* Thesis, Univ. London.

Davies, J., Dempsey, E. W. & Amoroso, E. C. (1961a). The sub-placenta of the guinea-pig. *J. Anat.* **95**: 457–473.

Davies, J., Dempsey, E. W. & Amoroso, E. C. (1961b). The sub-placenta of the guinea-pig: an electron microscope study. *J. Anat.* **95**: 311—324.

de Lange, D. (1938). Some remarks on the early development of *Ctenodactylus gundi* Pall. *Archs néerl. Zool.* **3**: 131–147.

Dempsey, E. W. & Wislocki, G. B. (1947). Further observations on the distribution of phosphatases in mammalian placentas. *Am. J. Anat.* **80**: 1–33.

Duval, M. (1892). La placenta des rongeurs. Le placenta du cochon d'Inde. *J. Anat. Physiol.* **28**: 58–98.

Enders, A. (1965). A comparative study of the fine structure of the trophoblast in several haemochorial placentas. *Am. J. Anat.* **116**: 29–68.

Fischer, T. V. & Mossman, H. W. (1969). The fetal membranes of *Pedetes capensis* and their taxonomic significance. *Am. J. Anat.* **124**: 89–116.

Hard, W. L. (1946). A histochemical and quantitative study of phosphatase in the placenta and fetal membranes of the guinea-pig. *Am. J. Anat.* **78**: 47–77.

Harman, M. T. & Prickett, M. (1932). The development of the external form of the guinea-pig (*Cavia cobaya*) between the ages of 11 days and 20 days of gestation. *Am. J. Anat.* **49**: 351–378.

Heap, R. B. & Illingworth, D. V. (1974). The maintenance of gestation in the guinea-pig and other hystricomorph rodents; changes in the dynamics of progesterone metabolism and the occurrence of progesterone-binding globulin (PBG). *Symp. zool. Soc. Lond.* No. 34: 385–415.

Hillemann, H. H. & Gaynor, A. I. (1961). The definitive architecture of the placenta of nutria, *Myocastor coypus* (Molina). *Am. J. Anat.* **109**: 299–318.

Hillemann, H. H., Tibbitts, F. D. & Gaynor, A. I. (1959). *Reproductive biology in chinchilla.* National Chinchilla Breeders of America Inc.

Huggett, A. St. G. & Widdas, W. F. (1951). The relationship between mammalian foetal weight and conception age. *J. Physiol., Lond.* **114**: 306–317.

Lavocat, R. (1974). What is an hystricomorph? *Symp. zool. Soc. Lond.* No. 34: 7–20.

McLaren, N. (1926). Development of *Cavia*: implantation. *Trans. R. Soc. Edinb.* **55**: 115–123.

Mossman, H. W. (1937). The comparative morphogenesis of the fetal membranes and accessory uterine structures. *Contrib. Embryol.* **26**: 129–246.

Mossman, H. W. (1967). The principal interchange vessels of the chorio-allantoic placenta of mammals. In *Organogenesis*: 771–786. de Haan, R. L. & Urspring, H. (Eds.). New York: Holt.

Newson, R. M. (1966). Reproduction in the feral coypu (*Myocastor coypus*). *Symp. zool. Soc. Lond.* No. 15: 323–334.

Perrotta, C. A. (1959). Fetal membranes of the Canadian porcupine, *Erethizon dorsatum. Am. J. Anat.* **104**: 35–59.

Roberts, C. M. (1973). *The embryology of certain hystricomorph rodents.* Ph.D. Thesis, University of London.

Roberts, C. M. & Weir, B. J. (1973). Implantation in the plains viscacha, *Lagostomus maximus. J. Reprod. Fert.* **33**: 299–307.

Rood, J. P. & Weir, B. J. (1970). Reproduction in female wild guinea-pigs. *J. Reprod. Fert.* **23**: 393–409.

Rowlands, I. W. (1957). Insemination of the guinea-pig by intraperitoneal injection. *J. Endocr.* **16**: 98–106.

Sansom, G. S. & Hill, J. P. (1931). Observations on the structure and mode of implantation of the blastocyst of *Cavia. Trans. zool. Soc. Lond.* **21**: 295–354.

Simpson, G. G. (1945). The principles of classification and a classification of mammals. *Bull. Am. Mus. nat. Hist.* **85**: 1–350.

Squier, R. R. (1932). The living egg and early stages of its development in the guinea-pig. *Contrib. Embryol.* **23**: 223–250.

Strahl, H. (1905). Eine Placenta mit einem Mesoplacentarium. *Anat. Anz.* **26**: 524–528.

Tibbitts, F. D. & Hillemann, H. H. (1959). The development and histology of the chinchilla placenta. *J. Morph.* **105**: 317–365.

Tripp, H. R. H. (1971). Reproduction in elephant shrews (Macroscelididae) with special reference to ovulation and implantation. *J. Reprod. Fert.* **26**: 149–159.

van der Horst, C. J. & Gillman, J. (1941). The number of eggs and surviving embryos in *Elephantulus. Anat. Rec.* **86**: 443–445.

von Spee, F. (1883). Beitrag zur Entwicklungsgeschichte der früheren Stadien des Meerschweinchens bis zur Vollendung der Keimblase. *Arch. Anat. Physiol.* **7**: 44–60.

Weir, B. J. (1967). *Aspects of reproduction in some hystricomorph rodents.* Ph.D. Thesis, University of Cambridge.

Weir, B. J. (1970). Chinchilla. In *Reproduction and breeding techniques for laboratory animals*: 209–223. Hafez, E. S. E. (ed.). Philadelphia: Lea & Febiger.

Weir, B. J. (1971a). The reproductive organs of the female plains viscacha, *Lagostomus maximus*. *J. Reprod. Fert.* **25**: 365–373.

Weir, B. J. (1971b). The reproductive physiology of the plains viscacha, *Lagostomus maximus*. *J. Reprod. Fert.* **25**: 355–363.

Weir, B. J. (1971c). Some observations on reproduction in the female agouti, *Dasyprocta aguti*. *J. Reprod. Fert.* **24**: 203–211.

Weir, B. J. (1974). Reproductive characteristics of hystricomorph rodents. *Symp. zool. Soc. Lond.* No. 34: 265–301.

Wise, P. H., Weir, B. J., Hime, J. M. & Forrest, E. (1972). The diabetic syndrome in the tuco-tuco (*Ctenomys talarum*). *Diabetologia* **8**: 165–172.

Wislocki, G. B., Deane, H. W. & Dempsey, E. W. (1946). Histochemistry of the rodent placenta. *Am. J. Anat.* **77**: 281–290.

Wislocki, G. B. & Dempsey, E. W. (1945). Histochemical reactions of the endometrium in pregnancy. *Am. J. Anat.* **77**: 365–378.

Wood, A. E. (1974). The evolution of the Old World and New World hystricomorphs. *Symp. zool. Soc. Lond.* No. 34: 21–54.

Wynn, R. M. (1964). Ultrastructure of the deciduo-trophoblastic junction of the guinea-pig. *Am. J. Obstet. Gynec.* **90**: 690–693.

Wynn, R. M. (1967). Comparative aspects of placental junctional zones. *Obstet. Gynec., N.Y.* **29**: 644–661.

DISCUSSION

SEATON (*London*): When I looked at the line you drew for the cube of the body-weight as a function of time, I noted a distinct hint of a curve and wondered if you have ever tried fitting an exponential to the data rather than a straight line? What I am suggesting is that the growth rather than being linear may be exponential. It only appears linear because you don't have enough points in that bottom region to indicate whether it is curved or straight.

ROBERTS (*London*): This is true, but the difficulty arises with the low points in early gestation because the embryos are so small it is just not possible to measure them.

PERRY (*Cambridge*): This formula is merely a device for estimating the gestation age from post-implantation stages of pregnancy, for which you do get a straight line. If you could do the bottom end of the curve, of course, you would see that it is, indeed, an exponential. Huggett & Widdas (1951)* showed that the t_0 intersect held fairly well for a certain range of gestation lengths so that you could put intermediate points on your curve and you would be able to give a reasonable estimate of the age of the fetus you were dealing with. It is a purely practical, empirical thing to use but, if you use it for a guinea-

---

* Huggett, A. St. G. & Widdas, W. F. (1951). *J. Physiol., Lond.* **114**: 306–317.

pig, or any other hystricomorph, you come unstuck using the theoretical levels of this intersect, simply because, instead of their growth being slow merely until implantation and a little after, it goes on being slow for quite a long time. But that in itself does not explain the very long gestation period; it is a combination of the time it takes for the embryo to get going and its subsequent slow growth.

ROBERTS: I have also plotted crown-rump lengths and you get more or less the same intersect. I have no doubt that the figures are correct and that the early growth, particularly the immediate post-implantation growth, of the embryo is very slow.

SEATON: I see. Can I just make one other point which I think is valid. You say that there are no points at the very bottom end. There is one very important point which often gets forgotten and that is the zero-zero point.

PERRY: The zero-zero point is very difficult to estimate in many of these animals, as Dr. Asibey said this morning.

ASIBEY (*Ghana*): I thought I had large enough samples of different stages of development to get some figures at the lower end of the scale but below a certain age the embryo is so tiny that you destroy the whole thing if you try to dissect it.

PERRY: As you saw from some of the last slides that Dr Roberts showed, we are not talking about grams of tissue, we are talking about a few hundred cells.

ASIBEY: That is right. Then you also find that from there on it seems to grow very fast. If my assumptions that birth of a 65 g fetus can occur are correct then it appears that, in the grasscutter, about one-third of gestation is spent at about 1 g and the rest of the growth occurs in the other two-thirds.

FRAZER (*London*): You found that, in the plains viscacha, implantation was earliest in the cervical ends of the uterine horns. In quite a different rodent, in the ordinary domestic rat, it was shown about 20 years ago (Frazer, 1955)* that when fewer than five eggs implanted they were usually at the cervical end of the horn. Has this part of the uterine horn some particular physiological property, or is it something to do with the blood supply at that end as compared with higher up the horn?

ROBERTS: Dr Weir has done some preliminary experiments in which the blood vessels at the cervical end of the horn were ligated. Implantation occurred normally just above the region where the vessels were ligated.

---

* Frazer, J. F. D. (1955). Foetal death in the rat. *J. Embryol. exp. Morphol.* **3**: 332–334.

Asibey: Are there any preferences of implantation in the left or right horns?

Roberts: In the viscacha? No, there appears to be no difference at all.

Lavocat (*Montpellier*): If you had no previous opinions about the possible primitive relationships between the different groups of rodents, could you deduce what is primitive by comparison of the embryology?

Roberts: I think there are generally accepted views by rodent embryologists that the Sciuromorpha represent the primitive condition. For instance, their implantation is superficial, that is, the embryo expands, it becomes quite large, it is visible to the naked eye and it does not embed itself in the uterine wall. You get trophoblastic processes which invade the uterine epithelium and the blastocyst expands in the lumen. The condition midway between sciuromorphs and our hystricomorphs is shown by the myomorphs which have eccentric or partly interstitial implantation whereby the embryo sits in a little cleft which then closes off. But the hystricomorphs actively burrow their way through the uterine epithelium.

Lavocat: And this condition of the hystricomorphs, could it have been evolved from the condition found in sciuromorphs or is it a different line of evolution from a common stem?

Roberts: Well, I would have thought there had been a progression along an evolutionary line such as this, but it is, of course, hypothetical.

CHAIRMAN'S INTRODUCTION: ENDOCRINOLOGY

I. W. ROWLANDS

Wellcome Institute of Comparative Physiology,
Zoological Society of London,
Regent's Park, London, England

A session of this title is a logical extension of this morning's meeting on the reproductive physiology of hystricomorph rodents. This afternoon before tea we are to listen to two papers on the hormonal mechanisms involved in the maintenance of the long gestation period characteristic of these animals, to which ample reference has already been made in this Symposium. The first paper, in particular, forms part of the original programme of endocrinological research at the Wellcome Institute, and both serve to complement our knowledge of ovarian structure that was described this morning. In recent years the guinea-pig has been used extensively to study these matters and the results form an important baseline for comparative purposes. However, it should not be overlooked that *Cavia porcellus*, the domestic guinea-pig, is a highly inbred species that no longer exists in the wild, so that its physiological mechanisms may have become modified by centuries of domestication.

The work described in the paper which is to be given after tea is probably one of the best examples we have to offer of an investigation of an event that arose quite spontaneously and unexpectedly. It is a study of considerable interest to chemists and endocrinologists in the pursuit of knowledge, and the findings could have far-reaching medical importance. This study of the chemical nature of hystricomorph insulin was entirely unforeseen, and was the direct outcome of the appearance of symptoms typical of diabetes mellitus and of infertility which arose during an attempt to establish a breeding colony of the tuco-tuco, *Ctenomys*, for the first time in captivity. Such serendipity is what one anticipates but rarely experiences in scientific work, but is surely more likely in studies involving 'new' species.

Symp. zool. Soc. Lond. (1974) No. 34, 363–384

THE SYNTHESIS OF PROGESTERONE
IN SOME HYSTRICOMORPH RODENTS

W. H. TAM*

*Wellcome Institute of Comparative Physiology
The Zoological Society of London,
Regent's Park, London, England
and
Department of Zoology, University of Western
Ontario, London, Ontario, Canada*

SYNOPSIS

The hystricomorph rodents show a number of structural and functional modifications which enhance the synthesis of progesterone for the maintenance of gestation. In the guinea-pig and the cuis, the placenta is an extra-ovarian site of progesterone synthesis, and in the chinchilla and the green acouchi, in which the ovary is probably the only source of progesterone, this organ is complex and characterized by an abundance of interstitial cells and the occurrence of numerous accessory corpora lutea (CL). In this investigation, the arterial levels of progesterone have been determined during gestation in the cuis and the chinchilla. The progesterone content of the ovary and its synthesis and metabolism *in vitro* by ovarian and placental tissues have also been studied to determine quantitatively the functional significance of these tissues in the maintenance of gestation. It was found that the progesterone content of the accessory CL of the green acouchi and chinchilla, and therefore probably secretory activity, increased progressively with gestation. This finding was also supported by the high *in vitro* rate of progesterone synthesis by the accessory CL of the chinchilla. The progesterone content and the rate of progesterone synthesis *in vitro* by the interstitial tissue of the chinchilla was also high relative to those of the two luteal tissues. The chinchilla luteal tissues were not capable of 17-hydroxylation, but the interstitial tissue produced a full complement of C_{21} and C_{19} metabolites and oestradiol-17β via the 4- and 5-unsaturated pathways. During pregnancy the 5-unsaturated pathway was demonstrated to be more active than the 4-unsaturated pathway so that androstenedione was mainly produced without the intermediary progesterone. This probably indicates the occurrence of a regulatory mechanism in the interstitium whereby either progesterone or oestrogen could be synthesized independently. Work on the guinea-pig yolk-sac and placenta *in vitro* revealed that the giant cells were not involved in steroid metabolism, that the yolk-sac was capable of synthesizing a small amount of progesterone throughout pregnancy, and that the syncytiotrophoblast produced large quantities of progesterone, showing a peak of activity at 35 days of gestation when the CL began to show signs of degeneration.

INTRODUCTION

Many hystricomorph rodents show a number of structural and functional modifications in the ovary and placenta to enhance the synthesis of progesterone for the maintenance of gestation. Thus, in the chinchilla

* Present address.

(*Chinchilla laniger*), the ovary is probably the only important source of progesterone as ovariectomy invariably resulted in abortion (B. J. Weir, pers. comm.). This organ is complex, contains much interstitial tissue, and possesses numerous primary (CL) and accessory (ACL) corpora lutea during pregnancy (Weir, 1967a). The ovary of the green acouchi (*Myoprocta pratti*) also possesses large numbers of ACL during gestation (Rowlands *et al.*, 1970), but the endocrine role of the placenta of this species is unknown. The ACL in these two species are formed from un-ovulated, luteinized follicles, and it is reasonable to suppose that they are capable of progesterone synthesis and that in the chinchilla, at least, the interstitial tissue also secretes progesterone. In the guinea-pig (*Cavia porcellus*) and the cuis (*Galea musteloides*), the placenta is believed to be an extra-ovarian site of progesterone synthesis during the later stages of gestation (Loeb, 1923; B. J. Weir, pers. comm.). Such adaptations of the ovary and placenta are not restricted to hystricomorph rodents. The ovary of the mare is known to possess secondary CL as the result of ovulation during pregnancy (Cole *et al.*, 1931; Amoroso *et al.*, 1948). The placentae of a considerable number of species, including man, are also probably sites of progesterone production (see Short, 1961).

In this communication, the plasma levels of progesterone and the progesterone content of ovarian tissues during gestation in some hystricomorph rodents are reported. The synthesis and metabolism of progesterone *in vitro* by the ovarian and placental tissues of the chinchilla and guinea-pig have also been investigated to provide a quantitative assessment of the functional significance of the different types of ovarian and placental tissues in the maintenance of gestation.

MATERIALS AND METHODS

Animals

Most of the animals used were taken from the colonies bred at the Wellcome Institute of Comparative Physiology. The management of the animals and observations on their oestrous cycles and gestation lengths were reported by Weir (1967b; 1970). The guinea-pigs used in the placental studies were obtained from the colony in the Department of Zoology, University of Western Ontario. They were kept at 22–25°C with 14 hours of light per day in winter and in natural lighting conditions during the summer months. Food and water were provided in excess. The oestrous cycle was approximately 16 days and the gestation period 68 days. They were observed through at least 3 cycles before being used for experimental purposes; they were then between 6 and 12 months of age and weighed 800–1 100 g.

Plasma sampling and tissue preparation

Blood was withdrawn from the left ventricle into a heparinized syringe from the animals under ether anaesthesia, and the plasma, separated by centrifugation, was stored at $-17°C$. The dissection of ovaries to give the major types of tissue components was performed using the method described by Rowlands & Short (1959) and the criteria for distinguishing primary and accessory CL were discussed by Tam (1971). Ovarian tissue remaining after dissection of the CL was referred to as 'residual tissue' and its composition, which varied according to the species and reproductive state, is described in the relevant sections below.

When the tissues were intended for experiments *in vitro*, they were stored over ice during collection and incubated separately within minutes after the animal had been killed. Tissues for the determination of steroids were stored in either methanol or ethanol at $-17°C$ until extraction. In the guinea-pig work, the yolk-sac and placenta were immersed in ice-cold $0·9\%$ NaCl solution immediately after removal from the animal, and the placenta was cut into 2 mm slices. Blood was removed with changes of $0·9\%$ NaCl solution. The superficial layers of the placenta were then dissected. In sections, the dissected outer portions of the placenta were seen to contain an outer monolayer of endodermal cells and an inner multi-layer of layered syncytiotrophoblast interspersed with a layer of giant cells. Samples taken near the subplacenta contained some connective tissue. These outer portions of the placenta are referred to as giant cell-rich tissue, and the inner core of the placenta, containing labyrinthine and spongy syncytiotrophoblast, is, for simplicity, called the syncytiotrophoblast.

Determination of steroids in plasma and ovarian tissue

Methods for the extraction and separation of progesterone in plasma and ovarian tissue by thin-layer chromatography (TLC), gas-liquid chromatographic quantitation with electron capture detection, and the calculation of recovery using radioactive internal standard have been described in detail by Tam (1973). The methods used to measure progesterone, 20α- and 20β-hydroxy-4-pregnen-3-one and oestradiol-17β in chinchilla plasma were given by Tam (1971). In the determination of progesterone, 20α- and 20β-hydroxy-4-pregnen-3-one, 18-hydroxy-progesterone, $17\alpha,20\alpha$-dihydroxy-4-pregnen-3-one, androstenedione, testosterone and oestradiol-17β in the ovarian tissues of the chinchilla, 15 000 d/min of $[4\text{-}^{14}C]$progesterone, 15 000 d/min of $[4\text{-}^{14}C]20\beta$-hydroxy-4-pregnen-3-one, 14 000 d/min of $[4\text{-}^{14}C]$androstenedione,

N

19 000 d/min of [4-^{14}C]testosterone and 15 000 d/min of [4-^{14}C]-oestradiol-17β were added to the preserved tissue just before extraction for the calculation of recovery. The recovery of 20α-hydroxy-4-pregnen-3-one was based on that of the 20β-epimer. The amounts of compounds in the radioactive internal standards were always subtracted from the final results. However, the experimental losses of 18-hydroxyprogesterone and 17α,20α-dihydroxy-4-pregnen-3-one were not accounted for as no labelled, authentic compounds were available. The tissue was then homogenized in 0·6 M-sodium hydroxide solution. This solution was then neutralized with acetic acid and the homogenate made up to 25 ml with distilled water. The extraction was then carried out by the method described by Brown (1955) and the steroids divided into the neutral and phenolic fractions. The lipids in the neutral fraction were removed by partition between light petroleum (b.p. 60–80°C) and 70% aqueous methanol (v/v). The neutral fraction was initially separated in the paper chromatographic system, light petroleum (b.p. 60–80°C) : methanol : water (4 : 3 : 1, by vol). The corresponding monochloro-acetates of progesterone (after enzymic 20β-reduction), the 20-hydroxy-4-pregnen-3-ones, androstenedione (after sodium borohydride reduction to testosterone) and testosterone were prepared and purified either in the TLC system, benzene : ethyl acetate (6 : 1, v/v) or in benzene : ethyl acetate (9 : 1, v/v). The 18-hydroxyprogesterone and 17α,20α-dihydroxy-4-pregnen-3-one fractions after the initial paper chromato-graphy were purified in the TLC system; benzene : ethyl acetate (1 : 1, v/v). The 18-hydroxyprogesterone was acetylated and the 17α,20α-dihydroxy-4-pregnen-3-one oxidised to androstenedione with chromic acid and re-developed in the same TLC system. The 18-hydroxypro-gesterone acetate was, however, hydrolysed with sodium carbonate before subsequent gas-liquid chromatography (GLC). The oestradiol-17β in the phenolic fraction was separated by TLC benzene : ethyl acetate (1 : 1, v/v), and the diacetate prepared and purified in the same TLC system before quantitative determination by GLC.

The preparation and GLC of steroid monochloroacetates with electron capture detection have been given in detail before (Tam, 1971). Flame ionization detection was used for the determination of 18-hyd-roxyprogesterone, 17α,20α-dihydroxy-4-pregnen-3-one (as androstene-dione) and oestradiol-17β diacetate. Chromatography was on a 0·42% (w/w) XE-60 column (Chromosorb G, 80–100 mesh). Nitrogen was used as carrier gas at the flow rate of 60 ml/min, and compressed air and hydrogen were supplied to the detector at 500 and 30 ml/min, respective-ly. The detector oven was maintained at 230°C. Column temperature was 225°C for the elution of 18-hydroxyprogesterone and oestradiol-17β

diacetate, and 215°C for androstenedione prepared from the $17\alpha,20\alpha$-dihydroxy-4-pregnen-3-one.

Incubation experiments

The procedure for the kinetic incubation of chinchilla ovarian tissues, using [7α-^{3}H]pregnenolone as substrate, was described by Tam (1972). The 4 h incubations of chinchilla primary CL, accessory CL and residual tissue were carried out in Krebs-Henseleit bicarbonate buffer (pH 7·4) at 37°C and in an atmosphere of 95% oxygen : 5% carbon dioxide. [7α-^{3}H]Pregnenolone (660·0 mCi/mmol) and [4-^{14}C]progesterone (58·5 mCi/mmol), purchased from Radiochemical Centre, Amersham, U.K., were used as substrates. The amounts of tissue (luteal, 1 mg/ml; residual, 2 mg/ml) and the concentrations of the substrates (0·14 μg/ml each) in the incubation medium were the same for all the incubations. The guinea-pig yolk-sac and placental tissues were incubated for only 2 h, but at the same temperature and in the same gaseous environment as the chinchilla ovarian tissues. In all these experiments, the amounts of tissue used were approximately 70 mg of yolk-sac, 200 mg of giant cell-rich tissue and 200 mg of syncytiotrophoblast per incubation, using 10 ml of Krebs-Henseleit buffer and 59·6 ng of [7α-^{3}H]pregnenolone (18·0 Ci/mmol; Amersham/Searle Corporation, Illinois) as substrate. The methods of extraction, separation and purification of steroid metabolites in the incubation medium by chromatography, identification by derivative formation, and the calculation of recovery (with one exception) by the addition of known quantities of carriers were the same as those described earlier (Tam, 1972). The recovery of progesterone in the guinea-pig was calculated by the addition of 22 200 d/min of [4-^{14}C]-progesterone (52·8 mCi/mmol; New England Nuclear, Massachusetts) to the incubation medium prior to extraction, instead of chemical carriers. The isolation and identification of 18-hydroxyprogesterone and $17\alpha,20\alpha$-dihydroxy-4-pregnen-3-one in the chinchilla incubation medium are reported in detail elsewhere (Tam, in preparation).

Determination of radioactivity

In all cases, radioactivity was detected on paper and thin-layer chromatogram with a Panax radiochromatogram scanner. Counting of carbon-14 and tritium was carried out either in a three-channel Packard Tri-Carb or a Nuclear-Chicago Unilux II liquid scintillation spectrometer.

RESULTS

Progesterone levels in plasma

Cuis (cycle 22 days, gestation 53 days)

In the cuis, the occurrence of oestrus is determined only by contact with a male (Rood & Weir, 1970; Weir, 1971), and the non-pregnant cuis used for the studies in this paper were those experiencing oestrus but not conceiving.

The progesterone concentrations in the arterial plasma of non-pregnant and pregnant cuis are given in Fig. 1a. The levels throughout

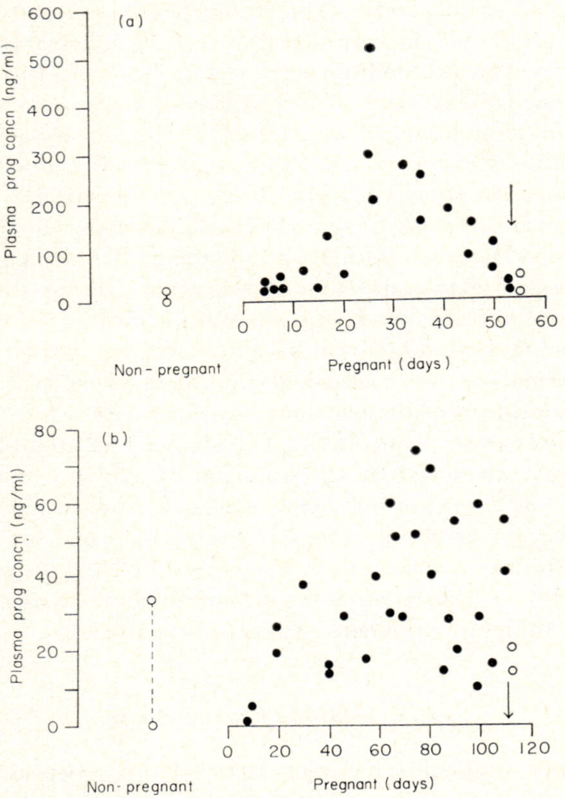

FIG. 1. The arterial levels of progesterone in (a) cuis* and (b) chinchilla. In the non-pregnant animals, only the range of progesterone concentration during the oestrous cycle is indicated. (○ = non-pregnant, ● = pregnant). Arrows indicate the time of parturition. * From Tam (1973).

the cycle and during the first 14 days of gestation are low compared with those found later in gestation; a maximum of 513 ng/ml was recorded on Day 26. From then on the level declined and just before parturition it varied between 50 and 100 ng/ml. It decreased further after parturition. The pattern of changes in progesterone levels during pregnancy is, therefore, one of relatively low concentrations in the first third of gestation, rising to a maximum at mid-term and then a gradual decrease towards parturition.

Chinchilla (cycle 40 days, gestation 111 days)

The changes in plasma progesterone concentration in the pregnant chinchilla (Fig. 1b) are quite different from those in cuis. Though there was an initial period of low levels, peak values (75 ng/ml) were much lower and occurred somewhat later (Days 60–90) than in cuis.

Progesterone content and synthesis by the ovary

Acouchi (cycle 30–55 days, gestation 100 days)

The concentrations and total content of progesterone in the primary CL and residual tissue (accessory CL and stroma: Rowlands, Tam & Kleiman, 1970) are given in Table I. As the acouchi ovary contains comparatively little, if any, interstitial tissue it is reasonable to assume that most, if not all the progesterone in the residual tissue was derived from the accessory CL. The total progesterone content in the accessory CL remained fairly constant from 20–75 days of pregnancy but that of

TABLE I

*Progesterone in primary CL and residual ovarian tissue (stroma and accessory CL) of the acouchi**

Pregnancy (days)	Primary CL		Residue	
	Total† (ng)	Concn. (ng/mg)	Total† (ng)	Concn (ng/mg)
20	740	74·0	254	6·7
29	911	45·6	227	17·9
45	649	37·5	458	12·9
60	142	34·6	272	4·0
75	768	47·1	236	8·2

* After Rowlands *et al.* (1970).
† Total progesterone content in the pair of ovaries.

the primary CL gradually declined during this time, so that as gestation advanced the contribution of the former became a progressively greater proportion of the total amount secreted by the ovary.

Chinchilla

Residual tissue in the chinchilla consisted of about equal amounts of interstitial tissue and stroma and probably included some small follicles and accessory CL (B. J. Weir, pers. comm.). The separate contribution of these tissues to progesterone production could not be determined, but as the interstitial tissue represented the most abundant steroidogenic element, it is assumed that most of the progesterone came from the interstitial cells. The results of these determinations, given in Table II, show that the content and concentration of progesterone in the accessory CL were as high as those in the primary CL throughout the pregnancy, except in the 8-day pregnant chinchilla in which they were low. The total content of progesterone in the residual tissue was considerable compared to both types of CL, but it did not show any appreciable variation.

The amounts of progesterone produced *in vitro* from endogenous and exogenous precursors by the three types of steroidogenic tissue in the ovaries of a 50-day-pregnant chinchilla were given by Tam (1972: Table III). The amounts of ovarian tissues used were as follows: residual tissue, 63·8 mg; primary CL, 6·1 mg and accessory CL, 2·3 mg. The quantities of progesterone synthesized by 1 mg of primary CL and accessory CL during 60 min of incubation were much greater than those produced by 1 mg of residual tissue. However, because of the large

TABLE II

Progesterone in primary CL, accessory CL and residual ovarian tissue (interstitial tissue, small follicles and accessory CL) of the chinchilla

	Primary CL		Accessory CL		Residue	
Pregnancy (days)	Total* (ng)	Concn (ng/mg)	Total* (ng)	Concn (ng/mg)	Total* (ng)	Concn (ng/mg)
8	195·2	78·1	8·1	3·7	113·8	3·0
62	—†	—	444·7	69·5	135·5	2·4
66	266·9	30·0	—	—	103·9	2·5
75	—	—	210·0	43·8	—	—
106	159·8	9·6	271·0	19·2	169·3	8·9

* Total progesterone content in the pair of ovaries.
† Not determined.

TABLE III

The percentage yield of progesterone from [7α-³H]pregnenolone
(18·0 Ci/mmol) by the yolk-sac and placental tissues of the
guinea-pig

Gestation (days)	Yolk sac	Giant cell-rich tissue	Syncytiotro-phoblast
25	2·62	26·45	34·48
	1·35	15·29	26·24
35	—	—	44·69
	0·35	13·86	46·49
40	6·49	15·66	17·67
51	3·36	19·48	14·29
	8·05	5·65	4·82
66	4·99	6·81	7·58

amount of interstitial tissue present, the total quantity of progesterone produced by this tissue *in vitro* was approximately equal to the total quantity synthesized by the two types of luteal tissue. This proportion agrees with that estimated by the determination of progesterone content in the luteal and residual tissues (Table II).

Progesterone synthesis in vitro *by the yolk-sac and placental tissues of the guinea-pig*

The syncytiotrophoblast was the most active tissue in the placenta for the transformation of [7α-³H]pregnenolone into progesterone (Table III). As early as Day 25 of gestation, the percentage yield of progesterone was already high, but the peak was not observed until Day 35. From Day 40 onwards the percentage yield was unexpectedly low compared with that in the first half of gestation. The yield from the giant cell-rich tissue was always much lower than that of the syncytiotrophoblast. A small amount of progesterone was synthesized *in vitro* by the yolk-sac.

Progesterone metabolism by the ovary of the chinchilla

In earlier investigations (Tam, 1971, 1972), using [7α-³H]pregnenolone as substrate and a kinetic incubation technique, the residual tissue was found to produce progesterone, 20α- and 20β-hydroxy-4-pregnen-3-one, 17α-hydroxyprogesterone, 17α-hydroxypregnenolone, dehydroepiandrosterone, androstenedione, testosterone and oestradiol-17β and that the synthesis of these compounds in the ovary during pregnancy involved both the 4- and 5-unsaturated pathways. It was also found that the luteal cells were incapable of 17-hydroxylation.

Metabolites present in arterial plasma and ovarian tissue

To ascertain that the steroid metabolites found in the incubations were not formed as an artefact of experimental conditions *in vitro*, the occurrence of these compounds was investigated in arterial plasma and ovarian tissues taken from animals in various stages of pregnancy. The results of these steroid determinations are given in Table IV. It can be seen that 20α- and 20β-hydroxy-4-pregnen-3-one and oestradiol-17β were present in considerable quantities in plasma and that 20β-hydroxy-4-pregnen-3-one and testosterone were detected only in the interstitial tissue. The concentration of 20α-hydroxy-4-pregnen-3-one was much higher in the luteal tissue than in the residual tissue.

Metabolism in vitro of steroids by the interstitial tissue

Using equimolar amounts of $[7\alpha\text{-}^3\text{H}]$pregnenolone and $[4\text{-}^{14}\text{C}]$-progesterone as substrates, 18-hydroxyprogesterone and $17\alpha,20\alpha$-dihydroxy-4-pregnen-3-one were detected in the incubation medium of the residual tissue in addition to the steroids isolated in the tritiated pregnenolone incubations mentioned above. Both compounds were

TABLE IV

The determination of steroids in arterial plasma (18·5 ml pooled from 50–110 days of pregnancy) and ovarian tissues (luteal 98 mg, residual tissue 384 mg; pooled from 4–110 days of pregnancy) of the chinchilla

Steroids	Plasma (ng/ml)	Luteal* (ng/mg)	Residue (ng/mg)
Progesterone	43·8	43·1	7·3
18-Hydroxyprogesterone†	—‡	N.D.	N.D.
20α-Hydroxy-4-pregnen-3-one	19·6	1·0	0·2
20β-Hydroxy-4-pregnen-3-one	5·6	N.D.	0.05
17α,20α-Dihydroxy-4-pregnen-3-one†	—	N.D.	N.D.
Androstenedione	—	Lost	Lost
Testosterone	—	N.D.	0·1
Oestradiol-17β	0·3	Lost	1·9

* Luteal tissue = a mixture of primary and accessory CL. Both luteal and residual tissue were taken from 21 ovaries fairly evenly spread from 50–110 days of pregnancy.
† Uncorrected for losses, but all other compounds have been corrected to 100% recovery.
‡ A blank indicates that no determination was attempted, but N.D. = not detectable.

TABLE V

The percentage yield of various steroids from 4-h incubation of ovarian tissues from cyclic and pregnant chinchilla with $[7\alpha\text{-}^3H]$pregnenolone (660 mCi/mmol) and $[4\text{-}^{14}C]$progesterone (58·5 mCi/mmol)

Steroids produced	Residue (20 days cycle)		Residue		Primary CL (74 days pregnant)		Accessory CL	
	^3H	^{14}C	^3H	^{14}C	^3H	^{14}C	^3H	^{14}C
Progesterone	26·63	55·83	13·60	30·38	28·02	41·70	17·16	42·12
18-Hydroxyprogesterone*	0·27	0·35	0·25	0·38	0·35	0·45	1·18	1·67
20α-Hydroxy-4-pregnen-3-one	1·43	1·49						
17α,20α-Dihydroxy-4-pregnen-3-one*	0·05	0·06	0·25	0·38				
17α-Hydroxyprogesterone	5·76	7·36	1·04	1·13				
17α-Hydroxypregnenolone			0·65	N.D.†				
Testosterone			0·09	0·12				
Androstenedione	14·28	17·67	4·33	5·03				

* Uncorrected for losses, but all other compounds have been corrected to 100% recovery.

† N.D. = not detectable. Blank spaces indicate that either no determination was attempted or the metabolites were not detectable. See text for detail.

TABLE VI

The between experiment comparison of the contribution (R) of [7α-³H]pregnenolone relative to [4-¹⁴C]progesterone in the production of various steroids by the residual tissue of the chinchilla ovary
(Mean ± S.D., n = number of experiments)

Steroids produced	Mean contribution $R*$	
	5–40 days of cycle	20–88 days pregnant
Progesterone	0·42 ± 0·08(5)	0·38 ± 0·08(5)
18-Hydroxyprogesterone	0·70 ± 0·02(3)	1·41(1)
20α-Hydroxy-4-pregnen-3-one	0·91 ± 0·09(2) †	—
17α,20α-Dihydroxy-4-pregnen-3-one	0·92 ± 0·13(4)	0·88 ± 0·04(2)
17α-Hydroxyprogesterone	0·72 ± 0·09(2)	0·70 ± 0·01(2)
Androstenedione	0·68 ± 0·13(5) †	0·90 ± 0·10(5)

Left column significance: Progesterone & 18-Hydroxyprogesterone $P < 0.001$; 17α,20α-Dihydroxy-4-pregnen-3-one & 17α-Hydroxyprogesterone $P < 0.05$; overall $P < 0.001$.
Right column significance: 17α,20α-Dihydroxy-4-pregnen-3-one & 17α-Hydroxyprogesterone $P < 0.05$; Androstenedione $P < 0.005$; overall $P < 0.001$.

$$* R = \frac{{}^{3}\text{H d/min} \times \text{specific activity of } [4\text{-}^{14}\text{C}]\text{progesterone}}{{}^{14}\text{C d/min} \times \text{specific activity of } [7\alpha\text{-}^{3}\text{H}]\text{pregnenolone}}$$

† = not significant, $P > 0.05$

labelled with tritium and carbon-14. However, $17\alpha,20\alpha$-dihydroxy-4-pregnen-3-one was not isolated from the incubation medium of the luteal tissues. Table V gives the percentage yields of these C_{21} and C_{19} steroids. No attempt was made to correlate the yield with the reproductive state of the animals. The percentage yield was presented to enable a comparison to be made of the amounts of different steroids produced. Progesterone was the principal product synthesized by all the ovarian tissues, but considerable amounts of 17α-hydroxyprogesterone and androstenedione were also produced by the residual tissue. Though 20β-hydroxy-4-pregnen-3-one was found, the $17\alpha,20\beta$-epimer of $17\alpha,20\alpha$-dihydroxy-4-pregnen-3-one could not be detected in any of the incubations.

The contribution of $[7\alpha\text{-}^{3}H]$pregnenolone relative to $[4\text{-}^{14}C]$progesterone towards the formation of any metabolite was expressed as the ratio R =

$$\frac{^{3}H \text{ d/min} \times \text{specific activity of } [4\text{-}^{14}C]\text{progesterone}}{^{14}C \text{ d/min} \times \text{specific activity of } [7\alpha\text{-}^{3}H]\text{pregnenolone}}.$$

Thus a large R value indicated a higher proportion of the metabolite having been derived from $[7\alpha\text{-}^{3}H]$pregnenolone than $[4\text{-}^{14}C]$progesterone, and *vice versa*. The mean R values for some C_{21} and C_{19} steroids isolated from the incubation medium of non-pregnant and pregnant residual tissue were tabulated in Table VI. The R values of all compounds were significantly higher than that of progesterone, showing the existence of biosynthetic pathways leading from pregnenolone to these metabolites without the intermediary progesterone. The means of the R values of 20α-hydroxy-4-pregnen-3-one and $17\alpha,20\alpha$-dihydroxy-4-pregnen-3-one, and 17α-hydroxyprogesterone and androstenedione in the non-pregnant animals were not significantly different. In the pregnant animals, however, the mean R values of androstenedione were very significantly greater than those of 17α-hydroxyprogesterone, and therefore the contribution of $[7\alpha\text{-}^{3}H]$pregnenolone towards the formation of androstenedione was higher than that of 17α-hydroxyprogesterone.

DISCUSSION

In some of the hystricomorphs so far studied there is an initial period of pregnancy, one-quarter to one-half of gestation, in which plasma progesterone concentrations are very low compared with those occurring at later stages of pregnancy. Thus, in the cuis (Fig. 1a and Tam, 1973), guinea-pig (Challis *et al.*, 1971) and coypu (Rowlands &

Heap, 1966) the levels of progesterone reached in this initial period do not exceed 60 ng/ml and are usually much lower, but, later, there is a dramatic increase in all three species to between 520 and 550 ng/ml for a variable period before recession, which is usually advanced by the time of parturition. There are, however, other species, such as the chinchilla (Fig. 1b) and acouchi (Rowlands *et al.*, 1970), in which these high concentrations of plasma progesterone are not attained, and, in fact, do not exceed the level found during the 'initial' period in the cuis, guinea-pig and coypu. Even so, this latter level (60 ng/ml) is greater than that found in the pregnant hamster (25 ng/ml; Leavitt & Blaha, 1970), ewe (15–20 ng/ml; Bassett *et al.*, 1969; Fylling, 1970) and cow (8 ng/ml; Donaldson *et al.*, 1970). One factor responsible for such high progesterone levels in the plasma of hystricomorph rodents could be the increase in plasma binding globulin (Illingworth *et al.*, 1973; Heap & Illingworth, 1974) which prolongs the half-life of the hormone in blood. The experimental evidence presented here shows that within this suborder at least two methods have been evolved of supplementing the production of progesterone by the primary CL. These are the formation of accessory CL and the assumption of an endocrine role by the placenta. In the acouchi, an increasing reliance on the accessory CL as a source of progesterone appears to occur as gestation progresses. The primary CL regressed to one-third of their original weight, while the number of accessory CL increased tenfold after the first 50 days of gestation (Rowlands *et al.*, 1970). The progesterone content of the primary CL tended to decrease, but that of the accessory CL remained reasonably constant (Table I). Weir (1967a) reported that with the formation of accessory CL, the total volume of luteal tissue in the chinchilla ovary increased throughout pregnancy and Fig. 1b illustrates the concomitant increase in plasma progesterone for at least the major part of gestation. The capacity of the accessory CL of the chinchilla to produce progesterone is indicated in Table III of Tam (1972). When expressed as amounts produced per mg of luteal tissue, the quantities of progesterone synthesized *in vitro* in a given time by both primary and accessory CL were closely similar. The total progesterone content in both types of luteal tissue in the ovary were also similar (Table II). Of course, the importance of the chinchilla interstitial tissue cannot be ignored. The two types of CL and the residual tissue appeared to provide equal contributions to the progesterone pool.

The first suggestion that the guinea-pig placenta might secrete progestational hormone came from the work of Loeb (1923) who demonstrated that ovariectomy did not result in abortion. Rowlands (1956) reported a slow regression of the CL beginning on Day 35 of gestation.

The results of Illingworth & Deanesly (1972) indicated a specific contribution of placental progesterone from Day 18 of pregnancy. The work reported here not only represents efforts to measure the capacity of the placenta to synthesize progesterone *in vitro* during different stages of gestation, but also attempts to show the type of placental tissue which was most active in the production of this hormone. The results (Table III) show that the syncytiotrophoblast was most active in the synthesis of progesterone from pregnenolone *in vitro* and that this activity reached a maximum at the 35th day of pregnancy. These results were most unlikely to have been masked by the further metabolism of the progesterone synthesized during the experiment. The other metabolites, tentatively identified as 5β-pregnane-3,20-dione, $3(?)$-hydroxy-5β-pregnan-20-one and $3(?),20\alpha$-dihydroxy-5β-pregnane, were identical for the yolk-sac and the two types of placental tissues and the amounts they produced, as revealed by a radiochromatogram scanner, also seemed to be proportionally equal. It is hoped that the detailed results of all the metabolites formed will be published later. The yolk-sac of the guinea-pig could also metabolize a small amount of tritiated pregnenolone. It is at the moment difficult to assign any physiological significance to the ability of the yolk-sac to synthesize progesterone. The amount produced was probably too small to be of importance. It is unlikely that the giant cells secrete any significant quantities of progesterone in view of the low percent yield of the hormone produced by these cells compared with that synthesized by an equal amount of syncytiotrophoblast. By the application of histochemical tests (Wattenberg, 1958), the reaction for the enzyme 5-ene-3β-hydroxysteroid dehydrogenase was shown to be only weakly positive in the endoderm of the yolk-sac, but much more intensive in the spongy zone syncytiotrophoblast (S. M. Burgess & W. H. Tam, unpublished results). The labyrinthine syncytiotrophoblast was also weakly positive, and the giant cells were negative to the test. Thus, the spongy zone syncytiotrophoblast is responsible for the synthesis of most of the progesterone in the guinea-pig placenta. Such is probably also the situation in the closely related species, the cuis, in which plasma progesterone levels (Fig. 1a) during pregnancy are comparable with those of the guinea-pig and ovariectomy during pregnancy does not cause abortion (B. J. Weir, pers. comm.).

The isolation of a whole spectrum of C_{21} and C_{19} steroids from the chinchilla residual tissue and incubation medium demonstrates the wide variety of steroid enzymes occurring in the interstitium and follicles which the residual tissue certainly contained. Some of these compounds were obviously produced in sufficient amounts to be present in blood. Though 18-hydroxyprogesterone was persistently formed *in vitro* by luteal and residual tissues and 17α,20α-dihydroxy-4-pregnen-3-one by just the residual tissue, these compounds were not detected in the tissue. This is probably because their concentrations were too low to be measured and in fact their percentage yield was the lowest in the *in vitro* experiments (Table V). However, the isolation of 18-hydroxylated metabolites of sex steroids is not without precedent. Loke *et al.* (1959) isolated 18-hydroxyoestrone from human pregnancy urine. Fukushima *et al.* (1962) isolated 18-hydroxyetiocholanolone and 18-hydroxyandrosterone from male human urine. These authors could not locate the origin of these compounds. It has been shown that 18-hydroxyprogesterone is synthesized by the residual and luteal tissues of the chinchilla ovary (Tam, in prep.). There have been many reports on the isolation of 17α,20α-dihydroxy-4-pregnen-3-one from *in vitro* systems and Dominguez (1961) isolated this compound from human spermatic venous blood. The fact that this compound was not produced by the luteal tissue supports the author's previous finding that 17α-hydroxylation did not occur in the granulosa lutein cells of the chinchilla (Tam, 1972). This also explains the much higher concentration of 20α-hydroxy-4-pregnen-3-one in luteal rather than in residual tissue (Table V) as the latter was able to metabolize it further to the 17α-hydroxylated form.

The interpretation of the results in Table VI depends on the principle that the contribution (R) of [7α-³H]pregnenolone relative to [4-¹⁴C]-progesterone towards the formation of a product (B) can only be smaller than its contribution to an *obligatory* intermediate (A) of the product (case I). When the relative contribution (R) of [7α-³H]pregnenolone to the product (B) is greater than that towards the *presumed* obligatory intermediate (A), a separate pathway exists (case II). With such consideration in mind and information from the literature, it is possible to expand the major 4- and 5-unsaturated routes, known from a previous investigation (Tam, 1972) to be similar to those of other animals, into detailed steroidogenic pathways occurring in these *in vitro* experiments. Thus, the presence of the carbon-14 compound and the significantly

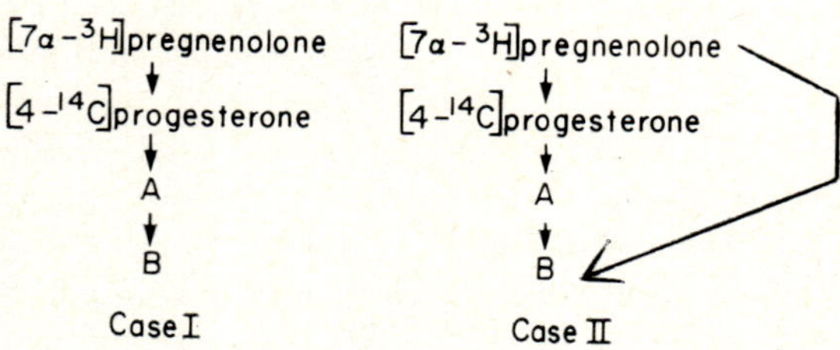

higher R value than that of progesterone (Table VI) suggest that 20α-hydroxy-4-pregnen-3-one can be formed from pregnenolone with and without intermediary progesterone (Fig. 2). The synthesis *in vitro* of 20α-hydroxy-4-pregnen-3-one from progesterone in human and sparrow testes and human ovary has been suggested by Dominguez (1961), Fevold & Eik-Nes (1963) and Forleo & Collins (1964). The formation *in vitro* of 20α-hydroxy-4-pregnen-3-one from pregnenolone was established in the rat corpus luteum (Kuhn & Briley, 1970). Following the same argument, the synthesis of 18-hydroxyprogesterone from two different origins probably also occurred. The almost identical R values of 20α-hydroxy-4-pregnen-3-one and 17α,20α-dihydroxy-4-pregnen-3-one in the cyclic chinchilla indicates that the 17α,20α-dihydroxy-4-pregnen-3-one was formed from 20α-hydroxy-4-pregnen-3-one as suggested by Dominguez (1961). The conversion of 17α,20α-dihydroxy-4-pregnen-3-one → 17α-hydroxyprogesterone → androstenedione → testosterone has been demonstrated by Dominguez (1966). The R value of 17α-hydroxyprogesterone is significantly higher than that of progesterone, but lower than that of 17α,20α-dihydroxy-4-pregnen-3-one (Table VI). The interpretation seems to be that 17α-hydroxyprogesterone could be produced by three different routes, *viz.* via 17α,20α-dihydroxy-4-pregnen-3-one, progesterone and 17α-hydroxypregnenolone as indicated in Fig. 2.

There is a strong possibility that most of the androstenedione in the residual tissue of the non-pregnant animal was formed from 17α-hydroxyprogesterone as the R values of these compounds are not significantly different (Table VI). However, the proportion of the tritiated compound in androstenedione was much higher in the pregnant animals. This additional tritiated androstenedione can be accounted for, if it was

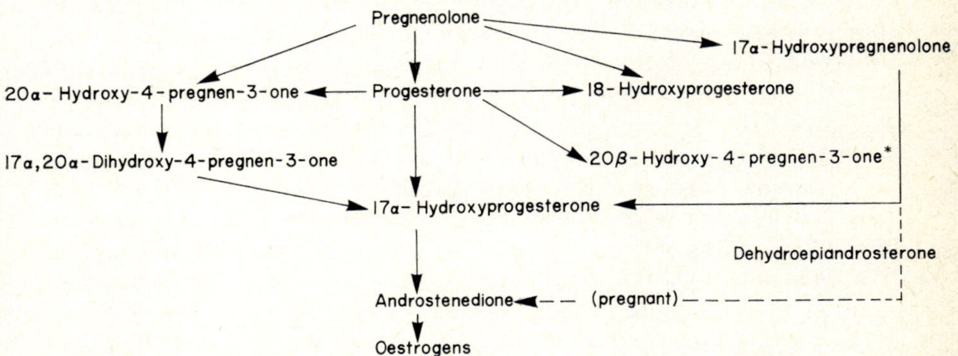

FIG. 2. Suggested steroid metabolic pathways *in vitro* in the residual tissue of the chinchilla ovary. *This step is after Tam (1972).

synthesized from pregnenolone via dehydroepiandrosterone, a pathway established earlier (Tam, 1972). It is not known whether the 4- and 5-unsaturated pathways for the formation of androstenedione took place in different cell types in the residual tissue, i.e. the interstitial cells or the thecal cells of the follicles which the residual tissue most probably contained. It is proposed here that the conversion of progesterone to 20α-hydroxy-4-pregnen-3-one and 18-hydroxyprogesterone in the corpora lutea and residual tissue of the chinchilla represents a means of inactivating any excessive progesterone synthesized as postulated by Wiest *et al.* (1968), Wiest (1968), and Hashimoto & Wiest (1969). It is also postulated that the *in vitro* biosynthetic pathway pregnenolone $\rightarrow 20\alpha$-hydroxy-4-pregnen-3-one $\rightarrow 17\alpha,20\alpha$-dihydroxy-4-pregnen-3-one $\rightarrow 17\alpha$-hydroxyprogesterone $\rightarrow$ androstenedione and the shift in emphasis from the 4-unsaturated to the 5-unsaturated pathway for the formation of androstenedione in the pregnant chinchilla interstitium may be bypasses by means of which C_{19} compounds and therefore probably oestrogens can be produced without the formation of progesterone, when the latter is not required. The synthesis of oestrogen independent of progesterone has important implications. Recently, oestradiol-17β has been shown to be luteolytic and to decrease circulating progesterone level in menstruating rhesus monkey (Karsch *et al.*, 1973).

Summary

Many hystricomorph rodents possess two ways of supplementing progesterone secretion by the primary CL to maintain pregnancy. In the green acouchi and chinchilla, the placenta is not thought to secrete progestational hormone, and the ovary during pregnancy contains numerous accessory CL. The accessory CL become progressively a more important source of progesterone in the pregnant animal. The ovaries of the guinea-pig do not possess accessory CL and the primary CL begin to regress after 35 days of pregnancy. However, *in vitro* experiments showed that the spongy zone syncytiotrophoblast is able to synthesize large amounts of progesterone once the placenta is established and that these cells reach a peak of activity at about the 35th day of gestation. The endocrine role of the placenta, therefore, represents the other method of supplementing the hormonal activity of the CL of conception. Residual tissue of the chinchilla ovary produced *in vitro* as much progesterone as the luteal tissue, and synthesized C_{19} steroids by the usual 4- and 5-unsaturated pathways, and also directly from the pregnenolone through 20α-hydroxy-4-pregnen-3-one and $17\alpha,20\alpha$-dihydroxy-4-pregnen-3-one. The 5-unsaturated pathway was more active in pregnant animals. The possibility that these biosynthetic pathways might act as bypasses and diversions in the regulation of progesterone synthesis is discussed.

ACKNOWLEDGMENT

The author is most grateful to Dr I. W. Rowlands and Dr B. J. Weir who provided and dissected all the tissue in the work performed in the Wellcome Institute of Comparative Physiology. The author also expresses his gratitude to the Journal of Endocrinology and Journal of Reproduction and Fertility for their permission to reproduce some of the author's published data. The work was supported by the Ford Foundation, The Medical Research Council (U.K.) and the National Research Council of Canada. The attendance at the Symposium was made possible by grants from the University of Western Ontario, Wellcome Trust and the Society for Endocrinology (U.K.).

REFERENCES

Amoroso, E. C., Hancock, J. L. & Rowlands, I. W. (1948). Ovarian activity in the pregnant mare. *Nature, Lond.* **161**: 353–356.

Bassett, J. M., Oxborrow, T. J., Smith, I. D. & Thorburn, G. D. (1969). The concentration of progesterone in the peripheral plasma of the pregnant ewe. *J. Endocr.* **45**: 449–457.

Brown, J. B. (1955). A chemical method for the determination of oestriol, oestrone and oestradiol in human urine. *Analyt. Biochem.* **19**: 234–242.

Challis, J. R. G., Heap, R. B. & Illingworth, D. V. (1971). Concentrations of oestrogen and progesterone in the plasma of non-pregnant, pregnant and lactating guinea-pigs. *J. Endocr.* **51**: 333–345.

Cole, H. H., Howell, C. E. & Hart, G. H. (1931). The changes occurring in the ovary of the mare during pregnancy. *Anat. Rec.* **49**: 199–209.

Dominguez, O. V. (1961). Biosynthesis of steroids by testicular tumors complicating congenital adrenocortical hyperplasia. *J. clin. Endocr. Metab.* **21**: 663–674.

Dominguez, O. V. (1966). Biosynthesis of androgens from C_{21}-steroids exhibiting differences in their side chain. *Steroids* **7**: 433–445.

Donaldson, L. E., Bassett, J. M. & Thorburn, G. D. (1970). Peripheral plasma progesterone concentration of cows during puberty, oestrous cycle, pregnancy and lactation, and the effects of undernutrition or exogenous oxytocin on progesterone concentration. *J. Endocr.* **48**: 599–614.

Fevold, H. R. & Eik-Nes, K. B. (1963). Progesterone metabolism by testicular tissue of the English sparrow (*Passer domesticus*). *Gen. comp. Endocr.* **3**: 335–345.

Forleo, R. & Collins, W. P. (1964). Some aspects of steroid biosynthesis in human ovarian tissue. *Acta endocr., Copenh.* **46**: 265–278.

Fukushima, D. K., Bradlow, H. L., Hellman, L. & Gallagher, T. F. (1962). Isolation and characterization of 18-hydroxy-17-ketosteroids. *J. biol. Chem.* **237**: 3359–3363.

Fylling, P. (1970). The effect of pregnancy, ovariectomy and parturition on plasma progesterone level in sheep. *Acta endocr., Copenh.* **65**: 273–283.

Hashimoto, I. & Wiest, W. G. (1969). Luteotrophic and luteolytic mechanisms in rat corpora lutea. *Endocrinology* **84**: 886–892.

Heap, R. B. & Illingworth, D. V. (1974). Progesterone in hystricomorph rodents. *Symp. zool. Soc. Lond.* No. 34: 385–415.

Illingworth, D. V., Ackland, N., Heap, R. B. & Weir, B. J. (1973). Progesterone-binding proteins: occurrence, capacity and binding affinity in hystricomorph rodents. *J. Endocr.* **58**: ii.

Illingworth, D. V. & Deanesly, R. (1972). Maintenance of pregnancy by synthetic progestagens in guinea-pigs ovariectomized before implantation; progesterone-binding protein and placental progesterone secretion. *J. Endocr.* **54**: 435–444.

Karsch, F. J., Krey, L. C., Weick, R. F., Dierschke, D. J. & Knobil, E. (1973). Functional luteolysis in the rhesus monkey: the role of estrogen. *Endocrinology* **92**: 1148–1152.

Kuhn, N. J. & Briley, M. S. (1970). The roles of pregn-5-ene-3β,20α-diol and 20α-hydroxy steroid dehydrogenase in the control of progesterone synthesis preceding parturition and lactogenesis in the rat. *Biochem. J.* **117**: 193–201.

Leavitt, W. W. & Blaha, G. C. (1970). Circulating progesterone levels in the golden hamster during the estrous cycle, pregnancy, and lactation. *Biol. Reprod.* **3**: 353–361.

Loeb, L. (1923). The mechanisms of the sexual cycle with special reference to the corpus luteum. *Am. J. Anat.* **32**: 305–343.

Loke, K. H., Marrian, G. F. & Watson, E. J. D. (1959). The isolation of a sixth Kober chromogen from the urine of pregnant woman and its identification as 18-hydroxyoestrone. *Biochem. J.* **71**: 43–48.

Rood, J. P. & Weir, B. J. (1970). Reproduction in female wild guinea-pigs. *J. Reprod. Fert.* **23**: 393–409.

Rowlands, I. W. (1956). The corpus luteum of the guinea-pig. *Ciba Fdn Collog Ageing* **2**: 69–83.

Rowlands, I. W. & Heap, R. B. (1966). Histological observations on the ovary and progesterone levels in the coypu, *Myocastor coypus*. *Symp. zool. Soc. Lond.* 15: 335–352.

Rowlands, I. W. & Short, R. V. (1959). The progesterone content of the guinea-pig corpus luteum during the reproductive cycle and after hysterectomy. *J. Endocr.* **19**: 81–86.

Rowlands, I. W., Tam, W. H. & Kleiman, D. G. (1970). Histological and biochemical studies on the ovary and of progesterone levels in the systemic. blood of the green acouchi (*Myoprocta pratti*). *J. Reprod. Fert.* **22**: 533–545.

Short, R. V. (1961). Progesterone. In *Hormones in blood*: 379–437. C. H. Gray & A. L. Bacharach. (Eds.) New York: Academic Press.

Tam, W. H. (1971). The production of hormonal steroids by ovarian tissues of the chinchilla (*Chinchilla laniger*). *J. Endocr.* **50**: 267–279.

Tam, W. H. (1972). Steroid metabolic pathways in the ovary of the chinchilla (*Chinchilla laniger*). *J. Endocr.* **52**: 37–50.

Tam, W. H. (1973). Progesterone levels during the oestrous cycle and pregnancy in the cuis (*Galea musteloides*). *J. Reprod. Fert.* **35**: 105–114.

Tam, W. H. (In preparation). The isolation and characterization of $17\alpha,20\alpha$-dihydroxy-4-pregnen-3-one and 18-hydroxyprogesterone from the chinchilla ovary.

Wattenberg, L. W. (1958). Microscopic histochemical demonstration of steroid-3β-ol dehydrogenase in tissue sections. *J. Histochem. Cytochem.* **6**: 225–232.

Weir, B. J. (1967a). *Aspects of reproduction in some hystricomorph rodents*. Ph.D. thesis, University of Cambridge.

Weir, B. J. (1967b). The care and management of laboratory hystricomorph rodents. *Lab. Anim.* **1**: 95–104.

Weir, B. J. (1970). The management and breeding of some more hystricomorph rodents. *Lab. Anim.* **4**: 83–97.

Weir, B. J. (1971). The evocation of oestrus in the cuis, *Galea musteloides*. *J. Reprod. Fert.* **26**: 405–408.

Wiest, W. G. (1968). On the function of a 20α-hydroxypregn-4-en-3-one during parturition in the rat. *Endocrinology* **83**: 1181–1184.

Wiest, W. G., Kidwell, W. R. & Balogh, Jr, K. (1968). Progesterone catabolism in the rat ovary: a regulatory mechanism for progestational potency during pregnancy. *Endocrinology* **82**: 844—859.

DISCUSSION

ROWLANDS (*London*): Tam, I would like to ask one or two questions. In some of your first tables you seemed to indicate that there was a difference between progesterone content of accessory and primary corpora lutea in the chinchilla. Am I right in thinking this?

TAM (*London, Ont.*): No, I don't think so. In one particular instance the content of progesterone in the accessory corpora lutea was 444 ng/mg, I think. All the others were around 200 ng/mg and these values agree with the progesterone content in the primary corpora lutea. So there is only one exception.

WEIR (*London*): Am I right in thinking you said there was 25% progesterone production in the residual tissue of the chinchilla ovary?

TAM: Right.

WEIR: Is it not possible that a large proportion of this 25% is luteal and not interstitial in origin? We dissected these ovaries and I am quite sure that we couldn't possibly have taken out every single corpus luteum and yet you seem to exclude the possibility of their contribution.

TAM: I wouldn't say a large proportion, though I wouldn't exclude that there is a certain proportion of it that comes from the luteal tissue that was left behind. If you stain the chinchilla ovary for the steroid enzyme Δ^5-3β-hydroxysteroid dehydrogenase, you can see that this enzyme is very active in the interstitial tissue.

WEIR: Yes, but it may not be the conversion enzyme for progesterone.

TAM: Why not?

WEIR: What about all the androgens that you say are produced by the interstitial tissue?

TAM: All that I am saying is that interstitial tissue produces a large amount of progesterone, and also much androgen, probably for the synthesis of oestrogen.

WEIR: I should feel happier if you could use for your biochemical work *in vitro* a species in which we know there are fewer accessory CL .

TAM: One thing I should like to see done is the application of a method for eliminating the accessory corpora lutea in a living animal to see whether gestation is disturbed.

WEIR: I have found just a few chinchilla which have managed without accessories.

PEARSON (*Berkeley*): I should like to know, are there comparative data for the progesterone content of residual tissue of non-hystricomorphs?

TAM: Yes. I think that it has been shown (Kraicer, Kisch & Nussbaum, 1971*) that the non-luteal tissue is probably very important for the synthesis of oestrogen, but I don't think any conclusion was reached about progesterone synthesis by this tissue. I think that in the chinchilla, in which the placenta is not producing any progesterone or oestrogen or may not be producing enough, the interstitial tissue must be very important for the synthesis of these two compounds.

ROWLANDS: The guinea-pig does not produce accessory CL and its ovaries contain little or no interstitial tissue. Its residual tissue is practically free of progesterone, isn't it?

TAM: Yes, this was shown by Dr Heap.

---

* Kraicer, P. F., Kisch, E. S. & Nussbaum, M. (1971). Role of non-luteal ovarian tissues in the maintenance of pregnancy. *Acta endocr., Copenh.* **66**: 462–470.

Symp. zool. Soc. Lond. (1974) No. 34, 385–415

THE MAINTENANCE OF GESTATION IN THE GUINEA-PIG AND OTHER HYSTRICOMORPH RODENTS: CHANGES IN THE DYNAMICS OF PROGESTERONE METABOLISM AND THE OCCURRENCE OF PROGESTERONE-BINDING GLOBULIN (PBG)

R. B. HEAP and D. V. ILLINGWORTH*

*A.R.C. Institute of Animal Physiology,
Babraham, Cambridge, England*

SYNOPSIS

Among the several endocrine control systems by which the progesterone requirements of pregnancy are met in eutherian mammals, recent studies of hystricomorph rodents have drawn attention to yet another variant, namely, a progesterone-conserving mechanism whereby the rate at which progesterone is removed from the circulation is much reduced in pregnancy compared with that in the non-pregnant animal. This phenomenon has been observed in the domestic guinea-pig and coypu (*Myocastor coypus*) and appears to be related to the synthesis in gestation of a plasma protein with a high affinity and a moderately high capacity for progesterone. This protein has been identified and characterized by several groups of workers as progesterone-binding globulin (PBG) or progesterone-binding protein (PBP).

The occurrence of PBG, or of related plasma proteins with a high affinity for progesterone, has been observed in five other hystricomorph species: *Proechimys guairae* (casiragua), *Galea musteloides* (cuis), *Chinchilla laniger* (chinchilla), *Lagostomus maximus* (plains viscacha) and *Ctenomys talarum* (tuco-tuco).

The stimulus for the synthesis of PBG in the pregnant guinea-pig is associated with the presence of a viable conceptus, and the initiation of PBG synthesis appears to be related temporally to the establishment of the definitive allantochorionic placenta. The time-course of synthesis of progesterone-binding protein(s) in other hystricomorph rodents has been compared with that found in the guinea-pig, and the probable physiological role of these binding proteins is discussed the context of the economy of pregnancy maintenance in a group of rodents whose gestation is unusually long relative to body weight.

INTRODUCTION

It has long been recognized that a basic requirement for viviparity in most, if not all, eutherian mammals is the secretion of an adequate amount of progesterone to sustain gestation. The means by which the progesterone requirements of pregnancy are supplied differ. In some

* Present address: *Division of Steroid Endocrinology, Department of Chemical Pathology, University of Leeds, England.*

species specialized endocrine mechanisms have been selected, notably, a prolongation of the life and secretory activity of the corpus luteum (as in the pig); an enhanced activity of the corpus luteum that is stimulated by a pituitary and/or placental luteotrophin (as in the rat); an accession of endocrine function by the placenta, thereby ensuring the local synthesis of progesterone within the gravid uterus (as in man); and the combination of several of these mechanisms (as may be found in many species including those already mentioned). In some species the progesterone requirements of pregnancy are met in different ways at different times during gestation. In the sheep, for example, the early requirements are met from a prolongation of the life and secretory activity of the corpus luteum that is supplemented later and replaced eventually by placental secretion. In contrast there are those species, such as the ferret and mink, where the pattern of plasma progesterone concentration in pregnancy resembles that of pseudopregnant females, presumably because of the similar duration and the similar secretory activity of the corpus luteum in both conditions.

Recent studies of hystricomorph rodents have drawn attention to yet another variant of these endocrine control systems. This concerns what has been referred to as a progesterone-conserving mechanism, whereby the rate at which progesterone is removed from the circulation during pregnancy is much reduced in comparison with that in the non-pregnant animal. In this way, the concentration of progesterone in blood is greatly increased, though the increase in the actual production of the hormone is comparatively modest. The synthesis of progesterone-binding globulin, PBG, a plasma protein of high affinity and moderately high capacity for progesterone, appears to play a key role in this alteration in the dynamics of progesterone metabolism and in this paper we shall examine the evidence for the occurrence of this protein in hystricomorph rodents and its probable contribution to the maintenance of gestation.

MATERIALS AND METHODS

Animals

Details of the guinea-pigs used in these experiments have been given previously (Heap *et al.*, 1967; Illingworth *et al.*, 1970; Challis *et al.*, 1971). The coypu (*Myocastor coypus*) and material from the other hystricomorph rodents, cuis (*Galea musteloides*), casiragua (*Proechimys guairae*), chinchilla (*Chinchilla laniger*), viscacha (*Lagostomus maximus*) and tuco-tuco (*Ctenomys talarum*), were provided from the breeding colonies at the Wellcome Institute (see Weir, 1970, 1972, 1973). The time of pregnancy when samples were collected was determined as follows. In the guinea-pig, the morning when a vaginal plug was found was denoted Day 1 of pregnancy. In the other species pregnancy was dated mostly from known conceptions (vaginal plug or spermatozoa in the vaginal smear), sometimes retrospectively from parturition dates since gestation lengths are known for all these species except the tuco-tuco (see Weir, 1974). The regression of the cube root of embryo weight on age (Huggett & Widdas, 1951) was used for animals of unknown gestational age or as confirmation of the other data as shown by Newson (1966) for coypu, Weir (1970) for chinchilla, and Roberts (1973) for some of the other species.

Analytical procedures

Progesterone was determined by fluorimetry (Heap, 1964), competitive protein binding (Challis *et al.*, 1971) or by radioimmunoassay (method A of Heap, Gwyn, Laing & Walters, 1973). The antiserum (S257 No. 2 from Dr G. Abraham, University of California) used in the radioimmunoassay had a high specificity for progesterone being raised in a ewe immunized against deoxycorticosterone-21-monohemisuccinate-human serum albumin. When compared with a cross-reaction of 100% against progesterone, the cross-reaction against deoxycorticosterone was 35% and that against 17α-hydroxyprogesterone, 1%. The values obtained for plasma progesterone concentrations in the cuis and casiragua during pregnancy were similar to those found with the competitive protein-binding technique.

The metabolic clearance rate (MCR) of progesterone was measured in the guinea-pig by the method described previously (Illingworth *et al.*, 1970). The MCR of progesterone was also determined in coypu anaesthetized with a fluothane:oxygen mixture supplied through an open mask. [1,2-^{3}H]progesterone (41 Ci/mmol) was infused for 4 h at about 0·076 ml/min, 0·25 μCi/min, through a P.T.F.E. catheter placed in the left jugular vein. Three or four blood samples were withdrawn from a catheter in the right carotid artery during the fourth hour of the infusion. The arterial cannula was also connected to a mercury manometer and blood pressure and heart rate were monitored throughout the experiment. The heparinized blood samples were centrifuged at 4°C and the plasma stored frozen at -10°C until analysed. A tracer quantity of [4-^{14}C]progesterone was added to the plasma to estimate procedural losses. The plasma was extracted three times with 5 volumes of diethyl ether, and the ether extract was evaporated and chromatographed by thin-layer chromatography in a solvent system, hexane:ethyl acetate, 1 : 1. The progesterone region of the chromatogram was eluted with diethyl ether, evaporated, acetylated and chromatographed a second time in the solvent system, benzene : ethyl acetate, 4: 6. The progesterone region was again eluted with diethyl ether and an aliquot taken to determine the amount of radioactivity. Derivative formation and further chromatography (20β-dihydroprogesterone and its acetate) showed that the ^{3}H : ^{14}C ratio was constant after the second chromatogram. In three instances the infusion was stopped and timed blood samples were taken during the next hour to measure the rate of removal of [^{3}H]progesterone from plasma.

The capacity and the association constant of high-affinity progesterone-binding proteins were estimated by a method similar to that of Pegg & Keane (1969). Progesterone or cortisol were dissolved overnight in 50 mM phosphate buffer, pH 7·4, at 37°C. A constant amount of [7α-^{3}H]progesterone (14·6 Ci/mmol) and increasing concentrations of progesterone (up to 4·5 μg) were dissolved in 0·5 ml buffer. Plasma was thawed, diluted with buffer 1 : 50 and portions (0·5 ml) were incubated with equal volumes of the dissolved steroid for 2 h at 4°C. Florisil (80 mg) was added and mixed briefly for 10 sec. (The florisil had previously been washed repeatedly with methanol to remove fine particles, dried quickly at 60°C, and stored over a desiccant.) After 30 min at 4°C, the samples were centrifuged briefly to complete the sedimentation of the florisil and the supernatant was removed to count the amount of protein-bound radioactivity. The quantities of bound and free steroid were calculated from the total of endogenous and added steroid, the results were plotted according to the method of Scatchard (1949), and the binding protein concentration and its affinity constant were calculated by the method described by Pegg & Keane (1969).

THE KINETICS OF PROGESTERONE METABOLISM

Guinea-pig

Hystricomorph rodents have a long gestation in relation to adult body weight and in two, the guinea-pig and coypu, of the seven species which have been studied in the course of our experiments, there is a striking change in the dynamics of progesterone metabolism during pregnancy. This feature was first noted and investigated in detail in the guinea-pig Illingworth *et al.* (1970). In this species the plasma progesterone concentration is very high during pregnancy. Whereas in non-pregnant guinea-pigs plasma progesterone levels increase by about ten-fold between oestrus (<0.2 ng/ml) and the mid-luteal phase of the cycle (3·0 ng/ml) the increase between the time of mating and mid-pregnancy (200 to 300 ng/ml) is more than 300-fold (Heap & Deanesly, 1966; Feder *et al.*, 1968; Challis *et al.*, 1971). High plasma progesterone concentrations are also found in pregnant women, in which the increase is related to an increasing placental secretion of the hormone from an early stage of gestation. In the pregnant guinea-pig, however, although progesterone secretion by the placenta apparently begins as early as Days 16–18, it accounts for only about 40% of the total progesterone production in mid-pregnancy. It is only during the last two weeks of gestation that almost all the progesterone produced is secreted by the gravid uterus (Fig. 1; Heap & Deanesly, 1966; Illingworth & Deanesly, 1972; Illingworth & Challis, 1973; Heap, Illingworth & Perry, 1973). Thus, unlike the human female, the high concentration of plasma progesterone in the guinea-pig from an early stage of pregnancy cannot be attributed solely to an increased placental secretion though the placenta can undoubtedly produce sufficient progesterone to maintain a viable pregnancy in the absence of the ovaries (Herrick, 1928) from about Day 20 in some individuals (Courrier *et al.*, 1929), and from the fourth week in a larger proportion of animals (Heap & Deanesly, 1966).

The question now arises whether the high progesterone concentration in arterial blood is the result of an increased secretion by the corpora lutea of pregnancy, or a decreased rate of removal of the hormone from

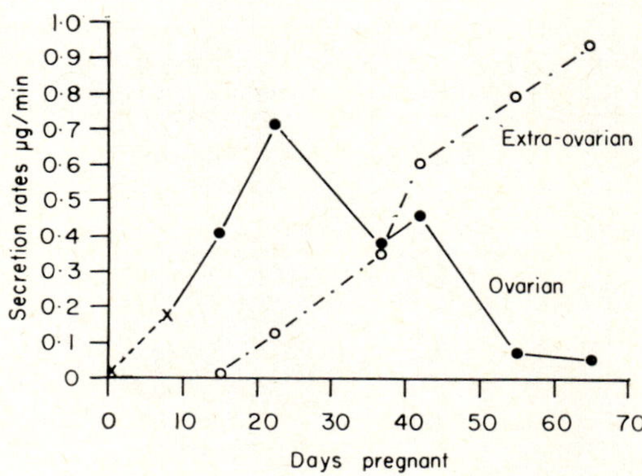

Fig. 1. The estimated secretion rate of ovarian and extra-ovarian (mainly placental) progesterone in the pregnant guinea-pig. Values calculated from the metabolic clearance rate (Illingworth, Heap & Perry, 1970) and production rate (Challis, Heap & Illingworth, 1971) of progesterone, and the plasma progesterone concentration in ovariectomized (Heap & Deanesly, 1966) and ovariectomized-norgestrel treated (Illingworth & Deanesly, 1972) pregnant guinea-pigs.

blood; actually both phenomena probably occur. Measurements of the production rate of progesterone at various times of the reproductive cycle (Illingworth *et al.*, 1970; Challis *et al.*, 1971; Heap, Perry & Challis, 1973) show that the production rate of progesterone in non-pregnant guinea-pigs rises more than ten-fold between oestrus (<0.01 μg/min) and the middle of the luteal phase (0·18 μg/min), and in pregnancy it increases from about 0·4 μg/min on Day 15 *post coitum* (*p.c.*) to about 0·8 μg/min during the next 10 days, and to a mean value of 1·32 μg/min during the second half of gestation (Table I). Similar changes probably occur in hypophysectomized pregnant guinea-pigs where the survival of viable conceptuses is associated with increases in plasma progesterone levels comparable with those of intact animals (Heap *et al.*, 1967). Thus, during the third week of gestation, the developing conceptus (rather than the maternal pituitary) apparently exerts a direct luteo-trophic influence by stimulating the secretory activity of luteal cells. However, this enhanced secretory activity of luteal cells after about day 15 *p.c.* only results in a three-fold increase in the production rate of progesterone, in contrast with a 100-fold increase in plasma progesterone concentration.

Table I

The production rate of progesterone in the guinea-pig during the reproductive cycle

Condition	Progesterone		
	Metabolic clearance rate (ml/min)	Plasma concentration (ng/ml)	Production rate (μg/min)
Non-pregnant			
Oestrous	71(4)	0·2(2)	0·01
Luteal phase	63(4)	2·8(9)	0·18
Pregnant			
Day 15	53(1)	7·6(3)	0·40
16	36(1)	11·4(2)	0·41
17	34(1)	17·5(2)	0·60
18	30(1)	46·3(4)	1·39
19	10(1)	83·2(4)	0·83
20	5(1)	131 (4)	0·66
21–25	5(1)	220 (11)	1·10
26–45	4(11)	329 (37)	1·32

The number of animals is given in parentheses.

A clue to the reason for the high plasma progesterone concentration in pregnant guinea-pigs came from experiments on ovariectomized females with progesterone tablets implanted subcutaneously (Heap & Deanesly, 1967). In non-pregnant animals bearing such implants plasma levels of progesterone were low. But in pregant guinea-pigs which had been ovariectomized in the first week after mating, the progesterone level was at least 20-fold greater by the 30th day *p.c.* This increase was not due solely to placental secretion which was far less than that of the ovary at this time; nor was it because of a difference in uptake of the hormone, since the absorption from progesterone tablets was similar in pregnant and non-pregnant guinea-pigs. We were forced to conclude that the rate of progesterone clearance from blood differed as between pregnant and non-pregnant animals. Direct measurements of progesterone clearance subsequently showed that the metabolic clearance rate, measured by a constant infusion technique in anaesthetized guinea-pigs, decreased by about 90% between the 15th and 20th day of gestation and remained low until after parturition (Table I). Thus MCR was 63–71 ml/min in the normal cycle, 53 ml/min on Day 15 of pregnancy and 5 ml/min on Day 20 *p.c.* Between the 26th and 15th day of pregnancy the average MCR was 4 ml/min (Illingworth *et al.*, 1970). It can be seen from Table I that the decrease in MCR was closely related to an increase in plasma progesterone concentration.

Coypu

The coypu has a gestation period of 130 to 135 days and a high concentration of plasma progesterone is found during the second half of pregnancy (Rowlands & Heap, 1966). Although implantation occurs at about Day 8 *p.c.* (Roberts, 1973), the embryo grows slowly during the first seven weeks, and the volume of the corpus luteum and the concentration of plasma progesterone resemble those of the non-pregnant animal. Thereafter, from about the eighth week, the growth rate of the embryo, the volume of the corpus luteum and the concentration of plasma progesterone increases markedly. The latter increase, as in the guinea-pig, is related to a decreased rate of progesterone clearance. In two anoestrous coypu the MCR of progesterone was 130 and 154 ml/min. The MCR was also high in a coypu 34 days pregnant (151 ml/min), but in animals 92 and 100 days pregnant it had decreased dramatically to values of 9 and 10 ml/min, respectively. The production rate of progesterone in anoestrous and early pregnant animals was about 0·7 μg/min (plasma progesterone concentration, 5 ng/ml). By Day 92 *p.c.* it had increased to about 2·8 μg/min (280 ng/ml plasma), and by Day 100 to 4·2 μg/min (448 ng/ml) (Fig. 2; Illingworth & Heap, 1971).

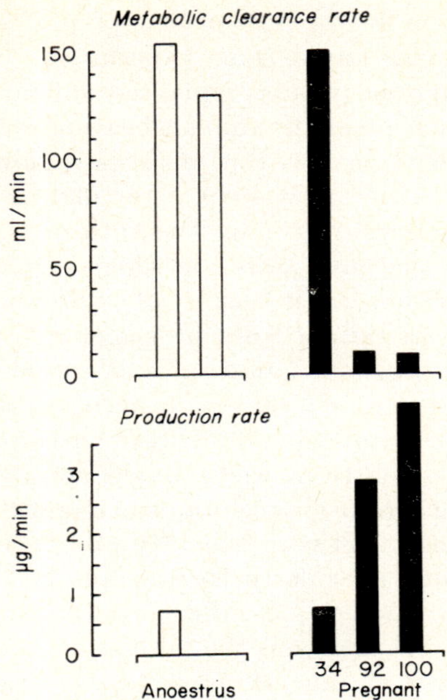

FIG. 2. The metabolic clearance rate and production rate of progesterone in the anoestrous and pregnant coypu.

It can be seen that in the coypu, as in the guinea-pig, the progesterone requirements of pregnancy appear to be supplied by a 3 to 6-fold increase in the production rate of the hormone accompanied by a 90% reduction of its clearance rate, resulting in a circulating concentration some 100-fold greater than in the non-pregnant animal. On the basis of bodyweight, the average production rate in the coypu and guinea-pig during the second half of gestation lies within the range of 0·5–2·0 µg/min/kg.

We have a reservation about the interpretation of these experiments concerning the effect of anaesthesia, which may account for the apparently high production rate of 0·7 µg/min in anoestrus owing to an enhanced adrenal response to operative stress. Studies on the kinetics of progesterone metabolism in anaesthetized animals are susceptible to the same criticism. In three coypu studied at anoestrus, Day 34 and Day 100 of pregnancy, the rate of removal of [³H]progesterone after the infusion of the isotope tracer had been stopped, differed according to the reproductive state of the animal. In one non-pregnant and one pregnant (Day 34) coypu, the concentration of [³H]progesterone in plasma fell to 50% of its equilibrium value within two minutes. At

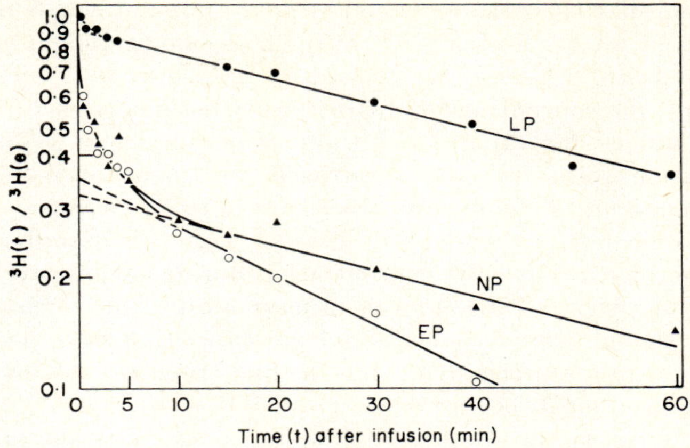

FIG. 3. The kinetics of progesterone metabolism in the coypu after stopping the continuous infusion of [³H]progesterone. NP, not pregnant; EP, early pregnancy (Day 34); LP, late pregnancy (Day 100).

100 days of pregnancy, the decrease took over 30 min (Fig. 3). In all three animals the half-life of the first exponent was similar, about one min, whereas the half-life of the second exponent was 20–46 min. The striking difference in these experiments was in the fractional rate constant of clearance which was approximately $0·5$ min^{-1} in the non-pregnant and early pregnant coypu, and $0·09$ min^{-1} at the 100th day of gestation (Illingworth & Heap, 1971).

PROGESTERONE-BINDING GLOBULIN (PBG), A HIGH-AFFINITY PLASMA PROTEIN

Evidence for its occurrence

Steroids have long been known to bind to plasma proteins (for a review of this subject see Westphal, 1971). For some time the role of plasma proteins has been seen as one of transport, though earlier work had suggested that they produced a biological inactivation of bound steroid. Among the proteins responsible for steroid-protein associations, serum albumin has been considered as one of the major macromolecules in plasma, possessing as it does a very large binding capacity, though a relatively low affinity, for steroids. Daughaday (1956, 1958) had earlier demonstrated that corticosteroids in plasma were also associated with a high-affinity, low capacity protein which was subsequently identified in the sera of all the vertebrate classes as corticosteroid binding globulin, CBG, or transcortin (see Seal & Doe 1961, 1966; Westphal, 1971).

This protein has been characterized as an α_1-globulin with a molecular weight about 50 000, and 26–29% of the molecule by weight consisting of carbohydrate.

In man, the concentration of CBG in plasma increases during pregnancy to values more than twice those of non-pregnant women. CBG concentrations also rise after oestrogen treatment of non-pregnant women (Sandberg & Slaunwhite, 1959) owing to an increased synthesis of the protein rather than to a decrease in turnover (Sandberg *et al.*, 1964). The increase in CBG concentration during pregnancy has been related to the high oestrogen levels in plasma especially during the last trimester, though more recent studies have shown that the concentration of CBG increases appreciably during the first trimester, before the time when oestrogen levels begin to rise sharply (Rosenthal *et al.*, 1969a). Of special interest is the finding that a high concentration of CBG is associated with a decreased rate of clearance of cortisol from blood (Tait & Burstein, 1964). Examples of a marked decrease in metabolic clearance rate of steroid hormones are not common, but where the phenomenon does occur, it is frequently associated with a change in the concentration of plasma proteins whose affinity for a specific hormone is of a high order. The binding of a steroid such as cortisol or testosterone to a specific plasma protein, CBG or testosterone binding globulin, apparently reduces its hepatic uptake, though it does not abolish it (Paterson, 1973, and see Baird *et al.*, 1969), thereby leading to a diminution in the rate of clearance of the hormone.

In the guinea-pig the concentration of CBG rises to very high values in pregnancy, some 30 times greater than the amount reported for the non-pregnant animal. Comparative studies in a further 16 species from eight mammalian orders have shown that this increase in pregnant guinea-pigs is many times greater than that in other species studied, though all species in which maternal blood bathes fetal tissue appear to show some increase (as little as 1·5- to as much as 30-fold) (Seal & Doe, 1966, 1967). Westphal (1967) found that in plasma from pregnant guinea-pigs the apparent association constant of CBG and cortisol (or corticosterone) was $0.5 - 1.1 \times 10^8 \text{M}^{-1}$ at 4°C, and $0.4 - 1.4 \times 10^7 \text{M}^{-1}$ at 37°C, with a concentration of CBG binding sites of $5.7 \times 10^{-7}\text{M}$. A notable feature was the high association constant of CBG for progesterone, a less polar steroid, the binding being approximately 100 times greater than that of the cortisol complex at 4° and 37°C (Westphal, 1967). These findings were confirmed by Rosenthal *et al.* (1969b) who also showed that whereas the CBG concentration increased in pregnant guinea-pigs, oestrogen treatment (oestradiol-3-benzoate, 25 μg daily for 15 days; or 50 μg diethyl stilboestrol feeding twice daily) failed to promote a consistent increase in CBG synthesis in all animals. Diethyl-stilboestrol administered orally for six weeks gave a consistent response in five of six guinea-pigs, but even so the CBG concentration only increased to about one-third that seen in pregnant guinea-pigs 50–60 days *p.c.* Similar results have been reported with oestrone given as daily injections of 1 mg/kg body weight for 12 weeks (Yudaev *et al.*, 1964) and with water-soluble oestrogens given intraperitoneally (Seal & Doe, 1961).

The possibility of the existence of a plasma protein differing from CBG yet with a high affinity for progesterone emerged when protein-bound progesterone was measured in plasma samples from non-pregnant and pregnant guinea-pigs (Heap, 1969). Endogenous progesterone and corticosteroids were removed from plasma proteins by gel filtration on Sephadex G25 at 45°C in 50 mM phosphate buffer, pH 7·4, according to the technique described by Westphal (1967). After dilution, 100 μl portions of the protein fraction were equilibrated with equimolar concentrations (2×10^{-6}M) labelled steroid, and free and protein-bound labelled steroid was separated on small columns of Sephadex G25 (200 mg) at 4° or 23°C. In plasma from non-pregnant guinea-pigs less than 1% of the radioactivity was bound to plasma proteins of high-affinity, whereas in plasma from pregnant guinea-pigs about 50% of [³H]progesterone and 30% of [³H]corticosterone was recovered in a protein fraction. After heating the plasma at 60°C for 20 min, which largely eliminates CBG binding of corticosterone and progesterone, there was still an appreciable amount of progesterone that remained bound to a high-affinity protein (Fig. 4; Heap, 1969). The contribution of albumin to these binding values was negligible. Diamond *et al.* (1969) also concluded that the high affinity of plasma proteins in guinea-pig serum for progesterone and cortisol was derived from two independent molecules, since binding values of plasma progesterone rose almost 100-fold during pregnancy whereas those of cortisol increased only four to five times (see Westphal, 1971). Subsequent work has verified the occurrence of progesterone-binding globulin, PBG, a plasma protein of high affinity for progesterone which is produced in appreciable quantity in pregnant guinea-pigs (Fig. 5; PBP—Milgrom *et al.*, 1970a; PBG—Burton *et al.*, 1971).

Physical properties

Physico-chemical studies have shown that the molecular weight of PBG determined by sedimentation analysis and polyacrylamide gel electrophoresis is 77,500 (Milgrom *et al.*, 1973) and that the protein contains a high proportion of carbohydrate (48·7%, Milgrom *et al.*,

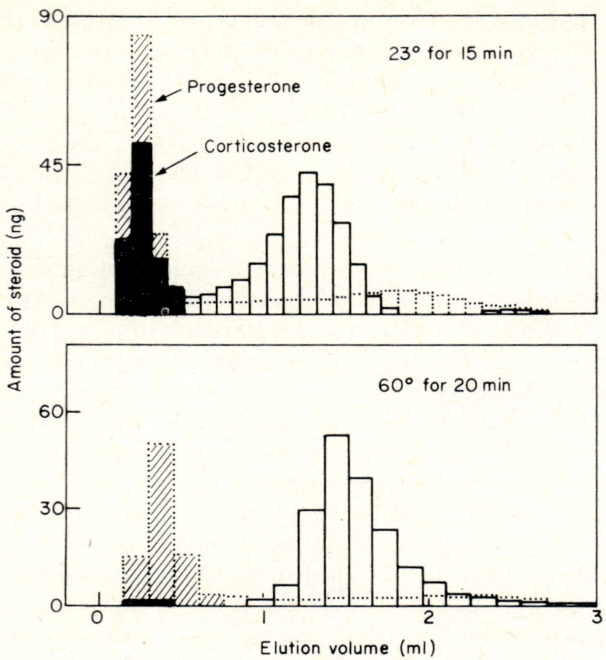

Fig. 4. The binding of [^{14}C]progesterone (300 ng, 0·06 μCi) and [^{3}H]corticosterone (319 ng, 0·14 μCi) to plasma from a pregnant guinea-pig, 48 days *post coitum*. A 100 μl volume of plasma stripped of endogenous steroids was incubated at 23°C for 15 min, or 60°C for 20 min with labelled steroids. Chromatography using Sephadex G25 (fine) micro-columns was carried out at 4°C. The protein fraction was eluted in 0·8 ml 0·1 M phosphate buffer, pH 7·4. and free steroid was eluted thereafter. Progesterone, hatched columns; corticosterone, black columns. Plasma binding attributable to progesterone binding globulin was retained after heating the sample at 60°C whereas the binding activity of corticosteroid binding globulin was lost (see Heap, 1969).

1973; 53%, MacLaughlin, Burton, Aboul-Hosn & Westphal, 1973) which explains its stability with regard to temperature and pH. Milgrom *et al.* (1973) have reported the separation of a homogeneous protein according to the criteria of sucrose gradient ultracentrifugation, polyacrylamide gel electrophoresis and sedimentation velocity characteristics. MacLaughlin *et al.* (1973), however, have reported the existence of PBG in two size classes —PBG I, mol. wt 117 000 and PBG II, mol. wt. 78 000—based on Sephadex G200 gel filtration. Both components were found to be similar in their carbohydrate content and binding characteristics, though they differed in their electrophoretic mobility. PBG II could be fractionated into a series of components of differing molecular weights; this was perhaps attributable to the separation techniques used rather than the nature of the material.

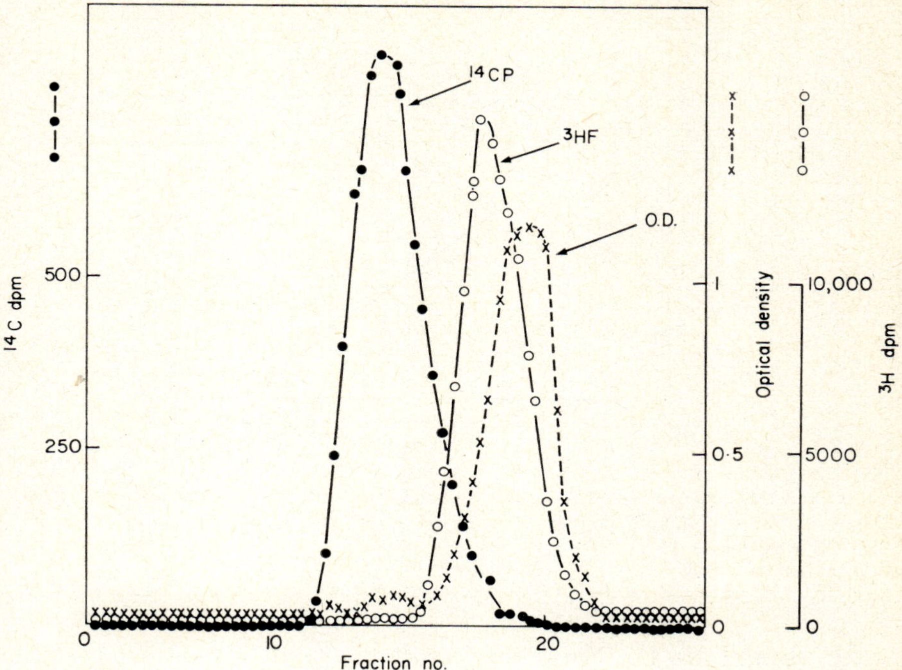

FIG. 5. Purification of PBG, Sephadex G-200 chromatography. Pregnant guinea-pig plasma (12 ml) was incubated 30 min at 37°C with 1 μM-[^{14}C]progesterone and 10 nM-[^{3}H]cortisol. The protein fraction (2 ml) was applied to a Sephadex G-200 column equilibrated with Tris-0·11 M-KCl buffer (diameter 5 cm, gel height 28 cm; 8-ml fractions, 5 fractions per hour; 0°C). In each fraction absorption at 280 nm was measured and an aliquot was used to measure radioactivity. ^{14}CP, [^{14}C]progesterone; ^{3}HF, [^{3}H]cortisol; O.D. optical density (from Milgrom, Allouch, Atger & Baulieu, 1973).

The high concentration of carbohydrate groups in the highly polar PBG molecule is reflected in its extremely acid iso-electric point. Milgrom *et al.* (1973) obtained a value of pH 3·6 for the isoelectrofocussing of [^{3}H]progesterone incubated with pregnant guinea-pig plasma, whereas MacLaughlin *et al.* (1973) claimed that the value was pH 2·0. Each molecule of PBG binds one molecule of progesterone (Fig. 6; Milgrom *et al.*, 1973), and in this respect PBG resembles CBG. PBG has been found to have a high affinity for steroids other than progesterone, including 5α-pregnan-3,20-dione, 20α-hydroxypregn-4-en-3-one, 21-hydroxypregn-4-en-3,20-dione, 17β-hydroxy-androstan-3-one and testosterone. Oestradiol-17β, cortisol and corticosterone, however, failed to compete with progesterone for binding to PBG (Milgrom

o

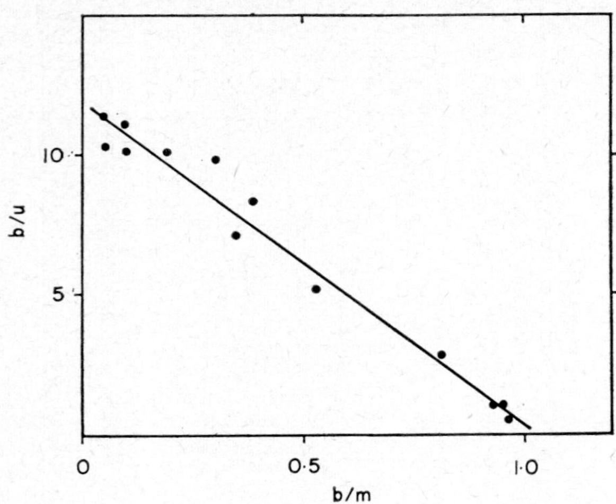

FIG. 6. Number of progesterone-binding sites per PBG molecule. Purified PBG solution from pregnant guinea-pigs was dialysed at 5°C for 46 hours with various amounts of [³H]progesterone in Tris-NaCl buffer. b/m = bound progesterone per mole of PBG (calculated by using $E^{1cm}_{1mg/ml} = 0.49$ and molecular weight $= 77,500$); (b), bound radio-activity; (u), unbound radioactivity (from Milgrom, Allouch, Atger & Baulieu, 1973).

et al., 1973; Table II). The affinity constant of diluted guinea pig plasma for androstenedione at the end of gestation was found to be comparable with that for progesterone, but the binding capacity for androstene-dione was approximately half that for progesterone (D. V. Illingworth, unpublished data).

In summary, it can be seen that PBG is distinguishable from CBG on several grounds; its larger molecular weight, its relative stability to temperature and pH changes, its affinity for 5α- and 20α-substituted steroids and for androgens, and its lack of affinity for the corticosteroids. PBG resembles sex steroid binding plasma protein (SBP, Mercier-Bodard *et al.*, 1970) only insofar as it shows some affinity for testo-sterone, but, unlike SBP, it lacks affinity for oestrogens such as oestra-diol-17β.

Determination in hystricomorph rodents

Following the identification of a plasma protein in pregnant guinea-pigs with a high affinity for progesterone and with binding properties that can be distinguished from CBG and other plasma proteins, a further question concerns the temporal relations between the concentra-tion of PBG in plasma and the change in the dynamics of progesterone

TABLE II

Steroid specificity of PBG (from Milgrom *et al.*, 1973)

Unlabelled steroid	Competing efficiency with [³H]progesterone
Progesterone	100
5α-Pregnan-3,20-dione	97
20α-Hydroxy-pregn-4-en-3-one	90
21-Hydroxy-pregn-4-en-3,20-dione	86
17β-Hydroxy-androstan-3-one	69
Testosterone	40
3β-Hydroxy-pregn-5-en-20-one (pregnenolone)	33
5β-Pregnan-3,20-dione	32
20β-Hydroxy-pregn-4-en-3-one	14
17α-Hydroxy-pregn-4-en-3,20-dione	11
Cortisol	0
Corticosterone	0
Estradiol	0

Pregnant guinea-pig plasma was diluted 80-fold with Tris-NaCl buffer and incubated 3 hours at 0°C with 10 nM [³H]progesterone and in some experiments 1 μM unlabelled steroid. The incubate (0·2 ml) was centrifuged 18 hours at 45,000 rpm in a SW 50·1 rotor. Bound radioactivity (B) was calculated. Competing efficiency of various steroids was compared to that of unlabelled progesterone by calculating

$$\frac{(B) \text{ with } [^3H]\text{progesterone alone} - (B) \text{ with } [^3H]\text{progesterone and competing steroid}}{(B) \text{ with } [^3H]\text{progesterone alone} - (B) \text{ with } [^3H]\text{progesterone and unlabelled progesterone}} \times 100.$$

metabolism during gestation. A simple method was devised to estimate high-affinity binding of plasma progesterone at different times of gestation and during lactation (Illingworth & Deanesly, 1972). The method was based on the finding that the extraction of progesterone by petroleum ether from samples of pregnant guinea-pig plasma was much lower than from non-pregnant females (Challis *et al.*, 1971). After the incubation of a tracer quantity of labelled progesterone with 0·5 ml guinea-pig plasma, the percentage extraction of radioactivity with petroleum ether was greater than 95% in samples taken before Day 15 of pregnancy. Thereafter the extraction decreased to values of about 30% by Day 25 *p.c.*, remaining low for the rest of gestation until after parturition when it returned to 95–100% extraction by Day 10 *post partum* (Fig. 7; Illingworth & Deanesly, 1972). These results indicated a

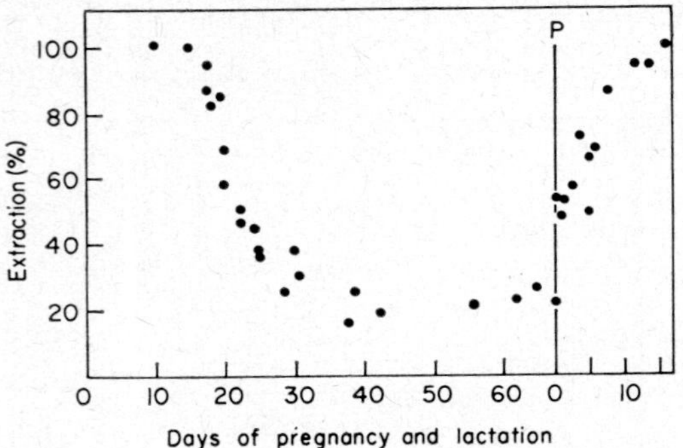

Fig. 7. Petroleum ether extraction of added labelled progesterone from guinea-pig plasma during pregnancy and lactation. The amount of labelled steroid extracted from plasma samples by petroleum ether is expressed as a percentage of the radioactivity added. P =day of parturition (from Illingworth & Deanesly, 1972).

significant increase in high-affinity plasma binding of progesterone after about the 15th day of gestation which is the time when the MCR of progesterone starts to decrease.

High-affinity progesterone binding in plasma has now been measured by an adsorbent technique using florisil. Plasma samples from pregnant guinea-pigs were obtained either by cardiac puncture under light ether anaesthesia or from conscious animals with an indwelling carotid catheter. After Day 13 of pregnancy the concentration of plasma progesterone and PBG increased in a parallel manner and by the fourth week of gestation the values were about 250 ng/ml and $10\text{--}15 \times 10^{-6}$M respectively. The association constant measured by the florisil technique was 41×10^{8}M^{-1} at 4°C (Table III). Milgrom *et al.* (1973) obtained values of $15\cdot1\text{--}17\cdot6 \times 10^{-6}$M for PBG concentration in a 50-day pregnant guinea-pig and an association constant of 9×10^{8}M^{-1} for pure PBG using equilibrium dialysis. The difference between the two sets of data probably relates to the techniques used. The florisil technique, for example, does not distinguish between PBG and CBG-binding of progesterone and since the association constant of progesterone and guinea-pig CBG at 4°C is 48×10^{8}M^{-1} (Westphal, 1967) and the concentration of CBG is about 14×10^{-6}M (Rosenthal *et al.*, 1969b), CBG-binding of progesterone probably contributes to the values obtained by this method. When plasma was heated to 60°C for 15 min, largely abolishing the binding of cortisol by maternal plasma in pregnancy, the values obtained for progesterone binding decreased by 20%. An additional feature of these determinations concerns the removal of endogenous steroids such as progesterone and corticosteroids from

TABLE III

Concentration of plasma progesterone, and the concentration and association constant of progesterone-binding globulin (PBG) in pregnant guinea-pigs

Days pregnant	Plasma progesterone (ng/ml)	PBG Concentration $\times 10^{-6}$M	PBG Association constant $\times 10^8$ M^{-1}
11	6	n.d.	—
12	10	n.d.	—
13	5	n.d.	—
14	16(2)	2·2(1)	34(1)
15	16(5)	0·6(5)	72(3)
16	21(4)	0·5(3)	53(2)
17	39(3)	0·7(2)	63(1)
18	57(5)	1·3(5)	70(4)
19	105(6)	1·9(6)	76(5)
20	120(5)	3·5(5)	43(4)
21	203(4)	4·2(4)	37(3)
22	162(4)	5·8(4)	47(3)
23	276(4)	7·8(4)	27(4)
24	259(5)	7·6(5)	36(5)
25	240(5)	8·1(5)	37(4)
26	311(4)	8·4(4)	37(4)
27	262(5)	9·5(5)	28(5)
28	328(4)	10·2(4)	38(4)
29	285(5)	11·1(5)	24(5)
30	304(2)	13·0(2)	23(2)
31	273(2)	9·2(2)	31(2)
32	278(2)	11·3(2)	31(2)
33	144(1)	10·9(1)	43(1)
34	312(2)	14·8(2)	26(2)
35	122(1)	10·5(1)	42(1)
36	260(1)	11·4(1)	36(1)
59	155(1)	24(1)	16(1)

Number of animals in parentheses. n.d., not detectable.

plasma before the concentration of binding proteins was determined. Westphal (1967) and Milgrom *et al.* (1973) treated plasma to eliminate the contribution of endogenous steroids, whereas in our experiments progesterone binding was measured in the presence of the normal endogenous levels of steroids and the results were corrected for the concentration of endogenous progesterone (or cortisol in the case of CBG).

The time when PBG was first detected in maternal plasma of pregnant guinea-pigs corresponds closely with that of the increased concentration of plasma progesterone (Fig. 8). Although in Fig. 8 the increase in PBG concentration appears to follow rather than precede the rising progesterone level, it should be noted that a PBG concentration of 3.5×10^{-6}M at Day 20 *p.c.* (Table III) represents a binding capacity for progesterone of 1099 ng/ml. Since the actual concentration of plasma progesterone at this time is 120 ng/ml of which more than 90% is bound to high-affinity proteins, it is apparent that even a low level of PBG has

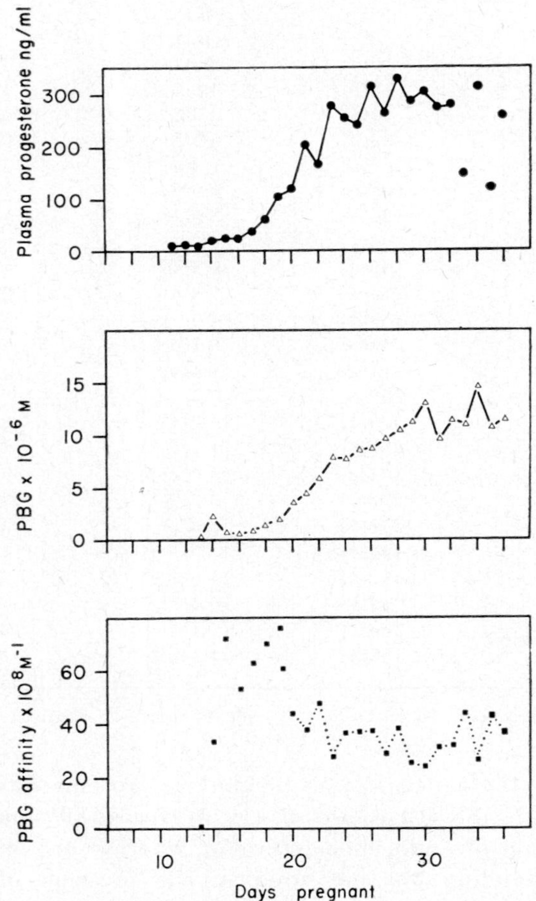

FIG. 8. The concentration of plasma progesterone and of progesterone-binding globulin (PBG), and the association constant of PBG, from Day 11 to Day 36 in pregnant guinea-pigs.

the capacity to bind all the progesterone present in plasma after about Day 14 of gestation. Further studies of PBG in the guinea-pig reveal that there is a positive correlation between the concentration of PBG and plasma progesterone levels (degrees of freedom, 76; correlation coefficient, 0·6144, $P < 0·05$). Furthermore the amount of binding protein does not decrease until after the delivery of the fetuses (Fig. 6) when MCR begins to rise (Illingworth & Heap, 1971). In contrast to CBG, which is present in both fetal and maternal blood, the larger molecule of PBG is excluded from the fetal compartment even when the maternal levels have reached their maximum values, indicating placental impermeability to the protein (see also Milgrom et al., 1973). It is notable that little or no progesterone can be found in fetal plasma during gestation (Illingworth & Heap, 1971).

A high-affinity progesterone binding protein has been detected so far in seven species of hystricomorph rodents (Table IV; Illingworth et al., 1973), but it has not been detected in plasma from murine rodents (rat and mouse) or from rabbit or women (MacLaughlin et al., 1972; Milgrom et al., 1973), ferret, pig or sheep (D. V. Illingworth & R. B. Heap, unpublished observations). A comprehensive study of PBG in two hystricomorphs, the cuis and casiragua, has been carried out on

TABLE IV

Progesterone binding protein(s) of high affinity in hystricomorph rodents

Species	Capacity $\times 10^{-6}$M	Affinity $\times 10^{-8}$M[1]	% reduction in binding after heating to 60°C
Cavia porcellus (guinea-pig)	13·6 ± 0·9	25·3 ± 2·7	19·2
Proechimys guairae (casiragua)	7·9 ± 1·1	32·7 ± 3·5	42·9
Galea musteloides (cuis)	7·9 ± 1·1	17·8 ± 2·2	30·1
Myocastor coypus (coypu)	15·2	15·0	—
Chinchilla laniger (chinchilla)	15·9	5·6	—
Lagostomus maximus (plains viscacha)	5·0	10·7	—
Ctenomys talarum (tuco-tuco)	3·2	34·5	—

plasma samples taken throughout pregnancy. The cuis (a wild guinea-pig) resembles the domestic guinea-pig in that the secretion of PBG begins in early pregnancy and continues till term (Fig. 9). In the casira-gua, which is believed to be a generalized hystricomorph, the pattern of binding protein production is quite different. Although the concentrat-ions of progesterone-binding protein and plasma progesterone both rise in gestation, the increase occurs relatively late in pregnancy (Fig. 9). The high-affinity binding protein has been detected in the coypu when plasma progesterone levels are high, and in the chinchilla where circu-lating progesterone levels rise to a lesser degree above those seen during the oestrous cycle (Tam, 1974).

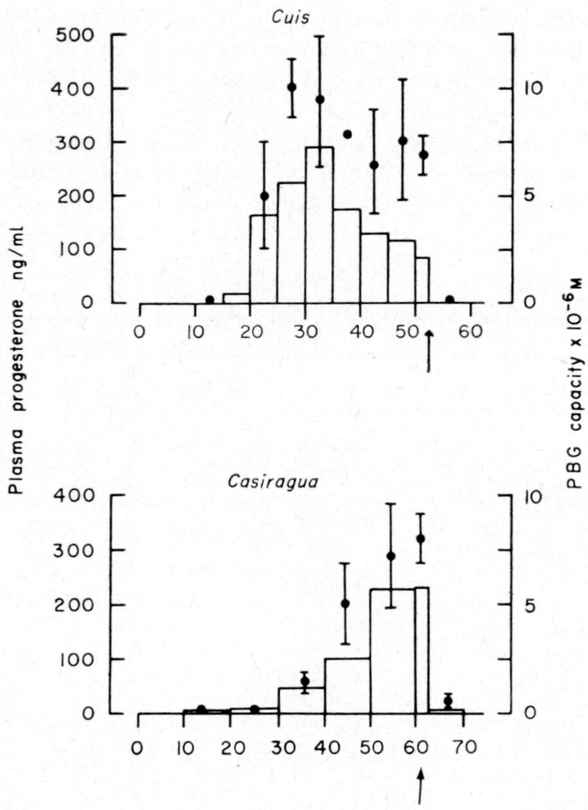

Fig. 9. The concentration of plasma progesterone (histogram) and high-affinity progesterone binding proteins (mean ± S.E.M.) in *Galea musteloides* and *Proechimys guairae*. Arrow denotes day of parturition.

Relationship to a uterine receptor for progesterone

In the guinea-pig, Milgrom *et al.* (1970b) detected a uterine cytosol receptor (progesterone binding protein, uterine PBP) that differs from PBG in that it is thermolabile when heated at 60°C for 30 min whereas PBG is thermostable; it is very sensitive to p-hydroxymercuribenzoate whereas PBG is not; and it has a sedimentation velocity of 6·7S compared with a value of 4·5S for PBG (Milgrom *et al.*, 1970b; Milgrom *et al.*, 1972). Recently, Kontula *et al.* (1972) have reported the partial purification of a progesterone receptor from the pregnant guinea-pig uterus with a sedimentation velocity of 5S and an acid isoelectric point of pH 2·5. At the present time there is some doubt about whether the progesterone receptor is related to the PBG molecule. Kontula *et al.* (1972) found that the binding specificity of the uterine receptor differed from that of PBG, and MacLaughlin *et al.* (1973) reported that the uterine receptor differed in sedimentation velocity and Stokes radius. However, the finding that the uterine receptor and PBG I share common properties e.g. electrophoretic mobility (MacLaughlin *et al.*, 1973), suggests that the two binding components may be derived from a similar source.

Stimulus for synthesis

The production of PBG in hystricomorph rodents is associated with pregnancy, and time will tell whether this group of rodents is unique in this respect. As yet, there is no convincing proof that PBG synthesis can be induced in the non-pregnant guinea-pig. Among the treatments tested have been injections of the steroid hormones, oestrogen, progesterone, and testosterone, and of fetal and placental homogenates (R. B. Heap, unpublished observations; Rosenthal *et al.*, 1969b; Milgrom *et al.*, 1973). The transplantation of placental tissue prolongs the life-span of the corpus luteum in the non-pregnant animal by its anti-luteolytic action (Bland & Donovan, 1969) but fails to induce PBG formation (D. V. Illingworth & C. A. Samuel, unpublished observations). The synthesis of PBG depends neither on the presence of the maternal pituitary, which can be removed early in gestation without preventing the continuation of pregnancy, nor on the secretory activity of the ovaries. However, PBG synthesis is dependent on the presence of a viable conceptus—a term used in this context to describe the products of conception, the fetus, fetal membranes and the placenta. When pregnant guinea-pigs were ovariectomized before implantation and pregnancy was maintained after the first week with implants of synthetic progestagens, an appreciable concentration of PBG was produced, though plasma progesterone concentrations were low. In those animals in which pregnancy was failing and fetuses were dying, presumably because of an inadequate supply of progestagen, no PBG could be detected (Illingworth & Deanesly, 1972). There is evidence from the studies of Diamond *et al.* (1969) that testosterone propionate (1 mg/day) injected from Day 18 *p.c.* will stimulate the progesterone binding activity of plasma in pregnant guinea-pigs (22–66 days *p.c.*). In addition to a possible role of testosterone in PBG production during pregnancy, the effect of a synthetic progestagen, or of progesterone itself, cannot be ignored. Ovariectomized guinea-pigs with progesterone implants had circulating levels of progesterone considerably greater than those of normal animals, even though the absorption rate from the implants was comparable with the amount normally produced in intact pregnant females (Heap & Deanesly, 1967; Illingworth & Deanesly, 1972).

As we have seen, the secretion of PBG into the maternal circulation of the pregnant guinea-pig begins on, or shortly after, Day 14 *p.c.* The synthesis of high-affinity binding proteins that play a part in the transport of steroid hormones in plasma is usually attributed to the liver. But there are features about the occurrence of PBG that suggest the gravid uterus may be closely involved in PBG synthesis, namely the necessity for the presence of a viable conceptus and the disappearance of PBG from the circulation after parturition or when the conceptus fails to develop owing to a progestagen deficiency. A deficiency in progesterone secretion from about Day 12 of pregnancy causes a slowing down in development, and the death of many embryos, from about Day 14 *p.c.* (Deanesly, 1963). More recent work (Deanesly, 1972) shows that the failure of pregnancy in ovariectomized guinea-pigs not receiving progesterone treatment is due to the total collapse of the decidua with massive haemorrhage, almost always on Day 16 *p.c.*, the approximate duration of the oestrous cycle and of artificially induced deciduomata and also about the time when PBG synthesis is initiated.

There is a close temporal relation between the regression of the corpus luteum in the non-pregnant animal, the luteotrophic effect of the conceptus, and the onset of PBG production in the pregnant guinea-pig (Heap, Illingworth & Perry, 1973). It is about this time of pregnancy (Day 15 *p.c.*) that the allantois makes contact with the chorion and the definitive allantochorionic placenta becomes established (Duval, 1892). This first point of contact is the central region where the sub-placenta becomes differentiated, a structure considered to be unique to this group of rodents (Mossman, 1937; Perrotta, 1959) and one which contains droplets of a PAS-positive, saliva-resistant material, probably glycoprotein or mucoprotein in nature (Davies *et al.*, 1961). It may be that this site is involved in the synthesis of PBG, a protein known to have a high carbohydrate content.

On the other hand, although the formation of the definitive allanto-chorionic placenta is closely correlated with the onset of PBG synthesis in the guinea-pig and the cuis, this is not so in the casiragua where the allantois meets the chorion at the 15th to 20th day but PBG synthesis is not detectable in significant amounts until Days 30–40 of pregnancy.

The absence of PBG in fetal plasma suggests that the protein is not secreted by tissues of the fetal placenta. However, it is possible that PBG synthesis may involve the interaction of several tissues within the gravid uterus, particularly in view of the isolation from the pregnant uterus of PBG I, a molecule related to PBG in plasma (MacLaughlin *et al.*, 1973). Preliminary results of incubations of guinea-pig conceptuses showed that a binding component was synthesized *in vitro* which resembled PBG in terms of its thermostability and binding affinity.

Physiological significance

From the evidence reviewed above it appears that pregnancy maintenance in hystricomorph rodents such as the guinea-pig, coypu, cuis and casiragua involves the development of a progesterone-conserv-ing mechanism almost certainly dependent on the increased synthesis of PBG which markedly reduces the MCR of progesterone. In the guinea-pig it seems probable that the conceptus provides a signal for PBG synthesis and thereby protects its own future. Whether such a signal occurs at a precise time in the development of the embryo is not yet known (Heap, Perry & Challis, 1973), but Deanesly (1963) showed that, in guinea-pigs ovariectomized three or four days *p.c.*, embryos usually survived for a further 10 days after which their survival depended on progesterone treatment. In this respect, it appears that progesterone itself is essential to sustain a chain of events that permit normal embryo development and the induction of PBG synthesis, a protein that subsequently protects the hormone from metabolism.

The role of PBG in shielding progesterone from metabolism can be deduced from the temporal relations between the induction of PBG synthesis and the decrease in MCR of progesterone. Furthermore, the half-life of progesterone in plasma is greatly increased from one minute in the non-pregnant guinea-pig to about 20 min in the pregnant animal (Illingworth, 1970; Heap, Illingworth & Perry, 1973). Calculations of the apparent sizes of the theoretical pools of progesterone distribution show that an inner compartment in the non-pregnant animal has an apparent volume approximately equivalent to that of extra-cellular fluid, whereas in the pregnant animal it is closely comparable with plasma volume. This indicates that the high progesterone concentrations are confined to plasma and leads to the supposition that PBG may be also confined to a vascular compartment. Thus, the high concentration of PBG in maternal blood may not only shield progesterone from very rapid metabolism in hepatic and extra-hepatic tissues, but may also protect extra-vascular tissues from the effects of high levels of progesterone. It is notable that endogenous progesterone is undetectable in fetal blood even when the concentration in maternal blood is high. In contrast corticosteroid and CBG levels are high on both sides of the placenta. The most probable explanation of these findings is that the placenta is impermeable to PBG, but not to CBG, or that the fetus is capable of producing CBG but not PBG. It is improbable that progesterone is transferred across the placenta and rapidly metabolized by the fetus for, when infused intravenously into the mother, the concentration of [^{3}H]progesterone or of ^{3}H-labelled compounds in fetal blood 3–4 h later are both very low compared with the amounts in maternal blood (Castrén et al., 1960; Illingworth & Heap, 1971). These results contrast with the findings for pregnant women in which the endogenous concentration of progesterone is extremely high in both fetal and maternal blood, and fetal metabolism of progesterone is appreciable.

Diamond et al. (1969) postulated that the high-affinity binding of androgens by plasma proteins protected the fetuses against excessive androgenization. Such effects in the mother have been observed in ovariectomized pregnant guinea-pigs treated with a synthetic progestagen, norgestrel, to maintain gestation, but only after the animals had delivered their young and PBG concentrations had fallen (Illingworth & Challis, 1973). Preliminary results indicated that during gestation norgestrel, like norethynodrel (MacLaughlin et al., 1972), was bound by a high-affinity plasma protein thus reducing its androgenic side-effect.

Further support for the hypothesis that high-affinity protein binding reduces the biological potency of progesterone has been derived from experiments showing that when progesterone was bound to a_1-acid glycoprotein (Westphal & Forbes, 1963), CBG, or to high concentrations of albumin (Hoffmann et al., 1969), biological activity, as measured in the mouse, was reduced or eliminated. In addition, progesterone bound to CBG and to high concentrations of human serum albumin was protected against metabolism by human placental 20α-hydroxysteroid dehydrogenase in vitro (Billiar et al., 1969; see Westphal, 1971).

Thus the physiological role of PBG in gestation may reside in its capacity to shield progesterone from the high rate of metabolism that normally occurs in the absence of this binding protein. Its presence in plasma also provides a relatively large concentration of circulating progesterone which is bound yet in dynamic equilibrium with that fraction which is in a free and biologically active form. It can be seen that in addition to its progesterone-conserving function PBG may also, by its buffering effect, serve to protect vulnerable tissues from high concentrations of potent steroids. Whether the occurrence of PBG is a unique feature of the economy of pregnancy maintenance in a group of rodents whose gestation is unusually long relative to their adult weight, or whether PBG has some specialized function in the intracellular mechanism of action of progesterone, are problems for future studies.

ACKNOWLEDGMENTS

We gratefully acknowledge the expert technical assistance of Mrs Nicola Ackland, and the generous supply of experimental material by Drs Barbara J. Weir, Christine Roberts and I. W. Rowlands from the colonies of hystricomorph rodents at the Wellcome Institute of Comparative Physiology. D. V. I. thanks the Lalor Foundation for the support of a research fellowship held during these studies.

RERERENCES

Baird, D. T., Horton, R., Longcope, C. & Tait, J. F. (1969). Steroid dynamics under steady-state conditions. *Recent Prog. Horm. Res.* **25**: 611–656.

Billiar, R. B., Tanaka, Y., Knappenberger, M. Hernandez, R. & Little, B. (1969). Influence of transcortin and albumin on the rate of reduction of progesterone by human placental 20α-hydroxysteroid dehydrogenase. *Endocrinology* **84**: 1152–1160.

Bland, K. P. & Donovan, B. T. (1969). Observations on the time of action and the pathway of the uterine luteolytic effect of the guinea-pig. *J. Endocr.* **43**: 259–264.

Burton, R. M., Harding, G. B., Rust, N. & Westphal, U. (1971). Steroid-protein interactions. XXIII. Nonidentity of cortisol-binding globulin and progesterone-binding globulin in guinea-pig serum. *Steroids* **17**: 1–16.

Castrén, O., Hirvönen, L., Närvänen, S. & Soiva, K. (1960). On the permeability of the guinea pig placenta to intravenously injected progesterone-4-^{14}C. *Acta endocr., Copenh.* **35**: 204–210.

Challis, J. R. G., Heap, R. B. & Illingworth, D. V. (1971). Concentrations of oestrogen and progesterone in the plasma of non-pregnant, pregnant and lactating guinea-pigs. *J. Endocr.* **51**: 333–345.

Courrier, R., Kehl, R. & Raynaud, R. (1929). Neutralisation de l'hormone folliculaire chez la femelle gestante castrée. *C. r. Séanc. Soc. Biol.* **100**: 1103–1105.

Daughaday, W. H. (1956). Evidence for two corticosteroid binding systems in human plasma. *J. Lab. clin. Med.* **48**: 799–800.

Daughaday, W. H. (1958). Binding of corticosteroids by plasma proteins. III. The binding of corticosteroid and related hormones by human plasma and plasma protein fractions as measured by equilibrium dialysis. *J. clin. Invest.* **37**: 511–518.

Davies, J., Dempsey, E. W. & Amoroso, E. C. (1961). The subplacenta of the guinea-pig: development, histology and histochemistry. *J. Anat.* **95**: 457–473.

Deanesly, R. (1963). Early embryonic growth and progestagen function in ovariectomized guinea-pigs. *J. Reprod. Fert.* **6**: 143–152.

Deanesly, R. (1972). Retarded embryonic development and pregnancy termination in ovariectomized guinea-pigs: progesterone deficiency and decidual collapse. *J. Reprod. Fert.* **28**: 241–247.

Diamond, M., Rust, N. & Westphal, U. (1969). High-affinity binding of progesterone, testosterone and cortisol in normal and androgen-treated guinea-pigs during various reproductive stages: relationship to masculinization. *Endocrinology* **84**: 1143–1151.

Duval, M. (1892). Le placenta des Rongeurs. III. Le placenta du Cochon d'Inde. *J. Anat. Physiol., Paris* **28**: 58–98.

Feder, H. H., Resko, J. A. & Goy, R. W. (1968), Progesterone concentrations in the arterial plasma of guinea-pigs during the oestrous cycle. *J. Endocr.* **40**: 505–513.

Heap, R. B. (1964). A fluorescence assay of progesterone. *J. Endocr.* **30**: 293–305.

Heap, R. B. (1969). The binding of plasma progesterone in pregnancy. *J. Reprod. Fert.* **18**: 546–548.

Heap, R. B. & Deanesly, R. (1966). Progesterone in systemic blood and placenta of intact and ovariectomized pregnant guinea-pigs. *J. Endocr.* **34**: 417–423.

Heap, R. B. & Deanesly, R. (1967). The increase in plasma progesterone levels in the pregnant guinea-pig and its possible significance. *J. Reprod. Fert.* **14**: 339–341.

Heap, R. B., Gwyn, M., Laing, J. A. & Walters, D. E. (1973). Pregnancy diagnosis in cows; changes in milk progesterone concentration during the oestrous cycle and pregnancy measured by a rapid radioimmunoassay. *J. agric. Sci., Camb.* **81**: 151–157.

Heap, R. B., Illingworth, D. V. & Perry, J. S. (1973). The secretory activity of the corpus luteum in the guinea-pig and its role in the establishment and maintenance of pregnancy. In *Le corps jaune*: 69–80. Denamur, R. & Netter, A. (eds). Paris: Masson.

Heap, R. B., Perry, J. S. & Challis, J. R. G. (1973). Hormonal maintenance of pregnancy. In *Handbook of physiology*, Section 7: Endocrinology. **2**: 217-260. Washington, D.C.: American Physiological Society.

Heap, R. B., Perry, J. S. & Rowlands, I. W. (1967). Corpus luteum function in the guinea-pig; arterial and luteal progesterone levels, and the effects of hysterectomy and hypophysectomy. *J. Reprod. Fert.* **13**: 537–553.

Herrick, E. H. (1928). The duration of pregnancy in the guinea-pig after removal and also after transplantation of ovaries. *Anat. Rec.* **39**: 193–200.

Hoffman, W., Forbes, T. R. & Westphal, U. (1969). Biological inactivation of progesterone by interaction with corticosteroid-binding globulin and with albumin. *Endocrinology* **85**: 778–781.

Huggett, A. St. G. & Widdas, W. F. (1951). The relationship between mammalian foetal weight and conception age. *J. Physiol., Lond.* **114**: 306–317.

Illingworth, D. V. (1970). Kinetics of progesterone metabolism in pregnant and non-pregnant guinea-pigs. *J. Physiol., Lond.* **210**: 99–100P.

Illingworth, D. V., Ackland, N., Heap,, R. B. & Weir, B. J. (1973). Progesterone binding proteins; occurrence, capacity and binding affinity in hystrico-morph rodents. *J. Endocr.* **58**: ii.

Illingworth, D. V. & Challis, J. R. G. (1973). Concentrations of oestrogens and progesterone in the plasma of ovariectomized and ovariectomized-Norgestrel-treated pregnant guinea-pigs. *J. Reprod. Fert.* **34**: 289–296.

Illingworth, D. V. & Deanesly, R. (1972). Maintenance of pregnancy by synthetic progestagens in guinea-pigs ovariectomized before implantation; progest-erone-binding protein and placental progesterone secretion. *J. Endocr.* **54**: 435–444.

Illingworth, D. V. & Heap, R. B. (1971). A decrease in the metabolic clearance rate of progesterone in the coypu during pregnancy. *J. Reprod. Fert.* **27**: 492–494.

Illingworth, D. V., Heap, R. B., & Perry, J. S. (1970). Changes in the metabolic clearance rate of progesterone in the guinea-pig. *J. Endocr.* **48**: 408–417.

Kontula, K., Jänne, O., Jänne, J. & Vihko, R. (1972). Partial purification and characterization of progesterone-binding protein from pregnant guinea pig uterus. *Biochem. Biophys. Res. Commun.* **47**: 596–603.

MacLaughlin, D. T., Burton, R. M., Aboul-Hosn, W. & Westphal, U. (1973). Progesterone-binding globulin (PBG) and uterine progesterone receptor in the pregnant guinea-pig. *Fedn Proc. Fedn Am. Socs exp. Biol.* **32**: 1962A.

MacLaughlin, D. T., Harding, G. B. & Westphal, U. (1972). Steroid-protein interactions. XXV. Binding of progesterone and cortisol in pregnancy sera; progesterone-binding globulin and uterine cytosol receptor in the pregnant guinea-pig. *Am. J. Anat.* **135**: 179–186.

Mercier-Bodard, C., Alfsen, A. & Baulieu, E.-E. (1970). Sex steroid binding plasma protein (SBP). In *Steroid assay by protein binding*: 204–221. Diczfalusy, E. (ed.). Karolinska Symposia on Research Methods in Reproductive Endo-crinology. Copenhagen: Bogtrykkeriet Forum.

Milgrom, E., Allouch, P., Atger, M. & Baulieu, E.-E. (1973). Progesterone-binding plasma protein of pregnant guinea-pig. *J. biol. Chem.* **248**: 1106–1114.

Milgrom, E., Atger, M. & Baulieu, E.-E. (1970a). Progesterone binding plasma protein (PBP). *Nature, Lond.* **228**: 1205–1206.

Milgrom, E., Atger, M. & Baulieu, E.-E. (1970b). Progesterone in uterus and plasma. IV. Progesterone receptor(s) in guinea-pig uterus cytosol. *Steroids* **16**: 741–754.

Milgrom, E.. Atger, M., Perrot, M. & Baulieu, E.-E. (1972). Progesterone in uterus and plasma. VI. Uterine progesterone *receptors* during the estrus cycle and implantation in the guinea-pig. *Endocrinology* **90**: 1071–1078.

Mossman, H. W. (1937). Comparative morphogenesis of the fetal membranes and accessory uterine structures. *Contr. Embryol.* **26**: 129–246.

Newson, R. M. (1966). Reproduction in the feral coypu (*Myocastor coypus*). *Symp. zool. Soc. Lond.* No. 15: 323–334.

Paterson, J. Y. F. (1973). The rate constants for the interaction of cortisol and transcortin, and the rate of dissociation of transcortin-bound cortisol in the liver. *J. Endocr.* **56**: 551–570.

Pegg, P. J. & Keane, P. M. (1969). The simultaneous estimation of plasma cortisol and transcortin binding characteristics by a competitive protein binding technique. *Steroids* **14**: 705–715.

Perrotta, C. A. (1959). Fetal membranes of the Canadian porcupine, *Erethizon dorsatum*. *Am. J. Anat.* **104**: 35–60.

Roberts, C. M. (1973). *The embryology of certain hystricomorph rodents*. Ph.D. Thesis, Univ. of London.

Rosenthal, H. E., Slaunwhite, W. R., Jr & Sandberg, A. A. (1969a). Transcortin: a corticosteroid-binding protein of plasma. X. Cortisol and progesterone interplay and unbound levels of these steroids in pregnancy. *J. clin. Endocr. Metab.* **29**: 352–367.

Rosenthal, H. E., Slaunwhite, W. R., Jr & Sandberg, A. A. (1969b). Transcortin: A corticosteroid-binding protein of plasma. XI. Effects of estrogens on pregnancy in guinea-pigs. *Endocrinology* **85**: 825–830.

Rowlands, I. W. & Heap, R. B. (1966). Histological observations on the ovary and progesterone levels in the coypu, *Myocastor coypus*. *Symp. zool. Soc. Lond.* No. 15: 335–352.

Sandberg, A. A. & Slaunwhite, W. R., Jr (1959). Transcortin: a corticosteroid-binding protein of plasma. II. Levels in various conditions and the effects of estrogens. *J. clin. Invest.* **38**: 1290–1297.

Sandberg, A. A., Woodruff, M., Rosenthal, H., Nienhouse, S. & Slaunwhite, W. R., Jr (1964). Transcortin: a corticosteroid-binding protein of plasma. VII. Half-life in normal and estrogen-treated subjects. *J. clin. Invest.* **43**: 461–466.

Scatchard, G. (1949). The attraction of proteins for small molecules and ions. *Ann. N.Y. Acad. Sci.* **51**: 660–672.

Seal, U.S. & Doe, R. P. (1961). Purification and properties of "transcortin", the corticosteroid-binding globulin. *Fedn Proc. Fedn Am. Socs exp. Biol.* **20**: 179.

Seal, U. S. & Doe, R. P. (1966). Corticosteroid-binding globulin: biochemistry, physiology and phylogeny. In *Steroid dynamics*: 63–88. Pincus, G., Nakao, T. & Tait, J. F. (eds). New York: Academic Press.

Seal, U. S. & Doe, R. P. (1967). The role of corticosteroid-binding globulin in mammalian pregnancy. Proc. 2nd Int. Congr. on Hormonal Steroids. *Excerpta Med. Int. Congr. Series* No. 132: 697–706.

Tait, J. F. & Burstein S. (1964). *In vivo* studies of steroid dynamics in man. In *The hormones* 5: 441–557. Pincus, G., Thimann, K. V. & Astwood, E. B. (eds). New York: Academic Press.

Tam, W. H. (1974). The synthesis of progesterone in some hystricomorph rodents. *Symp. zool. Soc. Lond.* No. 34: 363–384.

Weir, B. J. (1970). Chinchilla. In *Reproduction and breeding techniques of laboratory animals*: 209–223. Hafez, E. S. E. (Ed). Pennsylvania: Lea & Febiger.

Weir, B. J. (1972). Laboratory hystricomorph rodents other than the guinea-pig and chinchilla. In *The UFAW Handbook on the care and management of laboratory animals*: 278–286. 4th edn. UFAW (Ed.) Edinburgh and London: Churchill Livingstone.

Weir, B. J. (1973). Another hystricomorph rodent: keeping casiragua (*Proechimys guairae*) in captivity. *Lab. Anim.* **7**: 125–134.

Weir, B. J. (1974). Reproductive characteristics of hystricomorph rodents. *Symp. zool. Soc. Lond.* No. 34: 265–301.

Westphal, U. (1967). Steroid-protein interactions. XIII. Concentrations and binding affinities of corticosteroid-binding globulins in sera of man, monkey, rat, rabbit, and guinea-pig. *Archs Biochem.* **118**: 556–567.

Westphal, U. (1971). *Steroid-protein interactions*. Berlin: Springer-Verlag.

Westphal, U. & Forbes, T. R. (1963). Biological inactivation of progesterone by binding to alpha-1 acid glycoprotein (orosomucoid). *Endocrinology* **73**: 504–507.

Yudaev, N. A., Rozen, V. B. & Mikosha, A. S. (1964). Plasma binding of hydrocortisone in estrogen-treated guinea-pigs. *Prob. Endokrinol. Gormonoterap.* **10**: 73. Translated in *Fedn Proc. Fedn Am. Socs exp. Biol.* **23** Translation Suppl. T-1264-1266 (1964).

DISCUSSION

LAZARUS (*Dartford*): Could you tell me firstly, whether you have purified the protein, secondly, is it difficult to purify and, thirdly, have you ever used it as an assay for progesterone?

HEAP (*Cambridge*): The binding globulin has been used as an assay for progesterone by a number of workers. For instance, Pichon & Milgrom (1973)* have used it as a rapid assay of progesterone for women during pregnancy and the menstrual cycle. The protein is being currently investigated by a number of groups particularly those of Professor Westphal in the United States and Professors Milgrom and Baulieu in Paris. Professor Baulieu's group has recently published evidence relating to the purification and isolation of this protein, and the figures I have quoted for molecular weight are based on their observations (Milgrom *et al.*, 1973†). There is some controversy about the detailed physico-chemical properties of this protein because the reports from the laboratories of Professors Westphal and Milgrom differ. This is a field which obviously needs further investigation. One point that emerges from Professor Milgrom's work is that each molecule of progesterone-binding globulin combines with one molecule of steroid. Plasma contains about 1 mg/ml of this protein, which is an extraordinarily high concentration; the plasma concentration of progesterone in mid-pregnancy is approximately 300 ng/ml. There is, therefore, a very large excess of PBG for progesterone binding.

TAM (*London, Ont.*): Do you think changes in the uterus have got something to do with triggering this progesterone-binding protein?

HEAP: It would be interesting if that is the case. We have examined the time of production of this protein (PBG) with the formation of the definitive allantochorionic placenta in the guinea-pig. The protein occurred precisely at the time when the allantois appeared to fuse with the chorion (see p. 406). This coincidence holds for one or two other species, e.g. the cuis and possibly the coypu, but not the casiragua in which the definitive allantochorionic placenta forms between Days 16 and 20 of gestation and the protein is not produced until Days 30 to 35. Another possibility, hinted at this morning, is that the

---

* Pichon & Milgrom, E. (1973). *Steroids* **21**: 335–346.

† Milgrom, E., Allovch, P., Atger, M. & Baulieu, E.-E. (1973), *J. biol. Chem.* **248**: 1106–1114,

subplacenta may be involved. Professor Amoroso has described the presence of PAS-positive material in this structure and it is interesting to note that 48·7% of the progesterone-binding globulin is carbohydrate. This is an unusually high figure and obviously, the subplacenta cannot be ignored as a source of PBG synthesis. But, bssed on previous observations in the field of steroid-binding proteins, one would naturally expect this binding protein to be produced by the liver.

TAM: Have you measured the concentration of this progesterone-binding globulin in guinea-pigs with a deciduoma in the uterus?

HEAP: Yes, and we are unable to detect it.

TAM: So the source cannot be the uterus.

HEAP: It cannot be the decidua!

WEIR (*London*): One of your suggestions was that the binding protein acted as a protection against the high levels of circulating progesterone, but do all hystricomorphs have progesterone levels of the order of 500 ng/ml that you indicated for the guinea-pig?

HEAP: There are a number of species in which the level is only about 50–100 ng/ml, though this does not prove that PBG has no protective function in these species. The most direct information we have about a possible protective function of PGB is that from some recent work by Illingworth & Challis (1973)*. They implanted ovariectomized, pregnant guinea-pigs with Norgestrel, a synthetic progestagen that maintains gestation in the absence of the ovaries. Immediately after delivery, when the binding protein started to disappear, masculinization of the external genitalia of the mother occurred. We subsequently found that Norgestrel is bound by progesterone-binding globulin, and this probably explains why the Norgestrel revealed its androgenicity after parturition and not before. Such evidence suggests that the presence of a high-affinity binding protein may protect the fetus from very high steroid levels that may be circulating during gestation. In other hystricomorph species it may be that the relative order of concentration of PBG and progesterone is important.

WEIR: But they do all appear to have the protein?

HEAP: There is evidence for a high-affinity protein in all the species we have examined but its concentration is not as high as in the guinea-pig and the coypu.

GOODWIN (*London*): It is a bit small for an immunoglobulin, of course, and I gather from what you said you do not think it is an immunoglobulin because extracts from fetuses and other tissues do not provoke

---

* Illingworth, D. V. & Challis. J. R. G. (1973). *J. Reprod. Fert.* **34**: 289–296.

it. Is there any sign that the guinea-pig is tipped off about it? Does it come up faster in the second pregnancy?

HEAP: This is a fascinating point, and one to which we have given some thought. We have no evidence as yet that the protein is stimulated more rapidly in the second gestation or at a higher level of production. But it is interesting to recall the work of Seal & Doe (1966)* concerning corticosteroid-binding globulin. They have drawn attention to the fact that this plasma protein increases in concentration in pregnancy in all species studied so far in which maternal blood bathes fetal tissue. It may be that there is an interaction between fetal and maternal tissues which promotes the synthesis of steroid-binding proteins such as PGB.

* Seal, U. S. & Doe, R. P. (1966). In *Steroid dynamics:* pp. 63–88. New York: Academic Press.

Symp. zool. Soc. Lond. (1974) No. 34, 417–435

HYSTRICOMORPH INSULINS

R. W. J. NEVILLE, BARBARA J. WEIR* and
NORMAN R. LAZARUS

*Diabetes Research Unit, Wellcome Foundation,
Temple Hill, Dartford
and
*Wellcome Institute of Comparative Physiology,
Zoological Society of London,
Regent's Park, London, England*

BACKGROUND

Insulin is a hormone essential to the body if it is to use the various food elements, carbohydrates, fat and protein, in a normal manner for energy and growth (see Fig. 1).

Its primary action is to maintain glucose homeostasis by facilitating glucose utilization, and controlling the formation and storage of hepatic glycogen. Insulin also increases the conversion of glucose to fat for storage and to muscle glycogen for energy requirements, increases the rate of protein synthesis, and decreases the new formation of glucose from protein. It assists in all these activities and many others essential in the regulation of normal metabolism. Insulin lack (see Fig. 1), arising either through damage to the pancreas or by a change in the sensitivity of the body's response to insulin, results in increased glucose levels in the body fluids, the persistent elevation of which constitutes the primary clinical symptom of diabetes mellitus. There is an inability to convert glucose and store it in the liver as glycogen and the rate of glycogen breakdown to glucose is increased. Glucose utilization at the periphery is decreased and gluconeogenesis is increased. Thus, in the diabetic condition, hyperglycaemia, glycosuria and a general debility resulting from protein, carbohydrate and fat wastage are characteristic (Oakley, Pyke & Taylor, 1973).

Knowledge of diabetes is of great antiquity. It was recognized in Ancient Egypt and is described in the Sanscrit Vedic literature of India as "the passing of urine with honey" (see Papaspyros, 1964). Numerous descriptions of the disorder can be found in the works of Arabian, Chinese and Japanese writers throughout several centuries. The association of diabetes with changes or malfunction of the pancreas was recog-

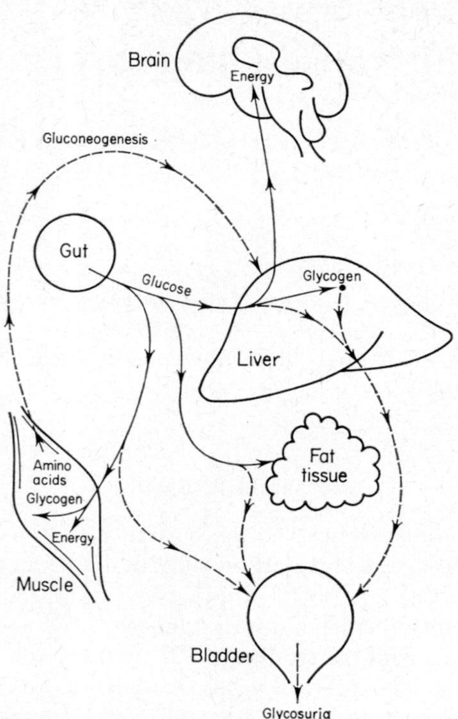

Fig. 1. Diagram of the most important activities of insulin. Solid lines represent glucose movement when insulin is present; broken lines represent glucose movement when insulin is absent.

nized at the end of the eighteenth century, but was confirmed only in 1889 when von Meering & Minkowski discovered that removal of the pancreas of a dog caused sugar to appear in the urine and resulted in a condition in the animal resembling diabetes mellitus in man. Before this, Langerhans (1869) had described particular cells, appearing as islets, in the pancreas and Schafer (1895) suggested that it was from these islets that the substance controlling the metabolism of carbohydrates was secreted. The extraction of insulin in a crude form by Banting, Best & Macleod (1922) in Canada and, independently, although rarely credited for it, Paulesco (1921) in Roumania, was an historic occasion and one which was to revolutionize the treatment of diabetes mellitus in man; it also uncovered many new problems relating to the disease and, despite the notable advance made meanwhile into the chemistry of ox insulin which is used to combat diabetes in man, many questions relating to its pharmacology still remain unanswered. Comparatively little is known about the chemical structure of insulin secreted by mammals other than the domesticated ones. Amongst the few hystricomorph species that have been studied (see below) there is evidence of variation in insulin structure so that an investigation of a range of these animals, available at the Wellcome Institute, seemed worth pursuing.

Our work is directed towards a clarification of those areas of the insulin molecule that are responsible for its biological and its immunological properties in some hystricomorph rodents.

CHEMISTRY

In 1955, Sanger and his co-workers announced the first complete analysis of the primary structure of a protein (Ryle, Sanger, Smith & Kitai, 1955). This protein was insulin and knowledge of the number and order of its amino-acids (Fig. 2) paved the way for studies of the relationships between structure and activity. Ox insulin is a protein of molecular weight 5 850 which consists of two chains, of 21 (A) and 30 (B) amino acid residues, linked by two disulphide bridges, and with an intra-chain disulphide bridge on the A-chain enclosing residues 8, 9 and 10.

Sanger's team then analysed the insulins of the pig, sheep (Brown, Sanger & Kitai, 1955), sperm whale (*Physeter catodon*) and horse (Harris, Sanger & Naughton, 1956). They found that residues 8, 9 and 10 within the intradisulphide bridge of the A-chain were the only positions involved in change (Table I). Smith (1966) later found that the C terminus of the B-chain of rabbit and human insulins was different from that of ox insulin and each other. Smith (1966) also showed that more than one insulin could exist in the same species; in the rat (Table I), positions 9 and 29 in the B-chain are occupied by proline and lysine respectively in one form of the specific insulin and by serine and methionine respectively in the other. The B-9 proline residue was detected by Clark & Steiner (1969) and later confirmed by Smith (1972).

Recently Markussen (1971) and Bünzli, Glatthaar, Kunz, Mülhaupt & Humbel (1972) have shown that the mouse has two insulins and that they are both identical to those of the rat. Balant and his co-authors (1971) obtained evidence for the presence of two insulins in the spiny mouse (*Acomys cahirinus*). However, Bünzli & Humbel (1972) reinvestigated the chemical structure of *Acomys* insulin. Two different strains, black (*A. cahirinus minous*) and yellow (*A. cahirinus dimidiatus*), were investigated and the amino acid compositions of their insulins were shown to be identical. The composition of the A- and B-chains were the same as those chains for the rat and rabbit, respectively (see Table I), and

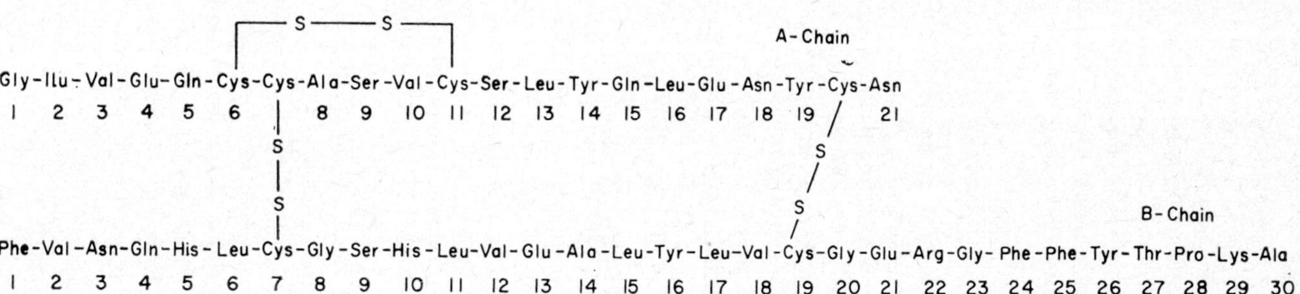

FIG. 2. The primary structure of ox insulin. Note the two chains, A and B, the one intra-chain and two cross-chain disulphide bonds, and the histidine residue at position 10 in the B-chain. (After Ryle *et al.*, 1955.)

Table I

Species differences in the amino acid sequences of some mammalian insulins

Species	Position								Authority
	A-chain				B-chain				
	4	8	9	10	3	9	29	30	
Ox	Glu	Ala	Ser	Val	Asn	Ser	Lys	Ala	Ryle *et al.*, 1955
Pig*, sperm whale†	O	Thr	O	Ilu	O	O	O	O	* Brown *et al.*, 1955
Horse†	O	Thr	Gly	Ilu	O	O	O	O	† Harris *et al.*, 1956
Sheep*	O	O	Gly	O	O	O	O	O	
Rabbit	O	Thr	O	Ilu	O	O	O	Ser	Smith, 1966
Man	O	Thr	O	Ilu	O	O	O	Thr	
Rat 1	Asp	Thr	O	Ilu	Lys	Pro	O	Ser	Smith, 1972
Rat 2	Asp	Thr	O	Ilu	Lys	Ser	Met	Ser	

further studies reinforce the assumption that the amino acid sequences of the chains are identical respectively with the species mentioned. So *Acomys* insulin differs from rabbit insulin by one residue at position A-4 where glutamic acid is replaced by aspartic acid. *Acomys* insulin may thus represent the hypothetical link between rabbit and rat insulins in the scheme of evolution suggested by Smith (1966).

Thus it appeared that although variation did occur, it was of a limited nature, suggesting that the major part of the insulin molecule was necessarily invariant. It was known, however, that antibodies to ox insulin could be raised in the guinea-pig (Moloney & Coval, 1955), indicating that the guinea-pig possessed a very different insulin, probably with major sequence changes, from that of the ox whose insulin was recognized as a foreign protein. This supposition was confirmed when Smith in 1966 and 1972 analysed the primary structure of guinea-pig insulin (Fig. 3). Changes were found at nine positions in the A-chain and at ten positions in the B-chain. This discovery prompted the investigation of the insulin structure of another hystricomorph, the coypu (*Myocastor coypus*). In coypu insulin (Fig. 4), there is a residue deleted at position 24 or 25 in the B-chain and an addition at the C-terminal of the A-chain (Smith, 1972). Thus the primary sequence of the insulin of the coypu differs from that of the ox in 22 residues and from that of the guinea-pig in 17 residues; 14 residues are different from those of guinea-pig and ox insulins.

These differences in the primary structures may explain the differences between the storage characteristics of the insulin of the guinea-pig and other mammals *in vivo*. It has been suggested that insulin is stored in the granules of the β-cells of the islets of Langerhans in the pancreas as small crystalline hexamers. Greider, Howell & Lacy (1969) have shown that some granules in rat pancreatic cells are hexagonal in shape (Fig. 5) and bear a striking resemblance to the rhombohedral crystals formed by zinc insulin *in vitro* (Fig. 6). In their work on the three-dimensional analysis of pig insulin, Hodgkin and her colleagues (Blundell *et al.*, 1972) have shown that the insulin hexamer is formed by the aggregation of three insulin dimers in the presence of zinc ions (Fig. 7). Further organization of the hexamers results in rhombohedral crystals. There are two important facts which suggest that the insulins of the guinea-pig and coypu are not stored in this highly organized and compact form. Firstly, there appears to be no zinc (Pihl, 1968) in the β-cells of these two species, thus precluding the aggregation of six insulin molecules around two zinc atoms to form the insulin hexamer. Secondly, the histidine residue at position 10 in the B-chain of ox and pig insulin is replaced by asparagine in the guinea-pig

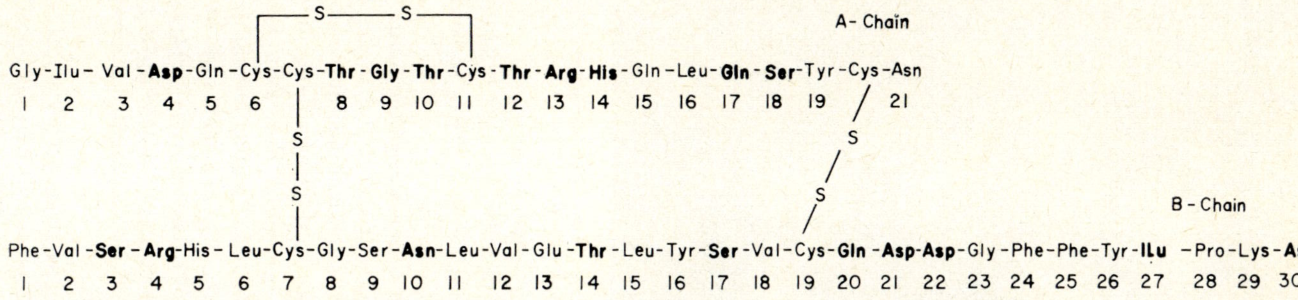

Fig. 3. The primary structure of guinea-pig insulin. (After Smith, 1972.) The 19 residues indicated in bold type are those that are different in ox insulin. The B-10 histidine residue has been replaced by an asparagine.

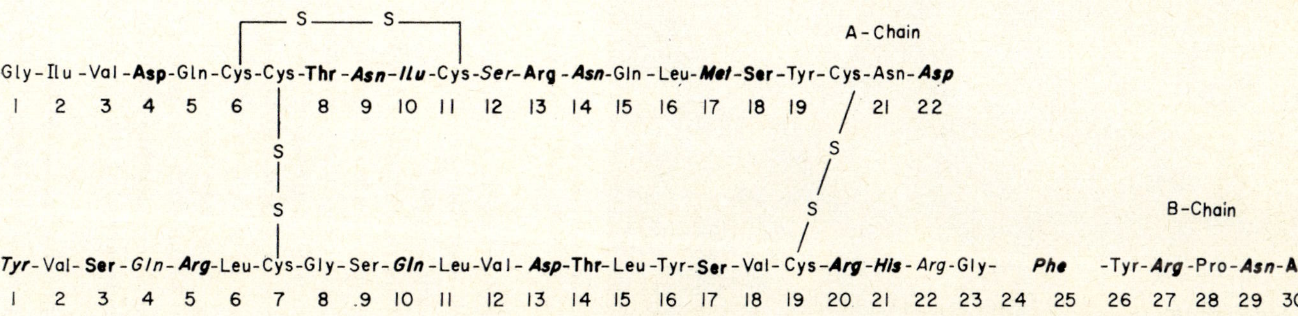

Fig. 4. The primary structure of coypu insulin. (After Smith, 1972.) The 17 amino acids in italic type indicate residues which are different from those in guinea-pig insulin (Fig. 3); the 22 residues indicated in bold type are those which are different in ox insulin. It can be seen that there are 14 residues in coypu insulin which are different from those of ox and guinea-pig insulin.

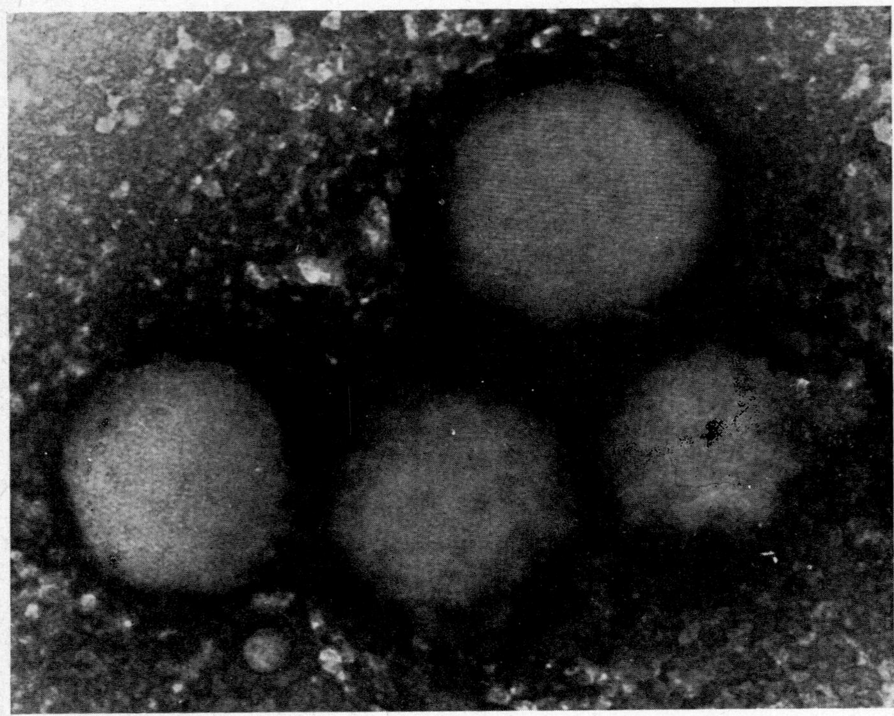

Fig. 5. Electron micrograph of insulin storage granules, some of which resemble insulin crystals, in the β-cells of rat pancreas. (From Greider *et al.*, 1969.)

and by glutamine in the coypu (Figs 3 and 4). Thus in the absence of zinc ions and B10 histidine residues hexamer formation is impossible in the guinea-pig and coypu.

It was this remarkable variation, from each other and other mammals, in the structure of two hystricomorph insulins that led to our interest in other members of the suborder. The biological activity of a hormone is dependent upon the preservation and integrity of specific areas of the molecule and not upon the molecule as a whole. For example, the removal of the N-terminal residue of adrenocorticotrophic hormone results in a complete loss of biological activity whereas the loss of residues 20–39 has little or no effect (Sayers, 1967). The study of hystricomorph insulins, therefore, seemed a good opportunity for attempting to define those areas of the insulin molecule that are responsible for its activity.

One other hystricomorph insulin has been analysed, that of the chinchilla (*Chinchilla laniger*). The primary sequence will be published

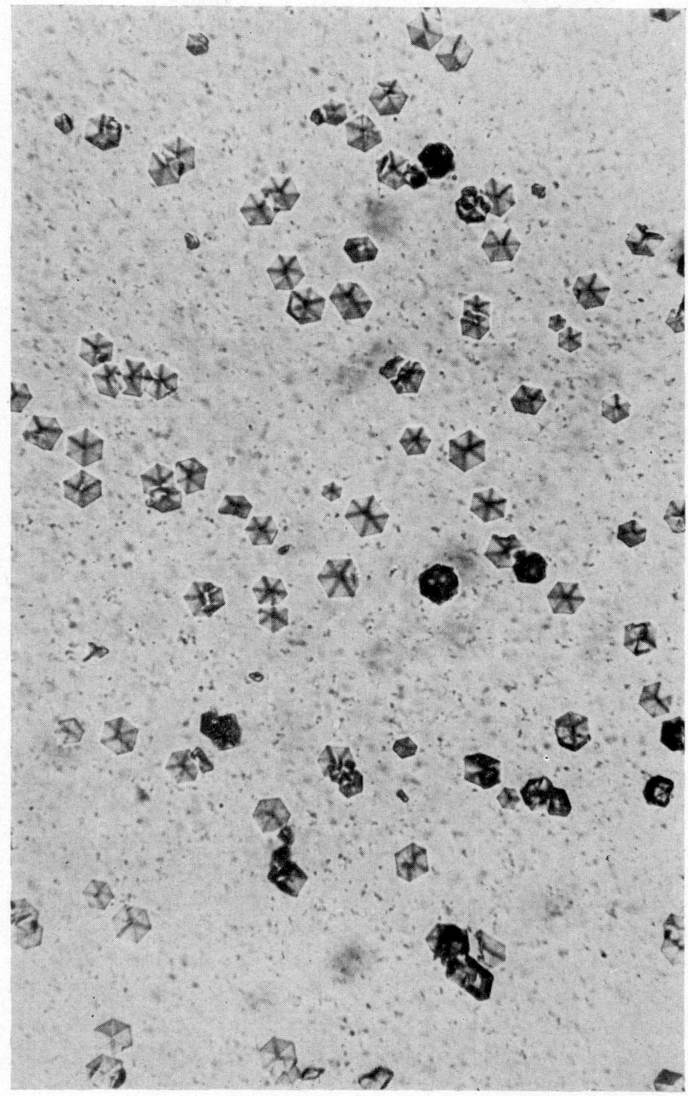

FIG. 6. Rhombohedral zinc insulin crystals. Note the similarity of the shape with the granule in Fig. 5.

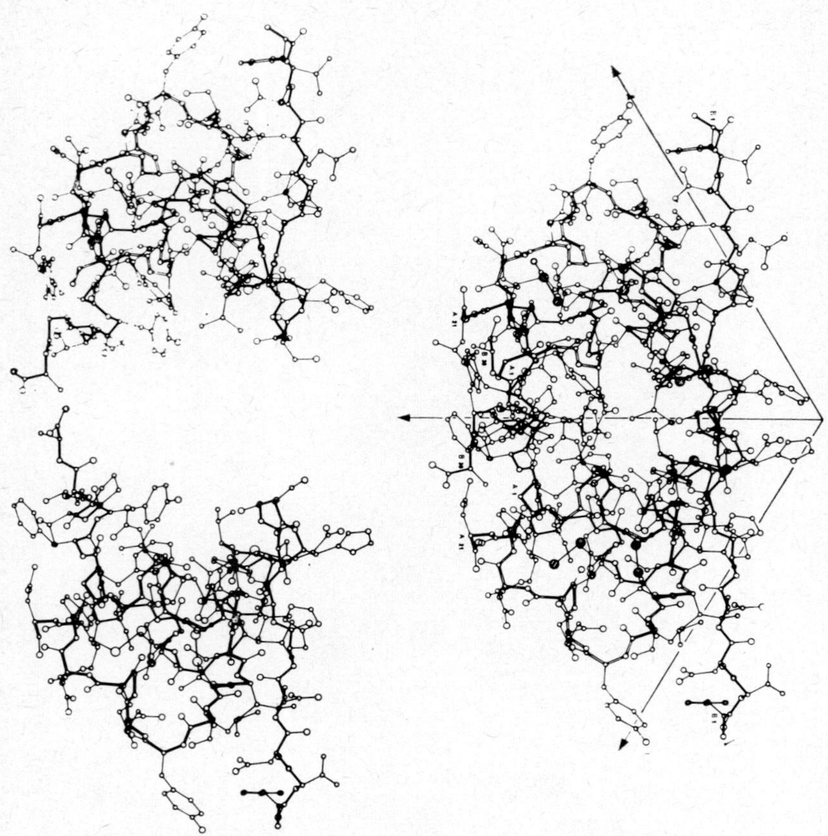

Fig. 7. The relationship between the monomer, dimer and hexamer of the insulin molecule. (From Blundell *et al.*, 1972).

in full elsewhere (D. G. Smythe, R. W. J. Neville & N. R. Lazarus, in preparation) but, as we expected from the biological activity and immunological properties of chinchilla insulin (see below), it is not as different from ox insulin as is guinea-pig or coypu insulin. Only eight changed residues have been identified and there is a histidine residue at position 10 on the B-chain, thus suggesting the potential to bind zinc ions and hexamerize in the manner described for other species.

IMMUNOLOGICAL PROPERTIES

Moloney & Coval (1955) showed that when guinea-pig anti-ox insulin serum was administered to mice a state of transient acute

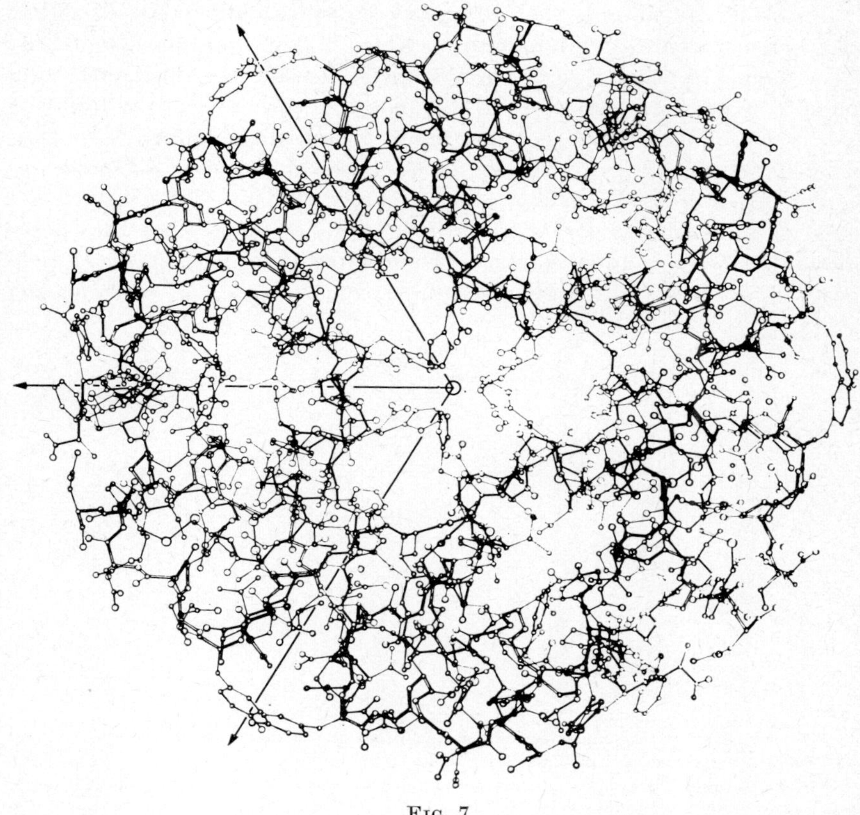

FIG. 7

diabetes was induced. This condition was characterized by raised blood glucose levels caused by inactivation of the endogenous insulin secreted by the mice. The inference from this neutralization of endogenous mouse insulin by an ox insulin antiserum was that the mouse insulin was likely to be structurally closely similar to ox insulin. This inference was subsequently shown to be valid and gave rise to the prediction that an insulin that is structurally dissimilar to ox insulin will not cross-react with anti-ox insulin serum. Moloney & Coval (1955) and Armin, Grant & Wright (1960) demonstrated normoglycaemia in guinea-pigs treated with ox insulin antiserum. In 1968, Davidson, Zeigler & Haist found that coypu insulin was not neutralized by antiserum *in vitro*, a result that was expected since coypu insulin differs from ox insulin. Davidson *et al.* (1969) also found that the insulin of the capybara (*Hydrochoerus hydrochaeris*) was not neutralized *in vitro*.

Since it was possible that all hystricomorphs might possess insulins that were structurally different from those of other mammals, we tested the response of other hystricomorph species to an ox insulin antiserum (Neville, Weir & Lazarus, 1973). The blood glucose levels were followed over four hours in at least seven animals of each of the following species: *Octodon degus* (degu), *Ctenomys talarum* (tuco-tuco), *Proechimys guairae* (casiragua), *Chinchilla laniger* (chinchilla) and *Galea musteloides* (cuis). Guinea-pigs and rats were included as controls since they were known to possess non-cross-reacting and cross-reacting insulins respectively. The results obtained are summarized in Fig. 8. The guinea-pig

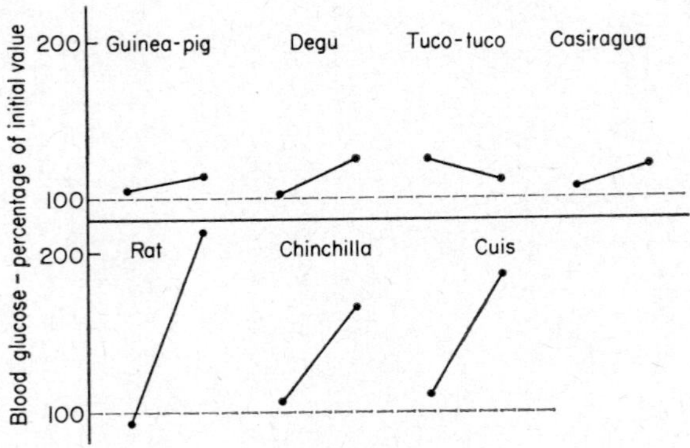

Fig. 8. Diagram showing the effect of guinea-pig anti-ox insulin serum on six hystricomorph species and the rat. The means only of at least seven animals in each group are given and are the means of 1-, 2- and 4-hour cumulated blood glucose values expressed as percentages of the initial values. The left point of each pair represents control animals treated with normal guinea-pig serum; the right point represents test animals. (After Neville *et al.*, 1973.)

and the rat behaved as expected but two of the other five hystricomorph species developed hyperglycaemia, thus indicating that the chinchilla and the cuis possess a cross-reacting insulin and that the degu, tuco-tuco and casiragua are like the guinea-pig in having a non-cross-reacting insulin.

Although the sample is small, it is apparent (see Fig. 9) that the possession of a cross-reacting insulin is not a familial characteristic. All the members of the superfamily Octodontoidea so far studied have not cross-reacted but one of the three cavioid species has cross-reacted.

The chinchilla and the cuis will probably be found to have insulins which have antigenic sites similar to those of other mammalian insulins, and which cross-react *in vitro* in the standard insulin immunoassay. The ability of various insulins to displace radioactive insulin from ox insulin

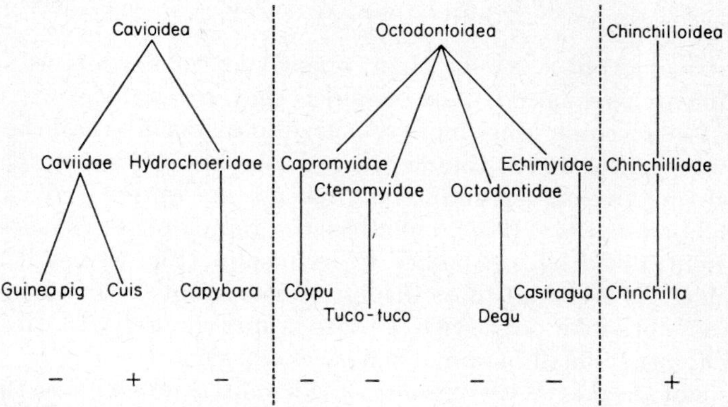

Fig. 9. Diagram summarizing the known cross-reactivity of hystricomorph insulins and the taxonomic relationships (after Simpson, 1945) of the species involved.

antibody is shown in Fig. 10. The amount of ox insulin required to displace a given quantity of radioactivity compared with that for chinchilla or guinea-pig is very small. The guinea-pig insulin cross-reacts very weakly but chinchilla insulin occupies an intermediate position. These results confirm the experiments *in vivo* and show that while chinchilla insulin is different from guinea-pig insulin it is far from being identical with ox insulin (see earlier). Thus not all hystricomorph species possess insulins with antigenic determinants markedly different from those of other mammals.

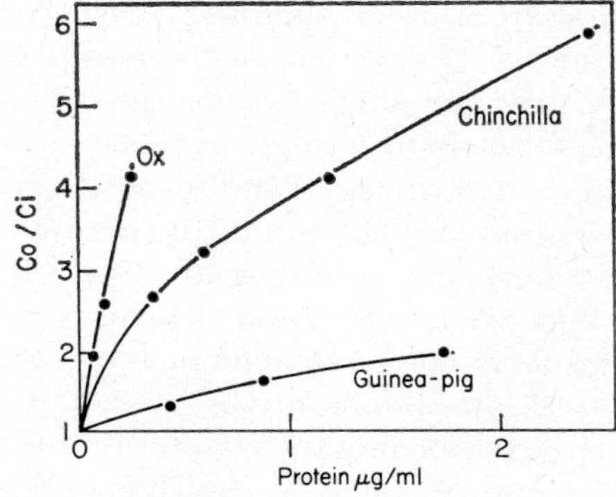

Fig. 10. The cross-reactivity of guinea-pig, chinchilla and ox insulins in an insulin radioimmunoassay.

P

BIOLOGICAL ACTIVITY

Hystricomorphs vary in their sensitivity to exogenous insulins. Chinchilla are as sensitive to ox insulin as they are to their own. Guinea-pigs are two or three times more sensitive to ox insulin than they are to their own (Smith, 1972; Zimmerman, Kells & Yip, 1972). The tuco-tuco, which develops spontaneous diabetes in captivity (Wise, Weir, Hime & Forrest, 1968, 1972), resembles the guinea-pig in that its insulin is not neutralized by antibodies to ox-insulin (Fig. 9) but it appears to be relatively resistant to ox insulin. A consistent fall in blood glucose level was obtained only with a dose approximately 16 times that needed by other small mammals (Wise *et al.*, 1972).

The biological activity of guinea-pig insulin is much lower than that of other mammalian insulins. Estimates of its relative potency (Smith, 1972; Zimmerman, Kells & Yip, 1972) using the standard mouse convulsion technique have given values of 2 to 4 units/mg compared with 25 units/mg generally obtained with ox and pig insulins. A biological assay of a pure sample of chinchilla insulin (S. Wood, R. W. J. Neville & N. R. Lazarus, in preparation) gave a potency of 17·3 units/mg. This confirms the intermediate position of chinchilla insulin between those of the guinea-pig and the ox.

DISCUSSION

The marked difference in structure of the insulins of the guinea-pig and the coypu from that of the insulins of other mammals, and the evidence from studies of coypu and capybara insulins *in vitro* led to our suggestion that the insulins of hystricomorphs might be different from each other as well as from other mammals. Our studies using several animals from each of five hystricomorph species indicate that this suggestion may not be true for all hystricomorphs. Chinchilla insulin at least appears to be intermediate between ox and guinea-pig insulins with respect to its antigenic determinants and its active sites. The cuis may also be found to react similarly but our antibody results (Fig. 8) indicate that the tuco-tuco, degu and casiragua are likely to possess insulins whose structure differs extensively from that of ox insulin.

That ox insulin is more potent than guinea-pig insulin in the guinea-pig suggests structural alterations in the guinea-pig insulin molecule have occurred without a concomitant adaptation by its receptor. Such a theory is difficult to prove. Initially, Smith (1972) suggested that, at some stage during evolution, insulin became less essential to the guinea-pig than it was to other mammals. The discovery in the coypu of another insulin which was structurally very different (Smith, 1972) from ox and also biologically active (Davidson *et al.*, 1968) has made this theory less tenable. The absence of a histidine residue at position B-10 in the guinea-pig and coypu insulins does not favour the formation of hexamers in these species. In addition a number of sequence changes unique to the guinea-pig and coypu, namely B-14, B-17, B-20, A-13, B-4 (guinea-pig only) and B-5 (coypu only), are situated at the hydrophobic dimer-dimer interface of the hexamer. These substituted residues are more hydrophilic, larger or both and in consequence they also do not favour hexamer formation. Blundell, Dodson, Dodson, Hodgkin & Vijayan (1971) have suggested that the high rate of mutation for guinea-pig insulin, which appears to be ten times greater than that for other insulins (Smith, 1966), may derive from the possibility that insulin hexamers have no role in this species. Thus the removal of certain constraints on the conformation of the molecule has favoured a high rate of mutation. Nevertheless, it is not readily apparent why hexamer formation should be so biologically advantageous since the guinea-pig and coypu are both successful species. The degree of replacement of amino acids, which are invariant residues in other insulins, would account for the different biological potencies and immunological activities of, and the variation of structure between, the insulins of the hystricomorph species. Assuming that the immunological activities are a reflection of the structural discrepancies, it is probable that there are no taxonomic correlations between the insulins of the hystricomorph species so far studied. None of the octodontids cross-reacted but these were only four out of possibly 80 species. The cavioids are divided because the guinea-pig and probably the capybara possess insulins which differ substantially from ox insulin whereas that of the cuis may be more similar to the insulins of other mammals. There is obviously no reason to suppose that this is the only violation of present taxonomy and it is possible that the substitution of residues in the insulin molecule is variable throughout the hystricomorphs. If insulin is not as essential to some of these rodents as it is to other mammals, it is necessary to explain how hystricomorphs do control their blood sugar levels.

Our primary aim was to isolate, analyse and evaluate biologically and immunologically the insulins of some hystricomorph rodents and our preliminary results are encouraging. In following this line of research we hope not only to improve our knowledge of the hystricomorphs but also to contribute towards identifying the areas of the insulin molecule responsible for its numerous biological activities.

Acknowledgments

We are grateful to Dr I. W. Rowlands for his constant encouragement; to many commercial breeders for chinchilla; to Dr O. A. Reig for the casiragua; to the Wellcome Trust, the Royal Society and the Medical Research Council for funds to capture and establish the cuis and tuco-tuco colonies; to Dr T. Blundell, Dr S. Howell, Mr S. Wood and Dr D. Smythe for allowing us to use some of their slides and material. The laboratory work of one of us (B. J. W.) was financed by the Ford Foundation.

References

Armin, J., Grant, R. T. & Wright, P. H. (1960). Acute insulin deficiency provoked by single injections of anti-insulin serum. *J. Physiol., Lond.* **153**: 131–145.

Balant, L., Burr, I. M., Stauffacher, W., Cameron, D. P., Bünzli, H. F., Humbel, R. E. & Renold, A. E. (1971). Insulin of spiny mice (*Acomys cahirinus*). Partial characterization and evidence for two insulins. *Endocrinology* **88**: 517–521.

Banting, F. G., Best, C. H. & Macleod, J. J. R. (1922). The internal secretion of the pancreas. *Am. J. Physiol.* **59**: 479.

Blundell, T. L., Cutfield, J. F., Cutfield, S. M., Dodson, E., Dodson, G. G., Hodgkin, D. C. & Mercola, D. A. (1972). Three-dimensional atomic structure of insulin and its relationship to activity. *Diabetes* **21**, Suppl. 2: 492–505.

Blundell, T. L., Dodson, G. G., Dodson, E., Hodgkin, D. C. & Vijayan, M. (1971). X-ray analysis and the structure of insulin. *Recent Prog. Horm. Res.* **27**: 1–40.

Brown, H., Sanger, F. & Kitai, R. (1955). The structure of pig and sheep insulins. *Biochem. J.* **60**: 556–565.

Bünzli, H. F., Glatthaar, B., Kunz, P., Mülhaupt, E. & Humbel, R. E. (1972). Amino-acid sequence of the insulins from mouse (*Mus musculus*). *Hoppe-Selyer's Z. physiol. Chem.* **253**: 451–458.

Bünzli, H. F. & Humbel, R. E. (1972). Isolation and partial structural analysis of insulin from mouse (*Mus musculus*) and spiny mouse (*Acomys cahirinus*). *Hoppe-Selyer's Z. physiol. Chem.* **353**: 444–450.

Clark, J. L. & Steiner, D. F. (1969). Insulin biosynthesis in the rat: demonstration of two proinsulins. *Proc. natn. Acad. Sci. U.S.A.* **62**: 278–285.

Davidson, J. K., Zeigler, M. & Haist, R. E. (1968). Failure of guinea pig antibody to beef insulin to neutralize coypu (nutria) insulin. *Diabetes* **17**: 8–12.

Davidson, J. K., Zeigler, M. & Haist, R. E. (1969). Failure of guinea pig antibodies to beef insulin, chicken insulin and cod insulin to neutralize capybara insulin. *Diabetes* **18**: 212–215.

Greider, M. H., Howell, S. L. & Lacy, P. E. (1969). Isolation and properties of secretory granules from the rat islets of Langerhans. II. Ultra-structure of the beta granule. *J. Cell Biol.* **41**: 162–166.

Harris, J. I., Sanger, F. & Naughton, M. A. (1956). Species differences in insulins. *Archs Biochem.* **65**: 427–438.

Langerhans, P. (1869). *Beitrage zur mikroskopischen Anatomie der Bauchspeichel-drusse.* Berlin.

Markussen, J. (1971). Mouse insulins—separation and structure. *Int. J. Prot. Res.* **3**: 149–155.

Moloney, P. J. & Coval, M. (1955). Antigenicity of insulin: diabetes induced by specific antibodies. *Biochem. J.* **59**: 179–185.

Neville, R. W. J., Weir, B. J. & Lazarus, N. R. (1973). Insulins of hystricomorph rodents. *Diabetes* **22**: 851–853.

Oakley, W. G., Pyke, D. A. & Taylor, K. W. (1973). *Diabetes and its management.* Oxford: Blackwells.

Papaspyros, N. S. (1964). *History of diabetes mellitus,* 2nd edn. Stuttgart: Thieme Verlag.

Paulesco, N. C. (1921). Recherche sur le rôle du pancréas dans l'assimilation nutritive. *Archs int. Physiol.* **17**: 85–109.

Pihl, E. (1968). An ultrastructural study of the distribution in the pancreatic islets as revealed by the sulfide silver method. *Acta path. microbiol. scand.* **74**: 145–160.

Ryle, A. P., Sanger, F., Smith, L. F. & Kitai, R. (1955). The disulphide bonds of insulin. *Biochem. J.* **60**: 541–556.

Sayers, G. (1967). Adrenocorticotrophin. In *Hormones in blood*: 169–194. 2nd edn. Gray, C. H. & Bacharach, A. L. (eds.) London and New York: Academic Press.

Schafer, E. (1895). On internal secretion. *Br. med. J.* **2**: 341–348.

Simpson, G. G. (1945). The principles of classification and a classification of mammals. *Bull. Am. Mus. nat. Hist.* **85**: 1–350.

Smith, L. F. (1966). Species variation in the amino acid sequence of insulin. *Am. J. Med.* **40**: 662–666.

Smith, L. F. (1972). Amino acid sequences of insulins. *Diabetes* **21**, Suppl. 2: 457–460.

von Mering, J. & Minkowski, O. (1889). Diabetes mellitus nach Pankreasextirpation. *Naunyn.-Schmiedebergs Arch. exp. Path. Pharmak.* **26**: 371–387.

Wise, P. H., Weir, B. J., Hime, J. M. & Forrest, E. (1968). Implications of hyperglycaemia and cataract in a colony of tuco-tucos (*Ctenomys talarum*). *Nature, Lond.* **219**: 1374–1376.

Wise, P. H., Weir, B. J., Hime, J. M. & Forrest, E. (1972). The diabetic syndrome in the tuco-tuco (*Ctenomys talarum*). *Diabetologia* **8**: 165–172.

Zimmerman, A. E., Kells, D. I. L. & Yip, C. C. (1972). Physical and biological properties of guinea-pig insulin. *Biochem. biophys. Res. Commun.* **46**: 2127–2133.

DISCUSSION

WOOD (*Sussex*): We have some preliminary evidence on the aggregation of chinchilla insulin using circular dichroism in the ultra-violet region and we have been able to show that, on the addition of zinc to dilute chinchilla insulin solutions, there is some aggregation which is very similar to that which occurs with ox insulin, and I wonder if this fits in with your preliminary ideas for the sequence of chinchilla insulin?

NEVILLE (*Dartford*): That is interesting because it does begin to tie up. It seems, now, that with the chinchilla we may have yet another exception to the rule in that the first 10 residues of the B chain do appear, on the first analysis, to contain two histidine residues. This strongly supports the suggestion that a histidine does exist at position 10 on the B-chain, and this could mean it would indeed be a zinc-binding molecule and as such would be the first of three hystricomorph insulins isolated to perhaps crystallise.

BLUNDELL (*Sussex*): Is there anything in the environment of the natural habitat of the guinea-pig and the coypu which would lead anybody to think that zinc wouldn't be so available to these animals?

PEARSON (*Berkeley*): I, of course, immediately started flipping through my mental card file of where we find guinea-pigs in the wild in South America and where we find coypu and clutching for some environmental parameter that might indicate zinc shortages or other items of interest. We have a problem here, in that I am not clear as to the ancestral origin of our domestic guinea-pig. I would hate to put my finger on the map where the ancestor of the domestic guinea-pig came from, and the existing numbers of the genus *Cavia* in South America cover 2 000 miles at least from Patagonia up to the Central and Northern Andes. The coypu covers a wide variety of habitats both geographically and ecologically, although they are always moist habitats, so that I am afraid I can't offer any light at the moment as to environmental correlates with the insulin picture.

BLUNDELL: As one of the people working on the three dimensional X-ray analysis of insulin structures I would just like to comment on one thing that Mr Neville mentioned. He said that the mutations may be random in hystricomorphs but, in fact, most of the mutations which occur in the coypu and the guinea-pig are on the one surface of the molecule which is involved in forming hexamers from dimers, so it is not really quite random. There appears to be quite a lot of consistency in the mutations; the residues get less hydrophobic, they also tend to become larger. This is consistent with their failure to aggregate and I wonder whether this has happened in the coypu and guinea-pig because for some reason zinc isn't available and these mutations have occurred in order to keep the insulin dimer soluble. This has resulted in a lower activity, but somehow the animals have managed to survive.

SIMPSON (*Tucson*): An evolutionist is very reluctant to admit that anything happens completely at random, and I certainly feel that further study is required when one seems to find something happening at random. As the history of evolutionary science is revealed it is more likely to show our ignorance than to add to our knowledge. Another point here is that I thought there was an implication of a definite correlation, possibly a causal relationship, between mutation rate and between replacement rate. I would like to suggest that this is not necessarily true, that the replacement rate is largely determined by the force of selection and that even a rather high mutation rate will not effect a change on an entire population to a uniform new condition unless it is favoured by selection. On the other hand, a high mutation rate will not cause a change at all if it is definitely opposed by selection, and the fact that insulins are, in general, among most animals, so uniform, suggests that indeed this has been acting against mutation rates, and that we do not really know what the mutation rates are. They may be even higher in other animals than in the so-called hystricomorphs.

Symp. zool. Soc. Lond. (1974) No. 34, 437–446

NOTES ON THE ORIGIN OF THE DOMESTIC GUINEA-PIG*

BARBARA J. WEIR

*Wellcome Institute of Comparative Physiology,
Zoological Society of London,
Regent's Park, London, England*

INTRODUCTION

All modern taxonomists are agreed that the cavies of the subfamily Caviinae comprise four genera: *Cavia* (cavy, guinea-pig, cobaye, meerschwein, aperea, préa, cori, cui, quiso), *Galea* (cuis), *Microcavia* (desert cavy) and *Kerodon* (moco). In the past, however, there have been numerous changes of species between and within genera and the tendency has been to split rather than to lump species (see Tate, 1935; Ellerman, 1940; Cabrera, 1957–1961). It is this confusion of the basic taxonomy of the whole group which seems to have led to the uncertainty of the origin of the domestic guinea-pig, *Cavia porcellus* (Gilmore, 1950).

HISTORY

The first mention of the Indian guinea-pig (del chanchito de la India) was by Oviedo in 1547 who described a common mammal called a 'cori' in Santo Domingo (Cabrera, 1953). However, no cavy is indigenous to Central and Caribbean America and if Oviedo actually saw guinea-pigs in Santo Domingo, they must have been taken there by the Spaniards themselves. The source of such an importation would almost certainly have been from the west coast of South America and probably from Peru, which the Conquistadores invaded in 1532. At the time of the Spanish invasion, guinea-pigs had already been domesticated by the Incas as a source of food, much as the Romans kept the edible dormouse (*Glis glis*), and for religious ceremonies. A study of guinea-pig mummies has led to the suggestion that domestication first occurred in the region of Lake Titicaca (Hückinghaus, 1961) but no date has yet been given. Ceramics portraying guinea-pigs are known from the period 500–1000 A.D. (Schmidt, 1929) and suggest a pre-Incan involvement of man with this species. Muñoz (1970) considers that domestication was probably effected by the indigenes of Peru, Colombia and Ecuador during the 'agricultural era' about 1000 B.C.

* (Not presented at the Symposium but included in the Proceedings because of its relevance–see Preface. Eds.)

NAME

The origin of the name 'guinea-pig' has often been questioned, and several suggestions have been proposed. The guinea-pig probably arrived in Europe about 1580 and Topsell (1607) first referred to it in English as the "Indian little pig coney". The word 'Indian' was presumably added because South America was at that time thought to be part of the West Indies but the association with 'pigs', and subsequently with 'guinea' is hard to explain. One suggestion (Cumberland, 1905) is that the Spaniards in Peru first saw these animals in markets and were reminded of sucking pigs. This seems unlikely since of all the European vernaculars, only the Spanish name (el conejillo de Indias) contains no reference to pigs. The association probably arose because the squeals made by cavies, and the domestic species particularly, can be thought to resemble those of pigs, hence 'little' pig. The simile has been furthered by referring to the male as a boar and the female as a sow (although no-one seems to have referred to a guinea-piglet). The appellation 'guinea' may have arisen because ships carrying these cavies stopped along the West African coast on the way to Europe. A more likely reason is that misunderstanding arose over the place of embarkation. This was probably from the northern coast of South America and associated by English writers with Guyana; this could easily be confused with the Old World settlement of Guinea (Cabrera, 1953; Muñoz, 1970). Although commonly called the 'guinea-pig' throughout the English-speaking world, the term 'cavy' is frequently used. The latter is derived from Quechuan names and was probably first used as a vernacular by Pennant (1781) who referred to the "restless cavy".

In 1648, Marcgrave wrote of the "*aperea brasiliensibus*" and called it "cavia cobaya" which was an adaptation of the name given by the Brazilian natives to this animal and was subsequently adopted by scientists. *Cavia* has become the generic name (Pallas, 1766), and in French the guinea-pig is called "le cobaye". The trivial name of *porcellus* derives from the 10th edition of Linneaus' *Systema Naturae* (1758) although he had called it *Mus cobaya* in 1747 and *Mus brasiliensis* in 1754. The binomial form *Cavia porcellus* was first used by Erxleben (1777). There is no doubt that the animals described by Marcgrave were of a domesticated species because they were multicoloured. Molina (1782) saw a similar type of coloured guinea-pig (el porcellino d'India) in Chile and it is probable that one or more domesticated varieties were widespread throughout the continent. Examples of this tamed species, *C. porcellus* (see Thomas, 1901), were brought to Europe at the end of the sixteenth century (Cabrera, 1953) and have been kept as curiosities, pets and laboratory species ever since.

ORIGIN

The conditions in which guinea-pigs are kept today in South America are not rigorous; the animals are left to scavenge in and around the huts of the Indians (Fig. 1) and it may be assumed that similar husbandry has always existed (Velasco, 1837*). Thus it is not surprising that *C. porcellus* may also occur as a feral species although there is no wild form of *C. porcellus*. Therefore, other wild cavy species must be

FiG. 1. Domestic guinea-pigs running freely in and around an Indian dwelling at Ulla Ulla, Bolivia (5,000 m above sea level).

considered as possible candidates for ancestry, on the assumption that they might not have changed greatly since the time *C. porcellus* was domesticated. Three possible areas have been suggested as the place of origin of the domestic guinea-pig.

* An early chronicle, c. 1560–1630, appears to have been used.

Peru

In 1835, Bennett described a single animal and called it *C. cutleri*. The crest of hairs that he reported on the nape is typical of wild cavies but as Thomas (1917) suggested, the rosette pattern of the cheek hairs and the length of the body hair indicate that Bennett's *C. cutleri* was probably a melanic form of the domestic guinea-pig. Tschudi (1845) also referred to animals from Ica, Perua as *C. cutleri*, but these were clearly different from that of Bennett (1835) and Fitzinger (1867) renamed Tschudi's *C. cutleri* as *C. tschudii*. Castle (in Castle & Wright, 1916) decided that his cavies, reputedly from Ica, were feral animals and that wild cavies from Arequipa were *C. cutleri* because they fitted Bennett's description. However, this confusion has probably arisen because throughout Peru cavies can be found which are *C. tschudii* (Hückinghaus, 1962), domestic cavies (*C. porcellus*) or feral *C. porcellus* which betray their origin by producing white spotted animals.

Argentina, Paraguay and southern Brazil

C. aperea Erxleben, 1777 is the most widespread cavy in these areas.

Eastern Brazil

C. rufescens was described by Lund (1841) but Thomas in 1901 considered that this was a synonym of *C. fulgida* Wagler, 1831, and in 1917 that *C. fulgida* was distinct from all the other, *aperea*-like, cavies.

INTERRELATIONSHIPS OF CAVIES

Waterhouse (1848) believed that *C. porcellus*, *C. aperea* and *C. cutleri* (Bennett's) were all the same species. Nehring (1889) considered *C. cutleri* (*C. tschudii*) the best ancestor and Thomas (1901) thought that *C. aperea* or *C. rufescens* could have been ancestral. Hückinghaus (1962) agrees that *C. fulgida* is a distinct species and concludes that *C. porcellus*, *C. rufescens* and *C. tschudii* are synonyms of *C. aperea*. However, Hückinghaus based his conclusions on an entirely morphological study of skull characteristics and gave no consideration to the work published at the beginning of the century on the genetics of various cavy crosses. Most of these studies (Detlefson, 1914; Castle & Wright, 1916) were concerned with coat colour but some information is given on the fertility of hybrids produced by crossing domestic and wild cavies.

Castle & Wright (1916) reported fertile hybrids of both sexes when *C. cutleri* (= *C. tschudii*) and *C. porcellus* are crossed.

Reciprocal crosses have been made between *C. rufescens* (= *C. fulgida*?) and *C. porcellus* by Ubisch & Mello (1940) although Detlefson

(1914) was unable to produce fertile hybrids between the domestic male and wild female. This was surprising since the offspring of the wild male and domestic female cross were larger than those of the pure-bred domestic guinea-pig. The hybrid females were mostly fertile but all the F_1 males were sterile. Detlefson (1914) considered that *C. rufescens* is more distantly related to *C. porcellus* than are *C. aperea* and *C. cutleri* (= *C. tschudii*).

When *C. aperea* males are mated to *C. porcellus* females, the offspring of both sexes are fertile (Nehring, 1893; Pictet & Ferrero, 1951; Rood & Weir, 1970), although Guyenot & Duszynska-Wietrzykowska (1935) reported that many females became infertile in the second to fifth generation. Since the exact contribution of the *C. aperea* male(s) in these colonies was difficult to elucidate, a study was made in our laboratory of the fertility of *C. aperea* × *C. porcellus* hybrids mated to hybrids of their own generation and backcrossed to the domestic guinea-pig (Fig. 2). The number of animals used was small but the results are of interest in view of the findings in other colonies. As in the

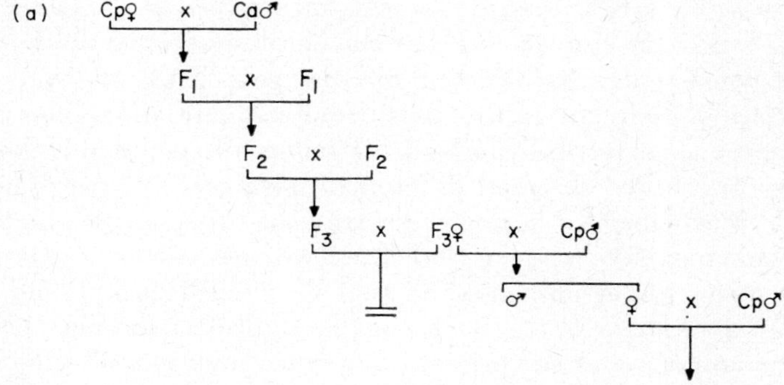

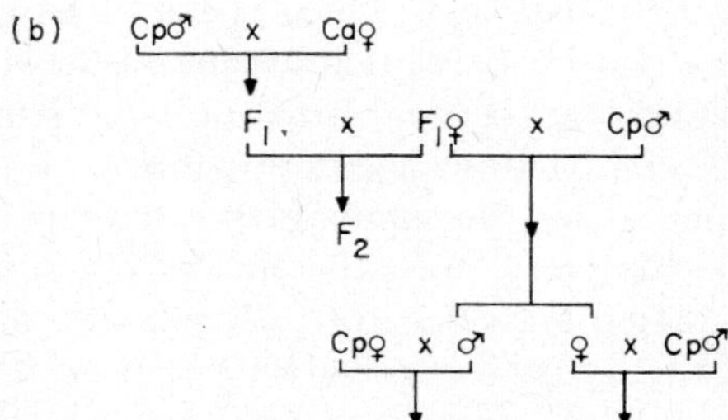

Fig. 2. Diagram showing the matings between *Cavia porcellus* (Cp) and *Cavia aperea* (Ca) and their hybrids.

case of the F_1 hybrids from the *C. rufescens* male and *C. porcellus* female mating, those from the cross between *C. aperea* males and *C. porcellus* females were large at birth (about 250 g), but later generations were similar in size to domestic guinea-pigs at birth. The progeny of matings between *C. porcellus* females and *C. aperea* males (Fig. 2a) were fertile to each other and to the parent species as far as the third generation except for the F_3 males who were unfortunately not tested. The F_3 hybrids could not be tested with *C. aperea* animals because they were required for other purposes at that time. The males resulting from the cross of F_3 females with *C. porcellus* males do not appear to be fertile with other hybrids or with *C. porcellus* females, although spermatozoa are produced. Fertile F_1 hybrids have been produced by the reciprocal cross, *C. porcellus* male and *C. aperea* female (Fig. 2b), but the F_2 offspring have not yet reached puberty. All these hybrids seem fertile with the *C. porcellus* stock. The block to continual fertility in hybrids with *C. aperea* blood seems to be at mating or fertilization since the males are spermatogenic and the females experience regular oestrous cycles. The fault may be genetic.

George *et al.* (1972) found that the karyotypes of *C. porcellus* and *C. aperea* were very similar and that the chromosomes of the F_1 hybrid matched consistently well except for one pair. Exaggeration of this difference in subsequent generations could result in anomalous pairing of chromosomes at fertilization and the failure of zygotic development.

There are other distinct differences between *C. aperea* and *C. porcellus*. Biochemical studies on variants of the phosphoglucose isomerase enzyme have shown that the two species have completely different forms of the enzyme, and that a mixed form is found in the hybrids (Carter *et al.*, 1972). Such enzyme hybridization may be found in other mammals and the original difference may not be sufficient to claim it as specifically diagnostic.

Darwin (1875: 145) said that *C. aperea* could not be the ancestor of *C. porcellus* because it harbours a different genus of louse. Patterns of social behaviour can be distinguished in the two species and hybrids display intermediate patterns (Rood, 1972). Unfortunately, nothing is known of the biochemistry, karyology, ectoparasites, or behaviour of the other cavy species, so further comparisons are impossible.

The only system for which there are comparative data is that of reproduction (see Table I). Contrary to expectation for a species selectively bred for food, gestation is longer in *C. porcellus* than it is in any of the wild species or in the hybrids of these with *C. porcellus*. The obvious effect of domestication in *C. porcellus*, apart from increase in body weight, has been to increase litter size, and an inverse correlation

TABLE I

Reproductive characteristics of some Cavia species and their hybrids

Species	No. of animals	Gestation length (days)		Litter size		Oestrous cycle (days)	Authors
		Mean	Range	Mean	Range		
C. porcellus	numerous	68·0	59–72	3·5	1–13	16·5	see Rood & Weir, 1970
C. aperea	>100	62·4 (40)	59–74	2·1 (53)	1–5	20·6 (67)	Rood & Weir, 1970; Weir, unpublished
C. tschudii		63·3 (19)	56–69	1·9 (53)	1–4		Castle & Wright, 1916
C. rufescens			62–65	1·35		9·0–25·0	Detlefson, 1914; Ubisch & Mello, 1940
C. rufescens ♂ × *C. porcellus* ♀			63–67				Detlefson, 1914
C. aperea ♂ × *C. porcellus* ♀	>5♂♂ >9♀♀	64·2 (15)	62–68	2·9 (16)	1–4		Rood & Weir, 1970; Weir, unpublished
F$_1$ hybrids	>3♂♂ >4♀♀	65·9 (11)	63–67	2·5 (21)	1–4	25·9 (9)	Rood & Weir, 1970; Weir, unpublished
F$_2$ hybrids	3♂♂ 10♀♀	64·5 (8)	63–67	1·7 (8)	1–3	18·7 (107)	Weir, unpublished
F$_3$ hybrids	6♀♀	64·3* (3)	64–65*	3·0* (3)	2–5*	17·1 (86)	Weir, unpublished
C. porcellus ♂ × *C. aperea* ♀	2♂♂ 1♀	63·3 (4)	61–67 e	1·2 (4)	1–2		Weir, unpublished
F$_1$ hybrids	1♂ 2♀♀	64·3 (4)	63–66	2·0 (4)	1–2	18·6 (5)	Weir, unpublished

* These are figures derived from matings between the F$_3$ hybrids and *C. porcellus* males. Figures in parentheses denote number of observations.

with gestation length is found in *C. porcellus* and in *C. aperea* (Rood & Weir, 1970). The mean length of the oestrous cycle is shorter in *C. porcellus* than in *C. aperea* though this can hardly have been a selected feature since this cycle is usually an artifact of laboratory conditions; most cavies mate at the post-partum oestrus and pregnancies are concurrent with lactation.

The white hair of the guinea-pig is also a trait of domestication. The only colour which breeds true is the white with pink eyes (albino) and it is generally considered that any free-living *Cavia* which has white in the pelage has domestic blood in it (see Castle & Wright, 1916). J. P. Rood (personal communication) did not find any evidence of white *C. aperea* in Argentina although every report of an albino was followed up. Some white fur has occurred in three lines of laboratory-kept *Galea musteloides*, but none of the animals so affected was fertile (Weir, unpublished observation). Many of the early descriptions of the guinea-pigs, taken to Europe and seen in South America, indicate that the hair was multicoloured and the implication must be that these cavies had been domesticated over many years before the Spanish conquest of Peru.

It is unfortunate that so little is known about species of wild cavy other than *C. aperea* as further information on their biology would permit a better assessment of the ancestry of *C. porcellus* than is now possible. It has been suggested (Cumberland, 1905) that guinea-pigs were domesticated in several places at once, and thus presumably from several species or at least races. This seems unlikely in view of the homogeneity of *C. porcellus* as it appears today.

It may be concluded that *C. porcellus* is not synonymous with *C. aperea*, *C. tschudii*, or *C. rufescens*. The latter species (as *C. fulgida?*) may not be synonymous with *C. aperea* or *C. tschudii* which may be conspecific. The origin of *C. porcellus* is probably to be found in these species and certainly *C. aperea* could have been ancestral to *C. porcellus*, but the likelihood of the progenitors having originated in Peru or anywhere else cannot be assessed.

ACKNOWLEDGMENTS

I am grateful to Dr I. W. Rowlands for his interest and encouragement in this subject. The wild cavies were caught during an expedition generously financed by The Royal Society, the Medical Research Council and the Wellcome Trust. The laboratory work was supported by the Ford Foundation.

REFERENCES

Bennett, E. T. (1835). On a new species of *Ctenomys* Blainv., and on other rodents
    collected near the Straits of Magellan by Capt. P. P. King, R.N. *Proc.
    zool. Soc. Lond.* **1835**: 189–191.

Cabrera, A. (1953). Los roedores argentinos de la familia "Caviidae". *Publnes
    Esc. Vet. Univ. B. Aires* **6**: 1–93.

Cabrera, A. (1957–1961). Catalogo de los mamíferos de America del Sur. *Revta
    Inst. nac. Invest. Cienc. nat. Mus. argent. Cienc. nat. Bernadino Rivadavia*
    4: 1–724.

Carter, N. R., Hill, M. D. & Weir, B. J. (1972). Genetic variation of phosphoglucose
    isomerase in some hystricomorph rodents. *Biochem. Genet.* **6**: 147–156.

Castle, W. E. & Wright, S. (1916). Studies of inheritance in guinea-pigs and rats.
    Publs Carnegie Instn No. 241: 1–129.

Cumberland, C. (1905). *The guinea-pig, or domestic cavy.* London: Upcott Gill.

Darwin, C. (1875). *The variation of animals and plants under domestication.* **2**.
    2nd edn. London: Murray.

Detlefson, J. A. (1914). Genetic studies on a cavy species cross. *Publs Carnegie
    Instn* No. 205: 1–134.

Ellerman, J. R. (1940). *The families and genera of living rodents.* **I**: 1–689. London:
    British Museum (Natural History).

Erxleben, J. C. P. (1777). *Systema Regni Animalis per classes, ordines, genera,
    species, varietates: cum synonymia et historia animalium. Classis I Mammalia.*
    Lipsiae.

Fitzinger, L. J. (1867). Versuch einer natürlichen Anordnung der Nagethiere
    (Rodentia). *Sber. K. Akad. Wiss. Wien.* (Math.-naturwiss.) **55**: 453–515;
    56: 57–168.

George, W., Weir, B. J. & Bedford, J. (1972). Chromosome studies in some mem-
    bers of the family Caviidae (Mammalia: Rodentia). *J. Zool., Lond.* **168**:
    81–89.

Gilmore, R. M. (1950). Fauna and ethnozoology of South America. In *Handbook
    of South American Indians* **6**: 345–464. J. H. Steward (Ed.) Washington:
    Smithsonian Institution, Bureau of American Ethnology, U.S. Government
    Printing Office.

Guyenot, E. & Duszynska-Wietrzykowska, J. (1935). Sterilité et virilisme chez
    des femelles de cobayes issue d'un croisement interspecifique. *Revue suisse
    Zool.* **42**: 341–388.

Hückinghaus, F. (1961). Zur Nomenklatur und Abstammung des Hausmeer-
    schweinchens. *Z. Säugetierk*, **26**: 108–111.

Hückinghaus, F. (1962). Vergleichende Untersuchungen über die Formenmannig
    faltigkeit der Unterfamilie-Caviinae Murray 1886 (Ergebnisse de Süd-
    amerika-expedition Herre/Rohrs 1956–57). *Z. wiss. Zool.* **166**: 1–98.

Linnaeus, C. (1747). *Wästgota-Resa, pa Rikseis hogoglige Ständers befallnig för-
    rättad ad 1746.* Stockholm.

Linnaeus, C. (1754). *Museum Adolphi Friderici Regis.* Stockholm.

Linnaeus, C. (1758). *Systema Naturae* (10th Ed.). Stockholm.

Lund, P. W. (1841). Blik paa Brasiliens Dyreverden for sidste Jordomvoeltning.
    K. dansk. Vidensk. Selsk. Skr. **8**: 282–283.

Marcgrave, G. (1648). *Historia rerum naturalium Brasiliae.*

Molina, T. (1782). *Saggio sulla storia naturale del Chile.* Bologna.

Muñoz, L. (1970). *Historia natural de conejillo de Indias*. Cauca: Talleres Editoriales del Departamento Popayan.

Nehring, A. (1889). Ueber die Herkunft des Haus-Meerschweinchens. *Sber. naturf. Ges. Berlin* **1**: 1–4.

Nehring, A. (1893). Über Kreuzungen von *Cavia aperea* und *Cavia cobaya*. *Sber. Ges. naturf. Freunde Berlin* **1893**: 249–252.

Oviedo, G. F. de (1547). *La Hystoria natural y general de las Indias yslas*. Toledo: Ramon de Petras.

Pallas, P. S. (1766). *Miscellanea zoologica*. (pp. 30–33.) Hagae Comitum.

Pennant, T. (1781). *History of quadrupeds*. London.

Pictet, A. & Ferrero, A. (1951). La déscendance d'un croisement interspecifique de cobayes (*Cavia aperea* D'Az × *Cavia cobaya* Marc) analysée durant 25 années. *Genetica* **25**: 357–515.

Rood, J. P. (1972). Ecological and behavioural comparisons of three genera of Argentine cavies. *Anim. Behav. Monogr.* **5**: 1–83.

Rood, J. P. & Weir, B. J. (1970). Reproduction in female wild guinea-pigs. *J. Reprod. Fert.* **23**: 393–409.

Schmidt, M. (1929). *Kunst und Kultur von Peru*. Berlin: Propyläen Verlag.

Tate, G. H. H. (1935). The taxonomy of the genera of the neotropical hystricoid rodents. *Bull. Am. Mus. nat. Hist.* **68**: 295–447.

Thomas, O. (1901). On mammals obtained by Mr. Alphonse Robert on the Rio Jordão, S.W. Minas Geraes. *Ann. Mag. nat. Hist.* (7) **8**: 526–536.

Thomas, O. (1917). Notes on the species of the genus *Cavia*. *Ann. Mag. nat. Hist.* (8) **19**: 152–160.

Topsell, E. (1607). *Historie of Foure-footed beastes . . . collected out of all the volumes of C. Gesner and all other writers to this present day*. London.

Tschudi, J. J. von (1845). *Fauna peruana*. St. Gallen: Suheitlim und Zollikofer.

Ubisch, G. & Mello, R. F. (1940). Genetic studies on a cavy species cross (*Cavia rufescens* Lund and *Cavia porcellus* Linné). *J. Hered.* **31**: (9) 389–398.

Velasco, J. de (1837). *Historia del reino de Quito en la America meridional*. Paris.

Wagler, J. (1831). Einige Mittheilungen über Thiere Mexicos. *Isis* **24**: 510.

Waterhouse, G. R. (1848). *A natural history of the Mammalia*. **2**. *Rodentia, or gnawing mammalia*. London: Baillière.

Symp. zool. Soc. Lond. (1974) No. 34, 447–453.

CONCLUDING REMARKS

E. C. AMOROSO

*A.R.C. Institute of Animal Physiology,
Babraham, Cambridge, England*

It was not, and it is not even now, evident to me why the organizers should have extended to me this invitation to undertake the very difficult task of summing up a symposium of such broad scope. It would be vain of me to hope to rise to the height of the opportunity, so let me say at the outset that if I venture on the task, I do so, not as a contestant, but as an observer, and with the certain knowledge that as an observer I have the advantage of emotional detachment and non-involvement. Hence, if any part of this summary fails in its allusion to major points while emphasising minor ones unduly, you must ascribe this to my own shortcomings and not to any departure from the strict neutrality that is usually expected of a judge's summing up. I would ask you on this occasion to regard the judge with friendly lenience as one who has been effectively separated from the cut-and-thrust of anatomical debate for several years and has never had any contact with fossil mammals.

Dr Rowlands, during his prefatory remarks, reminded us that the guinea-pig was the common laboratory animal long before the murine rodents took over. He might well have added that Lataste and his pupil Moreau, towards the end of the last century, first used the guinea-pig to demonstrate the cyclical changes in the vaginal epithelium and as a consequence provided Stockard and Papanicolaou in 1918 with the material for their distinguished researches. Dr Rowlands also called our attention to the important part which Leo Loeb played as a pioneer in the study of reproductive processes of the hystricomorphs. Indeed, Loeb will be remembered as the man who not only first opened to other explorers our knowledge of utero-ovarian interrelationships, but who first described, in the guinea-pig, the particular uterine transformation, the so-called deciduoma, which has provided reproductive endocrinology with some of its most recent and rapidly progressive developments.

This conference has given me, and I do not doubt that it has given to many of you, a vivid impression of the astonishing wealth of new knowledge that has accumulated on the biology of a group of mammals whose taxonomic relationships remain obscure. The papers which have been presented range from the evolutionary origin of these animals to their development and soft anatomy, from the endocrinological innovations during pregnancy and the perinatal period, and even beyond into the realms of their social behaviour and from the synthesis and secretion of steroids to the functional breakdown of the endocrine pancreas.

Were it the intention of the organizers to stimulate lively discussion, as I believe it was, it seems that the early discussants have not failed to answer the call. I am reminded at this point of such episodes as the common ancestry versus parallel evolution which became the basis of a preliminary skirmish early in our proceedings between Professor Lavocat and Professor Wood, and which was re-echoed to some extent in Professor Simpson's embarrassment when he felt that Dr Bugge was providing support for a viewpoint which he had long since abandoned. I can console Dr Simpson, however, by an utterance of Thomas Henry Huxley, who once assured us that advance came rather from errors clearly expressed than from a nebulous outline of the truth.

Dr Simpson in his opening remarks referred to the tendency to identify the sub-order Hystricomorpha with the presence of an infra-orbital canal and the attachment of the masseter muscle. Professor Wood also stated that the enlargement of the foramen went along with the invasion of the muscle. I seem to remember that it is the medial head of the muscle that is associated with the infra-orbital foramen and that the lateral head is a broad strip of muscle attached to the zygoma. Since, in association with these modifications, there must have been parallel changes in nerve supply, we might, with profit, look here for cues as giving support or negation for the views expressed by Professors Lavocat and Wood.

Professor Lavocat's main thesis was that all the important structures and combinations of structures found in the African fossil and extant families of Thryonomyoidea are present in the Hystricoidea of the Old World as well as in the New World Caviomorpha. He concluded, therefore, that the Phiomorpha, and the Hystricoidea and the Bathyergoidea are, with the Caviomorpha, members of one natural sub-order, which he calls Hystricognathi and considers to be of African origin. Professor Wood, on the other hand, reminded us that there is an extreme degree of parallelism in the evolution of features considered characteristic of the sub-order Hystricomorpha. This he regarded as providing strong support for a common ancestry for the hystricomorphs, but he insists that they must have been derived from non-hystricomorphous and probably non-hystricognathous northern hemisphere ancestors that had been evolving independently in the Old and New Worlds since late Palaeocene or early Eocene.

Dr Bugge used the distribution of the cephalic arteries of the Old and New World hystricomorphs to define their relationships but I must confess that it was a little difficult to decide whether he was on the side of Professor Lavocat or that of Professor Wood, or whether he adopted a position of neutrality. For my own part, I wish he had gone a little more slowly because I can think of few areas of the animal body whose development is beset with so many difficulties as is the remodelling of the blood vessels of the pharyngeal arches and with them the cephalic arterial supply. Dr Bugge's analysis of the cephalic arteries raises many exceedingly interesting and pertinent considerations that could have far-reaching implications.

The analysis of interspecific differences is prerequisite, I think, to any complete theory of evolution. The taxonomist generally distinguishes species on a morphological basis but in recent years karyology, the study of the nucleus, has provided an important adjunct to the more conventional methods of classification. A study of the nucleus may, in fact, give information of a different character from that derived from any other morphological study, and this was the importance of the study presented by Drs George and Weir on hystricomorph chromosomes. Their results do not always fit in tidily with those of either Professor Lavocat or Professor Wood, but they are, nevertheless, complementary to the studies of these two authors.

I pass now to those papers dealing with behaviour. Among the components of the total pattern of reproductive behaviour in the lower mammals are the prototypes of many of the behaviour patterns seen in man—aggression, courtship, mating and care of the young. It follows, therefore, that comparative study of the factors regulating and mediating the display of this behaviour could be as helpful in the clarification of the many problems in man as had been the comparative approach to the study of other vital activities. The observations of Dr Kleiman and Dr Eisenberg have provided the substance for a comparative consideration of many such problems.

While the speakers of the second session achieved their objective of painting a broad picture of the ecology and behaviour of the Old World and New World hystricomorphs, their remarks indicated how wide are the gaps in our knowledge that remain to be filled. The descriptions took us through many habitats and all were of value for the breadth they added to what would otherwise be laboratory studies in the narrower sense.

As I looked at the films and listened to the discussions, I was reminded of two sets of behavioural responses or reactions. I had read somewhere that chinchilla which have been mated for a long period are difficult to remate with other individuals and that fatal fighting often follows any attempt to establish a new partnership. I had also heard it said that male porcupines readily copulate with females which have been living in adjacent cages, but do not mount individuals with whom they are unfamiliar. And we learn today from Mr Asibey that certain features of these behaviour patterns are also evident in the grasscutter. Two questions then might be asked. Are such sexual partnerships as are shown by the chinchilla to be regarded as enduring mateships rather than temporary liaisons? And are these prohibitions observed in wild populations? If so, they may have arisen as a defence against the sexual rivalry which would develop if they were not enforced.

Each of the physiological papers generated its own interest, but the underlying theme as I saw it, is a self-evident one, namely that the business of animals is to stay alive until they have reproduced themselves and the business of the biologist is to try and understand how they do it, in terms which emphasize, at least qualitatively, the regulatory aspects of biological activity. This understanding demands the recognition that the structure and function are two indispensable adjuncts of animal organization, linked in patterns that have been determined by deduction and that are always worthy of speculation. The story of massive wastage of superovulated eggs in the plains viscacha and the formation of accessory corpora lutea in this and many other hystricomorphs described by Drs Weir and Rowlands, finds a ready place in the field of evolutionary discussions.

The study of ovarian control of gestation must embrace, in my view, an understanding, not only of the mechanisms behind the dramatic success of the unusual follicle that ovulates and the corpus luteum that replaces it, but also the background and physiological significance of the much more common process of atresia. This process is going on constantly through the normal cycle in adult ovaries but is more pronounced at certain times of the cycle. This, I believe, marks a point at which the endocrine mechanisms controlling gestation took several different trends towards further specialization in hystricomorph rodents. From the standpoint of comparative endocrinology the impression given is that gestation became an ecological problem involving both fetal and maternal adaptations for a common purpose. Although the end-result, in general, was the same, it was derived by different methods; similarities which appear being due notably to the sameness of the material that was used, namely, progesterone as discussed by Drs Heap and Illingworth and Dr Tam. Therefore, in attacking such problems in a particular group, one should hold only broad ideas as guides of approach to what might be found, rather than have fixed notions that the results should conform exactly to established observations in even closely related families.

The specializations of fetal and maternal tissue concerned with the promotion of gestation in hystricomorphs, as described by Dr Tam, are so diverse that it does not seem possible to arrange them in a series that might indicate more than an evolutionary trend. Yet, when reduced to its simplest form, it is conceivable that endocrine adoption of gonadal and pituitary function by the placenta became of major importance to sustain gestation in certain families (e.g. Caviidae) but, as far as we know now, it would seem that only in the domestic guinea-pig, among the hystricomorphs, has the placenta acquired a steroidogenic function. The restriction of this function to the primary corpus luteum of ovulation has been overcome in many species by the formation during pregnancy of an increasing array of luteinized follicles without the occurrence of ovulation. The development of supernumerary luteal tissue in such abundance and in such variety in animals such as the mountain and plains viscacha, chinchilla, Canadian porcupine and coypu supports a strong conviction that it is an hystricomorph device for prolonging gestation. But, whatever the origin of this steroidogenic tissue, the successful implantation of the fertilized egg in the endometrium and its survival as a developing fetus to term, are simply expressions of Nature's highly successful solution of the problems attendant upon the transplantation of one particular type of graft to one special type of bed, the decidua, a solution which is coeval with the origin of mammals.

Drs Roberts and Perry gave us a detailed account of the embryology of the hystricomorph rodents. I would refer only to the complex relationships of the placenta and sub-placenta to which they alluded, as I don't think we should adjourn in the belief that the pattern is replicated exactly in the guinea-pig, coypu and tuco-tuco. There are differences. I think it is essential to bear in mind that the important changes taking place as a result of the elaborate secretory efforts of the syncytiotrophoblast reported by Dr Tam, and the presence of a binding protein described by Drs Heap and Illingworth, may provide a physico-chemical barrier of great importance to the economy and survival of the young. Dr Tam expressed some surprise at having detected progesterone in the yolk sac—a finding which he suggested might be indicative of the steroidogenic activity of this fetal appendage. While this might well be true it is, of course, possible that the yolk-sac epithelium simply acts as a repository for the steroids known to be secreted by the chorion and transmitted by way of the yolk-sac and vitelline circulation.

Major advances have taken place in endocrinology following the initial development more than a decade ago of new micro-analytical methods having sensitivities of an order commensurate with the concentrations at which many compounds exert their effects in biological systems. These techniques exploit, on the one hand, the extreme delicacy of radioactivity measurements, and, on the other, the almost unique chemical specificity which characterizes many biochemical reactions, and allow fresh insight to be gained into the modes and sites of action of many hormones and other trace compounds. Many proteins bind steroids as we learnt, but the nature of the binding varies according to the protein and the steroid under consideration. The presence in the blood stream of macromolecules, or a macromolecular component that binds corticosteroids, seems to be a universal property of all vertebrates so far examined. Hence the importance of the studies that were presented by Drs Heap and Illingworth.

In the final paper, Drs Neville, Weir and Lazarus drew a broad picture of the functional activity of hystricomorph insulins. This study arose from the appearance of symptoms of diabetes and of infertility in consequence of the breakdown of the endocrine pancreas of the tuco-tuco in captivity. Of particular relevance, it seems to me, is their elucidation of the primary structure of the insulin molecule of the chinchilla. Since hystricomorphs also exhibit interspecific variations in their sensitivity to ox insulin and in the potency of their own insulin, it should cause us little surprise if, as more information becomes available, we are able to establish some sort of evolutionary series based on these biochemical profiles. Mention was made of the other secretion of the endocrine pancreas, glucagon. I should remind this audience that we celebrate year this the 50th anniversary of the discovery of glucagon, two years after that of insulin, but the contrast in the histories of these two polypeptides which originate from juxtaposed cells, the alpha- and beta-cells respectively, could not be more striking. Dr Neville reminded us that insulin was well accepted as a hormone even before its isolation. Glucagon, on the other hand, was of little interest to the discovers of insulin and Murlin and his associates, who separated it from insulin, gave it its name and first suggested its possible role as a gluco-regulatory hormone, but received little notice for their efforts. When glucagon was purified it served as an exceptional tool leading Sutherland and his co-workers to the discovery of $3',5'$-cyclic AMP, perhaps one of the most significant contributions to molecular biology in this half-century.

But it is time to stop an exposition that has reached the stage of merely thinking aloud, and only the fact that this is a privileged occasion can excuse speculation as tenuous as this. However, I am sure that you will all wish me, in conclusion, to voice a grateful acknowledgment of our indebtedness for the very special opportunity which we have all enjoyed at this meeting. The papers presented have all been of a very high order of excellence, and all have provided important information, in some cases from a wholly novel point of view. The fact that there is a community of problems has clearly emerged from your deliberations, and if answers have not been found to some of those that have engaged us, they will surely provide clues to the solution of others. But I need also to remind you that the collection of all this information has, at the same time, revealed how extensive are the gaps that remain in our knowledge. This meeting will have served a very useful purpose if, in the future, it focuses our attention on the many problems that have eluded us today. I think it is most regrettable, however, that an area of enquiry having such fundamental importance in both its practical and its theoretical aspects should have to be at risk of discontinuance. This information in this particular area must be gathered and it must be interpreted; and it is our common duty, the duty of everyone present here, to see that support for such an objective is forthcoming and that the colonies of animals that are now available should not be allowed to go to ruin.

To Dr Rowlands, to Dr Weir and their willing band of workers, we offer our most grateful thanks. We salute you as pioneers and we hope that in the years to come, when we again meet to discuss the biology of the hystricomorphs, both of you will be there to grace the occasion and to add to our pleasure as you have done during the course of the present meeting. To Dr Vevers, I say thanks on behalf of all your visitors for looking after us so splendidly. And, finally, I would like the organizers of this meeting to convey to the Lord Zuckerman our profound thanks for permission to hold this meeting here, so pleasantly and so appropriately in the premises of the Zoological Society of London.

Q

AUTHOR INDEX

Numbers in *italic* type indicate illustrations, numbers in **bold** type indicate pages in
the References at the end of each article

A

Aboul-Hosn, W. 396, 397, 405, 407,
411
Ackland, N. 376, **382**, 403, **411**
Adams, W. H. 201, **204**
Alfsen, A. 398, **411**
Allen, J. A. 277, **293**
Allouch, P. 395, 396, 397, *397*, 398,
398, 399, 400, 401, 403, 405, **411**,
413, **413**
Altman, P. L. 288, **294**
Alvarez, M. T. 290, **294**
Ambrosetti, J. B. 120, 126, **127**
Amoroso, E. C. 341, 344, 345, 348, 355,
355, **356**, 364, **381**, 406, **410**
Anderson, S. 2, **5**
Andrew, R. 211, 213, 218, **242**
Andrews, C. W. 33, **51**
Andrews, J. E. 22, **52**
Angulo, J. J. 290, **294**
Ansell, W. F. H. 277, **294**
Arrighi, F. E. 88, **106**
Armin, J. 427, **432**
Artunkal, T. 323, **327**
Asdell, S. A. 266, 283, 286, **294**
Asibey, E. O. A. 163, 167, **169**, 202, **204**,
251, **261**, 271, 277, 282, **294**, 352, **356**
Atger, M. 395, 396, 397, *397*, *398*, 399,
401, 403, 405, **411**, 413, **413**
Atwood, E. L. 201, **204**
Austin, C. 334, **356**

B

Bachmann, K. 94, **105**
Baird, D. T. 394, **409**
Baker, R. J. 94, **105**
Balant, L. 419, **432**
Balogh, K., Jr 380, **383**
Banks, E. M. 211, 242, **243**
Banting, F. G. 418, **432**
Barfield, R. J. 198, **204**, 236, **243**

Barlow, J. C. 121, **127**
Bassett, J. M. 376, **381**
Baulieu, E.-E. 396, 397, *397*, 398, *398*,
399, 400, 401, 403, 405, **411**, 413, **413**
Beach, F. A. 186, 187, 189, 190, **204**
Beauchamp, G. K. 181, **204**
Beçak, M. L. 85, **107**
Beçak, W. 85, **107**
Becher, H. 334, 345, **356**
Beddard, F. E. 16, **19**
Bedford, J. 80, 92, 94, *95*, 99, 100,
105, 442, **445**
Benirschke, K. 79, 80, 81, 83, 84, 85,
100, 104, **105**, **106**, **108**
Bennett, E. T. 134, **141**, 440, **445**
Berhaut, J. 150, **159**
Berry, D. L. 94, **105**
Best, C. H. 418, **432**
Binzstein, N. 114, 117, 118, **128**, 202,
207
Bignami, G. 186, 187, 189, **204**
Billewicz, W. Z. 273, **296**
Billiar, R. B. 408, **409**
Billington, W. D. 312, **327**, 345, **356**
Black, C. C. 31, 33, 38, 39, 42, 45, **51**
Blaha, G. C. 376, **382**
Bland, K. P. 325, **327**, 405, **409**
Blandau, R. J. 188, **205**, 268, **294**, 315,
327, 335, 337, 338, 339, **356**
Blandford, W. T. 143, **159**
Blatchley, F. R. 325, **327**
Bloom, R. T. 202, **206**, 282, **295**
Blundell, T. L. 422, *426–427*, 431, **432**
Blyth, E. 143, **159**
Bohlin, B. 39, 46, **52**
Boiry, L. 114, 117, 118, **128**, 202, **207**
Boling, J. L. 188, **205**
Bookhout, C. G. 314, **327**
Booth, A. H. 163, **170**, 259, **261**
Borgaonkar, D. S. 79, 84, **105**
Boyce, R. E. 22, **52**
Boyd, J. D. 334, **356**

Bradlow, H. L. 378, **382**
Brambell, F. W. R. 312, 317, **327**
Branca, A. 310, **327**
Brandt, J. K. 1, **5**, 44, **51**, 62, 69, **75**
Briley, M. S. 379, **382**
Brookes, J. 120, **127**
Brooks, R. J. 211, 242, **243**
Brown, E. 269, **294**
Brown, H. 421, **432**
Brown, J. B. 366, **381**
Bruce, H. M. 266, **294**
Bucher, G. C. 195, 197, 199, 201, **205**, 270, **294**
Bugge, J. 17, **19**, 61, *64*, 71, **75**
Bullard, R. W. 286, **294**
Bünzli, H. F. 419, **432**
Burge, B. L. 271, 287, **294**
Burr, I. M. 419, **432**
Burstein, S. 394, **412**
Burton, M. 148, **159**
Burton, R. M. 395, 396, 397, 405, 407, **409, 411**
Busch, C. 114, 117, 118, **128**, 202, **207**
Busnel, R. G. 211, **243**

C

Cabrera, A. 114, 116, 120, **127**, 132, 134, **141**, 202, **205**, 437, 438, **445**
Cameron, D. P. 419, **432**
Camus, L. 120, **127**, 290, **294**
Carter, C. O. 89, **106**
Carter, N. D. 442, **445**
Castle, W. E. 268, 275, 276, 286, **294**, 440, 443, 444, **445**
Castrén, O. 408, **409**
Challis, J. R. G. 324, **328**, 375, **381**, 387, 389, *389*, 390, 399, 407, 408, **409, 410**, 411, 414, **414**
Chester Jones, I. 321, **327**
Christie, G. A. 345, **356**
Cieca, P. de 131, **141**
Clark, J. L. 419, **432**
Clough, G. C. 201, **205**, 230, 234, **243**, 271, **295**
Cockerell, T. D. A. 46, **51**
Cohen, M. M. 80, **105**
Colby, D. R. 289, **298**
Cole, H. H. 364, **381**
Collias, N. 230, **243**

Collins, D. C. 310, **327**
Collins, L. R. 180, 184, 188, 189, 190, 200, **205**, 212, 213, 227, 233, 240, 242, **243**, 269, **294**
Collins, R. F. 314, **329**
Collins, W. P. 379, **382**
Cologne, R. A. 323, **327**
Comings, D. E. 98, **105**
Conaway, C. A. 285, **294**
Contreras, J. 118, **127**, 307, **327**, 347, **356**
Costello, D. F. 200, **205**
Coulon, J. 221, **243**
Courrier, R. 389, **409**
Coval, M. 422, 426, 427, **433**
Creighton, C. 344, **356**
Crespo, J. A. 120, 123, 126, **128**, 132, *133*, 134, 135, 136, **141**, 201, **206**, 284, **296**, 307, **328**
Crichton, E. 266, **294**, 355, **356**
Critch, P. J. 184, 186, **205**
Crooks, J. H. 334, 345, **356**
Cumberland, C. 438, 444, **445**
Cutfield, J. F. 422, *426–427*, **432**
Cutfield, S. M. 422, *426–427*, **432**

D

Dale, I. R. 150, **159**
Dansereau, P. 150, 151, **159**
Darwin, C. 109, 110, **111**, 114, 117, 118, 120, 123, **127**, 200, **205**, 442, **445**
Dathe, H. 47, **51**, 187, **205**, 289, **294**
Daughaday, W. H. 393, **409, 410**
Davidson, J. K. 427, 431, **432**
Davies, J. 344, 348, 355, **356**, 406, 410
Davis, D. E. 276, 282, **294**
Dawson, M. R. 40, **51**
d'Azara, F. 131, **141**
Deane, H. W. 345, **358**
Deanesly, R. 317, **327**, 377, 382, 389, *389*, 391, 399, *400*, 406, 407, **410, 411**
de Blainville, H. D. 120, **128**
Dekeyser, P. L. 3, **5**, 271, **294**
de Lange, D. 144, **159**, 249, **249**, 354, **356**
De la Torre, E. 63, 65, **75**
de Miranda-Ribeiro, A. 273, **294**
Dempsey, E. W. 344, 345, 348, 355, **356**, **358**, 406, **410**

SUBJECT INDEX

Numbers in *italic* type indicate illustrations